Birkhäuser Advanced Texts
Basler Lehrbücher

Edited by
Herbert Amann, Zürich University
Ranee Kathryn Brylinski, Penn State University

M. Holz
K. Steffens
E. Weitz

Introduction to Cardinal Arithmetic

Springer Basel AG

Authors:
M. Holz and K. Steffens
Institut für Mathematik
Universität Hannover
Welfengarten 1
30167 Hannover
Germany

E. Weitz
Bernadottestr. 33
22763 Hamburg
Germany

1991 Mathematics Subject Classification 04-01, 04A10, 03E10

A CIP catalogue record for this book is available from the
Library of Congress, Washington D.C., USA

Deutsche Bibliothek Cataloging-in-Publication Data
Holz, Michael:
Introduction to cardinal arithmetic / M. Holz ; K. Steffens ;
E. Weitz.
(Birkhäuser advanced texts)
ISBN 978-3-0346-0327-0 ISBN 978-3-0346-0330-0 (eBook)
DOI 10.1007/978-3-0346-0330-0

Originally published by Birkhäuser Verlag in 1999
Softcover reprint of the hardcover 1st edition 1999
Printed on acid-free paper produced from chlorine-free pulp. TCF ∞

ISBN 978-3-0346-0327-0

9 8 7 6 5 4 3 2 1

Contents

Preface

This book is an introduction to modern cardinal arithmetic, developed in the frame of the axioms of Zermelo-Fraenkel set theory together with the axiom of choice. It splits into three parts. Part one, which is contained in Chapter 1, describes the classical cardinal arithmetic due to Bernstein, Cantor, Hausdorff, König, and Tarski. The results were found in the years between 1870 and 1930. Part two, which is Chapter 2, characterizes the development of cardinal arithmetic in the seventies, which was led by Galvin, Hajnal, and Silver. The third part, contained in Chapters 3 to 9, presents the fundamental investigations in pcf-theory which has been developed by S. Shelah to answer the questions left open in the seventies. All theorems presented in Chapter 3 and Chapters 5 to 9 are due to Shelah, unless otherwise stated.

We are greatly indebted to all those set theorists whose work we have tried to expound. Concerning the literature we owe very much to S. Shelah's book [Sh5] and to the article by M. R. Burke and M. Magidor [BM] which also initiated our students' interest for Shelah's pcf-theory.

The enthusiasm of our students for cardinal arithmetic contributed much to the formation of this book. Our thanks are due to S. Shelah for answering several questions and to T. Jech for sending us some papers. Our thanks are also due to S. Neumann for many hints to Chapter 9 and to T. Espley and A. Truong for their contributions concerning the readability of this book.

Introduction

If M and N are sets[1] and if there exists a bijection[2] from M onto N, then we say that M and N are equinumerous, and write $M \approx N$. To measure the number of members of a set, we will introduce sets of comparison. With ω we denote the set of natural numbers; 0 is a natural number. If for example M is equinumerous to the set $\{n \in \omega : n < 25\}$, then we say that M has exactly 25 members, and $\{n \in \omega : n < 25\}$ is a set of comparison for M. If N is a set and if N and ω are equinumerous, then ω will be a set of comparison for N, and N will be called countably infinite or denumerable. A well known example for such a set is $N = \{n \in \omega : n \text{ is divisible by } 2\}$.

We will assign to each set M a set of comparison $|M|$, the so-called cardinality of M, such that the following holds.

(1) $M \approx N \Longleftrightarrow |M| = |N|$.

(2) $M \approx |M|$.

Chapter 1, § 1.5 shows that such an assignment $M \mapsto |M|$ is possible. Any set $|M|$ is called a cardinal number or a cardinal. If $|M|$ and $|N|$ are cardinal numbers, then we write $|M| \leq |N|$ iff[3] there is an injection[4] from M into N, and $|M| < |N|$ iff $|M| \leq |N|$ and $\neg(M \approx N)$. If we assume the axiom of choice, then any two cardinal numbers are comparable, i.e., we get $|M| \leq |N|$ or $|N| \leq |M|$ for any sets M and N; conversely it will not prove difficult to show that this assertion implies the axiom of choice. Thus the class CN of cardinal numbers is linearly ordered by $\leq$ and by $<$. These linear orderings have no greatest element. G. Cantor proved in 1873 that, for any set M, the so-called power set $\mathcal{P}(M)$ of M, i.e., the set $\{x : x \subseteq M\}$, satisfies $|M| < |\mathcal{P}(M)|$. In Section 1.5

[1]We assume that the reader has an intuitive knowledge about the mathematical notion of a set. $\{x : E(x)\}$ will be, as usual, the collection of all mathematical objects satisfying the property $E(x)$.

[2]That is a function f from M into N such that, for every $y \in N$, there is exactly one $x \in M$ with $y = f(x)$.

[3]We use the mathematical word *iff* as usual as an abbreviation for *if and only if*.

[4]That is a function f from M into N such that, for every $y \in N$, there is at most one $x \in M$ with $y = f(x)$.

we will see that CN is well-ordered by $<$, that means that every nonempty subset of CN has a $<$-least element. Consequently, for every cardinal number $|M|$, there exists a smallest cardinal number greater than $|M|$; we denote it by $|M|^+$. Every natural number will be shown to be cardinal, and ω is the set of finite cardinals. The infinite cardinals are enumerated by the so-called aleph function whose values $\aleph(\alpha)$ are usually written as $\aleph_\alpha$. The smallest infinite cardinal is ω, $\aleph_0 = \omega$. So we have $\aleph_1 = \aleph_0^+$, $\aleph_2 = \aleph_1^+$, and $\aleph_{n+1} = \aleph_n^+$ for every $n \in \omega$. A natural candidate for the next cardinal, the smallest cardinal greater than every $\aleph_n$, is $\bigcup\{\aleph_n : n \in \omega\}$. The following properties show that this set is the supremum of the set $\{\aleph_n : n \in \omega\}$, and we denote it by $\aleph_\omega$. Therefore we require further that, for any cardinals κ and μ,

(3) $\kappa < \mu \Longleftrightarrow \kappa \in \mu$.

(4) If K is a set of cardinal numbers, then $\bigcup K$ is a cardinal number.

(5) $\kappa < \mu \Longrightarrow \kappa \subseteq \mu$.

These properties will be realized, and thus CN will be completely ordered by $<$ in the sense that every nonempty set of cardinals has a supremum.

The name cardinal number would be inappropriate if one could not calculate with cardinals. We define the addition, multiplication and exponentiation of cardinals by

$$|M| + |N| := |M \cup N|, \text{ if } M \cap N = \emptyset,$$
$$|M| \cdot |N| := |M \times N|, \text{ and}$$
$$|M|^{|N|} := |{}^N M|,$$

where ${}^N M$ is the set of all functions from N into M. Each of these operations is a canonical extension of the corresponding operation for natural numbers. However, addition and multiplication satisfy a simple new rule: if one of the nonempty sets M and N is infinite, then

$$|M| + |N| = |M| \cdot |N| = \max\{|M|, |N|\}.$$

Thus we see that the course of values of addition and multiplication can be easily described. This is not at all the case for the exponentiation $(\kappa, \lambda) \mapsto \kappa^\lambda$ of cardinals. It is easy to see that $|{}^\omega 2| = |\mathcal{P}(\omega)|$, where $2 := \{0, 1\}$, and that ${}^\omega 2$ and the set of real numbers are equinumerous. With Cantor's result we get $\aleph_0 < 2^{\aleph_0}$. Cantor believed that there is no cardinal κ with $\aleph_0 < \kappa < 2^{\aleph_0}$, meaning that $2^{\aleph_0} = \aleph_0^+$. This assertion is called the continuum hypothesis, the more general assertion that $2^\kappa = \kappa^+$ for all infinite cardinals κ is called the generalized continuum hypothesis, written as (GCH). Under the assumption that Zermelo-Fraenkel set theory with the axiom of choice, denoted by ZFC, is consistent, K. Gödel proved that ZFC + (GCH), the theory ZFC with the additional axiom (GCH), is consistent, and P. J. Cohen proved, that ZFC + $\neg$(GCH) is consistent. If the reader wishes to accept (GCH), he need not study this book further, since the exponentiation of cardinals obeys in ZFC + (GCH) simple rules. The difficulties arise when cardinal exponentiation is studied in the axiom system ZFC, and we will describe some of them.

Let ICN be the class of infinite cardinal numbers. Clearly the so-called continuum function $(2^\kappa : \kappa \in \text{ICN})$ is increasing, i.e., we have $2^\kappa \leq 2^\mu$ for all $\kappa, \mu \in \text{ICN}$ with $\kappa < \mu$.

To formulate a further property of the continuum function, we need the notion of the cofinality of an infinite cardinal κ, written as $\text{cf}(\kappa)$. If $\kappa > \aleph_0$, then we call κ a successor cardinal iff there exists some cardinal μ with $\kappa = \mu^+$, and otherwise a limit cardinal. In this introduction let us define $\text{cf}(\aleph_0) = \aleph_0$, $\text{cf}(\mu^+) = \mu^+$ for any successor cardinal μ^+, and $\text{cf}(\lambda) = \min\{|K| : K \subseteq \{\nu \in \text{ICN} : \nu < \lambda\} \text{ and } \bigcup K = \lambda\}$ for limit cardinals λ. This minimum exists since, for $N = \{\nu \in \text{ICN} : \nu < \lambda\}$, we have $\bigcup N = \lambda$. Namely, (5) gives $\bigcup N \subseteq \lambda$, hence $|\bigcup N| \leq |\lambda|$. Furthermore, $\bigcup N$ is a cardinal by (4), and (1) and (2) yield $\bigcup N \leq \lambda$. The supposition $\bigcup N < \lambda$ leads to $(\bigcup N)^+ < \lambda$, since λ is a limit cardinal, and thus gives $(\bigcup N)^+ \in N$ and $(\bigcup N)^+ \subseteq \bigcup N$, hence $(\bigcup N)^+ \leq \bigcup N$, contradicting the definition of a cardinal successor. From $N \subseteq \lambda$ and thus $|N| \leq \lambda$, together with the other parts of the definition, we get $\text{cf}(\kappa) \leq \kappa$ for every infinite cardinal number κ.

If $\kappa \in \text{ICN}$, then we call κ regular iff $\text{cf}(\kappa) = \kappa$, and singular iff $\text{cf}(\kappa) < \kappa$. A well-known theorem of J. König says that $\kappa < \text{cf}(2^\kappa)$. Let Reg denote the class of regular cardinals. W. B. Easton proved the following theorem. Assume that $F : \text{Reg} \longrightarrow \text{ICN}$ is a function satisfying

(i) $\kappa < \mu \Longrightarrow F(\kappa) \leq F(\mu)$ and
(ii) $\kappa < \text{cf}(F(\kappa))$.

Then $\text{ZFC} + \forall \kappa \in \text{Reg}(2^\kappa = F(\kappa))$ is consistent, only if ZFC is consistent. This result shows that the course of values of the continuum function on the class of regular cardinals is only limited by the rules (i) and (ii). However, J. Silver proved for singular cardinals κ with $\aleph_0 < \text{cf}(\kappa)$: If $2^\mu = \mu^+$ for all infinite cardinals $\mu < \kappa$, then $2^\kappa = \kappa^+$. So we see that the values of the continuum function for singular cardinals κ can in fact depend on its course of values on the set of smaller cardinals. Further investigations on the exponentiation of cardinals showed that the so-called gimel function $(\kappa^{\text{cf}(\kappa)} : \kappa \in \text{ICN})$ is of outstanding importance. Already A. Tarski and K. Gödel remarked that one can reduce the study of cardinal arithmetic to the study of this function. If κ is regular, then we have $\kappa^{\text{cf}(\kappa)} = 2^\kappa$, and for singular κ we have $\kappa^{\text{cf}(\kappa)} \leq 2^\kappa$. Galvin and Hajnal found a nontrivial upper bound for $\kappa^{\text{cf}(\kappa)}$ if κ is singular and $\text{cf}(\kappa) > \aleph_0$. Until recently, one could not give a corresponding estimate for singular cardinals with countable cofinality. For this, Shelah developed a theory of its own, the so-called pcf-theory, which is complete in itself, allows to extend the Galvin-Hajnal theorem to singular cardinals with countable cofinality, and has other applications outside of cardinal arithmetic. For cardinal arithmetic, it yields for example the following theorem: If $2^{\aleph_n} < \aleph_\omega$ for every $n \in \omega$, then $2^{\aleph_\omega} < \aleph_{\aleph_4}$. Chapters 3 to 9 of this book give an introduction to the foundations of pcf-theory.

Chapter 1

Foundations

1.1 The Axioms of ZFC

The year of birth of set theory can be regarded as 1872. In this year, Georg Cantor introduced the notion of a transfinite (infinite) ordinal number. His investigations, which soon made set theory an independent area of mathematics, met with mistrust and sometimes open hostility from other mathematicians. At the start of the last decade of the century, set theory came into vogue and was being applied more and more extensively to analysis and geometry. Just when Cantor had found high recognition, and when he had put the finishing touches to his investigations, he himself discovered the first antinomy in his system, which was also found two years later, in 1897, by Burali-Forti. Neither of them could offer any solution, but they did not take the problem too seriously, since it arose in a pretty technical area, namely, the theory of well-ordered sets. They believed that a revision of proofs and theorems could settle the situation, as was often the case earlier, under similar circumstances. These hopes were radically dashed when, in 1902, Bertrand Russell published his famous antinomy which occurred just at the beginning of set theoretical investigations and demonstrated that something wasn't right in the foundations of that discipline. The disclosure of that antinomy led to a foundational crisis in mathematics. Several ways out of this crisis have been developed, the most successful one being without any doubt the axiomatic method. And amongst the various axiomatic systems of set theory, the system which was begun by Zermelo and completed by Fraenkel and Skolem is, adding the axiom of choice, perhaps the most known and the most practical one. The results of our book are set forth in this theory, denoted by ZFC, the axioms of which are represented below. We always begin with a naive formulation of the particular axiom of ZFC and will then make precise the notion of a "property" occurring in two axioms.

The axioms of ZFC are not **independent** in the sense that no axiom can be proved with the help of the other ones. But historical and practical reasons speak for choosing these axioms. The system ZF consists of the axioms (A1)–(A9); thus ZFC is the theory ZF together with the axiom of choice, (AC). We write $a \in A$ for the assertion "The set a is a member (or element) of the set A". The formal symbols we are using are explained later on when we will make precise the notion of a property.

(A1) **Axiom of Extensionality**
If any two sets have the same members, then they are equal.

(A2) **Empty Set Axiom**
There is a set having no members.
As usual, we denote this set by $\emptyset$; $\emptyset := \{x : x \neq x\}$.

(A3) **Comprehension Axioms** or **Subset Axioms**
If an arbitrary property is given, then to each set a there is a set b whose members are exactly those sets in a which satisfy the property.
If $E(x)$ is the considered property, then we write $b = \{x \in a : E(x)\}$.

(A4) **Pairing Axiom**
For any sets a and b, there is a set having as its members just a and b. We denote this (uniquely determined) set by $\{a, b\}$ and call it the **unordered pair** or **pair set** of a and b. This means $\{a, b\} := \{z : z = a \vee z = b\}$. Note that $\{a, a\} = \{a\}$ follows from (A1). This set is called the **singleton** $\{a\}$ or singleton of a.

(A5) **Union Axiom**
For any set a, there exists a set whose members are exactly the members of the members of a.
We donote this set by $\bigcup a$. So we have $\bigcup a := \{z : \exists u(u \in a \wedge z \in u)\}$. Furthermore we put $a \cup b := \bigcup\{a, b\}$ and $\{a_1, \ldots, a_n\} := \{a_1, \ldots, a_{n-1}\} \cup \{a_n\}$.

(A6) **Power Set Axiom**
For any set a, there exists a set whose members are exactly the subsets of a.
We denote this set by $\mathcal{P}(a)$ (i.e., $\mathcal{P}(a) := \{z : z \subseteq a\}$) and call it the **power set** of a. Hereby we say that a set z is a **subset** of a iff every member of z is a member of a, formally written as $\forall u(u \in z \Longrightarrow u \in a)$; of course, we abbreviate this by $z \subseteq a$.

(A7) **Infinity Axiom**
There is a set which has the empty set as a member, and has the member $b \cup \{b\}$ whenever b is a member of it.
Every set having this property is called an **inductive set**. Thus we can give a short formulation of the infinity axiom: "There is an inductive set".

(A8) **Replacement Axioms**
If the property $E(a,b)$ is **functional**, that means that for any sets a, b and c we can conclude $b = c$ from $E(a,b)$ and $E(a,c)$, then for each set A there is a set B whose members are exactly those sets b for which there exists a member a of A satisfying $E(a,b)$.
Very roughly speaking: "If the domain of a function is a set, then its range is also a set."[1]

(A9) **Regularity Axiom** or **Foundation Axiom**
Any nonempty set has a member which is minimal amongst its members with respect to the relation $\in$; or: Any nonempty set has a member which has no members in common with it.

(AC) **Axiom of Choice**
For any set A whose members all are nonempty, there is a function with domain A such that, for any member a of A, the value of this function at a is a member of a; or: For any set A whose members all are nonempty, there is a function with domain A "choosing" one member from any member of A.

If we define the **cartesian product** or the set of **choice functions** for A by

$$\prod A := \{f : f \text{ is a function from } A \text{ into } \bigcup A \text{ with } f(a) \in a \text{ for all } a \in A\},^{2}$$

then the axiom of choice, (AC), says:
If A is a set of nonempty sets, then $\prod A \neq \emptyset$.

To make the notion of a property precise, we introduce a **formal language** $\mathcal{L}_{\mathrm{ZF}}$ of set theory which differs from the so-called metalanguage or informal language, which is English. This formal language is a special case of a language of the **first-order predicate calculus** with equality, with one non-logical symbol $\in$, denoting membership.

The basic symbols of $\mathcal{L}_{\mathrm{ZF}}$, forming its **alphabet**, are **(sentential) connectives** $\neg$, $\wedge$, $\vee$, $\Longrightarrow$, and $\Longleftrightarrow$; **quantifiers** $\forall$ and $\exists$; **variables** v_j for each natural number j; $=$ and $\in$; parentheses (and) . Intuitively, $\neg$ means "no", $\wedge$ means "and", $\vee$ means "or", $\Longrightarrow$ means "implies"[3], $\Longleftrightarrow$ means "equivalent",

---

[1]Note that this assertion needs no axiom if a function of the form $f : A \longrightarrow B$ from the set A into a set B in the usual sense is given, since in this case its range is a set by a comprehension axiom. We don't need the replacements axioms before we assign, by a functional property, to each member of a given set a certain set, and do not know at the beginning that these "values" are members of another fixed set.

[2]If a and b are sets, then also $a \times b$ (see below) is a set. As a subclass of $\mathcal{P}(A \times \bigcup A)$, $\prod A$ is a set by (A5), (A6) and (A3). Note that $\prod \emptyset = \{\emptyset\}$.

[3]This usage of the word "implies" is a conventional part of mathematical jargon, but it is not exactly the way the rest of the world uses the word.

$\forall$ means "for all", and $\exists$ means "there exists". A **string of symbols** or **expression** is a finite sequence of symbols. The **concatenation** of strings of symbols is that sequence which results from writing the given strings one behind the other. Certain expressions, corresponding to meaningful assertions, will be called formulas. The **formulas of $\mathcal{L}_{ZF}$** are given by the following inductive definition:

(1) For any i and j, $(v_i \in v_j)$ and $(v_i = v_j)$ are formulas (so-called **atomic formulas**).

(2) If φ and ψ are formulas, so are $(\neg\varphi)$, $(\varphi \wedge \psi)$, $(\varphi \vee \psi)$, $(\varphi \implies \psi)$, $(\varphi \iff \psi)$, and $(\forall v_j \varphi)$ and $(\exists v_j \varphi)$ for any j.

Now a **property** simply is a formula of $\mathcal{L}_{ZF}$. If there is no danger of misunderstanding, we do not use parentheses as restrictively as in the definition of a formula; we will, in a formally incorrect manner, omit them for reasons of readability. The use of **bounded quantifiers** is very practical: $\forall v_i \in v_j\ \varphi$ is the formula $\forall v_i(v_i \in v_j \implies \varphi)$, $\exists v_i \in v_j\ \varphi$ is the formula $\exists v_i(v_i \in v_j \wedge \varphi)$. We also write $v_i \neq v_j$ for $\neg(v_i = v_j)$ and $v_i \notin v_j$ for $\neg(v_i \in v_j)$. A **sentence** is a formula with no free variable (see section 4.2 for further details), a **theory** or **axiom system** is a set of sentences which are called the **axioms of the theory**. If a formula φ is provable in a theory T, where for our purposes mostly $T = \mathrm{ZF}$ or $T = \mathrm{ZFC}$ will suffice, we write $T \vdash \varphi$.

As usual, we will introduce new function and relation symbols and so-called operations (as the intersection and union of sets). Furthermore it is very useful to speak about collections which need not be sets. Both aims can be realized by using **classes**: Every variable is a class; if $\varphi(x,p)$ is a formula, then $X := \{x : \varphi(x,p)\}$ is a class, where we allow φ to have free variables p etc. other than x, which are thought of as **parameters** upon which the class depends. Members of the class X are exactly those sets a which have the property $\varphi(a,p)$. This means that

$$a \in \{x : \varphi(x,p)\} \iff \varphi(a,p).$$

This is the main rule with whose help we can eliminate the new expressions $\{x : \varphi(a,p)\}$. We use capitel letters $A, B, \ldots$ to denote classes, also with indices, and define the **extended formulas** as those expressions which may contain the symbols $\{$, $\}$ and $:$ besides the symbols of $\mathcal{L}_{ZF}$ and which are defined in the same way as the formulas with the additional rule that $(A = B)$ and $(A \in B)$, for classes A and B, are also atomic formulas. If $A = \{x : \varphi(x)\}$ and $B = \{x : \psi(x)\}$, then $y \in A$ is the formula $\varphi(y)$, $A = B$ is the formula $\forall x(\varphi \iff \psi)$, $A = y$ is the formula $\forall x(\varphi \iff x \in y)$, $A \in B$ is the formula $\exists z(\psi(z) \wedge \forall u(u \in z \iff \varphi(u)))$, and $A \in y$ is the formula $\exists z(\forall x(\varphi \iff x \in z) \wedge z \in y)$, where the variable z can be chosen uniquely (and suitably). Thus any extended formula can be replaced by a formula of $\mathcal{L}_{ZF}$. We could go one step further and admit classes of the form $\{x : \Psi(x)\}$, where Ψ is allowed to contain further classes.

A. Levy proves in the appendix of his book [Le] that these extensions of the language can be eliminated in a formally correct way. In the first chapter of this book, the reader can find an explanation of our motivation and justification for our procedure of using classes.

We can now give a precise formulation of the **replacement axioms**, using formulae or, equivalently, using classes. The formalization of the comprehension axioms is left to the reader.

If $\varphi(x, y, p_1, \ldots, p_n)$, written as $\varphi(x, y)$, is a formula of $\mathcal{L}_{\mathrm{ZF}}$, then the following sentence is an axiom of ZF:

$$\forall p_1 \ldots \forall p_n (\forall u \forall v \forall w (\varphi(u, v) \wedge \varphi(u, w) \Rightarrow v = w) \\ \Rightarrow \forall x \exists y \forall v (v \in y \Leftrightarrow \exists u \in x\ \varphi(u, v))).$$

The class version says that, for any class F, the following formula is an axiom of ZF:

$$\mathrm{Func}(F) \implies \forall x \exists y (y = \mathrm{ran}(F \restriction x)).^4$$

We assume that the reader is familiar with the basic concepts of set theory as, for example, intersection, union, relation, function etc. Especially he should have seen a development of the elementary properties of the Boolean algebra of sets and should have learned to handle the "language of set theory" in the frame of an undergraduate course in mathematics. However, the basic properties of ordinals and cardinals will be described in detail, and principally no knowledge in set theory is required. We now give the definitions of all basic concepts that will be needed from the beginning.

$x, y, z, \ldots$ always denote sets, $A, B, C, \ldots$ denote classes; since $x = \{z : z \in x\}$ is provable[5], the definitions for classes also apply to sets.

The class $\mathrm{V} := \{x : x = x\}$ is the **universal class**. A is a **subclass** of B, written as $A \subseteq B$, iff $\forall z (z \in A \implies z \in B)$ (formally: $A \subseteq B$ is an abbreviation for this formula; we will also state this definition in the form $A \subseteq B :\iff \forall z (z \in A \implies z \in B)$). $A \subset B$ and $A \subsetneq B$ stand for the formula $A \subseteq B \wedge A \neq B$, to be read as "$A$ is a proper subclass of B". Any class A is called a **set** iff $\exists y \forall z (z \in y \iff z \in A)$ is provable. We call A a **proper class** iff $\forall y (y \neq A)$ is provable. $(x, y) := \{\{x\}, \{x, y\}\}$ is the **ordered pair** of x and y. The essential property of this set is $\forall x \forall y \forall u \forall v ((x, y) = (u, v) \iff x = u \wedge y = v)$. The class $A \times B := \{(x, y) : x \in A \wedge y \in B\}$ is the **cartesian product** or **product** of the classes A and B (where $\{(x, y) : \varphi(x, y)\} := \{z : \exists x \exists y (\ z = (x, y) \wedge \varphi(x, y))\}$). The classes $A \cup B := \{x : x \in A \vee x \in B\}$, $A \cap B := \{x : x \in A \wedge x \in B\}$ and $A \setminus B := \{x : x \in A \wedge x \notin B\}$ are respectively the **union**, the **intersection** and the **difference** of the classes A and B. We say that the classes A and B are

---

[4]For the definitions of Func, ran and $F \restriction x$, see below.

[5]By our conventions, this is the formula $\forall z (z \in x \iff z \in x)$.

disjoint iff $A \cap B = \emptyset$. $\bigcup A := \{x : \exists a(a \in A \wedge x \in a)\}$ is the **union of A** and $\bigcap A := \{x : \forall a(a \in A \implies x \in a)\}$ is the **intersection of A**. If A and B are sets, so are $A \setminus B$, $A \cup B$, $A \cap B$, $A \times B$ and $\bigcup A$; in the case that $A \neq \emptyset$, $\bigcap A$ is also a set. Note that $\bigcap \emptyset = \mathrm{V}$.

A **relation** is a class of ordered pairs: $\mathrm{Rel}(R) :\iff R \subseteq \mathrm{V} \times \mathrm{V}$. Instead of $(x, y) \in R$ we also write xRy. Assume that R is a relation. R is a **relation from A into B** iff $R \subseteq A \times B$. If $R \subseteq A \times A$, then R is a relation **on A** or a **binary relation on A**. Analogously, **n-ary relations** are defined as subclasses of the n-fold cartesian product of A. The relation $R \upharpoonright A := \{(x, y) \in R : x \in A\}$ is the **restriction** of the relation R to A, $R^{-1} := \{(y, x) : (x, y) \in R\}$ is the **inverse relation** of R. The **domain** of R is $\mathrm{dom}(R) := \{x : \exists y\ (x, y) \in R\}$, the **range** of R is $\mathrm{ran}(R) := \{y : \exists x\ (x, y) \in R\}$, and $\mathrm{fld}(R) := \mathrm{dom}(R) \cup \mathrm{ran}(R)$ is the **field** of the relation R. The **image of A under R** is $R[A] := \{y : \exists x \in A\ (x, y) \in R\}$. If S is a further relation, then the **composition of R and S** is the relation $R \circ S := \{(u, w) : \exists v((u, v) \in S \wedge (v, w) \in R)\}$.

We define F to be a **function**, written as $\mathrm{Func}(F)$, by

$$\mathrm{Func}(F) :\iff \mathrm{Rel}(F) \wedge\ \forall u \forall v \forall w((u, v) \in F \wedge (u, w) \in F \implies v = w).$$

If F is a function and $u \in \mathrm{dom}(F)$, then $F(u)$, the **value of F at u**, denotes the unique set v with $(u, v) \in F$. Let

$$F : A \longrightarrow B \quad :\iff \quad \mathrm{Func}(F) \wedge \mathrm{dom}(F) = A \wedge \mathrm{ran}(F) \subseteq B;$$

then F is a **function from A into B**. A function $F : A \longrightarrow \mathrm{V}$ is also called a **family** and written as $F =: (F(a) : a \in A)$. Often one introduces a new symbol, say B, and writes B_a instead of $F(a)$ and $(B_a : a \in A)$ instead of F. The **cartesian product** of a family $(F(a) : a \in A)$ is defined by

$$\prod_{a \in A} F(a) := \{f : f : A \longrightarrow \bigcup\{F(a) : a \in A\} \wedge \forall a \in A(f(a) \in F(a))\}.$$

Any function $F : A \longrightarrow B$ is a set, being a subclass of $A \times \mathrm{ran}(F)$, if A is a set. If A is a set, ${}^A B$ is the class of functions from A into B. id_A is the function $\{(a, a) : a \in A\}$, the so-called **identity function** or **identity** on A. A function F from A into B is called **surjective** or a **surjection** or a function from A **onto** B iff $\mathrm{ran}(F) = B$. Any function F is called **injective** or **one-to-one** or an **injection** iff $\forall u \forall v \forall w((u, v) \in F \wedge (w, v) \in F \implies u = w)$; in this case, F^{-1} is a function, the **inverse** of the function F. If F is an injective function from A onto B, then F is called **bijective** or a **bijection**.

Now assume that R and S are binary relations on the classes A and B, respectively. A function $F : A \longrightarrow B$ is a **homomorphism** (with respect to R and S) iff

$\forall x \forall y(xRy \implies F(x) S F(y))$; it is an **embedding** (w.r.t. R and S) iff it is injective and $\forall x \forall y(xRy \iff F(x)\, S\, F(y))$. Any surjective embedding is an **isomorphism**. If A and B – and thus also F, R and S – are sets, then we will use for such functions the familiar denotation $F : (A, R) \longrightarrow (B, S)$.

We say that a relation R is
reflexive iff $\forall x \in \operatorname{fld}(R)(xRx)$;
irreflexive iff $\forall x \in \operatorname{fld}(R)(\neg xRx)$;
antisymmetric iff $\forall x \forall y(xRy \wedge yRx \implies x = y)$;
asymmetric iff $\forall x \forall y(xRy \implies \neg yRx)$;
symmetric iff $\forall x \forall y(xRy \implies yRx)$;
connected or **trichotomic** iff $\forall x \in \operatorname{fld}(R)\, \forall y \in \operatorname{fld}(R)(xRy \vee x = y \vee yRx)$;
transitive iff $\forall x \forall y \forall z(xRy \wedge yRz \implies xRz)$;
an **equivalence relation** iff it is reflexive, symmetric, and transitive;
a **partial ordering** iff it is reflexive, antisymmetric, and transitive;
a **strict partial ordering** iff it is irreflexive and transitive[6];
a **linear ordering** iff R is a connected partial ordering;
well-founded, written as wf(R), iff

(i) $\forall x \in \operatorname{fld}(R) \exists y(x \in y \wedge R^{-1}[y] \subseteq y)$ and
(ii) $\forall x \subseteq \operatorname{fld}(R)(x \neq \emptyset \implies \exists y \in x \forall z \in x(\neg zRy))$;

a **well-ordering** iff R is a well-founded connected strict partial ordering.

If R is a relation, B is a subclass of fld(R) and $x \in B$, then x is a **greatest element** or **largest element** or **maximum** of B (w.r.t. R), denoted by $\max(B)$, iff $\forall z \in B(z \neq x \implies zRx)$;
least element or **smallest element** or **minimum** of B (w.r.t. R), denoted by $\min(B)$, iff $\forall z \in B(z \neq x \implies xRz)$;
maximal element (w.r.t. R) or ***R*-maximal element** of B iff $\neg \exists z \in B(z \neq x \wedge xRz)$;
minimal element (w.r.t. R) or ***R*-minimal element** of B iff $\neg \exists z \in B(z \neq x \wedge zRx)$.

If A is a class and R is a relation with $\operatorname{fld}(R) = A$, then we write $\boldsymbol{(A, R)}$. For example, if R is a partial ordering on A with $\operatorname{fld}(R) = A$, then we say that (A, R) is a partial ordering.

Assume that R is a relation on H with $\operatorname{fld}(R) = H$ and X is a subset of H. X is called a **chain** iff $(X, R \cap (X \times X))$ is a linear ordering (in particular, any two elements of X are **comparable** under R, i.e., for all $y, z \in X$ we have $yRz \vee y = z \vee zRy$). We say that X is an **antichain** iff any two different elements

[6]In most cases we will make no difference between partial orderings and strict partial orderings. For any partial ordering R, we get a strict partial ordering R' via $xR'y :\iff xRy \wedge x \neq y$, and conversely, any strict partial ordering R' yields a partial ordering R by $xRy :\iff xR'y \vee x = y$.

of X are **incomparable** under R, i.e., iff for all different $y, z \in X$ we have $\neg(yRz)$ and $\neg(zRy)$. Assume further that $x \in H$. x is an **upper bound** (**lower bound**) of X iff $sRx \vee s = x$ for all $s \in X$ ($xRs \vee x = s$ for all $s \in X$). We say that x is a **supremum** of X and write $x = \sup(X)$ iff x is a **least upper bound** of X; i.e., x is an upper bound of X, and $xRy \vee x = y$ for any upper bound y of X. Accordingly, x is an **infimum** of X, written as $x = \inf(X)$, iff x is a **greatest lower bound** of X.

If (H, R) is a partial ordering and a supremum or an infimum of X exists, then it is uniquely determined. X is called an **initial segment** of H iff, for all $s \in X$ and all $z \in H$, zRs implies $z \in X$. X is a **proper initial segment** of H iff X is an initial segment of H and different from H.

If $(H, \leq)$ and $(K, \preceq)$ are partial orderings and $f : H \longrightarrow K$ is a function, then f is called **increasing** or **monotone** or **monotonic** (**decreasing**) iff, for all $x, y \in H$, $x \leq y$ implies $f(x) \preceq f(y)$ ($f(y) \preceq f(x)$). We say that f is **strictly increasing** etc. (**strictly decreasing**) iff, for all $x, y \in H$, $x < y$ implies $f(x) \prec f(y)$ ($f(y) \prec f(x)$).

Note: If H is a well-ordering, then every strictly decreasing sequence (for the definition of a sequence, see section 1.4) of elements of H is finite, since its range has a least element.

Exercises

1) Let $x = \{\emptyset, \{\emptyset\}, \{\emptyset, \{\emptyset\}\}\}$. Write the following sets in the form $\{a_1, \ldots, a_k\}$: $\bigcup x$, $\bigcap x$, $\mathcal{P}(\{x\})$, $\mathcal{P}(x) \cap \bigcup\bigcap(x, \{x\})$.
2) Use the axioms of ZF to show for any sets x, y:

   a) $x \times y$, $\{\mathcal{P}(y) : y \in x\}$, $\{f : f : x \longrightarrow \mathcal{P}(x)\}$ are sets; if $x \neq \emptyset$, then $\bigcap x$ is a set.

   b) $\bigcap \emptyset$, $\{x : y \in x\}$, $\{x : \neg\exists y(x \in y \wedge y \in x)\}$ are not sets.

3) Prove without using (A9): There is not a set A such that $\mathcal{P}(A) \subseteq A$. Hint: Consider the class $\{x \in A : x \notin x\}$.
4) Show with the axiom of regularity, that there is not a finite sequence $(x_i : i \in n)$ such that $x_0 \in x_1 \in x_2 \in \ldots \in x_{n-1} \in x_0$.
5) The following sets $[x, y]$ have, for $x \neq y$, exactly two members. Which of them have the property of an ordered pair, i.e.,

$$[x, y] = [x', y'] \Longleftrightarrow x = x' \wedge y = y' \ ?$$

   a) $[x, y] := \{\{x\}, \{x, y\}\}$.

   b) $[x, y] := \{x, \{y\}\}$.

   c) $[x, y] := \{x, \{x, y\}\}$.

6) Prove that the pairing axiom follows from the other axioms of ZF. Hint: Observe that $\{\emptyset, \{\emptyset\}\}$ is a set by the power set axiom. Now let a and b be sets and apply a replacement axiom to the formula

$$(u = \emptyset \wedge v = a) \vee (u = \{\emptyset\} \wedge v = b).$$

7) Assume that a, b, c, d, e are different sets and

$$R = \{(a, b), (a, c), (a, d), (b, b), (b, c), (b, d), (c, c)\}.$$

Determine the following sets: $\mathrm{dom}(R)$, $\mathrm{ran}(R)$, $R[\{b, e\}]$, $R \upharpoonright \{b, e\}$, $R^{-1} \upharpoonright \{a, b\}$, $R^{-1}[\{a, b\}]$, $R \circ R$ and $\bigcup\bigcup R$. Is R reflexive, symmetric, antisymmetric, connected, transitive? Has $\mathrm{fld}(R)$ a minimal, maximal, greatest, smallest element?

8) In general, a function $F : A \longrightarrow B$ needs not be injective and F^{-1} does not exist. But prove that $(F^{-1})^* : \mathcal{P}(B) \longrightarrow \mathcal{P}(A)$, defined by $(F^{-1})^*(x) = F^{-1}[x]$ for any $x \in \mathcal{P}(B)$, is a function, and that $(F^{-1})^*$ is injective if F is surjective.

9) Let $f : A \longrightarrow B$ be a function, $x \subseteq A$ and $y \subseteq B$. Prove:

   a) $x \subseteq f^{-1}[f[x]]$.
   b) $f[f^{-1}[y]] \subseteq y$.
   c) If f is injective, then in a) equality holds.
   d) If f is surjective, then in b) equality holds.

   Give examples such that in a) and b) respectively equality does not hold.

10) Let X be a nonempty set. We call a set P a **partition** of X iff $\bigcup P = X$, $\emptyset \notin P$, and the members of P are pairwise disjoint. If (X, R) is an equivalence relation (we also say that R is an equivalence relation on X) and $x \in X$, then $\{y \in X : (x, y) \in R\}$ is the **equivalence class** of x (w.r.t. R).

   a) If R is an equivalence relation on X, then the set of equivalence classes is a partition of X.
   b) If P is a partition of X, and if the relation $R \subseteq X \times X$ is defined by $(a, b) \in R :\Longleftrightarrow \exists Y \in P(a \in Y \wedge b \in Y)$, then R is an equivalence relation on X.
   c) Assume that the set ω of natural numbers and $\mathbb{Z}$, the set of integers, are already known. For any $n \in \omega$ with $n \geq 2$ let R_n be defined by

   $$(a, b) \in R_n :\Longleftrightarrow n \text{ divides } b - a.$$

   Show that R_n is an equivalence relation on $\mathbb{Z}$, and determine the equivalence classes $\bar{a}$, $a \in \mathbb{Z}$, for $n = 6$. (Hint: Use division with remainder.)

d) Show that, for fixed $n \in \omega$,

$$G = \{(x, y) : \exists a \in \mathbb{Z} \exists c \in \mathbb{Z}(x = (\overline{a}, \overline{c}) \wedge y = \overline{a \cdot c})\}$$

is a function.

11) Assume that X is a nonempty set and $\mathrm{Part}(X)$ is the class of partitions of X. Let the relation $\preceq$ on $\mathrm{Part}(X)$ be defined by

$$P_1 \preceq P_2 :\Longleftrightarrow \forall Y \in P_1 \exists Z \in P_2 (Y \subseteq Z)$$

for $P_1, P_2 \in \mathrm{Part}(X)$. Prove:

a) $\mathrm{Part}(X)$ is a set.

b) $\preceq$ is a partial ordering on $\mathrm{Part}(X)$.

c) If $\mathcal{T}$ is a nonempty subset of $\mathrm{Part}(X)$, then $\mathcal{T}$ has an infimum under $\preceq$.
Hint: $\inf(\mathcal{T})$ is the partition
$\{Y \subseteq X : Y \neq \emptyset \wedge \exists D(D \subseteq \bigcup \mathcal{T} \wedge Y = \bigcap D \wedge \forall P \in \mathcal{T}\ |P \cap D| = 1)\}$.

d) If $\mathcal{T}$ is a nonempty subset of $\mathrm{Part}(X)$, then $\mathcal{T}$ has a supremum under $\preceq$.
Hint: $\sup(\mathcal{T}) = \inf\{P' \in \mathrm{Part}(X) : \forall P \in \mathcal{T}\ P \preceq P'\}$.

12) Assume that R is a reflexive transitive relation. Put $S := \{(x, y) : (x, y) \in R \wedge (y, x) \in R\}$. Show that S is an equivalence relation on $\mathrm{fld}(R)$. Further let $[a] := \{x \in \mathrm{fld}(R) : (a, x) \in S\}$ be the equivalence class of $a \in \mathrm{fld}(R)$. Prove that $T := \{([a], [b]) : (a, b) \in R\}$ is a partial ordering on the set $\{[a] : a \in \mathrm{fld}(R)\}$.

13) Assume that A and B are nonempty sets. Further let $(X_{(a,b)} : (a, b) \in A \times B)$ be a family of sets such that $X_{(a,b)} \cap X_{(a,c)} = \emptyset$ for all $a \in A$ and all $b, c \in B$ with $b \neq c$. Show with the axioms of ZF the existence of the families which are needed in the following equation and prove

$$\bigcap_{a \in A} \left(\bigcup_{b \in B} X_{(a,b)} \right) = \bigcup_{f \in {}^A B} \left(\bigcap_{a \in A} X_{(a, f(a))} \right).$$

14) Any two functions f and g are called **compatible** iff $f(x) = g(x)$ for all $x \in \mathrm{dom}(f) \cap \mathrm{dom}(g)$. Prove:

a) f and g are compatible iff $f \cup g$ is a function iff $f \restriction (\mathrm{dom}(f) \cap \mathrm{dom}(g)) = g \restriction (\mathrm{dom}(f) \cap \mathrm{dom}(g))$.

b) If F is a set of pairwise compatible functions, then $G := \bigcup F$ is a function with $\mathrm{dom}(G) = \bigcup\{\mathrm{dom}(f) : f \in F\}$ and $\mathrm{ran}(G) = \bigcup\{\mathrm{ran}(f) : f \in F\}$.

c) Let A and B be sets and $\mathrm{Func}(A, B)$ the class of all functions f with $\mathrm{dom}(f) \subseteq A$ and $\mathrm{ran}(f) \subseteq B$. Prove with the axioms of ZF that $\mathrm{Func}(A, B)$ is a set. We define a relation $\preceq$ on this set by

$$f \preceq g :\Longleftrightarrow f \subseteq g.$$

Show that $\preceq$ is a partial ordering.

d) Let $X \subseteq \mathrm{Func}(A, B)$. Show that X has a supremum under $\preceq$ in $\mathrm{Func}(A, B)$ iff X is a set of pairwise compatible functions.

1.2 Ordinals

In this section we will prove that every ordinal is well-ordered by $\in$. In Section 1.3 we will show that, for each well-ordered set $(A, <)$, there is a unique ordinal α such that $(A, <)$ is isomorphic to $(\alpha, \in)$. So it suffices to study the theory of ordinals to know the theory of well-orderings. Furthermore the principles of induction and recursion for natural numbers can be generalized for ordinals. So we get the fundamental tools of transfinite set theory. With the axiom of choice we will prove in Section 1.5 that every set can be well-ordered. This allows us to define for each set its cardinality or cardinal number as a special ordinal number.

Definition *A class A is* **transitive** *iff every member of A is a subset of A; this means formally:* $\forall y \forall z(z \in y \wedge y \in A \Longrightarrow z \in A)$. *An* **ordinal number** *or* **ordinal** *is a transitive set whose members are all transitive. We denote by* ON *the* **class of ordinals. Small greek letters** α, β, γ, δ, σ, τ, ξ, ζ **are variables for ordinals**. *The letters* κ, λ, μ, ν, ρ *are reserved for special ordinals, the cardinals or cardinal numbers.* **Instead of** $\sigma \in \tau$ **we also write** $\sigma < \tau$, *instead of* $\sigma < \tau \vee \sigma = \tau$ *we write* $\sigma \leq \tau$.

Example $\emptyset$ is an ordinal, and therefore $\mathrm{ON} \neq \emptyset$. Further no ordered pair is an ordinal. If we assume that $(a, b) = \{\{a\}, \{a, b\}\}$ is an ordinal, then $a \in (a, b)$, since every ordinal is transitive. Thus we get $a \in a$, from which we can infer that the set $\{a\}$ has no $\in$-minimal element, contradicting the axiom of regularity.

Lemma 1.2.1

a) *Any member of an ordinal is an ordinal.*
b) *Any nonempty class A of ordinals has an $\in$-minimal element (i.e.,* $A \subseteq \mathrm{ON} \wedge A \neq \emptyset \Longrightarrow \exists y(y \in A \wedge \forall z(z \in A \Longrightarrow z \notin y)))$*).*
c) $\sigma \notin \sigma$.[7]

[7]By the axiom of foundation, we get $x \notin x$ for every set x. The proof is exactly the same as that for σ.

Proof For a), let $x \in \sigma$. Since σ is transitive, we get $x \subseteq \sigma$. Therefore every member of x is transitive, since this holds for the members of σ. Finally, as a member of σ, x is transitive.

For b), let $\sigma \in A$. If $\sigma \cap A \neq \emptyset$, we can choose, by the axiom of foundation, an $\in$-minimal element of $\sigma \cap A$, which has no $\in$-predecessor in A, since σ is transitive.

To prove c), we assume that $\sigma \in \sigma$. The axiom of foundation implies that the nonempty subset $\{\sigma\}$ of σ has a member y satisfying $y \cap \{\sigma\} = \emptyset$. Certainly $y = \sigma$, and with $\sigma \in \sigma \cap \{\sigma\}$ we get a contradiction to $\sigma \cap \{\sigma\} = \emptyset$.

Corollary 1.2.2

a) ON *is not a set.*

b) Every class A of ordinals is partially ordered by $\in$.[8]

Proof To prove a), we suppose that there is a set y such that $y = \mathrm{ON}$. By Lemma 1.2.1, y is transitive. Furthermore, by definition, all members of y are transitive, and consequently y is an ordinal. This gives $y \in \mathrm{ON} = y$, contradicting Lemma 1.2.1 c).

Assertion b) follows immediately from Lemma 1.2.1 c) and the fact that all ordinals are transitive.

Lemma 1.2.3 $\in$ *is connected on* ON*; that means if σ and τ are ordinals, then*

$$\sigma \in \tau \vee \sigma = \tau \vee \tau \in \sigma.$$

Proof To get a contradiction, let us assume that the assumption is false. Then, by Lemma 1.2.1 b), the nonempty class

$$A := \{\sigma : \exists \tau (\sigma \notin \tau \wedge \sigma \neq \tau \wedge \tau \notin \sigma)\}$$

has an $\in$-minimal element, say σ_0. By the same reason, the nonempty class

$$B := \{\tau : \sigma_0 \notin \tau \wedge \sigma_0 \neq \tau \wedge \tau \notin \sigma_0\}$$

has an $\in$-minimal element, say τ_0. Thus $\sigma_0 \notin \tau_0$ and $\sigma_0 \neq \tau_0$ and $\tau_0 \notin \sigma_0$. The contradiction will have arrived if we show $\sigma_0 = \tau_0$. Every member of an ordinal is an ordinal. So let $\alpha \in \sigma_0$. Then $\alpha \notin A$, which implies $\alpha \in \tau_0$ or $\alpha = \tau_0$ or $\tau_0 \in \alpha$. From either of the last two assertions we can infer that $\tau_0 \in \sigma_0$, which is impossible. So we get $\alpha \in \tau_0$, hence $\sigma_0 \subseteq \tau_0$. In the same way we conclude successively from $\alpha \in \tau_0$ that $\alpha \notin B$, $\alpha \in \sigma_0$, and $\tau_0 \subseteq \sigma_0$. Altogether we have shown that $\sigma_0 = \tau_0$.

[8]If A does not have exactly one member, this means that $\in \cap (A \times A)$ is a partial ordering.

Corollary 1.2.4

a) Every class A of ordinals is well-ordered by $\in$.

b) If A is a transitive proper class and all members of A are transitive, then $A = \mathrm{ON}$.

c) $\sigma \in \tau \Longleftrightarrow \sigma \subsetneqq \tau$.

Proof Together with the preceding results, for the proof of a) we only have to verify the first condition in the definition of a well-founded relation. If $R := \in \cap (A \times A)$ and $\sigma \in \mathrm{fld}(R)$, then $y := \sigma \cup \{\sigma\}$ is a set satisfying $R^{-1}[y] \subseteq y$.

Next we prove part b). From the assumption we can easily infer that all members of A are ordinals. Consider, to get a contradiction, the ordinal $\alpha := \min(\mathrm{ON} \setminus A)$. Then A, being a proper class, is no subclass of α. So we can choose $\gamma \in A \setminus \alpha$. The previous lemma shows that $\alpha < \gamma$. But this yields $\alpha \in A$, since A is transitive, contradicting $\alpha \notin A$.

For the proof of c), we note that one implication of the equivalence holds by the definition of an ordinal and Lemma 1.2.1 c). For the converse, let $\sigma \subsetneqq \tau$. By Lemma 1.2.3 and $\tau \notin \tau$, we get $\sigma \in \tau$.

Definition *For any ordinals α and β such that $\alpha \leq \beta$, let*

$$(\alpha, \beta) := (\alpha, \beta)_{\mathrm{ON}} := \{x \in \mathrm{ON} : \alpha < x < \beta\},$$

$$(\alpha, \beta] := (\alpha, \beta]_{\mathrm{ON}} := \{x \in \mathrm{ON} : \alpha < x \leq \beta\}$$

be the **open interval, half open interval** *of ordinals with endpoints α and β, respectively. The intervals $[\alpha, \beta)_{\mathrm{ON}}$ and $[\alpha, \beta]_{\mathrm{ON}}$ are defined analogously, where $[\alpha, \beta]_{\mathrm{ON}}$ is called the* **closed interval** *with endpoints α and β.*

Lemma 1.2.5

a) If x is a set of ordinals, then $\bigcup x$ is an ordinal, and $\bigcup x = \sup(x)$.

b) If A is a nonempty class of ordinals, then $\bigcap A = \min(A)$.

Proof To prove a) we note that every member of $\bigcup x$ is a member of an ordinal, therefore is itself an ordinal and in particular transitive. In the same way one can show that $\bigcup x$ is transitive. Consequently, $\bigcup x$ is an ordinal. It is an upper bound of x, since $\tau \in x$ implies $\tau \subseteq \bigcup x$ and thus $\tau \leq \bigcup x$ by Corollary 1.2.4. If $\sigma < \bigcup x$, then σ is not an upper bound of x because there is an ordinal $\tau \in x$ such that $\sigma \in \tau$. So we have proved $\bigcup x = \sup(x)$.

For part b), Lemma 1.2.1 implies that A has an $\in$-minimal element σ_0. Obviously, this is the minimum of A. Certainly we have $\bigcap A \subseteq \sigma_0$. Conversely, if $\alpha < \sigma_0$, then $\alpha < \beta$ for all $\beta \in A$, and so we get $\alpha \in \bigcap A$. This proves $\sigma_0 = \bigcap A$.

Definition *If x is a set, then the set $S(x) := x \cup \{x\}$ is called the* **successor** *of x. An ordinal α is called a* **successor ordinal** *iff there is an ordinal σ such that $\alpha = S(\sigma)$. We also write $\sigma + 1$ instead of $S(\sigma)$. Let $0 := \emptyset$, called* **zero**. *An ordinal α is called a* **limit ordinal**, *written* $\mathrm{Lim}(\alpha)$, *iff $\alpha \neq 0$ and α is not a successor ordinal. We use* Lim *to denote the class of limit ordinals*[9].

Lemma 1.2.6

a) $S(\sigma)$ is an ordinal satisfying $\sigma < S(\sigma)$.
b) If $\sigma \leq \tau \leq S(\sigma)$, then $\tau = \sigma$ or $\tau = S(\sigma)$.
c) $\mathrm{Lim}(\alpha) \iff \alpha \neq 0 \wedge \bigcup \alpha = \alpha$.
d) Every ordinal is 0 or a successor ordinal or a limit ordinal.

Proof For part b), suppose that $\sigma < \tau < S(\sigma)$. Then $\sigma \in \sigma$, contradicting Lemma 1.2.1.

For part c), first let α be a limit ordinal and $\beta \in \alpha$. Then $\alpha \neq 0$ and, by part b), $\beta < \beta + 1 < \alpha$. So we get $\beta \in \bigcup \alpha$ and consequently $\alpha \subseteq \bigcup \alpha$. Since α is transitive, we have $\bigcup \alpha \subseteq \alpha$. Thus we have shown that $\bigcup \alpha = \alpha$. Conversely, if $\alpha \neq 0$ is a successor ordinal $\beta + 1$, then $\bigcup \alpha = \beta \neq \alpha$.

Lemma 1.2.7

a) There exists a limit ordinal.
b) If α_0 is the least limit ordinal and $C := \{x : x \text{ is inductive}\}$ is the class of inductive sets, then $\alpha_0 = \bigcap C$.

Proof First we prove part a). By the axiom of infinity, there is an inductive set y. Let the set x be given by an axiom of comprehension by $x := \{z \in y : z \in \mathrm{ON} \wedge z \subseteq y\}$. We want to show that x is an ordinal. First we have $0 \in x$, hence $x \neq \emptyset$, and every member of x is transitive. If $u \in z \in x$, we get $u \in y$ since $z \subseteq y$, and $u \in \mathrm{ON}$. Furthermore we have $u \subseteq y$, since $u \subseteq z$. This shows $u \in x$; x is transitive. Therefore x is an ordinal. Now suppose $x = S(\sigma)$. Then $S(\sigma) \subseteq y$, and $\sigma \in y$ leads to $S(\sigma) \in y$. This yields $x \in x$, contradicting Lemma 1.2.1.

To prove b), we note that α_0, as a limit ordinal, is an inductive set by Lemma 1.2.6. This implies $\bigcap C \subseteq \alpha_0$. To get a contradiction we assume that $\alpha_0 \setminus \bigcap C \neq \emptyset$. Then the ordinal $\sigma := \min(\alpha_0 \setminus \bigcap C)$ is a member of α_0 and thus must be a sucessor ordinal $S(\tau)$, since $0 \in \bigcap C$. Further we have $\tau \in \bigcap C$. This implies $\sigma = S(\tau) \in \bigcap C$, since any intersection of a nonempty class of inductive sets is inductive again, which contradicts the choice of σ.

---

[9] From the results in Section 1.3 we can easily infer that this is a proper class. If $\sigma \in \mathrm{ON}$ and ω is the set of natural numbers defined below, then we can define a function $(\sigma_n : n \in \omega)$ by $\sigma_0 := \sigma$ and $\sigma_{n+1} := \sigma_n + 1$; the ordinal $\tau := \sup\{\sigma_n : n \in \omega\}$ is a limit ordinal greater than σ.

Definition *We define*

$$\omega := \{x \in \mathrm{ON} : \ \forall z \leq x(\neg \mathrm{Lim}(z))\}.$$

ω is a set, since it is a subclass of the smallest limit ordinal. Further it is easy to show that ω is an inductive set. Now Lemma 1.2.7 says that ω is the least limit ordinal and the smallest inductive set. ω is called the **set of natural numbers**. *Each of its members is called a* **natural number**. *The first members of ω are* $0 := \emptyset$, $1 := \{0\}$, $2 := \{0, 1\}$, $\ldots$.

Corollary 1.2.8 (Induction Principle for ω, Complete Induction) *If A is an inductive subset of ω, then $A = \omega$. Formally:*

$$A \subseteq \omega \wedge 0 \in A \wedge \forall \alpha \in A(\alpha + 1 \in A) \implies A = \omega.$$

Exercises

1) Assume that a, b are sets and a is transitive. Prove:
   a) b is transitive iff $b \subseteq \mathcal{P}(b)$.
   b) If b is transitive, then $a \cup b$ is transitive.
   c) If b is transitive, then $a \cap b$ is transitive.
   d) If $S \subseteq \mathcal{P}(a)$, then $a \cup S$ is transitive.
   e) If every $x \in b$ is transitive, then $\bigcup b$ is transitive.
   f) Ist $a \neq \emptyset$, then $\bigcap a$ is transitive.
2) Let x be a set. Show that x is an ordinal iff $x = 1$ or

$$R := \{(y, z) : y \in x \wedge z \in x \wedge y \in z\}$$

   is a well-ordering with $\mathrm{fld}(R) = x$ such that $\forall y \in x(y = \{z : zRy\})$.
3) Show that the following assertions are equivalent for any set x:
   (i) x is an ordinal.
   (ii) x is transitive $\wedge \forall y \in x \forall z \in x(y \in z \vee y = z \vee z \in y)$.
   (iii) x is transitive and every transitive proper subset y of x is a member of x.
   (iv) $(x = 0 \vee 0 \in x) \wedge \forall y \in x(Sy = x \vee Sy \in x) \wedge \forall y(y \subseteq x \implies \bigcup y = x \vee \bigcup y \in x)$.

   Hint: We have shown above that (i) implies the other assertions. Therefore it is practical to prove that they are all equivalent to (i). For "(iii) $\implies$ (i)", consider the set $y := \{\alpha \in \mathrm{ON} : \alpha \in x\}$. For "(iv) $\implies$ (i)", consider $\alpha := \min(\mathrm{ON} \setminus x)$.

4) Let $\mathbb{R}$ be the **set of real numbers**. Show that any subset of $\mathbb{R}$ which is well-ordered by the usual ordering on $\mathbb{R}$ is countable.

   Hint: In a well-ordered set without a maximum, any member has an immediate successor.

5) Prove: If X is a nonempty set of ordinals without a maximum, then $\sup(X)$ is a limit ordinal.

1.3 Transfinite Induction and Recursion

We remind the reader that a relation R is called well-founded (written as wf(R)) iff

(i) $\forall x \in \text{fld}(R)\exists y(x \in y \wedge R^{-1}[y] \subseteq y)$ and

(ii) $\forall x \subseteq \text{fld}(R)(x \neq \emptyset \implies \exists y \in x \forall z \in x(\neg zRy))$.

In this definition, condition (i) has technical reasons. We will show that it can be replaced by the condition "the class of R-predecessors of each member x of fld(R) is a set", which intuitively is more intelligible. From this condition and the axiom of foundation it follows immediately that, for any class A, the relation $\in \cap A \times A$ is well-founded.

Lemma 1.3.1 *If R is a well-founded relation, then R is irreflexive, and any nonempty class $A \subseteq \text{fld}(R)$ has an R-minimal element.*

Proof If $x \in \text{fld}(R)$, then there is $y \in \{x\}$ such that $\neg zRy$ holds for all $z \in \{x\}$. Thus we get $\neg xRx$.

To show the second assertion, we take a member $x \in A$ and, by the above definition, a set y such that $x \in y$ and $R^{-1}[y] \subseteq y$. Then the set $y \cap A$ has an R-minimal element z. If uRz, then $u \in y$, since $R^{-1}[y] \subseteq y$, and thus $u \notin A$ by the choice of z. This shows that z is an R-minimal element of A.

Definition *If R is a relation and $x \in \text{fld}(R)$, we say that $\widehat{x} := \{y : yRx \wedge y \neq x\}$ is the* **class of strict R-predecessors of x**. *If R is well-founded, then $\widehat{x} = \{y : yRx\}$.*

Theorem 1.3.2 (Transfinite Induction Principle) *If R is a well-founded relation, then the following holds:*

$$\textit{If } \ \forall x \in \text{fld}(R)(\forall y \in \widehat{x}\ \varphi(y) \implies \varphi(x))\ , \textit{ then } \ \forall x \in \text{fld}(R)\ \varphi(x).$$

More precisely: If φ is a formula of $\mathcal{L}_{\text{ZF}}$, then

$$\text{ZF} \vdash \ \text{wf}(R) \wedge \forall x \in \text{fld}(R)(\forall y \in \widehat{x}\ \varphi(y) \implies \varphi(x)) \implies \forall x \in \text{fld}(R)\,\varphi(x).$$

Proof We assume, to get a contradiction, that the conclusion is false and the premiss is true. Then Lemma 1.3.1 says that there is an R-minimal element $x \in \mathrm{fld}(R)$ satisfying $\neg\varphi(x)$. So we have $\forall y \in \widehat{x}\varphi(y)$. From our assumption we get $\varphi(x)$, contradicting $\neg\varphi(x)$.

Remark 1.3.3 *Important special cases of Theorem 1.3.2 are:*

a) $R = \in \cap (\mathrm{ON} \times \mathrm{ON})$.
So we get the **Transfinite Induction Principle for ON***:*

$$\mathrm{ZF} \vdash \forall\alpha(\forall\beta \in \alpha\, \varphi(\beta) \Longrightarrow \varphi(\alpha)) \Longrightarrow \forall\alpha\, \varphi(\alpha).$$

This is also a direct consequence of Lemma 1.2.1.

b) $R = \in \cap\, \sigma \times \sigma$ *for* $\sigma \in \mathrm{ON}$.
In particular we obtain for $\sigma = \omega$ *the induction principle for* ω.

c) $R = \in$*: From the axiom of regularity and Lemma 1.3.8 (see below) we can conclude that the relation* $\in$ *is well-founded. Thus we can use in* ZF *the so-called*
$\in$**-Induction Principle***:*

$$\forall x(\forall y \in x\, \varphi(y) \Longrightarrow \varphi(x)) \Longrightarrow \forall x\varphi(x).$$

Corollary 1.3.4 (2nd Form of the Transfinite Induction Principle for ON)

$$\mathrm{ZF} \vdash \varphi(0) \land \forall\alpha(\varphi(\alpha) \Longrightarrow \varphi(\alpha+1)) \land \forall\alpha(\mathrm{Lim}(\alpha) \land \forall\beta \in \alpha\,\varphi(\beta) \\ \Longrightarrow \varphi(\alpha)) \Longrightarrow \forall\alpha\varphi(\alpha).$$

Proof This principle follows immediately from the transfinite induction principle for ON if one notices that every ordinal is either 0 or a successor ordinal or a limit ordinal.

Theorem 1.3.5 (Transfinite Recursion Theorem Schema) *If R, G and F are classes, then let* $\mathrm{rec}(F, R, G)$ *be an abbreviation for the formula*

$$\mathrm{Func}(F) \land \mathrm{dom}(F) = \mathrm{fld}(R) \land \forall x \in \mathrm{fld}(R)(F(x) = G(F \restriction \widehat{x})).$$

If R, F' and G are classes, then we can construct a class F, such that the following formula is provable in ZF*:*

$$(\mathrm{Rel}(R) \land \mathrm{wf}(R) \land G : \mathrm{V} \longrightarrow \mathrm{V}) \Longrightarrow \mathrm{rec}(F, R, G) \land (\mathrm{rec}(F', R, G) \Longrightarrow F = F').$$

We say that the function F is **defined by transfinite induction** *or* **defined by recursion** *or* **defined recursively** *with respect to the well-founded relation R and the function G.*

Proof The uniqueness of F can be easily shown, using Theorem 1.3.2. Namely, if $x \in \operatorname{fld}(R)$ and $F(z) = F'(z)$ for all $z \in \widehat{x}$, then $F \restriction \widehat{x} = F' \restriction \widehat{x}$, and so we get $F(x) = G(F \restriction \widehat{x}) = G(F' \restriction \widehat{x}) = F'(x)$. From Theorem 1.3.2 we can conclude that $\forall x \in \operatorname{fld}(R)\ F(x) = F'(x)$. Now we define the class F and prove that it has the desired properties. Let $F := \bigcup K$, where

$$K := \{f : \operatorname{Func}(f) \land \operatorname{dom}(f) \subseteq \operatorname{fld}(R) \land \forall x \in \operatorname{dom}(f)(\widehat{x} \subseteq \operatorname{dom}(f) \\ \land f(x) = G(f \restriction \widehat{x}))\}.$$

Claim 1 F is a function.
For this we show $f(x) = g(x)$ for any $f, g \in K$ and $x \in \operatorname{dom}(f) \cap \operatorname{dom}(g)$. We use transfinite induction, i.e., we apply Theorem 1.3.2, and assume that the assertion is true for all $z \in \widehat{x}$. Since $\widehat{x} \subseteq \operatorname{dom}(f) \cap \operatorname{dom}(g)$, this gives $f(z) = g(z)$ for all $z \in \widehat{x}$. Thus we get $f \restriction \widehat{x} = g \restriction \widehat{x}$ and, since G is a function, $f(x) = G(f \restriction \widehat{x}) = G(g \restriction \widehat{x}) = g(x)$.

Claim 2 $\forall x \in \operatorname{dom}(F)(F(x) = G(F \restriction \widehat{x}))$.
Let $x \in \operatorname{dom}(F)$, that means $x \in \operatorname{dom}(f)$ for some $f \in K$. Then $\widehat{x} \subseteq \operatorname{dom}(f) \subseteq \operatorname{dom}(F)$, hence $f \restriction \widehat{x} = F \restriction \widehat{x}$ and $F(x) = f(x) = G(f \restriction \widehat{x}) = G(F \restriction \widehat{x})$.

Claim 3 $\operatorname{dom}(F) = \operatorname{fld}(R)$.
By the definition of F, we have $\operatorname{dom}(F) \subseteq \operatorname{fld}(R)$. Now let $x \in \operatorname{fld}(R)$ and assume, for an application of Theorem 1.3.2, that $z \in \operatorname{dom}(F)$ for all $z \in \widehat{x}$, i.e., $\widehat{x} \subseteq \operatorname{dom}(F)$. Since R is well-founded, there is a set y such that $x \in y$ and $R^{-1}[y] \subseteq y$. From $\{x\} \subseteq y$ we get $R^{-1}[\{x\}] \subseteq R^{-1}[y] \subseteq y$, hence $\widehat{x} \subseteq y$. Put $f := F \restriction y$. If $x \in \operatorname{dom}(f)$, then $x \in \operatorname{dom}(F)$, and we are done.

So assume that $x \notin \operatorname{dom}(f)$. Since F is a function, f is also a function, and furthermore f is a set by the axiom of replacement. Obviously we have $\operatorname{dom}(f) = y \cap \operatorname{dom}(F)$ and $\widehat{x} \subseteq \operatorname{dom}(f)$. Therefore, $f' := f \cup \{(x, G(f \restriction \widehat{x}))\}$ is a function such that $\operatorname{dom}(f') \subseteq \operatorname{fld}(R)$. Now one easily verifies $\widehat{z} \subseteq \operatorname{dom}(f')$ and $f'(z) = G(f' \restriction \widehat{z})$ for any $z \in \operatorname{dom}(f')$. Namely, if $z = x$, then $\widehat{x} \subseteq \operatorname{dom}(f) \subseteq \operatorname{dom}(f')$, hence $f \restriction \widehat{x} = f' \restriction \widehat{x}$ and $f'(x) = G(f \restriction \widehat{x}) = G(f' \restriction \widehat{x})$. On the other hand, if $z \neq x$, then $z \in \operatorname{dom}(f) \subseteq y$, which gives $\widehat{z} \subseteq R^{-1}[y] \subseteq y$. Furthermore $\widehat{z} \subseteq \operatorname{dom}(F)$, since $z \in \operatorname{dom}(F)$. So we can conclude that $\widehat{z} \subseteq \operatorname{dom}(f) \subseteq \operatorname{dom}(f')$ and $f'(z) = f(z) = F(z) = G(F \restriction \widehat{z}) = G(f \restriction \widehat{z}) = G(f' \restriction \widehat{z})$. From this, we get $f' \in K$ and $x \in \operatorname{dom}(F)$. Since $x \in y$, we have $x \in \operatorname{dom}(f)$. This contradiction completes the proof of the theorem.

Remark

1) Let us call each member of the class K defined in the proof of Theorem 1.3.5 an **approximation** of F. Then it follows from the preceding arguments that

   - For each $x \in \operatorname{fld}(R)$, there is an approximation f of F with $F(x) = f(x)$.

- If $x \in \text{fld}(R)$ and $f \in \text{V}$ is an arbitrary function satisfying $\widehat{x} \subseteq \text{dom}(f)$ and $F(z) = f(z)$ for all $z \in \widehat{x}$, then $F \upharpoonright \widehat{x} = f \upharpoonright \widehat{x}$, hence $F(x) = G(F \upharpoonright \widehat{x}) = G(f \upharpoonright \widehat{x})$. If in addition R is transitive, then $f \upharpoonright \widehat{x}$ is an approximation of F.

2) Theorem 1.3.5 is a metatheorem, since it says something about the existence of classes. Given the formulae $\chi(y,z)$ and $\psi(y,z)$, defining the classes R and G respectively, we can find a formula $\varphi(y,z)$ defining the class F such that that formula of $\mathcal{L}_{\text{ZF}}$ which results from the formula in Theorem 1.3.5 by the elimination of classes is provable in ZF. For example, under the above premisses, for the formula $\forall x \in \text{fld}(R)(F(x) = G(F \upharpoonright \widehat{x}))$ (in the extended sense) there is a corresponding formula of ZF:
$\forall x(\exists z(\chi(x,z) \vee \chi(z,x)) \implies \exists y(\varphi(x,y) \wedge$
$\forall f(\text{Func}(f) \wedge \text{dom}(f) = \widehat{x} \wedge \forall z \in \widehat{x}\ \varphi(z, f(z)) \implies (\varphi(x,y) \iff \psi(f,y)))))$,
where $z \in \widehat{x}$ is the formula $\chi(z,x) \wedge z \neq x$ and $\text{dom}(f) = \widehat{x}$ is the formula $\forall z(z \in \text{dom}(f) \iff \chi(z,x) \wedge z \neq x)$.

3) The proof of Theorem 1.3.5 can be transferred word for word if we formulate Theorem 1.3.5 for functions G whose defining formulae contain parameters. Formally, the above formula $\psi(y,z)$ defining G may contain further free (set) variables in addition to y and z. In this connection we never mention explicitly the parameters. Well known examples are the sum, the product and the exponentiation of ordinals which will be defined in the example after the next corollary.

4) A further important example is the case that we want to define the function F recursively by $F(x) = G(x, F \upharpoonright \widehat{x})$ for all $x \in \text{fld}(R)$. The proof of Theorem 1.3.5 works for this case if we replace in the definition of the class K the equation $f(x) = G(f \upharpoonright \widehat{x})$ by $f(x) = G(x, f \upharpoonright \widehat{x})$.

Corollary 1.3.6 (Transfinite Recursion Principle for ON)

a) If G is a class, then we can construct a class F such that
$\text{ZF} \vdash G : \text{V} \longrightarrow \text{V} \implies \text{Func}(F) \wedge \text{dom}(F) = \text{ON} \wedge \forall \sigma(F(\sigma) = G(F \upharpoonright \sigma))$.

b) If a, H_1 and H_2 are classes, then we can construct a class F such that
$\text{ZF} \vdash H_1 : \text{V} \longrightarrow \text{V} \wedge H_2 : \text{V} \longrightarrow \text{V} \wedge \exists y(y = a) \implies$
$\text{Func}(F) \wedge \text{dom}(F) = \text{ON} \wedge F(0) = a \wedge \forall \sigma(F(\sigma + 1) = H_1(F(\sigma)) \wedge (\text{Lim}(\sigma) \implies F(\sigma) = H_2(F \upharpoonright \sigma)))$.

Proof Part a) follows from Theorem 1.3.5, if we take into account that the relation $\in \cap (\text{ON} \times \text{ON})$ is well-founded and satisfies $\widehat{\sigma} = \{z : z \in \text{ON} \wedge z \in \sigma\} = \sigma$.

For the proof of b), we note again that $\in \cap (\text{ON} \times \text{ON})$ is well-founded and has the field ON. To apply Theorem 1.3.5, we define the function G from V into V by $G(0) := a$; $G(x) := H_1(x(\sigma))$, if x is a function such that $\text{dom}(x) =$

$\sigma + 1$ for some $\sigma \in \mathrm{ON}$; $G(x) := H_2(x)$, if x is a function whose domain is a limit ordinal; $G(x) := 0$ in all other cases. Let F be the class given by Theorem 1.3.5. Then we get: $F(0) = G(F \upharpoonright 0) = G(0) = a$; $F(\sigma + 1) = G(F \upharpoonright \sigma + 1) = H_1(F(\sigma))$, since $F \upharpoonright \sigma + 1$ is a function with domain $\sigma + 1$; $F(\sigma) = G(F \upharpoonright \sigma) = H_2(F \upharpoonright \sigma)$ for limit ordinals σ, since $F \upharpoonright \sigma$ is a function whose domain is the limit ordinal σ. This completes the proof of the corollary.

Examples 1.3.7

a) From Corollary 1.3.6 and Theorem 1.3.5 we obtain by an obvious choice of H_1, H_2 and of G, respectively, the usual recursion principles for ω, taking $f := F \upharpoonright \omega$:

   For each function $g : \omega \longrightarrow \omega$ and each $a \in \omega$ there exists a unique function $f : \omega \longrightarrow \omega$ satisfying $f(0) = a$ and $f(n+1) = g(f(n))$ for all $n \in \omega$.

   For each function g with an appropriate domain, there is a unique function f such that $f(n) = g(f \upharpoonright n)$ for all $n \in \omega$.

b) Depending of the parameter $\alpha \in \mathrm{ON}$, we will define in Section 1.4 for every $\beta \in \mathrm{ON}$ the sum $\alpha + \beta$, the product $\alpha \cdot \beta$ and the exponentiation α^β; that means that we will apply the principle of transfinite recursion.

c) **von Neumann's Hierarchy:** For $a := \emptyset$ and the functions H_1 and H_2, given by $H_1(x) := \mathcal{P}(x)$ and $H_2(x) := \bigcup \mathrm{ran}(x)$, where $\mathrm{ran}(x) := \{v : \exists u\ (u, v) \in x\}$, we consider the function F from Corollary 1.3.6 and put $\mathrm{V}_\alpha := F(\alpha)$. Then

$$\mathrm{V}_0 = \emptyset; \quad \mathrm{V}_{\alpha+1} = \mathcal{P}(\mathrm{V}_\alpha); \quad \mathrm{V}_\gamma = \bigcup\{\mathrm{V}_\beta :\ \beta < \gamma\} \text{ for limit ordinals } \gamma\ .$$

The sets V_α ($\alpha \in \mathrm{ON}$) are called the **levels of the von Neumann hierarchy**, the function ($\mathrm{V}_\alpha : \alpha \in \mathrm{ON}$) is called **von Neumann's hierarchy** or **cumulative hierarchy**.

We use the principle of $\in$-induction to prove that $\bigcup\{\mathrm{V}_\alpha : \alpha \in \mathrm{ON}\}$ is the universal class. First, it is easy to see by transfinite induction that every set V_α is transitive and $\mathrm{V}_\alpha \subseteq \mathrm{V}_\beta$ for any α and β with $\alpha < \beta$. If $x \in \mathrm{V}$ is an arbitrary set, and if we assume that $y \in \bigcup\{\mathrm{V}_\alpha : \alpha \in \mathrm{ON}\}$ for every member y of x, then, for each $y \in x$, there is a least ordinal α_y such that $y \in \mathrm{V}_{\alpha_y+1}$. By an axiom of replacement and Lemma 1.2.5, there exists the ordinal $\alpha := \sup\{\alpha_y : y \in x\}$. Now we can conclude that $x \subseteq \mathrm{V}_\alpha$, hence $x \in \mathrm{V}_{\alpha+1}$. This implies

$$\mathrm{V} = \bigcup\{\mathrm{V}_\alpha : \alpha \in \mathrm{ON}\}.\text{[10]}$$

[10] We cannot prove this assertion without the axiom of regularity.

The least ordinal α with $x \in \mathrm{V}_{\alpha+1}$ is called the **rank** of the set x, written as $\mathrm{rk}(x)$. Sometimes it is practical to use induction on the rank of a set. If $y \in x \in \mathrm{V}_{\alpha+1}$, then $y \in x \subseteq \mathrm{V}_\alpha$, and so we get $\mathrm{rk}(y) < \mathrm{rk}(x)$.

Lemma 1.3.8 *Let R be a relation.*

a) The following assertions are equivalent:

(i) For every $x \in \mathrm{fld}(R)$, $\widehat{x}$ is a set.

(ii) For each $x \in \mathrm{fld}(R)$, there is a set y such that $x \in y$ and $R^{-1}[y] \subseteq y$.

b) R is well-founded iff

(1) For every $x \in \mathrm{fld}(R)$, $\widehat{x}$ is a set and

(2) There is not a family $(x_n : n \in \omega)$ of members of $\mathrm{fld}(R)$ satisfying $x_{n+1} R x_n$ for all $n \in \omega$.

Proof For the proof of part a), we note first that the implication "(ii) $\Longrightarrow$ (i)" follows from an axiom of subsets, since, for $x \in y$, $\widehat{x} \subseteq R^{-1}[\{x\}] \subseteq R^{-1}[y] \subseteq y$.

Now let us assume (i). Then for every set z the class $\{\widehat{u} : u \in z \wedge u \in \mathrm{fld}(R)\}$ is a set, since it is the range of the restriction of the function $F := \{(u, \widehat{u}) : u \in \mathrm{V}\}$ to the set $z \cap \mathrm{fld}(R)$. Consequently, the class $\bigcup\{\widehat{u} : u \in z \cap \mathrm{fld}(R)\} \cup z$ is also a set, and the class $R^{-1}[z]$ is a subclass of it. Thus $R^{-1}[z]$ is a set. Take $x \in \mathrm{fld}(R)$. Then Corollary 1.3.6 b) tells us that there is a function $f : \omega \longrightarrow \mathrm{V}$ satisfying $f(0) = \{x\}$ and $f(n+1) = R^{-1}[f(n)]$ for every $n \in \omega$. Put $y := \bigcup \mathrm{ran}(f) = \bigcup\{f(n) : n \in \omega\}$. Then we have $x \in f(0) \subseteq y$. Furthermore, if $u \in R^{-1}[y]$, then there is $v \in y$ with uRv. Since $v \in f(m)$ for some $m \in \omega$, our construction gives $u \in R^{-1}[f(m)] = f(m+1) \subseteq y$.

Next we turn to the proof that every well-founded relation satisfies properties (1) and (2). For (1), this is clear from part a). If there was a family $(x_n : n \in \omega)$ of members of $\mathrm{fld}(R)$ satisfying $x_{n+1} R x_n$ for all $n \in \omega$, then the range $\{x_n : n \in \omega\}$ of this function would not have a minimal element under R.

Conversely let R be a relation satisfying (1) and (2). Then a) implies that the first condition in the definition of a well-founded relation is fulfilled. For the second condition, we need (AC). Assume, to get a contradiction, that there is a nonempty subset X of $\mathrm{fld}(R)$ without a minimal element. Let g be a choice function for the set $\mathcal{P}(X) \setminus \{\emptyset\}$. By the recursion theorem, there is a function $(x_n : n \in \omega)$ such that $x_0 \in X$ and $x_{n+1} = g(\{y \in X : y R x_n\})$. It is easy to verify that we have obtained a function whose existence contradicts (2).

Definition and Example Let x be an arbitrary set. If there exists a transitive superset of x, then we obtain, taking the intersection of all these sets, an $\subseteq$-smallest set with this property. The recursion theorem provides us directly with this set. There is a function f satisfying $f(0) = x$ and $f(n+1) = \bigcup f(n)$ for every $n \in \omega$. Put $y := \bigcup \mathrm{ran}(f)$. Then clearly $x \subseteq y$. If $u \in v \in y$, then $v \in f(m)$ for some $m \in \omega$, hence $u \in \bigcup f(m) = f(m+1) \subseteq y$. Thus y is transitive. Furthermore we can conclude, using complete induction, that y is a subset of every transitive superset of x. It is called the **transitive closure** of x and is denoted by $\mathrm{tc}(x)$.

Lemma 1.3.9 *If A_1 and A_2 are transitive classes and $\pi : A_1 \longrightarrow A_2$ is an $\in$-isomorphism, then $A_1 = A_2$ and $\pi = \mathrm{id}_{A_1} = \{(x, x) : x \in A_1\}$.*

Proof Consider $x \in A_1$. We will show that $\pi(x) = \{\pi(y) : y \in x\}$. Then we can use transfinite induction on the well-founded relation $\in \cap (A_1 \times A_1)$ and obtain immediately $\pi(x) = x$ from the inductive hypothesis $\forall y \in x(\pi(y) = y)$, which proves the lemma by Theorem 1.3.2. The inclusion "$\supseteq$" holds by assumption. For the reverse inclusion let $z \in \pi(x)$. Then $z \in A_2$, since A_2 is transitive. Thus $z = \pi(u)$ for some $u \in A_1$. This gives $u \in x$, and thus $z \in \{\pi(y) : y \in x\}$, as desired.

Definition and Example *A relation R is called* **extensional** *iff*

$$\forall x, y \in \mathrm{fld}(R)(\widehat{x} = \widehat{y} \Longrightarrow x = y).$$

We call a class A **extensional** *iff $\forall x, y \in A(x \cap A = y \cap A \Longrightarrow x = y)$.*

The relation $\in$ is extensional – as stated by the axiom of extensionality. Any transitive class is extensional. Further every linear ordering $<$ is extensional; if we assume $\widehat{x} = \widehat{y}$ and $x \neq y$, then connectivity implies that, without loss of generality, $x < y$. Then $x \in \widehat{y} = \widehat{x}$, and thus we have $x < x$ which is a contradiction.

Theorem 1.3.10 (Mostowski's Collapsing Lemma)
If the relation R is well-founded and extensional, then there is a unique transitive class A and a unique isomorphism $\pi : \mathrm{fld}(R) \longrightarrow A$ with respect to R and $\in$.

Proof First we will prove uniqueness. If $\pi_1 : \mathrm{fld}(R) \longrightarrow A_1$ and $\pi_2 : \mathrm{fld}(R) \longrightarrow A_2$ are isomorphisms satisfying the properties in the theorem, then $\pi_2 \circ \pi_1^{-1} : A_1 \longrightarrow A_2$ is an $\in$-isomorphism. From Lemma 1.3.9 we know that $A_1 = A_2$ and $\pi_2 \circ \pi_1^{-1} = \mathrm{id}_{A_1}$, hence $\pi_1 = \pi_2$.

The existence of π follows from the recursion theorem, applied to the well-founded relation R. Let the function G be defined by $G(x) := \mathrm{ran}(x) := \{z : \exists y\ (y, z) \in x\}$. Then Theorem 1.3.5 yields a function $\pi : \mathrm{fld}(R) \longrightarrow \mathrm{V}$ such that

$$\pi(x) = G(\pi \restriction \widehat{x}) = \mathrm{ran}(\pi \restriction \widehat{x}) = \{\pi(y) : yRx\}.$$

(π is called the **collapsing** or **Mostowski isomorphism** for R.)
We put $A := \pi[\mathrm{fld}(R)]$ and show that π and A have the desired properties.

Claim 1 π is injective.

Assume, to get a contradiction, that there are $x, y \in \text{fld}(R)$ such that $x \neq y$ and $\pi(x) = \pi(y)$. By Lemma 1.3.1, there is an R-minimal element x_0 of the class $\{x \in \text{fld}(R) : \exists y \in \text{fld}(R)(\pi(x) = \pi(y) \wedge x \neq y)\}$. In the same way we get an R-minimal element y_0 of $\{y \in \text{fld}(R) : \pi(x_0) = \pi(y) \wedge x_0 \neq y\}$. Since R is extensional, we have $\widehat{x_0} \neq \widehat{y_0}$. So consider, without loss of generality, $z \in \widehat{x_0} \setminus \widehat{y_0}$. Then $\pi(z) \in \pi(x_0)$. If we suppose that $\pi(z) \in \pi(y_0)$, i.e., $\pi(z) = \pi(u)$ for some $u \in \widehat{y_0}$, then the choice of x_0 and $z \in \widehat{x_0}$ yield $z = u$. This gives $z \in \widehat{y_0}$, a contradiction. Therefore $\pi(z) \in \pi(x_0) \setminus \pi(y_0)$, contradicting $\pi(x_0) = \pi(y_0)$.

Claim 2 π is an isomorphism w.r.t. R and $\in$.

By Claim 1, π is injective, and by the definition of A, π is a surjective function from $\text{fld}(R)$ onto A. If yRx, then $\pi(y) \in \pi(x)$ by the above properties of π. Conversely, if $\pi(y) \in \pi(x)$, then there exists u such that uRx and $\pi(y) = \pi(u)$. Claim 1 shows that $y = u$, hence yRx.

Claim 3 A is transitive.

If $z \in y \in A$, then $y = \pi(x)$ for some $x \in \text{fld}(R)$. Thus $z = \pi(u)$ for some u with uRx, and consequently $z \in A$.

Corollary 1.3.11

a) *Any extensional class B is $\in$-isomorphic to a transitive class A.*

b) *If $(x, \prec)$ is a well-ordering, then there is a unique ordinal α and a unique isomorphism $\pi : (x, \prec) \longrightarrow (\alpha, \in)$.*

c) *If B is a proper class, well-ordered by the relation $\prec$, then B is isomorphic to* ON *w.r.t. $\prec$ and $\in$.*

Proof For the proof of part a), we observe first that for $B = \emptyset$ we can take $A = \emptyset$ and for $B = \{x\}$ we can take $A = \{0\}$. So we can assume that B has at least two members. By the remark at the beginning of this section, the relation $R := \in \cap (B \times B)$ is well-founded, by assumption it is extensional, and clearly $\text{fld}(R) \subseteq B$. To show that $\text{fld}(R) = B$, fix $x \in B$ and assume, to get a contradiction, that $x \notin \text{fld}(R)$. The class $B \setminus \{x\}$ has an $\in$-minimal element y. By the assumption on x, the set of $\in$-predecessors of y in B is empty, and the same holds for x. Since B is extensional, we get $x = y$, contradicting $y \in B \setminus \{x\}$. So we have shown that $B = \text{fld}(R)$, and Theorem 1.3.10 provides us with an isomorphism π from B onto a transitive class.

For b), assume without loss of generality that x has at least two elements and choose A and π according to Theorem 1.3.10. Then A is a transitive set. Furthermore it is easy to see that every member of A is transitive. Namely, if $u \in v \in z$ and $z \in A$, then $u \in A$ and $v \in A$, since A is transitive. Since π is an isomorphism and the relation $\prec$ is transitive, we get $u \in z$. This shows that A is an ordinal.

For c), choose again A and π according to Theorem 1.3.10. As in the proof of b) we can conclude that every member of A is transitive. Furthermore, A is a proper class, since otherwise $B = \pi^{-1}[A]$ would be a set. Now Corollary 1.2.4 tells us that $A = \mathrm{ON}$.

Definition *Let $(B, \prec)$ be a well-ordering. The unique isomorphism given by Corollary 1.3.11 is called the* **order isomorphism** *of $(B, \prec)$. If B is a set, then we call the ordinal characterized by Corollary 1.3.11 b) the* **ordinal number** *or* **order type** *of $(B, \prec)$ and write it as* $\mathrm{otp}(B, \prec)$.

Exercises

1) Prove the following induction principles:

   a) (**Finite Induction**) If $k \in \omega$, then

   $$\varphi(0) \wedge \forall n \in k(\varphi(n) \Longrightarrow \varphi(n+1)) \Longrightarrow \forall n \leq k\ \varphi(n).$$

   b) If $\forall m, n \in \omega(\forall k, l \in \omega((k \in m \vee (k = m \wedge l \in n)) \Longrightarrow \varphi(k, l)) \Longrightarrow \varphi(m, n))$, then $\forall m, n \in \omega\ \varphi(m, n)$.
   Hint: Define a suitable well-founded relation on $\omega \times \omega$.

2) a) Why are the relations $\mathrm{ON} \times \{0\}$ and $\{(\alpha+1, \alpha) : \alpha < \tau\}$, for any limit ordinal τ, not well-founded?

   b) Let α and β be infinite ordinals and the relation $\prec$ on $\alpha \times \beta$ be defined by $(\sigma_1, \sigma_2) \prec (\tau_1, \tau_2) \Longleftrightarrow \sigma_1 < \tau_1 \wedge \sigma_2 < \tau_2$. Is $\prec$ well-founded? Is it extensional?

3) Give an example of a linear ordering $<$ on a set A and of a $\subseteq$-chain K of subsets of A which are well-ordered by $<$, such that $\bigcup K$ is not well-ordered by $<$.

4) For $\alpha, \beta \in \mathrm{ON}$ let $\mathrm{fin}(\beta, \alpha)$ be the set of all functions f from β into α for which the set $\{\sigma < \beta : f(\sigma) \neq 0\}$ is finite. Now we define the **generalized Hebrew lexicographic ordering**; for $f, g \in \mathrm{fin}(\beta, \alpha)$ we put

   $$f \prec g \ :\Longleftrightarrow\ f \neq g \wedge (\tau = \max\{\sigma \in \beta : f(\sigma) \neq g(\sigma)\} \Longrightarrow f(\tau) < g(\tau)).$$

   Prove that $(\mathrm{fin}(\beta, \alpha), \prec)$ is a well-ordering.
   Hint: It is easy to verify that $\prec$ is a linear ordering. To show that every nonempty subset X of $\mathrm{fin}(\beta, \alpha)$ has a smallest element, use induction on β. Choose $f \in X$, $f \not\equiv 0$. Consider $\tau := \max\{\sigma < \beta : f(\sigma) \neq 0\}$ and $\gamma := \min\{g(\tau) : g \prec f \wedge g \in X\}$.

5) Assume that $(A, <)$ is a well-ordering. Prove:

   a) If $f : (A, <) \longrightarrow (A, <)$ is a homomorphism, then $a \leq f(a)$ for all $a \in A$.

   b) $(A, <)$ is not isomorphic to any of its proper initial segments.

6) Assume that $(A, <)$ and $(B, \prec)$ are well-orderings. As usual we write $(C, <)$ instead of $(C, < \cap (C \times C))$ for substructures C of A. For $u \in A$ we put $A_u := \{x \in A : x < u\}$, the **initial segment** of A **determined by** $\boldsymbol{u}$. Prove without using Corollary 1.3.11:
   Either $(A, <)$ is isomorphic to $(B, \prec)$, or there is $v \in B$ such that $(A, <)$ is isomorphic to $(B_v, \prec)$, or there is $u \in A$ such that $(A_u, <)$ is isomorphic to $(B, \prec)$.
   Hint: Show that
   $$F := \{(x, y) \in A \times B : (A_x, <) \text{ is isomorphic to } (B_y, \prec)\}$$
   is an order preserving function or, alternatively, use the recursion theorem.

7) Assume that $(W_\alpha : \alpha \in \mathrm{ON})$ is a function such that $W_\gamma = \bigcup\{W_\sigma : \sigma < \gamma\}$ for limit ordinals γ and $W_\alpha \subseteq W_{\alpha+1}$ for all $\alpha \in \mathrm{ON}$. Prove that $W_\alpha \subseteq W_\beta$ whenever $\alpha \in \beta \in \mathrm{ON}$.
   Hint: Use transfinite induction on β.

8) Prove the following properties of the cumulative hierarchy $(\mathrm{V}_\alpha : \alpha \in \mathrm{ON})$.

   a) V_α is transitive.

   b) $\alpha < \beta \Longrightarrow \mathrm{V}_\alpha \subseteq \mathrm{V}_\beta$ (Use the previous question.)

   c) $\mathrm{V}_\alpha \cap \mathrm{ON} = \alpha$.

   d) $\mathrm{rk}(x) = \sup\{\mathrm{rk}(y) + 1 : y \in x\}$.

   e) $\mathrm{rk}(\alpha) = \alpha$.

   Hint: Use transfinite induction on α.

9) Prove: If $x, y \in \mathrm{V}_\alpha$, then $\mathcal{P}(x), \bigcup x, x \times y, {}^x y \in \mathrm{V}_{\alpha+4}$.

10) We call a set x **hereditarily finite** iff its transitive closure is finite. Prove:
   $$\mathrm{V}_\omega = \{x : x \text{ is hereditarily finite}\}.$$

11) Prove: If α is a limit ordinal, then there is a set x such that $\mathrm{rk}(x) = \alpha$, x is transitive, and $x \cap \mathrm{ON} = \{0, 1\}$.

12) Assume that A, B and C are classes, $C \subseteq A$, B and C are transitive, and $\pi : A \longrightarrow B$ is an $\in$-isomorphism. Prove that $\pi \restriction C = \mathrm{id}_C$.

13) Assume that R is a well-founded relation and $\mathrm{fld}(R) = P$. Prove that there is a function $\rho_R : P \longrightarrow \mathrm{ON}$ such that

    a) $xRy \Longrightarrow \rho_R(x) < \rho_R(y)$ for all $x, y \in P$.

    b) $\rho_R(x) = \sup\{\rho_R(y) + 1 : yRx\}$.

    c) $\rho_R(x) = \min\{\alpha \in \mathrm{ON} : \forall y \in \widehat{x}\ \alpha > \rho_R(y)\}$.

    The sets $P_\alpha := \{x \in P : \rho_R(x) = \alpha\}$ are the ***R*-levels** of P, $\rho_R(x)$ is the ***R*-rank** of x.

    Hint: Use the recursion theorem.

14) Prove with the recursion theorem that there is a unique function $\rho : \mathrm{V} \longrightarrow \mathrm{ON}$ such that $\rho(x) = \sup\{\rho(y) + 1 : y \in x\}$. Consequently we have $\rho(x) = \mathrm{rk}(x)$ for all sets x.

15) Assume that $(A, <)$ and $(B, \prec)$ are well-orderings. Prove: If $g : (A, <) \longrightarrow (B, \prec)$ is a homomorphism, then $\mathrm{otp}(A, <) \leq \mathrm{otp}(B, \prec)$.

1.4 Arithmetic of Ordinals

Definition *Let A be an ordinal or the class* ON *of ordinals. If $F : A \longrightarrow \mathrm{V}$ is a function, then we call F a* **sequence**. *Notions as upper bound, supremum etc. of a sequence are understood as upper bound, supremum etc. of the range of the sequence. In later chapters we will extend the notion of a sequence to functions whose domain is a club, a stationary set, or a set of regular cardinals.*

F is called **continuous** *iff for every limit ordinal $\gamma \in A$ we have*

$$F(\gamma) = \bigcup\{F(\sigma) : \sigma < \gamma\};$$

we also say that F is a **continuous sequence** *of sets. Furthermore every function from A into* ON *that is continuous and strictly increasing under $<$ is called a* **normal function** *or* **normal sequence**.

More general, a function from an arbitrary set A into ON *is called an* **ordinal function** *on A.*

Remark In most applications of the notion of continuity, the domain A of the function is a limit ordinal or the class ON of ordinals. If $\mathrm{ran}(F) \subseteq \mathrm{ON}$, then the continuity of F is equivalent to

$$\forall \gamma \in \mathrm{dom}(F) \cap \mathrm{Lim}\ (F(\gamma) = \sup\{F(\sigma) : \sigma < \gamma\}).$$

If the last property is satisfied and if in addition $F(\alpha) < F(\alpha + 1)$ for every $\alpha \in A$, then F already is a normal function. This can be seen as follows: By transfinite induction on $\beta \in \mathrm{dom}(F)$ we show $\alpha < \beta \Longrightarrow F(\alpha) < F(\beta)$. For $\beta = 0$ there is nothing to prove. If β is a limit ordinal, then $\alpha < \beta$ implies $\alpha < \sigma$ for some $\sigma < \beta$. The inductive hypothesis and the continuity of F imply $F(\alpha) < F(\sigma) \leq \sup\{F(\sigma) : \sigma < \beta\} = F(\beta)$. Finally, if $\beta = \sigma + 1$, then we get $\alpha \leq \sigma$, $F(\alpha) \leq F(\sigma)$ by the inductive hypothesis, and consequently $F(\alpha) \leq F(\sigma) < F(\sigma + 1)$.

Further we note that normal functions always map limit ordinals onto limit ordinals.

Lemma 1.4.1 *Assume that* $A \in \mathrm{ON}$ *or* $A = \mathrm{ON}$ *and* $F : A \longrightarrow \mathrm{ON}$ *is a strictly increasing function. Then*

a) F *is continuous iff, for every nonempty subset* b *of* A *with* $\sup(b) \in A$,

$$F(\sup(b)) = \sup\{F(\alpha) : \alpha \in b\} = \sup(F[b]).$$

b) $\alpha \leq F(\alpha)$ *for all* $\alpha \in A$.

Proof For the proof of a) we note that, by the definition of continuity and since $\sup(\gamma) = \gamma$ for all limit ordinals γ, we only have to show one direction of the equivalence. So let F be a normal function and b be a nonempty subset of A such that $\sup(b) \in A$. If b has a maximum, then $\sup(b) \in b$, and the desired equation is obvious. Thus let us assume that b has no maximum. Then $\sup(b)$ is a limit ordinal, since otherwise its immediate predecessor would be an upper bound of b, too. This gives $F(\sup(b)) = \sup\{F(\sigma) : \sigma < \sup(b)\} \geq \sup\{F(\alpha) : \alpha \in b\}$. If $\gamma < F(\sup(b))$, then there is $\sigma < \sup(b)$ such that $\gamma < F(\sigma)$. Furthermore there is $\alpha \in b$ with $\sigma < \alpha$. By assumption, we have $F(\sigma) < F(\alpha)$ and consequently $\gamma < F(\alpha)$. Thus every ordinal $\gamma < F(\sup(b))$ is not an upper bound of the set $\{F(\alpha) : \alpha \in b\}$, which was still to be shown.

For an indirect proof of b), let $\alpha \in A$ be minimal with $F(\alpha) < \alpha$. Then $F(\alpha) \in A$, and therefore $F(F(\alpha)) < F(\alpha)$ since F is strictly increasing. This contradicts the choice of α, and part b) is shown.

The normal functions we are now going to define will play an important role in the sequel. They form a proper class of normal functions, since for every ordinal α the function $\beta \mapsto \alpha \diamond \beta$, where $\beta \in \mathrm{ON}$, is normal, when $\diamond$ denotes the addition, multiplication or exponentiation of ordinals.

Definition *If* α *is an ordinal, then we define by transfinite recursion on* $\beta \in \mathrm{ON}$ *the* **ordinal sum** $\alpha + \beta$*, the* **ordinal product** $\alpha \cdot \beta$*, and the* **ordinal exponentiation** α^β *as follows:*

$\alpha + 0 = \alpha$; $\alpha + (\sigma + 1) = (\alpha + \sigma) + 1$; $\mathrm{Lim}(\beta) \Longrightarrow \alpha + \beta = \sup\{\alpha + \sigma : \sigma < \beta\}$.
$\alpha \cdot 0 = 0$; $\alpha \cdot (\sigma + 1) = (\alpha \cdot \sigma) + \alpha$; $\mathrm{Lim}(\beta) \Longrightarrow \alpha \cdot \beta = \sup\{\alpha \cdot \sigma : \sigma < \beta\}$.
$\alpha^0 = 1$; $\alpha^{\sigma+1} = \alpha^\sigma \cdot \alpha$; $\mathrm{Lim}(\beta) \Longrightarrow \alpha^\beta = \sup\{\alpha^\sigma : \sigma < \beta\}$.

Remark 1.4.2

a) In the case of addition note remark 3) after Theorem 1.3.5 and choose, for an application of 1.3.5 (we could apply Corollary 1.3.6 as well – and will do this when defining the multiplication of ordinals), $G(0) := \alpha$; $G(x) := S(x(\sigma))$, if x is a function such that $\operatorname{dom}(x) = \sigma + 1$ for some $\sigma \in \mathrm{ON}$; $G(x) := \bigcup \operatorname{ran}(x)$, if x is a function whose domain is a limit ordinal; $G(x) := 0$ in all other cases. Theorem 1.3.5 says that there is a function $F : \mathrm{ON} \longrightarrow \mathrm{V}$ such that $F(\beta) = G(F \restriction \beta)$. Now define $\alpha + \beta := F(\beta)$.

   In the case of multiplication we apply Corollary 1.3.6: $a := 0$; $H_1(x) := x + \alpha$, if x is an ordinal, $H_1(x) := 0$ otherwise; $H_2(x) := \bigcup \operatorname{ran}(x)$.

b) From the definition of continuity and Lemma 1.4.3 a) (iii) it follows that the function $\beta \mapsto \alpha + \beta$ is normal for each $\alpha \in \mathrm{ON}$. Lemma 1.4.3 b) (iii) implies the corresponding assertion for the multiplication of ordinals if $\alpha > 0$, and Lemma 1.4.3 c) (iii) for the exponentiation of ordinals if $\alpha > 1$.

c) By complete induction it can be easily seen that the restrictions to the set ω of the functions defined above yield the usual addition, multiplication and exponentiation of natural numbers. (These operations also could have been obtained by applying a recursion theorem for ω.) The well known properties of these operations are proved using complete induction.

d) Intuitively speaking, the sum of ordinals corresponds to that order type which results from putting one well-order behind the other. Corollary 1.3.11 tells us that every well-ordered set is isomorphic to exactly one ordinal bearing the relation $\in$. We called this ordinal the order type of the well-ordered set. Now, if $(A, <_1)$ and $(B, <_2)$ are well-ordered sets of order type α and β respectively, satisfying $A \cap B = \emptyset$, then we obtain a new well-ordering $(A \cup B, <_3)$ as follows:

$$<_3 := <_1 \cup <_2 \cup \{(a, b) : a \in A \wedge b \in B\}$$

   ("first $(A, <_1)$ and then $(B, <_2)$"). One can show that this is a well-ordering of order type $\alpha + \beta$. Analogously, $\alpha \cdot \beta$ is the order type of that well-ordering $\prec$ that results from lining up β-times well-orderings $(A_\sigma, <_\sigma)$, $\sigma < \beta$, of order type α. More precisely: If $((A_\sigma, <_\sigma) : \sigma < \beta)$ is a family of well-orderings of order type α satisfying $A_\sigma \cap A_\tau = \emptyset$ for $\sigma \neq \tau$, then we can define a well-ordering $\prec$ of order type $\alpha \cdot \beta$ on the set $B := \bigcup\{A_\sigma : \sigma < \beta\}$ by

$$a \prec b :\Longleftrightarrow \exists\sigma\exists\tau((a \in A_\sigma \wedge b \in A_\tau \wedge \sigma < \tau) \vee (a \in A_\sigma \wedge b \in A_\sigma \wedge a <_\sigma b)).$$

It is simpler to well-order the set $\alpha \times \beta$ by the so-called **Hebrew lexicographic ordering**: $(\sigma_1, \tau_1) < (\sigma_2, \tau_2)$ iff $\tau_1 < \tau_2 \vee (\tau_1 = \tau_2 \wedge \sigma_1 < \sigma_2)$. Then $(\alpha \times \beta, <)$ has order type $\alpha \cdot \beta$. Shortly speaking, "we take α, β times".

e) The "visual" introduction of addition and multiplication facilitates the understanding of the following special features which immediately follow from the definitions. We have $1 + \omega = \sup\{1 + n : n \in \omega\} = \omega$. Therefore $1 + \omega < \omega + 1$, the addition of ordinals is not commutative. Further $2 \cdot \omega = \sup\{2 \cdot n : n \in \omega\} = \omega$, and consequently $2 \cdot \omega < \omega \cdot 2 = \omega + \omega$, the multiplication of ordinals is also not commutative. In addition, we have shown that $(1+1) \cdot \omega \neq 1 \cdot \omega + 1 \cdot \omega$ – the corresponding distributive law is false. The other distributive law, $\alpha \cdot (\beta + \gamma) = \alpha \cdot \beta + \alpha \cdot \gamma$, is true; furthermore the associative laws and certain monotonicity properties hold. We collect all these properties in the following lemma.

f) For the sake of completeness, let us mention how to introduce the ordinal exponentiation as an order type of a well-ordered set. For $\alpha, \beta \in \mathrm{ON}$ let $\mathrm{fin}(\beta, \alpha)$ be the set of all functions f from β into α for which the set $\{\sigma < \beta : f(\sigma) \neq 0\}$ is finite. Now the set $\mathrm{fin}(\beta, \alpha)$ will be well-ordered by a generalized Hebrew lexicographic ordering; for $f, g \in \mathrm{fin}(\beta, \alpha)$ we define

$$f \prec g \;:\Longleftrightarrow\; f \neq g \wedge (\tau = \max\{\sigma \in \beta : f(\sigma) \neq g(\sigma)\} \Longrightarrow f(\tau) < g(\tau)).$$

One can show that $(\mathrm{fin}(\beta, \alpha), \prec)$ is a well-ordering of order type α^β. The key for the proof is Theorem 1.4.6.

Lemma 1.4.3 *If α, β and γ are ordinals, then the following laws hold.*

a)	*(i)*	$0 + \alpha = \alpha = \alpha + 0$	*(ii)*	$\beta \leq \alpha + \beta$
	(iii)	$\beta < \gamma \Longrightarrow \alpha + \beta < \alpha + \gamma$	*(iv)*	$\alpha \leq \beta \Longrightarrow \alpha + \gamma \leq \beta + \gamma$
	(v)	$(\alpha + \beta) + \gamma = \alpha + (\beta + \gamma)$		
b)	*(i)*	$0 \cdot \alpha = 0 = \alpha \cdot 0$	*(ii)*	$\alpha \cdot 1 = \alpha = 1 \cdot \alpha$
	(iii)	$0 < \alpha \wedge \beta < \gamma \Longrightarrow \alpha \cdot \beta < \alpha \cdot \gamma$	*(iv)*	$\alpha \leq \beta \Longrightarrow \alpha \cdot \gamma \leq \beta \cdot \gamma$
	(v)	$(\alpha \cdot \beta) \cdot \gamma = \alpha \cdot (\beta \cdot \gamma)$	*(vi)*	$\alpha \cdot (\beta + \gamma) = \alpha \cdot \beta + \alpha \cdot \gamma$
c)	*(i)*	$\beta \neq 0 \Longrightarrow 0^\beta = 0$	*(ii)*	$1^\beta = 1$
	(iii)	$1 < \alpha \wedge \beta < \gamma \Longrightarrow \alpha^\beta < \alpha^\gamma$	*(iv)*	$\alpha \leq \beta \Longrightarrow \alpha^\gamma \leq \beta^\gamma$
	(v)	$1 < \alpha \Longrightarrow \beta \leq \alpha^\beta$	*(vi)*	$\alpha^{\beta+\gamma} = \alpha^\beta \cdot \alpha^\gamma$
	(vii)	$(\alpha^\beta)^\gamma = \alpha^{\beta \cdot \gamma}$		

Proof The assertions are verified by transfinite induction on the "argument on the right", and their proofs are thought of as exercises for the reader. We indicate some proofs and present all ideas which are necessary for the other ones.

For a) (iii), note that there is nothing to prove if $\gamma = 0$. If $\gamma = \sigma + 1$ and $\beta < \gamma$, then $\beta \leq \sigma$, and the inductive hypothesis and the definition of the sum of ordinals give $\alpha+\beta \le \alpha+\sigma < (\alpha+\sigma)+1 = \alpha+(\sigma+1) = \alpha+\gamma$. Finally, if γ is a limit ordinal, then $\gamma = \sup(\gamma)$, and $\beta < \gamma$ means $\beta < \sigma$ for some $\sigma < \gamma$. By the inductive hypothesis and by definition, we get $\alpha+\beta < \alpha+\sigma \le \sup\{\alpha+\sigma : \sigma < \gamma\} = \alpha+\gamma$.

For a) (v): If $\gamma = 0$, then $(\alpha+\beta)+0 = \alpha+\beta = \alpha+(\beta+0)$. Furthermore, from the inductive hypothesis and the definition of ordinal addition we can conclude that

$$\begin{aligned}(\alpha+\beta)+(\sigma+1) &= ((\alpha+\beta)+\sigma)+1 = (\alpha+(\beta+\sigma))+1\\ &= \alpha+((\beta+\sigma)+1) = \alpha+(\beta+(\sigma+1)).\end{aligned}$$

For limit ordinals γ we obtain $(\alpha+\beta)+\gamma = \sup\{(\alpha+\beta)+\sigma : \sigma < \gamma\} = \sup\{\alpha+(\beta+\sigma) : \sigma < \gamma\}$. Now we apply Lemma 1.4.1 to the normal function $F_\alpha : \mathrm{ON} \longrightarrow \mathrm{ON}$, given by $F_\alpha(\tau) := \alpha+\tau$, and to the set $b := \{\beta+\sigma : \sigma < \gamma\}$. Since $\sup(b) = \beta+\gamma$, it tells us that $\alpha+(\beta+\gamma) = F_\alpha(\sup(b)) = \sup\{\alpha+(\beta+\sigma) : \sigma < \gamma\}$. This completes the proof.

Completely analogously, the associative law of multiplication is shown, and in a similar way one obtains the distributive law.

For c) (vii): If $\alpha = 0$, then the assertion follows from the proceding ones by considering the cases $\gamma = 0$; $\beta = 0, \gamma \neq 0$; $\beta \neq 0, \gamma \neq 0$. Now let $\alpha \neq 0$. We prove (vii) using transfinite induction on γ. If $\gamma = 0$, then $\alpha^{\beta\cdot 0} = \alpha^0 = 1 = (\alpha^\beta)^0$. If $\gamma = \sigma+1$, then we can conclude from b) (vi), c) (vi), the inductive hypothesis and the definition of exponentiation that

$$\alpha^{\beta\cdot(\sigma+1)} = \alpha^{\beta\cdot\sigma+\beta} = \alpha^{\beta\cdot\sigma}\cdot\alpha^\beta = (\alpha^\beta)^\sigma\cdot\alpha^\beta = (\alpha^\beta)^{\sigma+1} = (\alpha^\beta)^\gamma.$$

Finally, if γ is a limit ordinal, then again we apply Lemma 1.4.1 to the continuous function G_α, defined by $G_\alpha(\tau) := \alpha^\tau$, and obtain with the inductive hypothesis:

$$\alpha^{\beta\cdot\gamma} = \alpha^{\sup\{\beta\cdot\sigma:\sigma<\gamma\}} = \sup\{\alpha^{\beta\cdot\sigma} : \sigma < \gamma\} = \sup\{(\alpha^\beta)^\sigma : \sigma < \gamma\} = (\alpha^\beta)^\gamma.$$

Lemma 1.4.4 *Let α and β be ordinals.*

a) **(Substraction Lemma)** *If $\alpha \le \beta$, then there is a unique ordinal γ such that $\alpha+\gamma = \beta$.*

b) **(Division Lemma)** *If $\beta \neq 0$, then there are unique ordinals ξ and η satisfying $\alpha = \beta\cdot\xi+\eta$ and $\eta < \beta$.*

c) **(Logarithm Lemma)** *If $\alpha \neq 0$ and $\beta > 1$, then there are unique ordinals σ, τ and γ (the logarithm, the coefficient and the remainder) such that*

$$\alpha = \beta^\sigma\cdot\tau+\gamma \wedge 1 \le \tau < \beta \wedge \gamma < \beta^\sigma\,.$$

Proof For part a) we note, that the uniqueness of γ is obvious if one uses the monotonicity properties of ordinal addition in Lemma 1.4.3. Next we prove existence. Since $\beta \leq \alpha+\beta$, there is a least ordinal γ with $\beta \leq \alpha+\gamma$. If equality holds, we are done. Otherwise we have $\gamma \neq 0$. Furthermore γ cannot be a limit ordinal, because $\beta < \alpha + \gamma$ yields, by the definition of ordinal addition, for limit ordinals γ the existence of an ordinal σ such that $\sigma < \gamma$ and $\beta < \alpha + \sigma$, contradicting the choice of γ. Thus we obtain $\gamma = \eta + 1$ for some η and, again by the choice of γ, $\alpha + \eta < \beta < \alpha + \eta + 1$, which contradicts Lemma 1.2.6 b).

For the proof of part b), we state first that the uniqueness of ξ and η follows again from a) and Lemma 1.4.3. Now we prove existence. Lemma 1.4.3 tells us that $\alpha \leq \beta \cdot \alpha$, and so there is a least ordinal γ such that $\alpha \leq \beta \cdot \gamma$. If equality holds, we are done taking $\xi := \gamma$ and $\eta := 0$. Suppose that $\alpha < \beta \cdot \gamma$. Then we get in complete analogy to the proof of a), that γ is a successor ordinal $\xi + 1$. Hence $\beta \cdot \xi < \alpha < \beta \cdot (\xi + 1) = \beta \cdot \xi + \beta$. From a) follows the existence of a unique ordinal η satisfying $\beta \cdot \xi + \eta = \alpha$. Since $\alpha = \beta \cdot \xi + \eta < \beta \cdot \xi + \beta$, Lemma 1.4.3 a) gives $\eta < \beta$.

Finally, we turn to the proof of c). Having studied the previous proofs, the reader will be familiar with the following arguments. To prove existence, we note first that $\alpha \leq \beta^\alpha$ by Lemma 1.4.3 c). So there is a least ordinal δ with $\alpha \leq \beta^\delta$. If equality holds, put $\sigma := \delta$, $\tau := 1$, and $\gamma := 0$. In the other case the continuity of exponentiation and the choice of δ ensure again that δ cannot be a limit ordinal. Thus $\delta = \sigma + 1$ for some $\sigma \in \mathrm{ON}$. Now we have $\beta^\sigma < \alpha$, and by part b) there are unique ordinals τ and γ satisfying $\alpha = \beta^\sigma \cdot \tau + \gamma$ and $\gamma < \beta^\sigma$. Since $\alpha < \beta^\delta = \beta^{\sigma+1} = \beta^\sigma \cdot \beta$, Lemma 1.4.3 b) tells us that $1 \leq \tau < \beta$. Uniqueness can be proved again with Lemma 1.4.3; this is left to the reader.

Corollary 1.4.5 *Let α and β be ordinals.*

a) α is a limit ordinal iff there is an ordinal $\xi \neq 0$ with $\alpha = \omega \cdot \xi$.

b) The operation $\alpha \mapsto \omega \cdot (1 + \alpha)$ defines the unique $\in$-isomorphism from ON onto the class of limit ordinals.

c) If α is a limit ordinal and $1 \leq n < \omega$, then $n \cdot \alpha = \alpha$.

d) If $n < \omega$ and $\omega \leq \alpha$, then $n + \alpha = \alpha$.

e) $\alpha \cdot \beta$ is a successor ordinal iff α and β are both successor ordinals.

Proof First we prove a). By Lemma 1.4.4 we get, for α and ω, unique ordinals ξ and η such that $\eta < \omega$ and $\alpha = \omega \cdot \xi + \eta$. If α is a limit ordinal, then $\eta = 0$ and $\xi \neq 0$. Conversely, if $\eta = 0$ and $\xi \neq 0$ then, by Remark 1.4.2 b), $\alpha = \omega \cdot \xi$ is a limit ordinal if this holds for ξ, and also in the case $\xi = \sigma + 1$, because ω is a limit ordinal and then $\alpha = \omega \cdot \sigma + \omega$.

For b), let the equation $G(\alpha) := \omega \cdot (1 + \alpha)$ define the function G. By part a), $G(\alpha)$ is a limit ordinal, and Lemma 1.4.3 says that G is strictly monotonic.

If α is a limit ordinal and $\sigma < \alpha$, then we get with d) $1+\sigma \le \sigma+1 < \alpha$ and $1+\alpha = \alpha$. Therefore

$$\begin{aligned} G(\alpha) = \omega\cdot\alpha &= \sup\{\omega\cdot\sigma : \sigma < \alpha\} \\ &= \sup\{\omega\cdot(1+\sigma) : \sigma < \alpha\} = \sup\{G(\sigma) : \sigma < \alpha\}, \end{aligned}$$

and thus G is continuous. Finally, G is surjective. Consider a fixed limit ordinal τ. By a), we can represent τ as $\tau = \omega\cdot\alpha$ for some $\alpha \neq 0$. If $\alpha \ge \omega$, then $G(\alpha) = \tau$, since $1+\alpha = \alpha$. On the other hand, if $\alpha = \sigma+1$ is finite, then $G(\sigma) = \tau$.

For c), the assumption and part a) give $\alpha = \omega\cdot\xi$ for some $\xi \neq 0$. Further, by the definition of ordinal multiplication, $n\cdot\omega = \sup\{n\cdot m : m < \omega\} = \omega$, since $n > 0$. The associative law gives $n\cdot\alpha = n\cdot(\omega\cdot\xi) = \omega\cdot\xi = \alpha$.

Part d) can be proved analogously to c), using Lemma 1.4.4 a). Since $\alpha \ge \omega$, there is an ordinal β with $\alpha = \omega+\beta$.

For the proof of e), let first α and β be successor ordinals. Then, by the definition of ordinal multiplication, $\alpha\cdot\beta$ is a successor ordinal. Now let conversely $\alpha\cdot\beta$ be a successor ordinal $\gamma+1$. Then Lemma 1.4.3 says that both α and β are nonzero. By Lemma 1.4.4, we obtain a unique representation $\gamma = \alpha\cdot\xi+\eta$ with $\eta < \alpha$. Hence $\gamma+1 = \alpha\cdot\xi+\eta+1 = \alpha\cdot\beta$. The assertion $\eta+1 < \alpha$ would contradict the uniqueness property in Lemma 1.4.4 b). Thus $\eta+1 = \alpha$ and $\gamma+1 = \alpha\cdot\xi+\alpha = \alpha\cdot(\xi+1) = \alpha\cdot\beta$. For the same reasons we get $\beta = \xi+1$, as desired.

Theorem 1.4.6 (Cantor Normal Form for the base β) *Assume that α and β are ordinals with $\alpha \neq 0$ and $\beta > 1$. Then there are unique ordinals $\sigma_1,\dots,\sigma_n$ and $\tau_1,\dots,\tau_n$ such that*

$$\alpha = \beta^{\sigma_1}\cdot\tau_1+\ldots+\beta^{\sigma_n}\cdot\tau_n,\quad \sigma_1 > \ldots > \sigma_n \quad\text{and}\quad 1 \le \tau_i < \beta \text{ for } i = 1,\dots,n.$$

In particular, for $\beta = \omega$, there are unique ordinals $\alpha_1,\dots,\alpha_n$ satisfying

$$\alpha = \omega^{\alpha_1}+\ldots+\omega^{\alpha_n} \quad\text{and}\quad \alpha_1 \ge \ldots \ge \alpha_n.$$

Proof We use transfinite induction on $\alpha \in \mathrm{ON}$. Most work has already been done in Lemma 1.4.4 c). This lemma provides us, for $\alpha \neq 0$ and $\beta > 1$, with unique ordinals σ_1, τ_1 and γ such that $1 \le \tau_1 < \beta$, $\gamma < \beta^{\sigma_1}$ and $\alpha = \beta^{\sigma_1}\cdot\tau_1+\gamma$. Since $\gamma < \alpha$, we can represent γ, if $\gamma \neq 0$, by the inductive hypothesis in the desired form with ordinals $\tau_2,\dots,\tau_n$, satisfying $1 \le \tau_i < \beta$ for $i = 1,\dots,n$ and $\sigma_2 > \ldots > \sigma_n$. Now we have $\beta^{\sigma_2} \le \gamma < \beta^{\sigma_1}$, from which we can conclude together with Lemma 1.4.3 that $\sigma_1 > \sigma_2$.

If $\beta = \omega$, then all ordinals τ_i are natural numbers, and $\omega^{\sigma_i}\cdot\tau_i$ is the τ_i-fold sum of ω^{σ_i}. From this, we get the second assertion of the theorem.

We have learnt that the operations for ordinals extend the usual operations on ω but satisfy no longer all of its rules. The representation of an ordinal in Theorem 1.4.6 permits the definition of a so-called natural sum, satisfying the most important laws of finite arithmetic.

Definition *The representation of an ordinal $\alpha \neq 0$, given by the first part of Theorem 1.4.6, is called the* **Cantor normal form of α for the base β** *or the* **Cantor normal form of α for the base ω**, *if $\beta = \omega$. The Cantor normal form of 0 is $\omega^0 \cdot 0$.*

Now assume that $\alpha \neq 0$ and β are represented in Cantor normal form for the base ω. If we add powers ω^σ of ω, occurring only in one representation, to the other one as the summand $\omega^\sigma \cdot 0$, then we can assume without loss of generality that α and β are uniquely given as $\alpha = \sum_{i=1}^n \omega^{\sigma_i} \cdot n_i$ and $\beta = \sum_{i=1}^n \omega^{\sigma_i} \cdot m_i$ such that $\sigma_1 > \ldots > \sigma_n$, $0 \leq n_i < \omega$, $0 \leq m_i < \omega$, and $m_i + n_i \neq 0$ for $i = 1, \ldots, n$. Then we can define the **natural sum** *$\alpha \# \beta$ of the ordinals α and β by*

$$\alpha \# \beta := \sum_{i=1}^{n} \omega^{\sigma_i} \cdot (m_i + n_i).$$

We call an ordinal γ a **principal number of addition** *iff $\gamma \neq 0$ and, for all $\alpha < \gamma$, $\alpha + \gamma = \gamma$. γ is called* **additively decomposable** *iff there are ordinals $\alpha < \gamma$ and $\beta < \gamma$ such that $\alpha + \beta = \gamma$; otherwise we say that γ is* **additively indecomposable**.

Remark The natural sum of ordinals satisfies the associative law and the commutative law. Later on we will need the fact that the equation $\alpha \# \beta = \gamma$ has, for a fixed ordinal γ, only finitely many solutions (α, β). This is true, since there are only finitely many nonzero coefficients in the Cantor normal form of γ and, for every natural number l, the equation $x + y = l$ has only finitely many solutions $(x, y) \in \omega \times \omega$.

In complete analogy to the notion of a principal number of addition one can define principal numbers of multiplication and of exponentiation. It turns out that $\{2\} \cup \{\omega^{(\omega^\sigma)} : \sigma \in \mathrm{ON}\}$ is the set of principal numbers of multiplication (see Exercise 1.4.7) and $\{\omega\} \cup \{\alpha \in \mathrm{ON} : \omega^\alpha = \alpha\}$ is the set of principal numbers of exponentiation (see Exercise 1.4.8). Clearly 1 is the only principal number of addition less than ω, and Corollary 1.4.5 d) says that ω is the least principal number of addition greater than 1. The next lemma characterizes the principal numbers of addition.

Lemma 1.4.7

a) An ordinal $\gamma \neq 0$ is a principal number of addition iff it is additively indecomposable.

b) The principal numbers of addition are exactly the ordinals of the form ω^α, where $\alpha \in \mathrm{ON}$.

c) $(\omega^\alpha : \alpha \in \mathrm{ON})$ *is the unique* $\in$*-isomorphism from* ON *onto the class of principal numbers of addition.*

Proof For a), let γ be a principal number of addition, and let α and β be less than γ. Then, by Lemma 1.4.3, $\alpha+\beta < \alpha+\gamma = \gamma$, and γ is additively indecomposable. Conversely, if this is the case for a fixed $\gamma \neq 0$, then take $\alpha < \gamma$ and choose, according to Lemma 1.4.4, an ordinal $\beta \leq \gamma$ with $\alpha+\beta = \gamma$. Then by definition $\beta = \gamma$, and γ is a principal number of addition.

For the proof of b), let $\alpha \in \mathrm{ON}$. Clearly, $\omega^\alpha \neq 0$. If $0 < \beta < \omega^\alpha$ and $\omega^{\beta_1}+\ldots+\omega^{\beta_n}$ is the Cantor normal form of β for the base ω, then Lemma 1.4.3 tells us that $\beta_i < \alpha$ for $i = 1,\ldots,n$. If we can show that $\sigma < \tau$ implies $\omega^\sigma + \omega^\tau = \omega^\tau$, then we obtain $\beta + \omega^\alpha = \omega^\alpha$, and ω^α is a principal number of addition. So assume that $\sigma < \tau$. By Lemma 1.4.4 a), there is an ordinal $\gamma \neq 0$ with $\tau = \sigma + \gamma$. Then $\omega^\gamma \geq \omega$, and Corollary 1.4.5 d) says that $1 + \omega^\gamma = \omega^\gamma$. By Lemma 1.4.3, we get

$$\omega^\sigma + \omega^\tau = \omega^\sigma + \omega^{\sigma+\gamma} = \omega^\sigma + \omega^\sigma \cdot \omega^\gamma = \omega^\sigma \cdot (1+\omega^\gamma) = \omega^\sigma \cdot \omega^\gamma = \omega^\tau.$$

To prove the converse, let, for any α, $\gamma \neq 0$ be not of the form ω^α. Then in the Cantor normal form $\omega^{\gamma_1} + \ldots + \omega^{\gamma_n}$ of γ occur at least two summands, so we have $\omega^{\gamma_1} < \gamma$, and the uniqueness assertion of Theorem 1.4.6 implies that $\omega^{\gamma_1} + \gamma \neq \gamma$. Thus γ is not a principal number of addition.

Part c) is easy to show, since $\alpha \mapsto \omega^\alpha$ defines, by Lemma 1.4.3, a normal function, which is surjective by part b).

Exercises

1) If τ is a limit ordinal and $(\alpha_\xi : \xi < \tau)$ is a strictly increasing sequence of ordinals, then $\sup\{\alpha_\xi : \xi < \tau\}$ is a limit ordinal.
2) If τ is a principal number of addition, $(\eta(\xi) : \xi < \tau)$ is a normal sequence of ordinals, and $\sigma < \tau$, then $\eta(\xi) + \sigma \leq \eta(\xi + \sigma)$ for all $\xi < \tau$.
3) Use Lemma 1.4.4 to show that the relations $<_3$ on $\alpha \times \{0\} \cup \alpha \times \{1\}$ and the Hebrew lexicographic ordering on $\alpha \times \beta$ from Remark 1.4.2 d) are well-orderings of order type $\alpha + \beta$ and $\alpha \cdot \beta$, respectively.
4) Prove: If $\sigma_1 > \ldots > \sigma_n$ and, for all $i \in \{1,\ldots,n\}$, $1 \leq \tau_i < \alpha$, then

$$\alpha^{\sigma_1} \cdot \tau_1 + \ldots + \alpha^{\sigma_n} \cdot \tau_n < \alpha^{\sigma_1+1}.$$

5) Consider the relation defined in Remark 1.4.2 f). It is a well-ordering by Exercise 1.3.4. Use the Cantor normal form theorem to show that this well-ordering has order type α^β.

Hint: If $f \in \text{fin}(\beta, \alpha)$, $\{\sigma \in \beta : f(\sigma) \neq 0\} = \{\sigma_1, \ldots, \sigma_n\}$ and $\sigma_1 > \ldots > \sigma_n$, let

$$\Phi(f) := \alpha^{\sigma_1} \cdot f(\sigma_1) + \alpha^{\sigma_2} \cdot f(\sigma_2) + \ldots + \alpha^{\sigma_n} f(\sigma_n).$$

Then we know from a previous question that $\Phi(f) < \alpha^{\sigma_1 + 1}$, hence $\Phi(f) \in \alpha^\beta$. Prove that Φ is strictly increasing and a function onto α^β.

6) Let $\prec$ be the canonical well-ordering of ON $\times$ ON, defined in Section 1.5, and let $ot(\alpha)$ be the order type of $(\alpha \times \alpha, \prec)$. Prove:

   a) $\{(\sigma, \tau) : (\sigma, \tau) \prec (0, \beta)\} = \beta \times \beta$.
   b) $ot(\alpha + 1) = ot(\alpha) + \alpha + \alpha + 1$.
   c) $ot(\omega) = \omega$ and $ot(\omega \cdot 2) = \omega \cdot \omega$.
   d) $ot(\alpha) \leq \omega^\alpha$.
   e) If $\alpha = \omega^\alpha$, then $ot(\alpha) = \alpha$.

   Hint: If A is a set, $(A, <)$ is a well-ordering, and $A_u = \{x \in A : x < u\}$ is the initial segment of $(A, <)$ determined by u, then $\text{otp}(A, <) = \sup\{\text{otp}(A_u, <) + 1 : u \in A\}$.

7) An ordinal α is called a **principal number of multiplication** iff $\alpha > 1$ and $\beta \cdot \alpha = \alpha$ for all β with $1 \leq \beta < \alpha$. Prove:

   a) 2 and ω are principal numbers of multiplication. If $2 < n < \omega$, then n is not a principal number of multiplication.
   b) If $\tau \neq 0$, then ω^{ω^τ} is a principal number of multiplication.
      Hint: Use the Cantor normal form theorem for the base ω and a previous exercise.
   c) If γ is nonzero and not a principal number of addition, then ω^γ is not a principal number of multiplication.
   d) If $\alpha_1 \neq 0$, $n > 1$, and $\alpha = \omega^{\alpha_1} + \cdots + \omega^{\alpha_n}$ with $\alpha_1 \geq \ldots \geq \alpha_n$, then α is not a principal number of multiplication.
   e) $\{\alpha : \alpha \text{ is a principal number of multiplication}\} = \{2\} \cup \{\omega^{\omega^\tau} : \tau \in \text{ON}\}$.
      Hint: If $\alpha \neq 2$ and $\alpha \neq \omega^{\omega^\tau}$ for all τ, represent α as $\alpha = \omega^{\alpha_1} + \ldots + \omega^{\alpha_n}$ with $\alpha_1 \geq \alpha_2 \geq \ldots \geq \alpha_n$.

8) ε is called a **principal number of exponentiation** iff $\varepsilon > 2$ and $\alpha^\varepsilon = \varepsilon$ for all α with $2 \leq \alpha < \varepsilon$. Prove:

   a) If $2^\gamma = \gamma$, then γ is a principal number of exponentiation.
      Hint: Prove successively that γ is a limit ordinal, a principal number of addition (using $\gamma = \omega \cdot \xi$ for some $\xi > 0$), and a principal number of multiplication.

b) $\{\omega\} \cup \{\alpha : \omega^\alpha = \alpha\}$ is the class of principal numbers of exponentiation.

c) If $f : \omega \longrightarrow \mathrm{ON}$ is given by $f(0) = \alpha$ and $f(n+1) = \omega^{f(n)}$, then $\gamma := \sup\{f(n) : n \in \omega\}$ has the property $\omega^\gamma = \gamma$. So the class of principal numbers of exponentiation is unbounded in ON. For $\alpha = 0$ we obtain the smallest principal number of exponentiation which is greater than ω.

1.5 Cardinal Numbers and their Elementary Properties

Theorem 1.5.1 *The following statements are equivalent (in* ZF*).*

(i) Axiom of choice.

(ii) **Well-ordering Principle**: *For every set x, there exists a relation $\prec$ on x such that $(x, \prec)$ is a well-ordering.*
(Shortly speaking: Every set can be well-ordered.)

(iii) **Zorn's Lemma**[11]: *If H is a set and $\leq$ is a partial ordering on H such that every subset of H linearly ordered by $\leq$, i.e., every chain in H, has a $\leq$-upper bound in H, then H possesses a $\leq$-maximal element.*

Proof For the proof of "(ii)$\Longrightarrow$(i)", assume that x is a set of nonempty sets and $\prec$ is a well-ordering on $\bigcup x$, and let the function f be defined by

$$f := \{(y, z) : y \in x \text{ and } z \text{ is the } \prec\text{-smallest element of } y\}.$$

Then obviously f is a choice function for x.

Next we prove "(i)$\Longrightarrow$(iii)". By transfinite recursion, we define a function corresponding to the following "visual" proof. Take $a_0 \in H$. If a_0 is maximal with respect to $\leq$, then we are ready. Otherwise, there is a member y of H such that $a_0 \leq y$ and $a_0 \neq y$ – choose a_1 to be such a member. If we have chosen $a_n \in H$ for each $n \in \omega$ without finding a maximal element, then our assumption tells us that the chain $\{a_n : n \in \omega\}$ has an upper bound in H. Take one of these upper bounds and call it a_ω. In this manner we go on and obtain an ascending chain of members of H. The process must stop at some ordinal yielding a $\leq$-maximal element of H, since otherwise, and this is the crucial point of the construction, we would get an injection of ON into H, contradicting the fact that H is a set. Let us make these ideas precise. Take f as a choice function for $\mathcal{P}(H) \setminus \{\emptyset\}$, and let for each $x \subseteq H$ the set $s(x)$ be defined as the set of

[11]This is only a lemma in the axiom system ZFC. Mostly it is used as an axiom. For historical and habituary reasons, we will keep this name.

upper bounds of x with respect to $\leq$ which are not a member of x. Further take $b \notin H$, which is possible since $H \neq V$. The transfinite recursion theorem provides us with a function $F : \mathrm{ON} \longrightarrow \mathrm{V}$ such that for all $\sigma \in \mathrm{ON}$ we have, putting $F(\sigma) =: a_\sigma$:

$$a_\sigma = \begin{cases} f(s(\{a_\tau : \tau < \sigma\})) & \text{if } s(\{a_\tau : \tau < \sigma\}) \neq \emptyset, \\ b & \text{otherwise.} \end{cases}$$

It is easy to see that $F \restriction \{\sigma : a_\sigma \neq b\}$ is an injective function. Thus there is smallest σ_0 with $a_{\sigma_0} = b$, since, assuming the contrary, we would get $F[\mathrm{ON}] \subseteq H$. Consequently, by the axiom of comprehension, $F[\mathrm{ON}]$ would be a set, and the axiom of replacement would imply that this would also hold for $F^{-1}[F[\mathrm{ON}]] = \mathrm{ON}$. This would contradict Corollary 1.2.2.

The assumptions of Zorn's lemma give that σ_0 is not a limit ordinal. So let $\sigma_0 = \tau_0 + 1$. Since $s(\{a_\tau : \tau < \sigma_0\}) = \emptyset$, a_{τ_0} is a $\leq$-maximal element of H, as desired.

Finally, for the proof of "(iii)$\Longrightarrow$(ii)", let x be a set. Note that x is well-orderable iff there is a bijection from an ordinal onto x: If this is the case, then each such bijection induces a well-ordering on x by the well-ordering $\in$ on this ordinal; the reverse implication follows from Corollary 1.3.11. This motivates the definition of

$$H := \{f : \exists\sigma(f : \sigma \longrightarrow x \text{ and } f \text{ is one-to-one})\}.$$

We take the inclusion $\subseteq$ as the partial ordering $\leq$ on H. If H satisfies the assumptions of Zorn's Lemma and g is a maximal element of H, say $g : \alpha \longrightarrow y$, then $y = x$. Otherwise we could choose $z \in x \setminus y$ and define $h := g \cup \{(\alpha, z)\}$. Then h would be a one-to-one function from $\alpha + 1$ into x, hence $h \in H$, and a proper extension of g, contradicting the fact that g is maximal. Let us check therefore whether H satisfies the assumptions of (iii). First we have $H \neq \emptyset$, since the empty function $\emptyset$ is a member of H. If f_0 is any union of a chain $\mathcal{K}$ of members of H, then f_0 is a function and one-to-one, since $\mathcal{K}$ is a chain. The domain of f_0 is $\bigcup\{\mathrm{dom}(g) : g \in \mathcal{K}\}$, hence is an ordinal by Lemma 1.2.5, being the union of a set of ordinals. Finally we state that $\mathrm{ran}(f_0) \subseteq x$, and thus $f_0 \in H$.

It remains to be shown that H is a set (for proper classes H, Zorn's Lemma generally fails – take $(H, \leq) := (\mathrm{ON}, \subseteq)$). Each $f \in H$ induces a well-ordering

$$R_f = \{(a, b) : a, b \in \mathrm{ran}(f) \wedge f^{-1}(a) < f^{-1}(b)\}$$

of the subset $\mathrm{ran}(f)$ of x. For each well-ordering $(\mathrm{ran}(f), R_f)$ there exists, by Corollary 1.3.11, a unique ordinal τ_f and a unique isomorphism, namely f, from $(\tau_f, \in)$ onto the given well-ordering. Thus the function $F := \{(f, (\mathrm{ran}(f), R_f)) :$

$f \in H\}$ is one-to-one. Its range is a set as a subset of the set $\mathcal{P}(x) \times \mathcal{P}(x \times x)$. Now the axiom of replacement, applied to the function F^{-1}, tells us that H is a set.

Remark There are a lot of assertions that are equivalent in ZF to the axiom of choice. Furthermore there are extensive investigations on weaker versions of this axiom and how they implicate one another. The reader is referred to the literature, especially to Th. Jech, The Axiom of Choice, [Je1], and to H. Rubin and J. E. Rubin, Equivalents of the Axiom of Choice, II, [RR]. We give examples for some further equivalent assertions; other examples can be found in the exercises.

(iv) **(Cardinal Comparability)**
$\forall x \forall y \exists f(f : x \longrightarrow y$ is injective $\vee f : y \longrightarrow x$ is injective).

(v) $\forall x \forall y(x \neq \emptyset \wedge y \neq \emptyset \Longrightarrow$
$\exists f(f : x \longrightarrow y$ is surjective $\vee f : y \longrightarrow x$ is surjective)).

(vi) $\forall x(\neg \exists n \in \omega(|x| = |n|) \Longrightarrow |x \times x| = |x|)$, where $|a| = |b|$ means: There is a bijection from a onto b.

(vii) Zorn's lemma restricted to sets partially ordered by the relation $\subseteq$.

(viii) Every vector space has a basis.

Definition *A* **cardinal number** *or* **cardinal** *is an ordinal that cannot be mapped one-to-one onto a smaller ordinal.*

If x is a set, then from (AC), *Theorem 1.5.1, and Corollary 1.3.11 we obtain an ordinal that can be mapped one-to-one onto x, since x can be well-ordered. The smallest ordinal with this property is called the* **cardinality** *or* **cardinal number** *of x, written as $|x|$.*

Any bijection from $|x|$ onto x is called an **enumeration** *of x.*

Remark The cardinality of any set is obviously a cardinal number. Furthermore we have, by definition, $|\alpha| \leq \alpha$, and for cardinal numbers α holds $|\alpha| = \alpha$. If α and β are cardinal numbers, and if there is a bijection from α onto β, then $\alpha = \beta$.

If the cardinal numbers $|x|$ and $|y|$ exist in ZF, then it is easy to obtain the following equivalences. They can be used to formulate in ZF many statements on cardinalities of sets which are of importance for the working mathematician and can be applied without explicitly defining the object $|x|$. Any two sets with the property in a) are called **equinumerous**.

a) $|x| = |y| \Longleftrightarrow \exists f(f : x \longrightarrow y \wedge f$ is bijective).

b) $|x| \leq |y| \Longleftrightarrow \exists f(f : x \longrightarrow y \wedge f$ is one-to-one).

Lemma 1.5.2 (ZF) *If z is a set and $h : (\mathcal{P}(z), \subseteq) \longrightarrow (\mathcal{P}(z), \subseteq)$ is a homomorphism, then h has a fixed point (i.e., there is $w \subseteq z$ such that $h(w) = w$).*

Proof Put $y := \{u \in \mathcal{P}(z) : u \subseteq h(u)\}$ and $w := \bigcup y$. By the definition of y, and since h is a homomorphism, we have $\bigcup y \subseteq \bigcup\{h(u) : u \in y\} \subseteq h(w)$. So we get $w \subseteq h(w)$ and thus $h(w) \subseteq h(h(w))$, which gives $h(w) \in y$. But this implies $h(w) \subseteq w$, and w is the desired fixed point.

Theorem 1.5.3 (Cantor-Schröder-Bernstein) (ZF) *If there are injective functions $f : x \longrightarrow y$ and $g : y \longrightarrow x$, then there exists a bijection from x onto y.*

Proof Let the function $h^* : \mathcal{P}(x) \longrightarrow \mathcal{P}(x)$ be defined by

$$h^*(u) := x \setminus g[y \setminus f[u]].$$

From $u \subseteq v$ we get $h^*(u) \subseteq h^*(v)$. Thus Lemma 1.5.2 says that there is a subset w of x such that $w = x \setminus g[y \setminus f[w]]$. Now the desired bijection $h : x \longrightarrow y$ is given by $h(z) := f(z)$, if $z \in w$, and $h(z) := g^{-1}(z)$, if $z \notin w$. The easy verification is left to the reader.

Corollary 1.5.4

a) $|x| = |y| \Longleftrightarrow \exists f(f : x \longrightarrow y \wedge f$ *is bijective*$)$.

b) $|x| \leq |y| \Longleftrightarrow \exists f(f : x \longrightarrow y \wedge f$ *is one-to-one*$)$.

c) $x \neq \emptyset \Longrightarrow (|x| \leq |y| \Longleftrightarrow \exists f(f : y \longrightarrow x \wedge f$ *is onto* $x))$.

Lemma 1.5.5 *If K is a set of cardinal numbers, then* $\sup(K)$ $(= \bigcup K)$ *is a cardinal number.*

Proof First we note that K is a set of ordinal numbers. Thus $\sup(K)$ is an ordinal, and we have $\sup(K) = \bigcup K$ by Lemma 1.2.5, and furthermore $|\bigcup K| \leq \bigcup K$. If $|\bigcup K|$ is an upper bound of K, then $\bigcup K = \sup(K) \leq |\bigcup K|$ and $\bigcup K$ is a cardinal, since $\bigcup K = |\bigcup K|$. To show that property, let $\kappa \in K$. Then $\kappa \subseteq \bigcup K$, and $\kappa = |\kappa| \leq |\bigcup K|$ follows from Corollary 1.5.4. Thus $|\bigcup K|$ is an upper bound of K.

Lemma 1.5.6

a) **(Cantor)** *If x is a set, then $|x| < |\mathcal{P}(x)|$ and $|\mathcal{P}(x)| = |{}^x 2|$.*

b) **(Hartogs' Theorem)**(ZF) *If κ is a cardinal number, then there is a cardinal number greater than κ.*[12]

Proof For the proof of a), we note first that $\{(z, \{z\}) : z \in x\}$ is an injection from x into $\mathcal{P}(x)$. We will show that there is no surjection from x onto $\mathcal{P}(x)$ and assume, to get a contradiction, that f is such a surjection. Let $y := \{z \in x : z \notin f(z)\}$. By assumption, there is $u \in x$ such that $f(u) = y$. Now

[12] The smallest cardinal number with this property, constructed in the proof, is called the **Hartogs number** of κ.

we can argue closely to Russell's antinomy and conclude that $u \in f(u) \Longleftrightarrow u \notin f(u)$. This contradiction, together with the previous remark, proves $|x| < |\mathcal{P}(x)|$. For the proof of the second assertion we map each subset v of x onto its so-called characteristic function from x into $\{0,1\}$, which has the value 1 at the argument $z \in x$ iff $z \in v$. This yields a bijection from $\mathcal{P}(x)$ onto ${}^x2$.

Next we turn to the proof of part b). In ZFC, the power set of each set can be well-ordered. Therefore b) follows at once from a). Since we have carried out the main steps for a proof of this assertion in ZF already in the proof of Theorem 1.5.1, we give Hartogs' proof of this assertion in ZF. Let

$$A := \{\alpha : \exists f(f : \alpha \longrightarrow \kappa \wedge f \text{ is one-to-one})\}.$$

When proving the well-ordering theorem with the help of Zorn's lemma, we showed in ZF that the class $B := \{f : \exists\alpha(f : \alpha \longrightarrow x \text{ and } f \text{ is one-to-one})\}$ is a set for each set x. Since, for $x = \kappa$, $A = \{\text{dom}(f) : f \in B\}$, we obtain with an axiom of replacement that A is also a set. Thus there is a smallest ordinal λ such that $\lambda \notin A$. It is easy to show that $\kappa \in A$ and $\forall\beta\forall\gamma(\beta \in A \wedge \gamma < \beta \Longrightarrow \gamma \in A)$. If we could map λ one-to-one onto a smaller ordinal β, then we would get $\lambda \in A$, since $\beta \in A$. This contradiction shows that λ is a cardinal greater than κ, as desired. Furthermore, it is the smallest cardinal greater than κ.

Corollary 1.5.7 *The class* CN *of cardinal numbers is a proper class. The same holds for* $\text{CN} \setminus \omega$*, the class of infinite cardinal numbers.*

Proof Assume that CN is a set y. Then, by Lemma 1.5.5, $\kappa := \sup(y) \in \text{CN}$. From Lemma 1.5.6 we obtain a cardinal $\lambda > \kappa$. Together with $\lambda \in y$, this yields the contradiction $\lambda < \lambda$.

Definition *Let* $\text{CN} := \{\kappa \in \text{ON} : \kappa \textit{ is a cardinal}\}$ *be the class of cardinal numbers. By complete induction it is easy to see that every natural number is a cardinal number. The members of* ω *are called* **finite cardinal numbers**. *Thus we call a cardinal number* κ **infinite** *or* **transfinite** *iff* $\kappa \geq \omega$. *Note that* ω *is the smallest infinite cardinal. We write* ICN *for the class of infinite cardinal numbers.*

Lemma 1.5.6 says that for each cardinal κ *there is a smallest cardinal greater than* κ. *We denote it by* κ^+ *and call it the* **cardinal successor** *or* **successor** *of* κ. *Every infinite cardinal of the form* κ^+ *is called a* **successor cardinal**. *These cardinal numbers are exactly the values* $\aleph_{\alpha+1}$ *of the aleph function (see below) at arguments that are successor ordinals. Any infinite cardinal number which is uncountable and not a successor cardinal is called a* **limit cardinal**.

A set x *is called* **finite** *if there is a natural number* n *and a bijection from* n *onto* x. *Otherwise we call* x **infinite**. $\mathcal{P}_{\text{fin}}(x)$ *denotes the set of finite subsets of the set* x. *We say that* x *is* **denumerable** *or* **countably infinite** *iff there is a bijection from* ω *onto* x. x *is called* **countable** *iff it is finite or countably infinite.*

By Corollary 1.5.7 and Corollary 1.3.11, there is exactly one $\in$*-isomorphism from* ON *onto the class* ICN *of infinite cardinal numbers. Usually, it is denoted by* $\aleph$ *and called the* **aleph function**. *Instead of* $\aleph(\alpha)$ *we write* $\aleph_\alpha$. *Especially,* $\aleph_0$*, that means* ω*, is the smallest infinite cardinal, and* $\aleph_1$*, that means* $(\aleph_0)^+$*, is the smallest uncountable cardinal (ordinal). Often in the literature, part c) of the following lemma is taken as a recursive definition (in* ZF *– see above) of the aleph function. For each* $\alpha \in$ ON*, the members of the set* $\aleph_{\alpha+1} \setminus \aleph_\alpha$ *are exactly the ordinal numbers, and hereby the representatives of well-ordered sets, of cardinality* $\aleph_\alpha$. *The smallest element of this set,* $\aleph_\alpha$*, often is called an* **initial ordinal**. *If we want to emphasize this order-theoretical character of* $\aleph_\alpha$*, and also for historical reasons, we use the denotation* ω_α *instead of* $\aleph_\alpha$.[13]

If μ *is an infinite cardinal,* $\mu = \aleph_\alpha$*, and* β *is an ordinal, then let the cardinal* $\mu^{+\beta}$ *be defined by* $\mu^{+\beta} := \aleph_{\alpha+\beta}$.

In analogy to intervals of ordinals we define, for any cardinals κ *and* μ *such that* $\kappa \leq \mu$*, the interval*

$$(\kappa, \mu)_{\mathrm{CN}} := (\kappa, \mu)_{\mathrm{ON}} \cap \mathrm{CN}$$

of cardinals. The intervals $(\kappa, \mu]_{\mathrm{CN}}$*,* $[\kappa, \mu)_{\mathrm{CN}}$*, and* $[\kappa, \mu]_{\mathrm{CN}}$ *are defined in the same way.*

Lemma 1.5.8

a) If κ *is an infinite cardinal, then there is an ordinal* α *with* $\kappa = \aleph_\alpha$.

b) If $\alpha < \beta$*, then* $\aleph_\alpha < \aleph_\beta$.

c) $\aleph_0 = \omega$*,* $\aleph_{\alpha+1} = \aleph_\alpha^+$ *and* $\aleph_\gamma = \sup\{\aleph_\alpha : \alpha < \gamma\}$ *for limit ordinals* γ.

Proof Part a) is true by the definition of the aleph function. b) and c) follow from the fact that $\aleph : \mathrm{ON} \longrightarrow \mathrm{ICN}$ is an $\in$-isomorphism.

Definition (Sum, product and exponentiation of cardinal numbers)
If κ *and* λ *are cardinal numbers, then the sets* $\kappa \times \{0\} \cup \lambda \times \{1\}$ *and* $\kappa \times \lambda$ *can be well-ordered (for example by the relation* $\prec$ *defined in Theorem 1.5.11). We call the cardinal numbers*

$$\kappa + \lambda := |\kappa \times \{0\} \cup \lambda \times \{1\}| \quad \text{and} \quad \kappa \cdot \lambda := |\kappa \times \lambda|$$

the (cardinal) sum *and the* **(cardinal) product** *of the cardinal numbers* κ *and* λ *(not to be confused with the sum and product of ordinal numbers).*

Furthermore ${}^x y = \{f : f : x \longrightarrow y\}$ *is a set, if* x *and* y *are sets. We define* $\kappa^\lambda := |{}^\lambda \kappa|$ *as the* **exponentiation** *or* **power** *of the cardinal numbers* κ *and* λ. *This definition makes sense in* ZFC.

[13] A typical example is the denotation $\aleph_{\omega_\alpha}$ instead of $\aleph_{\aleph_\alpha}$.

By Lemma 1.5.6, we have $\kappa < 2^\kappa$ for every cardinal κ, hence $\kappa^+ \leq 2^\kappa$. The **continuum hypothesis** *is the assertion $2^{\aleph_0} = \aleph_1$, the* **generalized continuum hypothesis**, *abbreviated by* (GCH), *is the assertion $2^{\aleph_\alpha} = \aleph_{\alpha+1}$ for all $\alpha \in$* ON.

Remark Usually, one introduces the real numbers as follows: First the set $\mathbb{Z}$ of integers is constructed from the set ω of natural numbers, then the set $\mathbb{Q}$ of rational numbers, the set of all fractions of integers, is constructed from $\mathbb{Z}$. A **Dedekind cut in** $\mathbb{Q}$ is an ordered pair (A, B) of nonempty subsets of $\mathbb{Q}$ satisfying $A \cup B = \mathbb{Q}$, $\forall a \in A\, \forall b \in B(a < b)$ and $\inf(B) \in B$, if the infimum exists. Then we can define the set $\mathbb{R}$ of real numbers, the so-called **continuum**, as the set of all Dedekind cuts in $\mathbb{Q}$. Clearly we have $|\mathbb{R}| \leq |\mathcal{P}(\mathbb{Q})| = 2^{|\mathbb{Q}|} = 2^{\aleph_0}$ by Lemma 1.5.6. If we can specify a subset of $\mathbb{R}$ of cardinality $2^{\aleph_0}$, then it is shown that $|\mathbb{R}| = 2^{\aleph_0}$. We get an injection from the set ${}^\omega 2$ into $\mathbb{R}$, if we assign to each sequence $f \in {}^\omega 2$ the ternary fraction $\sum_{n=0}^{\infty} 2f(n)3^{-n-1} \in [0,1] := \{x \in \mathbb{R} : 0 \leq x \leq 1\}$.

Thus the continuum hypothesis can be reformulated equivalently as

$$\neg \exists X \subseteq \mathbb{R}(\omega < |X| < |\mathbb{R}|).$$

Lemma 1.5.9

a) If x and y are sets with $|x| = \kappa$ and $|y| = \lambda$, then $\kappa + \lambda = |x \times \{0\} \cup y \times \{1\}|$, $\kappa \cdot \lambda = |x \times y|$, and $\kappa^\lambda = |{}^y x|$.

b) For natural numbers, the cardinal sum, product and exponentiation coincide with the corresponding ordinal operations defined in Section 1.4.

c) If $\diamond$ is the addition, multiplication, or exponentiation of ordinals, and if α and β are nonzero ordinals, one of them being infinite, then

$$|\alpha \diamond \beta| \leq \max\{|\alpha|, |\beta|\}.$$

In particular, every infinite cardinal number κ is a principal number of addition, i.e., we have $\alpha + \kappa = \kappa$ for all $\alpha < \kappa$.

Proof In a), we immediately obtain the desired bijections with the help of the given bijections from x onto κ and from y onto λ. Part b) is proved by complete induction, and the first assertion in part c) by transfinite induction according to the recursive definitions of the ordinal operations and by Theorem 1.5.14. The second assertion follows from Lemma 1.4.7.

Lemma 1.5.10 *For cardinal numbers κ, λ, μ and ν the following laws hold.*

a) *(i)* $(\kappa + \lambda) + \mu = \kappa + (\lambda + \mu)$ *(ii)* $\kappa + \lambda = \lambda + \kappa$

b) *(i)* $(\kappa \cdot \lambda) \cdot \mu = \kappa \cdot (\lambda \cdot \mu)$ *(ii)* $\kappa \cdot \lambda = \lambda \cdot \kappa$

(iii) $\kappa \cdot 0 = 0$ *(iv)* $\kappa \cdot 1 = \kappa$

(v) $\kappa \cdot (\lambda + \mu) = \kappa \cdot \lambda + \kappa \cdot \mu$

c) (i) $\kappa^{\lambda+\mu} = \kappa^{\lambda} \cdot \kappa^{\mu}$ (ii) $(\kappa^{\lambda})^{\mu} = \kappa^{\lambda \cdot \mu}$
(iii) $(\kappa \cdot \lambda)^{\mu} = \kappa^{\mu} \cdot \lambda^{\mu}$ (iv) $\kappa^0 = 1,\ 0^{\kappa} = 0$ *for* $\kappa \neq 0$
(v) $\kappa^1 = \kappa,\ 1^{\kappa} = 1$ (vi) $\kappa^2 = \kappa \cdot \kappa$

d) *If* $\kappa \leq \mu$ *and* $\lambda \leq \nu$, *then*:
(i) $\kappa + \lambda \leq \mu + \nu$ (ii) $\kappa \cdot \lambda \leq \mu \cdot \nu$
(iii) $\lambda \neq 0 \Longrightarrow \kappa^{\lambda} \leq \mu^{\nu}$

Proof We leave it as an exercise for the reader to find suitable bijections and injections. We indicate the proof of (ii) in part c). If $f : \mu \longrightarrow {}^{\lambda}\kappa$ is a function, then for each $\beta \in \mu$ the value $f(\beta)$ is a function from λ into κ. Let the function $\phi(f) : \lambda \times \mu \longrightarrow \kappa$ be defined by $\phi(f)(\alpha, \beta) := f(\beta)(\alpha)$. Then ϕ is a bijection from ${}^{\mu}({}^{\lambda}\kappa)$ onto ${}^{\lambda\times\mu}\kappa$.

Theorem 1.5.11 *There is a well-ordering* $\prec$ *on* ON $\times$ ON *such that, for every* $\alpha \in$ ON, *the well-ordering* $(\aleph_\alpha \times \aleph_\alpha, \prec)$ *is an initial segment of* $\prec$ *which is isomorphic to* $(\aleph_\alpha, \in)$. *In particular we have* $|\aleph_\alpha \times \aleph_\alpha| = \aleph_\alpha$, *that means* $\aleph_\alpha \cdot \aleph_\alpha = \aleph_\alpha$.

Proof We define the relation $\prec$ on ON $\times$ ON by $(\alpha, \beta) \prec (\sigma, \tau) :\Longleftrightarrow$

$$\max\{\alpha,\beta\} < \max\{\sigma,\tau\} \vee (\max\{\alpha,\beta\} = \max\{\sigma,\tau\} \wedge (\alpha < \sigma \vee (\alpha = \sigma \wedge \beta < \tau))).$$

That means that we well-order the pairs of ordinal numbers by their maxima and, if their maxima are equal, lexicographically. It is easy to verify that $\prec$ is a well-ordering on ON $\times$ ON, and that $(\aleph_\alpha \times \aleph_\alpha, \prec)$ is a proper initial segment of this well-ordering for every $\alpha \in$ ON. Assume, contrary to our hopes, that α is the smallest ordinal for which $(\aleph_\alpha \times \aleph_\alpha, \prec)$ is not isomorphic to $(\aleph_\alpha, \in)$. By Corollary 1.3.11, there is a unique ordinal γ and an isomorphism f from $(\gamma, \in)$ onto $(\aleph_\alpha \times \aleph_\alpha, \prec)$. Then $\gamma \geq |\gamma| = |\aleph_\alpha \times \aleph_\alpha| \geq \aleph_\alpha$, and thus, by assumption, $\gamma > \aleph_\alpha$. We conlude that the function $f \restriction \aleph_\alpha$ is an isomorphism from $(\aleph_\alpha, \in)$ onto a proper initial segment of $(\aleph_\alpha \times \aleph_\alpha, \prec)$. Since $\prec$ orders ON $\times$ ON first by the maxima of the pairs, the set of first as well as the set of second components of $f[\aleph_\alpha]$ must be bounded in $\aleph_\alpha$; take $\sigma < \aleph_\alpha$ such that $f[\aleph_\alpha] \subseteq \sigma \times \sigma$. Now σ is infinite, and so there is $\beta < \alpha$ with $|\sigma| = \aleph_\beta$. Then we get $\aleph_\beta < \aleph_\alpha$ and, by the choice of α, $\aleph_\beta \cdot \aleph_\beta = \aleph_\beta$. On the other hand,

$$\aleph_\alpha = |f[\aleph_\alpha]| \leq |\sigma \times \sigma| = |\sigma| \cdot |\sigma| = \aleph_\beta \cdot \aleph_\beta = \aleph_\beta.$$

This contradiction shows that such an ordinal α does not exist, as desired.

Corollary 1.5.12 *If* κ *and* λ *are cardinal numbers, one of them nonzero and the other infinite, then*

$$\kappa + \lambda = \kappa \cdot \lambda = \max\{\kappa, \lambda\}.$$

In particular,

$$\aleph_\alpha + \aleph_\beta = \aleph_\alpha \cdot \aleph_\beta = \max\{\aleph_\alpha, \aleph_\beta\} = \aleph_{\max\{\alpha,\beta\}}.$$

Proof Without loss of generality let $0 < \kappa \leq \lambda$ and $\lambda \geq \omega$. Then the definitions of the cardinal operations, Corollary 1.5.4, and Theorem 1.5.11 yield $\lambda \leq \kappa + \lambda \leq \kappa \cdot \lambda \leq \lambda \cdot \lambda = \lambda$. This gives the first assertion. The second follows from the monotonicity of the aleph function.

Definition *If a is a set, then we write $[a]^\lambda := \{x \subseteq a : |x| = \lambda\}$ for the set of those subsets of a which have cardinality λ.*[14]

Corollary 1.5.13 *Let a be an infinite set of cardinality κ and λ be a cardinal. Then $|[a]^\lambda| = |a|^\lambda = \kappa^\lambda$, if $\lambda \leq \kappa$, and $|[a]^\lambda| = 0$, if $\lambda > \kappa$.*

Proof The second assertion is obvious. So assume without loss of generality that $0 < \lambda \leq \kappa = |a|$. Then $|[a]^\lambda| = |[\kappa]^\lambda|$. Using (AC), we can assign uniquely to each subset X of κ of cardinality λ a bijection from λ onto X, that means a member of ${}^\lambda\kappa$. Clearly this defines an injective function, and thus $|[\kappa]^\lambda| \leq \kappa^\lambda$. For the proof of the reverse inequality we observe that every function from λ into κ is a subset of $\lambda \times \kappa$ with cardinality λ, and that we have $|\lambda \times \kappa| = \kappa$ by assumption and by Corollary 1.5.12. Thus we can conclude that

$$\kappa^\lambda = |{}^\lambda\kappa| \leq |[\lambda \times \kappa]^\lambda| = |[\kappa]^\lambda| \leq \kappa^\lambda,$$

as claimed.

Theorem 1.5.14 *The following assertions hold. They are generally sufficient to get cardinality results in everyday mathematics.*

a) (ZF) The Cantor-Schröder-Bernstein theorem.

b) If, for any set a and $n \in \omega \setminus \{0\}$, a^n is defined by $a^1 := a$ and $a^{k+1} := a^k \times a$, then we have for any infinite set a:

$$|a| = |a \times a| = |a^n| \quad \textit{and} \quad |{}^n a| = |a|.$$

c) If $(A_i : i \in I)$ is a family of sets, then

$$|\bigcup\{A_i : i \in I\}| \leq |I| \cdot \sup\{|A_i| : i \in I\}.$$

d) If a is an infinite set, then $|\mathcal{P}_{\text{fin}}(a)| = |a|$.

Proof The first two statements of b) follow directly from Corollary 1.5.12. Hereby a^n is the set of all n-tupels of members of a. We will also work with the set ${}^n a$ whose members can be regarded as n-tupels, too.[15] If we assume as

[14] The sets $[a]^{<\lambda}$, $[a]^{\leq\lambda}, \ldots$ are defined analogously in Section 1.6.

[15] Then we have two different definitions of an ordered pair, since the set (c, d) is not equal to the function $\{(0, c), (1, d)\}$.

inductive hypothesis that g is a bijection from ${}^n a$ onto a, then the equation $h(f) := (g(f \restriction n), f(n))$ defines a bijection from ${}^{n+1}a$ onto $a \times a$, and this gives $|{}^{n+1}a| = |a \times a| = |a|$.

For c), let $\kappa := \sup\{|A_i| : i \in I\}$, and let $<$ be a well-ordering of I. The function $f : \bigcup\{A_i : i \in I\} \longrightarrow \bigcup\{A_i \times \{i\} : i \in I\}$, given by $f(a) = (a, i)$ where i is $<$-minimal with $a \in A_i$, is injective. So we obtain

$$|\bigcup\{A_i : i \in I\}| \leq |\bigcup\{A_i \times \{i\} : i \in I\}| =$$

$$|\bigcup\{|A_i| \times \{i\} : i \in I\}| \leq |\bigcup\{\kappa \times \{i\} : i \in I\}| = |\kappa \times I| = |I| \cdot \kappa.$$

Finally, we prove d). With the axiom of choice it is easy to show for nonempty sets x and y that there is an injection from x into y iff there is a surjection from y onto x. From c) and b) we get $|\bigcup\{{}^n a : n \in \omega\}| = |a|$. If we assign to each member of $\bigcup\{{}^n a : n \in \omega\}$ its range, then we obtain a surjective function from this set onto $\mathcal{P}_{\text{fin}}(a)$.

Example We want to show that any two bases of a vector space W over a field K have the same cardinality.

If one of the bases is finite, then the exchange theorem of Steinitz says that the other basis is finite, too, and that the bases have the same number of members. So assume that B_1 and B_2 are infinite bases of W. For each vector $x \in B_1$, there is a finite subset b_x of B_2 such that x is a linear combination of the elements of b_x. Let $C := \bigcup\{b_x : x \in B_1\}$, and suppose that $C \neq B_2$. Take $y \in B_2 \setminus C$. y is a linear combination of certain members $x_1, \ldots, x_n$ of B_1, and each x_i is a linear combination of b_{x_i}. Thus y is a linear combination of $\bigcup\{b_{x_i} : i \in \{1, \ldots, n\}\}$, and our supposition implies that y is not a member of this set. This contradicts the fact that B_2 is linearly independent, hence $B_2 = \bigcup\{b_x : x \in B_1\}$. Theorem 1.5.14 c) says that $|B_2| \leq |B_1| \cdot \aleph_0 = |B_1|$. By symmetry, we have $|B_1| \leq |B_2|$. Now part a) of the theorem yields $|B_1| = |B_2|$, as desired.

Definition *Let R be a relation on A with* $\text{dom}(R) = A$. *If $B \subseteq A$, then B is called* **unbounded** *or* **cofinal in A** *under R iff, for each $a \in A$, there is a member $b \in B$ such that aRb. We define the* **cofinality of (A, R)**, *denoted by* $\text{cf}(A, R)$, *by*

$$\text{cf}(A, R) := \min\{|C| : C \subseteq A \wedge C \text{ is cofinal in } A \text{ under } R\}.$$

Thus $\text{cf}(A, R)$ *is the smallest cardinality of any subset of A that is cofinal in A under the relation R. Because of* $\text{dom}(R) = A$, *A itself has this property; so the cardinal number* $\text{cf}(A, R)$ *always exists.*

If α is an ordinal, then the **cofinality of α**, *written as* $\text{cf}(\alpha)$, *is the smallest ordinal μ with the property that there is a function $f : \mu \longrightarrow \alpha$ such that*

$\operatorname{ran}(f)$ *is cofinal in* α *under the relation* $\leq$ *on ordinals. Obviously,* $\operatorname{cf}(\alpha) \leq \alpha$. *Furthermore* $\operatorname{cf}(\alpha) \leq \operatorname{cf}(\alpha, \leq)$, *since for each cofinal subset* C *of* α *satisfying* $\operatorname{cf}(\alpha, \leq) = |C|$ *there is a bijection from* $|C|$ *onto* C *whose range is cofinal in* α; *an analogous argument shows that* $\operatorname{cf}(\alpha)$ *is a cardinal. Lemma 1.5.15 and Lemma 1.5.16 tell us that* $\operatorname{cf}(\alpha) = \operatorname{cf}(\alpha, \leq)$ *holds.*

We call a limit ordinal α **regular** *iff* $\operatorname{cf}(\alpha) = \alpha$; *we have just shown that regular ordinals are cardinals. A limit ordinal* α *is called* **singular** *iff* $\operatorname{cf}(\alpha) < \alpha$. *With* Reg *we denote the class of regular cardinal numbers and let* $(\kappa, \mu)_{\mathrm{reg}} := (\kappa, \mu)_{\mathrm{CN}} \cap \mathrm{Reg}$ *etc.*

Lemma 1.5.15 *Let* α *be an ordinal. Then*

a) $\operatorname{cf}(0) = 0$ *and* $\operatorname{cf}(\alpha + 1) = 1$.

b) If α *is a limit ordinal, then there exists a normal function* $h : \operatorname{cf}(\alpha) \longrightarrow \alpha$ *whose range is cofinal in* α.
The last condition is equivalent to $\sup(\operatorname{ran}(h)) = \alpha$, *i.e., to* $\bigcup \operatorname{ran}(h) = \alpha$.

Proof The range of the empty function is cofinal in 0, and the range of the function $\{(0, \alpha)\}$ is cofinal in $\alpha + 1$. This proves a).

Next we turn to the proof of part b). If $\alpha \in \mathrm{Lim}$ and $f : \gamma \longrightarrow \alpha$ is a function, then we get from the definition of a limit ordinal

$$\operatorname{ran}(f) \text{ is unbounded in } \alpha \iff \sup(\operatorname{ran}(f)) = \alpha \iff \bigcup \operatorname{ran}(f) = \alpha.$$

If $f : \operatorname{cf}(\alpha) \longrightarrow \alpha$ is a function such that $\sup(\operatorname{ran}(f)) = \alpha$, then, using transfinite recursion, we obtain a strictly increasing function $g : \operatorname{cf}(\alpha) \longrightarrow \alpha$ with $\sup(\operatorname{ran}(g)) = \alpha$ via

$$g(\beta) := \max\{f(\beta), \sup\{g(\sigma) + 1 : \sigma < \beta\}\}.$$

Note that, by the definition of $\operatorname{cf}(\alpha)$, the ordinal $\sup\{g(\sigma) + 1 : \sigma < \beta\}$ is smaller than α for any $\beta < \operatorname{cf}(\alpha)$.

For continuity, we define the function $h : \operatorname{cf}(\alpha) \longrightarrow \alpha$ by $h(0) := g(0)$, $h(\gamma) := \sup\{h(\beta) : \beta < \gamma\}$ for limit ordinals $\gamma < \alpha$ and $h(\beta + 1) := g(\beta + 1)$. Then h is continuous. Using transfinite induction, it is easy to verify that $h(\beta) \leq g(\beta)$ for every $\beta < \operatorname{cf}(\alpha)$. From this we get immediately that h is strictly increasing, and clearly we have $\sup(\operatorname{ran}(h)) = \alpha$.

Lemma 1.5.16

a) If α *is a limit ordinal, then* $\operatorname{cf}(\operatorname{cf}(\alpha)) = \operatorname{cf}(\alpha)$, *hence* $\operatorname{cf}(\alpha)$ *is a regular cardinal. Furthermore we have* $\operatorname{cf}(\alpha, \in) = \operatorname{cf}(\alpha)$.

b) $\aleph_0$ *is regular. Furthermore, every infinite successor cardinal* κ^+ *is regular.*

c) If α is a limit ordinal, then $\mathrm{cf}(\alpha) = \mathrm{cf}(\aleph_\alpha)$.

d) $\aleph_\omega$ *is the smallest singular cardinal number.*

e) The class of singular cardinal numbers is unbounded in ON.

Proof For part a), we take strictly increasing functions $f : \mathrm{cf}(\mathrm{cf}(\alpha)) \longrightarrow \mathrm{cf}(\alpha)$ and $g : \mathrm{cf}(\alpha) \longrightarrow \alpha$ such that $\sup(\mathrm{ran}(g)) = \alpha$ and $\sup(\mathrm{ran}(f)) = \mathrm{cf}(\alpha)$. Then the function $g \circ f$ is strictly increasing. To show that its range is unbounded in α, consider a fixed $\beta < \alpha$. First there is an $\gamma < \mathrm{cf}(\alpha)$ with $\beta < g(\gamma)$. Furthermore there is $\sigma < \mathrm{cf}(\mathrm{cf}(\alpha))$ such that $\gamma < f(\sigma)$. Thus we have, by the monotonicity of g, $\beta < g(f(\sigma))$. This gives $\sup(\mathrm{ran}(g \circ f)) = \alpha$, and we can conclude that $\mathrm{cf}(\alpha) \leq \mathrm{cf}(\mathrm{cf}(\alpha))$. The reverse inequality is trivial. So the first assertion in a) is shown.

Since we can find a strictly increasing function $f : \mathrm{cf}(\alpha) \longrightarrow \alpha$ whose range is cofinal in α, we get $\mathrm{cf}(\alpha, \in) \leq |\mathrm{ran}(f)| = |\mathrm{cf}(\alpha)| = \mathrm{cf}(\alpha)$. From this together with the remark after the definition of $\mathrm{cf}(\alpha)$ we obtain the second assertion of part a).

For the proof of b), we note first that obviously $\aleph_0$ is regular, since no finite subset of ω can be cofinal in ω. For the second assertion, consider a function $f : \mu \longrightarrow \kappa^+$. Then, by Theorem 1.5.14 c),

$$|\sup(\mathrm{ran}(f))| = |\bigcup \mathrm{ran}(f)| = |\bigcup \{f(\beta) : \beta < \mu\}| \leq |\mu| \cdot \kappa,$$

since $|f(\beta)| < \kappa^+$ for every $\beta < \mu$, hence $|f(\beta)| \leq \kappa$. Now, if $\mu \leq \kappa$, then $|\sup(\mathrm{ran}(f))| < \kappa^+$ and thus $\sup(\mathrm{ran}(f)) < \kappa^+$. So we get $\mathrm{cf}(\kappa^+) \geq \kappa^+$. Since $\mathrm{cf}(\kappa^+) \leq \kappa^+$, we have $\mathrm{cf}(\kappa^+) = \kappa^+$, and thus κ^+ is regular.

For the proof of c), we note first that $\aleph_\alpha = \sup\{\aleph_\beta : \beta < \alpha\}$ for any limit ordinal α. If $h : \mathrm{cf}(\alpha) \longrightarrow \alpha$ has the property that $\sup(\mathrm{ran}(h)) = \alpha$, then the range of the function $h' : \mathrm{cf}(\alpha) \longrightarrow \aleph_\alpha$, given by $h'(\beta) = \aleph_{h(\beta)}$, satisfies $\sup(\mathrm{ran}(h')) = \aleph_\alpha$, hence $\mathrm{cf}(\aleph_\alpha) \leq \mathrm{cf}(\alpha)$. Conversely, let $g : \mathrm{cf}(\aleph_\alpha) \longrightarrow \aleph_\alpha$ be fixed, such that $\mathrm{ran}(g)$ is cofinal in $\aleph_\alpha$. Let $g' : \mathrm{cf}(\aleph_\alpha) \longrightarrow \alpha$ be defined by $g'(\gamma) = \min\{\sigma < \alpha : g(\gamma) \leq \aleph_\sigma\}$. If $\tau < \alpha$, then there is $\gamma < \mathrm{cf}(\aleph_\alpha)$ such that $\aleph_\tau \leq g(\gamma)$. Then $\tau \leq g'(\gamma)$, hence $\mathrm{ran}(g')$ is cofinal in α and $\mathrm{cf}(\alpha) \leq \mathrm{cf}(\aleph_\alpha)$.

Next we turn to d). By definition, $\aleph_\omega = \sup\{\aleph_n : n \in \omega\}$. Thus $\mathrm{cf}(\aleph_\omega) = \omega = \aleph_0 < \aleph_\omega$, $\aleph_\omega$ is singular. By part b), the cardinal numbers $\aleph_n$ $(n \in \omega)$ are regular. Consequently, $\aleph_\omega$ is the smallest singular cardinal.

For e), we remind the reader of the fact that the ordinal sum is continuous in the second argument. Thus each cardinal $\aleph_{\alpha+\omega}$ is singular, since it is the supremum of the countable set $\{\aleph_{\alpha+n} : n \in \omega\}$, where $+$ is the ordinal sum, and furthermore $\alpha \leq \aleph_\alpha < \aleph_{\alpha+\omega}$. From this we also get that the class of singular cardinals is unbounded in ON; therefore it is a proper class.

Remark Further properties of regular and singular cardinal numbers are presented in Lemma 1.6.11, when we have infinite sums of cardinals at our disposal, and of course in the section on clubs and stationary sets. Those readers which are meeting these concepts for the first time, may always imagine $\aleph_\omega$ as the prototype for a singular cardinal and $\aleph_1$ as the prototype for a regular cardinal. The regularity of $\aleph_1$ also follows from the well-known fact that any union of countably many countable sets is countable (see Theorem 1.5.14). The corresponding generalization for arbitrary regular cardinal numbers can be found in Lemma 1.6.11.

Exercises

Assume that $a, b, c, \ldots, A, B, C, \ldots$ etc. are sets.

1) If $n \in \omega$, then there is no injection from n onto a proper subset A of n.
   Hint: Use complete induction. For the induction step, consider the cases $n \notin A$ and $n \in A$.
2) Prove that ω and every member of ω are cardinals.
3) Assume that B and C have the same finite cardinality n. Let $f : B \longrightarrow C$ be injective (surjective). Prove that f is surjective (injective).
   Hint: Use complete induction.
4) Prove that the following assertions are equivalent:

   (i) A is finite.

   (ii) Every nonempty subset Y of $\mathcal{P}(A)$ has a $\subseteq$-maximal element.

   (iii) $\forall f(f : A \longrightarrow A \implies (f \text{ is injective } \iff f \text{ is surjective}))$.

   Hint: For (i) $\implies$ (iii), use Exercise 3.
5) Prove that ordinal and cardinal addition (multiplication, exponentiation) coincide on $\omega \times \omega$.
6) If A is a finite set all of whose elements are nonempty, then A has a choice function.
7) (ZF) If B is a finite set and each member of B is countable, then $\bigcup B$ is countable. In ZFC, this assertion is provable for countable sets B.
8) ${}^{\omega}\{0,1\}$ is not countable.
   Hint: Suppose that ${}^{\omega}\{0,1\} = \{f_n : n \in \omega\}$ and construct a new function $f : \omega \longrightarrow \{0,1\}$.
9) $f : \omega \times \omega \longrightarrow \omega$, defined by $f(x,y) := \frac{1}{2}(x+y+1)\cdot(x+y)+x$, is bijective.
10) Assume that A is a nonempty set and B is a set. Prove: There is an injection from A into B iff there is a surjection from B onto A. For which implication is (AC) not needed?
11) Assume that A has at least three elements. Prove that there is no surjective function from A onto $\mathcal{P}(A) \setminus \{\{a\} : a \in A\}$.
    Hint: The proof is very similar to Cantor's proof in 1.5.6 a).

12) Prove that every infinite set has a countably infinite subset. Did you need (AC)? (Yes!)

13) Prove in ZF that (AC) is equivalent to the following assertion:

$$\forall x \, \forall y \, \exists f (f \text{ is injective } \wedge (f : x \longrightarrow y \vee f : y \longrightarrow x)).$$

Hint: One implication can be found above. Furthermore we have shown without using (AC) that $\{\sigma \in \mathrm{ON} : \exists f (f : \sigma \longrightarrow A \wedge f \text{ is injective})\}$ is a set if A is a set. Use this to show that the above assertion implies that A can be well-ordered.

14) a) If there is a surjection f from an ordinal α onto A, then A can be well-ordered.

Hint: For $a \in A$ put $\mathrm{od}(a) := \min\{\sigma \in \alpha : f(\sigma) = a\}$. Define $a \prec b :\Longleftrightarrow \mathrm{od}(a) < \mathrm{od}(b)$ for any $a, b \in A$.

b) Prove in ZF that (AC) is equivalent to the following assertion:

$$\forall x \, \forall y (x \neq \emptyset \wedge y \neq \emptyset$$
$$\Longrightarrow \exists f (f : x \longrightarrow y \text{ surjective} \vee f : y \longrightarrow x \text{ surjective})).$$

Hint: Use part a) for one of the implications and show that the class $\{\sigma : \exists f (f : A \longrightarrow \sigma \wedge f \text{ is surjective})\}$ is a set. Observe that every surjection from A onto $\sigma \in \mathrm{ON}$ yields an injection from σ into $\mathcal{P}(A)$.

15) (Vaught) Let $\mathrm{pd}(y)$ abbreviate the assertion $\forall u \in y \, \forall v \in y (u \neq v \Longrightarrow u \cap v = \emptyset)$. Prove in ZF that (AC) is equivalent to the following assertion:

$$\forall x \exists y (y \subseteq x \wedge \mathrm{pd}(y) \wedge \forall z (z \subseteq x \wedge \mathrm{pd}(z) \Longrightarrow \neg \, y \subsetneqq z)).$$

16) The **set version** of Zorn's lemma says:

If X is a set and $M \subseteq \mathcal{P}(X)$ is nonempty, such that $\bigcup K \in M$ for every nonempty $\subseteq$-chain $K \subseteq M$, then M has a $\subseteq$-maximal element.

Show in ZF that this assertion is equivalent to Zorn's lemma.

Hint: If $(H, \leq)$ is a partial ordering that satisfies the assumptions of Zorn's lemma, then define

$$M := \{X \subseteq H : X \text{ is a } \leq\text{-chain}\}.$$

17) Assume that X is a set and $M \subseteq \mathcal{P}(X)$. We say that M **has finite character** iff any subset Y of X is a member of M if and only if every finite subset of Y is an member of M. The **Teichmüller-Tukey Lemma** says:

> If $M \subseteq \mathcal{P}(X)$ has finite character and $M \neq \emptyset$, then M has a $\subseteq$-maximal element.

Show in ZF that this assertion is equivalent to Zorn's lemma.

Hint: For one implication, the proof of the nontrivial direction of the previous question is quite useful.

18) **Hausdorff's Maximality Principle** says: For every nonempty partial ordering $(H, \leq)$, where H is a set, there is a $\subseteq$-maximal $\leq$-chain in H.

    a) Prove in ZF that this principle follows from the Teichmüller-Tukey lemma.

    b) Prove in ZF that Zorn's lemma follows from this principle.

19) The **Principle of Dependent Choices** says that, for every nonempty set A and every relation R on A satisfying $\forall x \in A \, \exists y \in A \; xRy$, there is a sequence $f : \omega \longrightarrow A$ such that $(f(n), f(n+1)) \in R$ for all $n \in \omega$. Prove that this principle follows from (AC), and prove in ZF that it implies that every countable set of nonempty sets has a choice function. The last assertion is called the axiom of **countable choice**.

20) (Kelley) Let X be a topological space. An **open cover** of X is a set $\mathcal{U}$ of open sets such that $\bigcup \mathcal{U} = X$. X is called **compact** iff every open cover of X has a finite open subcover $\mathcal{U}' \subseteq \mathcal{U}$. In lectures on topology, one usually proves with (AC) the following assertion, known as **Tychonoff's Theorem**:

    > If I is a set and $(X_i : i \in I)$ is a family of compact topological spaces, then the topological space $\prod_{i \in I} X_i$ with the standard product topology is compact.

    Prove in ZF that (AC) follows from this assertion.

    Hint: Let $(X_i : i \in I)$ be a family of nonempty sets. We want to show that $\prod_{i \in I} X_i \neq \emptyset$. Choose a fixed new set a and put $Y_i := X_i \cup \{a\}$. Let a topology $\mathcal{T}_i$ on Y_i be defined by

    $$\mathcal{T}_i := \{\emptyset, \{a\}\} \cup \{Z \subseteq Y_i : Y_i \setminus Z \text{ is finite}\}.$$

    Now use the fact that X is compact iff every family of closed sets with the finite intersection property has a nonempty intersection, and consider the family $(A_i : i \in I)$, where $A_i := \{f \in \prod_{k \in I} Y_k : f(i) \in X_i\}$.

21) Prove without using (AC):

    a) If $B \cap C = \emptyset$, then there is a bijection from ${}^{B \cup C}A$ onto ${}^{B}A \times {}^{C}A$.

    b) There is a bijection from ${}^{B \times C}A$ onto ${}^{C}({}^{B}A)$.

c) If S is a countable subset of $\mathbb{R}$, then $|\mathbb{R} \setminus S| = |\mathbb{R}|$.

Hint: If S is infinite, assume w.l.o.g. that $\omega \setminus S$ is infinite, and map this set bijectively onto $\omega \cup S$.

d) There is a bijection from $\mathbb{R}^n$ onto $\mathbb{R}$.

e) There is a bijection from $\mathbb{R}$ onto the set of all sequences of real numbers.

f) There is a bijection from ${}^{\mathbb{R}}\mathbb{R}$ onto $\mathcal{P}(\mathbb{R})$.

22) (ZF) Show that, for any $a, b \in \mathbb{R}$ with $a < b$, $|\mathbb{R}^n| = |\mathbb{R}| = |[a,b]| = |]a,b[| = |]a,b]|$. Consequently, the continuum hypothesis is trivially true for open sets: Every nonempty open set is equinumerous to $\mathbb{R}$.

23) Show that, if $C = \{f \in {}^{\mathbb{R}}\mathbb{R} : f \text{ is continuous}\}$ ("continuous" in the usual meaning of calculus), then $|C| = |\mathbb{R}|$.

Hint: For any $f, g \in C$, $f \restriction \mathbb{Q} = g \restriction \mathbb{Q}$ implies $f = g$.

24) The set $C = \{\sum_{i=0}^{\infty} a_i 3^{-i-1} : (a_i : i \in \omega) \in {}^{\omega}\{0,2\}\}$ is called the **Cantor set**. Show:

a) The function $f : [0,1] \longrightarrow \{0,1\}$, defined by $f(x) = 0 \iff x \in C$, is Riemann integrable.

Hint: By the integrability criterion of Riemann, it suffices to specify a sequence of partitions of $[0,1]$, such that the corresponding lower Riemann sums converge to 1.

b) $|\{f \in {}^{[0,1]}[0,1] : f \text{ is Riemann integrable}\}| = 2^{|\mathbb{R}|}$.

Hint: Use the hint in a).

25) Prove that the set of open subsets of $\mathbb{R}$ (open with respect to the standard topology on $\mathbb{R}$) is equinumerous to $\mathbb{R}$.

26) If $A \subseteq \mathbb{R}$, then $a \in A$ is called an **isolated point** of A if there is $\varepsilon > 0$ such that $A \cap (a - \varepsilon, a + \varepsilon) = \{a\}$. Show that any subset of $\mathbb{R}$ has only countably many isolated points.

27) Assume that $(A_\alpha : \alpha < \gamma)$ is a sequence of closed subsets of $\mathbb{R}$ (with respect to the standard topology on $\mathbb{R}$) which is strictly decreasing under $\subseteq$. Prove that γ is countable.

28) **(Cantor-Bendixson Theorem)** Prove that every closed subset of $\mathbb{R}$ can be represented as $A = P \cup S$, where P is **perfect** (i.e., closed and without isolated points) and S is countable.

Hint: Put $A_0 := A$, $A_{\alpha+1} := A'_\alpha := \{x \in \mathbb{R} : x \text{ is an accumulation point of } A_\alpha\}$ and $A_\gamma := \bigcap\{A_\sigma : \sigma < \gamma\}$ for limit ordinals γ. Show that there is a countable ordinal β such that $A_\sigma = A_\beta$ for all $\sigma \geq \beta$, and apply the result from a previous question.

29) Prove that every perfect subset P of $\mathbb{R}$ has the cardinality $2^{\aleph_0}$.

Hint: Assign to each finite $\{0,1\}$-sequence s a member a_s of P such that (1) $\mathrm{dom}(s) \neq \mathrm{dom}(t) \implies a_s \neq a_t$, (2) $f \in {}^{\omega}\{0,1\} \implies a_f := \lim_{n\to\infty} a_{f\restriction n}$ exists. Then show that the function $(a_f : f \in {}^{\omega}\{0,1\})$ is injective.

30) Prove that the continuum hypothesis is true for closed sets: For every closed subset A of $\mathbb{R}$, we have $|A| \leq \aleph_0$ or $|A| = 2^{\aleph_0}$.

Hint: Use the results from the previous questions.

31) Prove:

a) $|\{X \subseteq \mathbb{R} : |X| \leq \aleph_0\}| = |\mathbb{R}|$.

b) $|\{X \subseteq \mathbb{R} : X \text{ is closed }\}| = |\{X \subseteq \mathbb{R} : X \text{ is perfect}\}| = |\mathbb{R}|$.

32) Prove: If $|\mathbb{R}| = \aleph_1$, then $\mathbb{R}^2$ can be partitioned into two sets X and Y such that every straight line parallel to the x-axis contains only countably many points of X, and every straight line parallel to the y-axis contains only countably many points of Y.

Hint: The assumption yields a well-ordering $<$ of $\mathbb{R}$ of order type $\aleph_1$. Put $X := \{(x,y) \in \mathbb{R}^2 : x \leq y\}$ and $Y := \{(x,y) \in \mathbb{R}^2 : y < x\}$.

33) Assume that W is a vector space over the field F. Prove:

a) If W has the finite dimension $k \in \omega$, then

$$|W| = \begin{cases} |F| & \text{if } |F| \geq \aleph_0, \\ m^k & \text{if } |F| = m \in \omega. \end{cases}$$

b) If B is a basis of W, $W \neq \{0\}$, and if one of the sets F and B is infinite, then

$$|W| = \max\{|F|, |B|\}.$$

Hint: Every $w \in W \setminus \{0\}$ is uniquely representable as $w = \sum_{i=1}^{n} \lambda_i b_i$ for some $\lambda_1, \ldots, \lambda_n \in F$ and $b_1, \ldots, b_n \in B$. Show first that the equations $\Phi(w) := \{(\lambda_1, b_1), \ldots, (\lambda_n, b_n)\}$ and $\Phi(0) := \emptyset$ define an injection from W into $\mathcal{P}_{\mathrm{fin}}(F \times B)$. Now use Theorem 1.5.14.

34) If F is a field, then $F[x] := \{f \in {}^{\omega}F : \exists n_0 \forall n \geq n_0\ f(n) = 0\}$ is the set of **polynomials** over F. Prove that $|F[x]| = \max\{\aleph_0, |F|\}$.

35) Assume that κ is infinite and $\lambda \leq \kappa$. Let $T = \{f \in {}^{\kappa}\lambda : f \text{ is surjective}\}$. Then $|T| = \lambda^{\kappa}$.

Hint: If $\lambda \neq 0$, consider the set $Y := \{f \in {}^{\kappa\times\lambda}\lambda : f \text{ is surjective}\}$ and assign to each function $f \in {}^{\kappa}\lambda$ one-to-one a member of Y. Observe that $|(\kappa \times \lambda) \setminus f| = \kappa$, if $\lambda \geq 2$.

36) Assume that κ is infinite and $\lambda \leq \kappa$. Let $T = \{f \in {}^{\lambda}\kappa : f \text{ is injective}\}$. Then $|T| = \kappa^{\lambda}$.
Hint: If $\lambda \neq 0$, then $\lambda \cdot \kappa = \kappa$. Therefore $|T| = |T'|$, where $T' := \{g \in {}^{\lambda}(\lambda \times \kappa) : g \text{ is injective}\}$. Find an injection from ${}^{\lambda}\kappa$ into T'.
37) Assume that κ is infinite. Let $T = \{f \in {}^{\kappa}\kappa : f \text{ is bijective}\}$. Then $|T| = 2^{\kappa}$.
Hint: First κ can be partitioned into two disjoint sets B and C of cardinality κ. Every injection from B into B can be extended to a bijection from A onto A. Now apply the previous exercise.
38) If A is infinite, then there are $2^{|A|}$ equivalence relations on A.
39) Assume that H is a set and $(H, \leq)$ is a linear ordering. Prove that $\mathrm{cf}\,(H, \leq)$ equals 0 or 1 or is a regular cardinal.
40) Give an example of a partial ordering which has singular cofinality.
41) Assume that τ is a limit ordinal, $(\alpha_\xi : \xi < \tau)$ is a strictly increasing sequence of ordinals, and $\alpha = \sup\{\alpha_\xi : \xi < \tau\}$. Prove that $\mathrm{cf}\,(\alpha) = \mathrm{cf}\,(\tau)$.
42) Prove that $|\alpha| \leq \alpha < |\alpha|^+$ for every ordinal α.
43) Assume that α is infinite. Show that $\alpha + 1$ is not a cardinal.
44) Prove: If a is an infinite set of ordinals, then $|a| \leq \sup(a)$.
45) Let κ be a cardinal. Prove that $\alpha < \kappa^+$ iff $|\alpha| \leq \kappa$.
46) Prove: If $f : B \longrightarrow \mathrm{V}$ is a function, then $|\mathrm{ran}(f)| \leq |B|$.
47) Assume that κ is an infinite cardinal and $\prec$ is the lexicographical ordering on ${}^{\kappa}\{0,1\}$. Prove that there is no strictly increasing or strictly decreasing sequence of length κ^+.
Hint: Assume that $\bar{f} = (f_\alpha : \alpha < \kappa^+)$ is strictly increasing under $\prec$. For each $\alpha < \kappa$ there is an ordinal $\alpha^* < \kappa$ such that $f_\alpha \restriction \alpha^* = f_{\alpha+1} \restriction \alpha^*$, $f_\alpha(\alpha^*) = 0$ and $f_{\alpha+1}(\alpha^*) = 1$, since $f_\alpha \prec f_{\alpha+1}$. Use the regularity of κ^+ to obtain $\gamma < \kappa$ such that $|A_\gamma| = \kappa^+$ for $A_\gamma := \{\alpha < \kappa : \alpha^* = \gamma\}$. Let $o(\bar{f})$ be the least ordinal γ with this property and take $\bar{f}$ such that $o(\bar{f})$ is minimal. If $\alpha, \beta \in A_{o(\bar{f})}$ and $\alpha < \beta$, then $f_{\alpha+1}(o(\bar{f})) = 1$, $f_\beta(o(\bar{f})) = 0$ and $\alpha + 1 < \beta$. So consider the subsequence $\bar{g}$ of $\bar{f}$ having the members f_α and $f_{\alpha+1}$ for all $\alpha \in A_{o(\bar{f})}$ to obtain the contradiction $o(\bar{g}) < o(\bar{f})$.
48) Prove: If $2^{\aleph_0} > \aleph_\omega$, then $\aleph_\omega^{\aleph_0} = 2^{\aleph_0}$.
49) Assume that κ is a cardinal, and define $\kappa_0 := \kappa$, $\kappa_{n+1} := \aleph_{\kappa_n}$ for each $n \in \omega$, and $\mu := \sup\{\kappa_n : n \in \omega\}$. Show that μ is the smallest fixed point of the aleph function which is greater or equal than κ. In particular, for $\kappa = \omega$, we obtain the smallest fixed point of the aleph function.

1.6 Infinite Sums and Products

Definition *Let $(\kappa_i : i \in I)$ be a family of cardinal numbers. Generalizing the definition in Section 1.5, we define the* **sum** *and the* **product** *of this family as follows:*

$$\sum(\kappa_i : i \in I) := \sum_{i \in I} \kappa_i := |\bigcup\{\kappa_i \times \{i\} : i \in I\}|,$$

$$\prod(\kappa_i : i \in I) := \prod_{i \in I} \kappa_i := |\{f : \ f : I \longrightarrow \bigcup\{\kappa_i : i \in I\} \wedge \forall i \in I\ f(i) \in \kappa_i\}|.$$

The product of a family $(\kappa_i : i \in I)$ of cardinal numbers is defined as the cardinality of the cartesian product of the family $(\kappa_i : i \in I)$ of sets. Thus we obtain a certain ambiguity of the denotation $\prod_{i \in I} \kappa_i$, with which we will put up, since it will not lead to serious confusions. The context will make plain the intended meaning. So it should be clear that in "$f \in \prod_{i \in I} \kappa_i$" the cartesian product is meant.

The following lemma summarizes some laws for the above cardinal operations. It is left as an exercise for the reader to find the suitable bijections and injections.

Lemma 1.6.1 *Let $(\kappa_i : i \in I)$ and $(\lambda_i : i \in I)$ be families of cardinal numbers. Then*

a) If, for every $i \in I$, x_i is a set with $|x_i| = \kappa_i$, then

$$\sum_{i \in I} \kappa_i = |\bigcup\{x_i \times \{i\} : i \in I\}| \text{ and } \prod_{i \in I} \kappa_i = |\prod_{i \in I} x_i|.$$

b) $\sum_{i \in \{0,1\}} \kappa_i = \kappa_0 + \kappa_1$, $\sum_{i \in \emptyset} \kappa_i = 0$, $\prod_{i \in \{0,1\}} \kappa_i = \kappa_0 \cdot \kappa_1$, $\prod_{i \in \emptyset} \kappa_i = 1$.

c) If $\kappa_i \leq \lambda_i$ for all $i \in I$, then $\sum_{i \in I} \kappa_i \leq \sum_{i \in I} \lambda_i$ and $\prod_{i \in I} \kappa_i \leq \prod_{i \in I} \lambda_i$.

d) $\sum_{i \in I} \kappa = \kappa \cdot |I|$ and $\prod_{i \in I} \kappa = \kappa^{|I|}$.

Lemma 1.6.2 *Let $(\kappa_i : i \in I)$ be a family of cardinal numbers, and let λ be a cardinal number. Then the following laws hold.*

a) If $(X_j : j \in J)$ is a partition of the index set I, then

(i) (General commutative and associative laws)

$$\sum_{i \in I} \kappa_i = \sum_{j \in J} \sum_{i \in X_j} \kappa_i \quad \text{and} \quad \prod_{i \in I} \kappa_i = \prod_{j \in J} \prod_{i \in X_j} \kappa_i .$$

(ii) (Distributive law)

$$\prod_{j\in J}\sum_{i\in X_j}\kappa_i = \sum\Big(\prod_{j\in J}\kappa_{f(j)} : f\in\prod_{j\in J}X_j\Big).$$

b) $\lambda\cdot\sum_{i\in I}\kappa_i=\sum_{i\in I}\lambda\cdot\kappa_i$.

c) $\lambda^{\sum_{i\in I}\kappa_i}=\prod_{i\in I}\lambda^{\kappa_i}$.

d) $(\prod_{i\in I}\kappa_i)^\lambda=\prod_{i\in I}\kappa_i^\lambda$.

Proof Again it is left to the reader to find the desired bijections. We will only prove the distributive law and will specify a bijection ϕ from $\prod_{j\in J}B_j$ onto the set $\bigcup\{C_f\times\{f\} : f\in\prod_{j\in J}X_j\}$, where $B_j:=\bigcup\{\kappa_i\times\{i\} : i\in X_j\}$ and $C_f:=\prod_{j\in J}\kappa_{f(j)}$. Then the assertion follows from Lemma 1.6.1. Fix $F\in\prod_{j\in J}B_j$. Then, for each $j\in J$, there is a unique $i\in X_j$, say $f(j)$, such that $F(j)\in\kappa_i\times\{i\}$. Put $\phi(F):=(h,f)$ where $h\in\prod_{j\in J}\kappa_{f(j)}$ is given by: $h(j)$ is the first component of $F(j)$.

Lemma 1.6.3 *Let $(\kappa_i : i\in I)$ be a family of cardinal numbers. Then*

a) $\sup\{\kappa_i : i\in I\}\le\sum_{i\in I}\kappa_i$.

b) If $\kappa_i>0$ for all $i\in I$, $I\neq\emptyset$ and at least one of the cardinals $|I|$ and κ_i $(i\in I)$ is infinite, then

(i) $\sum_{i\in I}\kappa_i=\max\{|I|,\sup\{\kappa_i : i\in I\}\}=|I|\cdot\sup\{\kappa_i : i\in I\}$.

(ii) If in addition $|I|\le\sup\{\kappa_i : i\in I\}$, then

$$\sum_{i\in I}\kappa_i=\sup\{\kappa_i : i\in I\}.$$

In particular this holds if the given family is injective.

Proof The second equation in b) (i) holds by Corollary 1.5.12. The assertion "$\le$" in the first equation follows from Theorem 1.5.14. Since $\kappa_i>0$ for all $i\in I$, we have $|I|=\sum_{i\in I}1\le\sum_{i\in I}\kappa_i$. Furthermore it is clear by definition that $\kappa_j\le\sum_{i\in I}\kappa_i$ for every $j\in I$. This gives "$\ge$" and also proves part a).

The first equation in b) (ii) follows immediately from (i). Now assume that $\kappa_i\neq\kappa_j$ for all $i,j\in I$ with $i\neq j$. Let f be the order isomorphism from the order type α of the well-ordered set $K:=\{\kappa_i : i\in I\}$ onto this set. Then $|I|\le\alpha$ and $f(\gamma)\ge\gamma$ for every $\gamma<\alpha$. From this we can conclude that $\sup\{\kappa_i : i\in I\}=\sup(\mathrm{ran}(f))\ge\alpha\ge|I|$, if α is a limit ordinal. If $\alpha=\beta+1$ and $\alpha<\omega$, then $\sup\{\kappa_i : i\in I\}=f(\beta)\ge\omega>\alpha=|I|$. Otherwise we have $\alpha=\beta+1$, $\alpha\ge\omega$ and $\sup(\mathrm{ran}(f))=f(\beta)\ge\beta\ge|\alpha|\ge|I|$.

Corollary 1.6.4 *Let α and β be ordinals.*

a) If β is a limit ordinal, $(\sigma_\xi : \xi < \beta)$ is a strictly increasing sequence of ordinals and $\alpha = \sup\{\sigma_\xi : \xi < \beta\}$, then $\sum_{\xi<\beta} \aleph_{\sigma_\xi} = \aleph_\alpha$. In particular, $\sum_{\xi<\beta} \aleph_\xi = \aleph_\beta$.

b) $\sum_{\xi<\alpha+1} \aleph_\xi = \aleph_\alpha$.

Definition *If κ and λ are cardinal numbers, then put*

$$\kappa^{<\lambda} := \sup\{\kappa^\nu : \nu \in \mathrm{CN} \wedge \nu < \lambda\}.$$

If a is a set and λ is a cardinal number, then let $[a]^{<\lambda} := \{x \subseteq a : |x| < \lambda\}$ be the set of all subsets of a whose cardinality is less than λ. In complete analogy the sets $[a]^{\leq\lambda}$, $[a]^{>\lambda}$ and $[a]^{\geq\lambda}$ are defined.

The set $[a]^\lambda$ has already been defined in Section 1.5. The next corollary establishes the relation between $\kappa^{<\lambda}$ and $[a]^{<\lambda}$ analogously to Corollary 1.5.13.

Corollary 1.6.5

a) If κ and λ are cardinal numbers such that $\kappa \geq 2$ and $\lambda \geq \omega$, then $\kappa^{<\lambda} \geq \lambda$.

b) If a is an infinite set, $|a| = \kappa$ and λ is a cardinal number with $2 \leq \lambda \leq \kappa$, then for all cardinals $\nu_0 < \lambda$

$$\kappa^{<\lambda} = |[a]^{<\lambda}| = \sum_{\nu\in\lambda\cap\mathrm{CN}} \kappa^\nu = \sum_{\nu_0\leq\nu\in\lambda\cap\mathrm{CN}} \kappa^\nu.$$ [16]

Proof For the proof of part a), first let λ be a successor cardinal μ^+. Then $\kappa^{<\lambda} = \kappa^\mu \geq 2^\mu > \mu$ by Lemma 1.5.6. So we get $\kappa^{<\lambda} \geq \mu^+ = \lambda$. Now let λ be a limit cardinal. If $\nu \in \lambda \cap \mathrm{CN}$, then $\kappa^\nu \geq 2^\nu > \nu$. From this we can conclude that $\kappa^{<\lambda} = \sup\{\kappa^\nu : \nu \in \lambda \cap \mathrm{CN}\} \geq \sup\{\nu : \nu \in \lambda \cap \mathrm{CN}\} = \lambda$.

Next we prove b). Corollary 1.5.13 says that the cardinality of $[a]^\nu$ is κ^ν for cardinal numbers $\nu \leq \kappa$ and 0 for cardinal numbers $\nu > \kappa$. Therefore $|[a]^{<\lambda}| = \sum_{\nu\in\lambda\cap\mathrm{CN}} \kappa^\nu$. For each cardinal ν_0 with $\nu_0 < \lambda$, it follows from Lemma 1.6.3 and $2 \leq \lambda \leq \kappa$ that this sum equals $\sup\{\kappa^\nu : \nu_0 \leq \nu \in \lambda \cap \mathrm{CN}\}$. In particular, for $\nu_0 = 0$, we get $\kappa^{<\lambda} = \sup\{\kappa^\nu : \nu \in \lambda \cap \mathrm{CN}\} = |[a]^{<\lambda}|$.

Corollary 1.6.6 *If a is an infinite set and λ is a cardinal number with $\lambda \leq |a|$, then*

a) $|[a]^\lambda| = |a|^\lambda$.

b) $|[a]^{<\lambda}| = \sup\{|a|^\nu : \nu \in \mathrm{CN} \cap \lambda\}$; in particular, $|[\omega]^{<\omega}| = \aleph_0$.

c) $|[a]^{\leq\lambda}| = |a|^\lambda$.

[16]In the literature, also the cardinal number $\sum_{\nu\in\lambda\cap\mathrm{CN}} \kappa^\nu$ instead of $\kappa^{<\lambda}$ is investigated. Further results on κ^λ can be found in Section 1.7.

Proof Part a) is Corollary 1.5.13, part b) can be found in Corollary 1.6.5 b), and part c) follows at once from b) and c) using $[a]^{\leq\lambda} = [a]^{\lambda} \cup [a]^{<\lambda}$.

Theorem 1.6.7 *Assume that* $(\kappa_i : i \in I)$ *and* $(\lambda_i : i \in I)$ *are families of cardinal numbers. Then the following holds.*

a) If $\lambda_i \geq 2$ *and* $\kappa_i \leq \lambda_i$ *for all* $i \in I$*, then* $\sum_{i\in I} \kappa_i \leq \prod_{i\in I} \lambda_i$.

b) **(J. König's Lemma)**
If $\kappa_i < \lambda_i$ *for all* $i \in I$*, then* $\sum_{i\in I} \kappa_i < \prod_{i\in I} \lambda_i$.

Proof First we prove a) and the relation "$\leq$" of b).

Let $I_1 := \{i \in I : \lambda_i \geq \omega\}$ and $I_2 := \{i \in I : \lambda_i < \omega\}$. Since $\sum_{i\in I} \kappa_i = |\bigcup\{\kappa_i \times \{i\} : i \in I\}|$, we assign uniquely to each pair (α, j) with $\alpha \in \kappa_j$ and $j \in I$ a function $f_{\alpha,j} \in \prod_{i\in I}(\lambda_i + 1)$ by

$$f_{\alpha,j}(i) := \kappa_i, \quad \text{if } i \neq j; \ f_{\alpha,j}(j) := \alpha \, .$$

It is easy to verify that the functions $(f_{\alpha,j} : \alpha \in \kappa_j \wedge j \in I)$ and $(f_{\alpha,j} \restriction I_1 : \alpha \in \kappa_j \wedge j \in I_1)$ are injective. This gives

$$\sum_{i\in I_1} \kappa_i \leq |\prod_{i\in I_1}(\lambda_i + 1)| = \prod_{i\in I_1} \lambda_i \, .$$

Furthermore the same argument, applied to $(\kappa_i : i \in I)$ using the first function, proves the assertion "$\leq$" in b), since $\kappa_i < \lambda_i$ for all $i \in I$ implies that $f_{\alpha,j} \in \prod_{i\in I} \lambda_i$. If I_2 is infinite, then, by assumption,

$$\sum_{i\in I_2} \kappa_i \leq \aleph_0 \cdot |I_2| = |I_2| < 2^{|I_2|} \leq |\prod_{i\in I_2} \lambda_i| = \prod_{i\in I_2} \lambda_i.$$

If I_2 is finite, then it follows from $\lambda_i \geq 2$ for all i that $\sum_{i\in I_2} \kappa_i \leq \prod_{i\in I_2} \lambda_i$. Altogether we obtain from Lemma 1.5.10:

$$\sum_{i\in I} \kappa_i = \sum_{i\in I_1} \kappa_i + \sum_{i\in I_2} \kappa_i \leq \prod_{i\in I_1} \lambda_i + \prod_{i\in I_2} \lambda_i \leq \prod_{i\in I} \lambda_i \, .$$

For the proof of b) we assume, to get a contradiction, that there is a bijection h from $\bigcup\{\kappa_i \times \{i\} : i \in I\}$ onto $\prod_{i\in I} \lambda_i$. Then this cartesian product is the union of the pairwise disjoint sets $A_i := h[\kappa_i \times \{i\}]$, $i \in I$. Put $B_i := \lambda_i \cap \{f(i) : f \in A_i\}$. Since $|B_i| \leq |A_i| = \kappa_i < \lambda_i$, B_i is a proper subset of λ_i for every $i \in I$. Therefore the axiom of choice yields a function $g \in \prod_{i\in I} \lambda_i$ such that $g(i) \in \lambda_i \setminus B_i$ for every $i \in I$. So we get the contradiction $g \notin \bigcup\{A_i : i \in I\}$.

Corollary 1.6.8 *Let* β *be an infinite ordinal and* $(\kappa_\xi : \xi < \beta)$ *be a sequence of nonzero cardinal numbers.*

a) If $(\kappa_\xi : \xi < \beta)$ is strictly increasing and β is a limit ordinal, then

$$\sum_{\xi<\beta} \kappa_\xi < \prod_{\xi<\beta} \kappa_\xi \quad \text{and} \quad \sum_{\xi<\beta} \kappa_\xi < (\sum_{\xi<\beta} \kappa_\xi)^{|\beta|}.$$

b) If $\kappa_\xi \geq 2$ for every $\xi < \beta$, then

$$(\sum_{\xi<\beta} \kappa_\xi)^{|\beta|} = (\prod_{\xi<\beta} \kappa_\xi)^{|\beta|}.$$

Proof The first inequality in a) follows from König's lemma, if we put $\lambda_\xi := \kappa_{\xi+1}$ and take into consideration that the factor $\kappa_0 \geq 1$ does not make the product $\prod_{1\leq\xi<\beta} \kappa_\xi$ smaller. Since $\kappa_\xi \leq \sum_{\xi<\beta} \kappa_\xi$ for each $\xi < \beta$, we get $\prod_{\xi<\beta} \kappa_\xi \leq (\sum_{\xi<\beta} \kappa_\xi)^{|\beta|}$. Now the second inequality in a) follows from the first one.

With the help of part a) of Theorem 1.6.7 it can be shown, using the same argument, that $\sum_{\xi<\beta} \kappa_\xi \leq \prod_{\xi<\beta} \kappa_\xi \leq (\sum_{\xi<\beta} \kappa_\xi)^{|\beta|}$. Hereby we obtain

$$(\sum_{\xi<\beta} \kappa_\xi)^{|\beta|} \leq (\prod_{\xi<\beta} \kappa_\xi)^{|\beta|} \leq (\sum_{\xi<\beta} \kappa_\xi)^{|\beta|\cdot|\beta|},$$

and with $|\beta| \cdot |\beta| = |\beta|$ part b) is proved.

Theorem 1.6.9 *Let κ be a cardinal number.*

a) If $\kappa \geq 2$, then $\aleph_\alpha < \mathrm{cf}\,(\kappa^{\aleph_\alpha})$.
b) If κ is infinite, then $\kappa < \kappa^{\mathrm{cf}(\kappa)}$.

Proof For a), assume that $\gamma \leq \aleph_\alpha$ and f is a function from γ into $\kappa^{\aleph_\alpha}$. Then we can infer from Theorem 1.6.7 and Lemma 1.5.10 that

$$|\sup(\mathrm{ran}(f))| \leq \sum_{\beta<\gamma} |f(\beta)| < \prod_{\beta<\gamma} \kappa^{\aleph_\alpha} = (\kappa^{\aleph_\alpha})^{|\gamma|} = \kappa^{\aleph_\alpha \cdot |\gamma|} = \kappa^{\aleph_\alpha}.$$

From this we get at once $\mathrm{cf}\,(\kappa^{\aleph_\alpha}) > \aleph_\alpha$.

For b), assume that $f : \mathrm{cf}\,(\kappa) \longrightarrow \kappa$ is a function such that $\sup(\mathrm{ran}(f)) = \kappa$. As above, we obtain

$$\kappa \leq \sum_{\beta<\mathrm{cf}(\kappa)} |f(\beta)| < \prod_{\beta<\mathrm{cf}(\kappa)} \kappa = \kappa^{\mathrm{cf}(\kappa)}.$$

Corollary 1.6.10

a) For any $\alpha \in \mathrm{ON}$, $\mathrm{cf}\,(2^{\aleph_\alpha}) > \aleph_\alpha$. In particular we have $2^{\mathrm{cf}(\aleph_\alpha)} \neq \aleph_\alpha$.
b) If α is a limit ordinal and $\aleph_\beta \geq \mathrm{cf}\,(\alpha)$, then $\aleph_\alpha \neq \aleph_\gamma^{\aleph_\beta}$ for all ordinals γ; in particular we have $\aleph_\alpha < \aleph_\alpha^{\aleph_\beta}$.

Proof Part a) follows immediately from the previous theorem. So let us prove b). Since $\mathrm{cf}(\aleph_\alpha) = \mathrm{cf}(\alpha)$, it follows from Theorem 1.6.9 that

$$\aleph_\alpha < \aleph_\alpha^{\mathrm{cf}(\aleph_\alpha)} = \aleph_\alpha^{\mathrm{cf}(\alpha)} \leq \aleph_\alpha^{\aleph_\beta}.$$

The assumption $\aleph_\alpha = \aleph_\gamma^{\aleph_\beta}$ would yield the contradiction $\aleph_\alpha^{\aleph_\beta} = \aleph_\gamma^{\aleph_\beta} = \aleph_\alpha$.

Remark Corollary 1.6.10 tells us that the assertion $\aleph_0 < \mathrm{cf}(2^{\aleph_0})$ is provable in ZFC. Essentially this is the only limitation which is set by the axioms of ZFC for the cardinality of the continuum. The continuum function $(2^\kappa : \kappa \in \mathrm{ICN})$ satisfies two essential laws:

(i) $\kappa < \mu \Longrightarrow 2^\kappa \leq 2^\mu$ (monotonicity) and

(ii) $\kappa < \mathrm{cf}(2^\kappa)$.

If $F : \mathrm{Reg} \longrightarrow \mathrm{ICN}$ is an arbitrary increasing function satisfying $\kappa < \mathrm{cf}(F(\kappa))$ for all $\kappa \in \mathrm{Reg}$, then by the results of W.B. Easton [Ea] it is consistent with the axioms of ZFC that $2^\kappa = F(\kappa)$ for all $\kappa \in \mathrm{Reg}$. This means that the axioms of ZFC set no limitations on the continuum function on the regular cardinals except the conditions (i) and (ii).

Lemma 1.6.11 *Let κ be an infinite cardinal number.*

a) The following assertions are equivalent.

(i) κ is regular.

(ii) For any sequence $(S_\alpha : \alpha < \gamma)$ of sets, satisfying $\gamma < \kappa$ and $|S_\alpha| < \kappa$ for all $\alpha < \kappa$, we have $|\bigcup\{S_\alpha : \alpha < \kappa\}| < \kappa$ (i.e., κ is not a union of less than κ sets, all of which are of cardinality less than κ).

(iii) κ is not a sum of less than κ cardinal numbers, all of which are less than κ.

b) Assume that κ is an infinite cardinal. Then κ is a sum of $\mathrm{cf}(\kappa)$ cardinals less than κ. If κ is a limit cardinal, then there is a strictly increasing sequence $(\lambda_\xi : \xi < \mathrm{cf}(\kappa))$ of cardinal numbers less than κ such that

$$\kappa = \sum_{\xi < \mathrm{cf}(\kappa)} \lambda_\xi = \sup\{\lambda_\xi : \xi < \mathrm{cf}(\kappa)\}\ .$$

For singular cardinals κ we can arrange that $\mathrm{cf}(\kappa) < \lambda_\xi < \kappa$ for all $\xi < \mathrm{cf}(\kappa)$. If in addition $\mathrm{cf}(\kappa) > \aleph_0$, then there also exists a normal sequence of singular cardinals with these properties.

Proof For the proof of a), "(ii)$\Longrightarrow$(i)", assume that κ is a singular cardinal and $f : \mathrm{cf}(\kappa) \longrightarrow \kappa$ is a function such that $\bigcup \mathrm{ran}(f) = \kappa$. Put $S_\alpha := f(\alpha)$ for all $\alpha < \mathrm{cf}(\kappa)$. Then $(S_\alpha : \alpha < \mathrm{cf}(\kappa))$ is a sequence of less than κ sets, all of which are of cardinality $< \kappa$. The union of these sets is κ.

For "(i)$\Longrightarrow$(iii)", let $(\kappa_i : i < \gamma)$ be a sequence of cardinal numbers less than κ with domain $\gamma < \kappa$. Then Theorem 1.5.14 yields $\sum_{i<\gamma} \kappa_i \le |\gamma| \cdot \sup\{\kappa_i : i < \gamma\} < \kappa$, since κ is regular.

For "(iii)$\Longrightarrow$(ii)" note that the cardinality of $\bigcup\{S_\alpha : \alpha < \gamma\}$ is at most $\sum_{\alpha<\gamma} |S_\alpha|$.

Next we turn to the proof of b). Observe that for regular cardinals κ and any $\nu \in \mathrm{CN} \setminus \{0\} \cap \kappa$ we have $\kappa = \sum_{\xi<\kappa} \nu$.

Now let κ be a limit cardinal. If, according to Lemma 1.5.15, $f : \mathrm{cf}(\kappa) \longrightarrow \kappa$ is a normal function, i.e., if f is strictly increasing and continuous, such that $\sup(\mathrm{ran}(f)) = \kappa$, then the set $c := \{|f(\beta)| : \beta < \mathrm{cf}(\kappa)\}$ is cofinal in κ. Otherwise the regular cardinal number $(\sup(c))^+$ would be smaller than κ and thus be dominated by some value $f(\sigma)$, and also by $|f(\sigma)|$, which would contradict the definition of c. Therefore we obtain by transfinite recursion a strictly increasing continuous sequence $(\lambda_\xi : \xi < \mathrm{cf}(\kappa))$ of members of c with supremum κ; from the definition of cofinality we can conclude that it has length $\mathrm{cf}(\kappa)$. Using Lemma 1.6.3, we get $\kappa = \sum_{\xi<\mathrm{cf}(\kappa)} \lambda_\xi$, as desired.

If κ is singular, the same proof works if we replace c by the set $c := \{|f(\beta)| : \beta < \mathrm{cf}(\kappa) \wedge |f(\beta)| > \mathrm{cf}(\kappa)\}$. If in addition we have $\mathrm{cf}(\kappa) > \aleph_0$, then the cardinals λ_ξ can be chosen as singular. Namely, let $\lambda_{\xi+1}$ for $\xi < \mathrm{cf}(\kappa)$ be the supremum of some strictly increasing sequence in c of length ω having $\max\{\lambda_\xi, |f(\xi)|\}$ as its first member, let λ_0 be defined analogously, and put, for limit ordinals $\gamma < \mathrm{cf}(\kappa)$, $\lambda_\gamma := \sup\{\lambda_\xi : \xi < \gamma\}$. Then we have, for any limit ordinal $\gamma < \mathrm{cf}(\kappa)$, $\mathrm{cf}(\lambda_\gamma) \le \gamma < \mathrm{cf}(\kappa) < \lambda_0 < \lambda_\gamma$, and λ_γ is singular. Since $\mathrm{cf}(\kappa) > \aleph_0$, all cardinals λ_ξ are singular. The construction shows that the sequence $(\lambda_\xi : \xi < \mathrm{cf}(\kappa))$ is normal and that $\sup\{\lambda_\xi : \xi < \mathrm{cf}(\kappa)\} = \kappa$.

Theorem 1.6.12 *Let α and β be ordinals.*

a) If $\alpha \le \beta$, then $\aleph_\alpha^{\aleph_\beta} = 2^{\aleph_\beta}$.

b) **(Hausdorff's Formula)**

$$\aleph_{\alpha+1}^{\aleph_\beta} = \aleph_{\alpha+1} \cdot \aleph_\alpha^{\aleph_\beta} .$$

c) **(Tarski's Formula)** *If $|\gamma| \le \aleph_\beta$, then*

$$\aleph_{\alpha+\gamma}^{\aleph_\beta} = \aleph_{\alpha+\gamma}^{|\gamma|} \cdot \aleph_\alpha^{\aleph_\beta} .$$

Proof For part a), we can conclude from Lemma 1.5.10 and Corollary 1.5.12 that $2^{\aleph_\beta} \le \aleph_\alpha^{\aleph_\beta} \le (2^{\aleph_\alpha})^{\aleph_\beta} = 2^{\aleph_\alpha \cdot \aleph_\beta} = 2^{\aleph_\beta}$.

For b), let first $\beta \le \alpha$. Then $\aleph_\beta \le \aleph_\alpha$, and since $\aleph_{\alpha+1}$ is regular, every function from $\aleph_\beta$ into $\aleph_{\alpha+1}$ is already a function from $\aleph_\beta$ into γ for some $\gamma < \aleph_{\alpha+1}$. This gives

$$\aleph_{\alpha+1}^{\aleph_\beta} = |^{\aleph_\beta}\aleph_{\alpha+1}| = |\bigcup\{^{\aleph_\beta}\gamma : \gamma < \aleph_{\alpha+1}\}| \le \sum_{\gamma<\aleph_{\alpha+1}} |\gamma|^{\aleph_\beta} \le \aleph_{\alpha+1} \cdot \aleph_\alpha^{\aleph_\beta} \le \aleph_{\alpha+1}^{\aleph_\beta}.$$

If $\alpha < \beta$, then $\aleph_{\alpha+1} \le \aleph_\beta < 2^{\aleph_\beta}$, and the assertion follows from a) and Corollary 1.5.12.

Now we prove part c) by transfinite induction on γ. If $\gamma = 0$, there is nothing to show. For the successor step, we can infer from part b), the inductive hypothesis, and Corollary 1.5.12 that

$$\begin{aligned}\aleph_{\alpha+\gamma+1}^{\aleph_\beta} = \aleph_{\alpha+\gamma}^{\aleph_\beta} \cdot \aleph_{\alpha+\gamma+1} &= \aleph_\alpha^{\aleph_\beta} \cdot \aleph_{\alpha+\gamma}^{|\gamma|} \cdot \aleph_{\alpha+\gamma+1} \\ &= \aleph_\alpha^{\aleph_\beta} \cdot \aleph_{\alpha+\gamma}^{|\gamma+1|} \cdot \aleph_{\alpha+\gamma+1} = \aleph_\alpha^{\aleph_\beta} \cdot \aleph_{\alpha+\gamma+1}^{|\gamma+1|}.\end{aligned}$$

Now let γ be a limit ordinal. Then from Theorem 1.6.7 and the continuity of ordinal addition we can conclude that

$$\aleph_{\alpha+\gamma} = \sup\{\aleph_{\alpha+\sigma} : \sigma < \gamma\} \le \sum_{\sigma<\gamma} \aleph_{\alpha+\sigma} \le \prod_{\sigma<\gamma} \aleph_{\alpha+\sigma}.$$

From the inductive hypothesis, Lemma 1.5.10, and Lemma 1.6.1 we can infer that

$$\begin{aligned}\aleph_{\alpha+\gamma}^{\aleph_\beta} &\le (\prod_{\sigma<\gamma} \aleph_{\alpha+\sigma})^{\aleph_\beta} = \prod_{\sigma<\gamma} \aleph_{\alpha+\sigma}^{\aleph_\beta} = \prod_{\sigma<\gamma} (\aleph_\alpha^{\aleph_\beta} \cdot \aleph_{\alpha+\sigma}^{|\sigma|}) \\ &\le (\aleph_\alpha^{\aleph_\beta})^{|\gamma|} \prod_{\sigma<\gamma} \aleph_{\alpha+\sigma}^{|\sigma|} \le \aleph_\alpha^{\aleph_\beta \cdot |\gamma|} \cdot \aleph_{\alpha+\gamma}^{|\gamma| \cdot |\gamma|} = \aleph_\alpha^{\aleph_\beta} \cdot \aleph_{\alpha+\gamma}^{|\gamma|} \le \aleph_{\alpha+\gamma}^{\aleph_\beta}.\end{aligned}$$

The last equation and the last inequality hold because, by assumption, $|\gamma| \le \aleph_\beta$.

Corollary 1.6.13 *If $n \in \omega$, then the following holds.*

a) **(Generalized Hausdorff Formula)** $\aleph_{\alpha+n}^{\aleph_\beta} = \aleph_\alpha^{\aleph_\beta} \cdot \aleph_{\alpha+n}$.

b) **(Bernstein's Formula)** $\aleph_n^{\aleph_\beta} = 2^{\aleph_\beta} \cdot \aleph_n$.

c) If $\alpha \le \aleph_\beta$, then $\aleph_\alpha^{\aleph_\beta} = 2^{\aleph_\beta} \cdot \aleph_\alpha^{|\alpha|}$.

Examples 1.6.14

a) $\aleph_1^{\aleph_0} = 2^{\aleph_0} \cdot \aleph_1$ follows from Corollary 1.6.13, and thus $\aleph_1^{\aleph_0} = 2^{\aleph_0}$.

b) $\aleph_\omega^{\aleph_1} = 2^{\aleph_1} \cdot \aleph_\omega^{\aleph_0}$.

c) If $2^{\aleph_1} = \aleph_2$, then $\aleph_\omega^{\aleph_0} \neq \aleph_{\omega_1}$. To see this, we conclude with Example b) from $2^{\aleph_1} = \aleph_2$ that $\aleph_\omega^{\aleph_0} = \aleph_\omega^{\aleph_1}$. Theorem 1.6.9 gives $\mathrm{cf}(\aleph_{\omega_1}) = \aleph_1 < \mathrm{cf}(\aleph_\omega^{\aleph_1})$. Now the assertion follows.

d) For any α with $\omega \leq \alpha < \aleph_1$, we have $\aleph_\alpha^{\aleph_\beta} = 2^{\aleph_\beta} \cdot \aleph_\alpha^{\aleph_0}$.

e) If $n \in \omega$ and $\aleph_{\omega_n} = \aleph_\gamma^{\aleph_\beta}$, then $\beta \in \{0,\dots,n-1\}$. This follows from Theorem 1.6.9, because, under the given assumptions, we have $\mathrm{cf}(\aleph_{\omega_n}) = \aleph_n > \aleph_\beta$, hence $n > \beta$. In particular $\aleph_\omega^{\aleph_m} \neq \aleph_{\omega_n}$ for all $m \geq n$ and, in view of the surprising role of $\aleph_{\omega_4}$ in Shelah's Theorem 8.1.4 in Chapter 8, $\aleph_\omega^{\aleph_m} \neq \aleph_{\omega_4}$ for all $m \geq 4$.

f) $2^{\aleph_\omega} \neq \aleph_{\omega_4}$ follows from Corollary 1.6.10.

Lemma 1.6.15 (The four golden rules of cardinal arithmetic)

a) (Tarski) If $\nu \geq \aleph_0$ is a cardinal number and $(\kappa_\xi : \xi < \nu)$ is an increasing sequence of infinite cardinals, then

$$\prod_{\xi<\nu} \kappa_\xi = (\sup\{\kappa_\xi : \xi < \nu\})^\nu .$$

b) If λ and κ are cardinal numbers such that $\kappa \geq \aleph_0$ and $\lambda \geq \mathrm{cf}(\kappa)$, then

$$\kappa^\lambda = (\sup\{\nu^\lambda : \nu \in \mathrm{CN} \cap \kappa\})^{\mathrm{cf}(\kappa)} .$$

c) (Tarski) If λ and κ are cardinal numbers such that $\kappa \geq \aleph_0$ and $0 < \lambda < \mathrm{cf}(\kappa)$, then

$$\kappa^\lambda = \kappa \cdot \sup\{\nu^\lambda : \nu \in \mathrm{CN} \cap \kappa\} = \kappa \cdot \sum_{\nu \in \kappa \cap \mathrm{CN}} \nu^\lambda .$$

d) If κ is an infinite cardinal number, then $2^\kappa = (2^{<\kappa})^{\mathrm{cf}(\kappa)}$.

Proof First we prove part a). Since $|\nu \times \nu| = \nu$, we can represent ν as a disjoint union $\nu = \bigcup\{B_\alpha : \alpha < \nu\}$ of subsets B_α of ν such that $|B_\alpha| = \nu$ for all $\alpha < \nu$. Then we get $\mu := \sup\{\kappa_\xi : \xi < \nu\} = \sup\{\kappa_\xi : \xi \in B_\alpha\}$ for every $\alpha < \nu$, and $\mu \leq \prod_{\xi \in B_\alpha} \kappa_\xi$. Consequently, Lemma 1.6.2 yields

$$\mu^\nu = \prod_{\alpha<\nu} \mu \leq \prod_{\alpha<\nu} \prod_{\xi \in B_\alpha} \kappa_\xi = \prod_{\xi<\nu} \kappa_\xi \leq \prod_{\xi<\nu} \mu \leq \mu^\nu .$$

For b), we can choose a sequence $(\kappa_\xi : \xi < \mathrm{cf}(\kappa))$ of cardinals κ_ξ with $2 \leq \kappa_\xi < \kappa$ satisfying $\kappa = \sum_{\xi<\mathrm{cf}(\kappa)} \kappa_\xi$; this is possible by Lemma 1.6.11. Then we can conclude for $\mu := \sup\{\nu^\lambda : \nu \in \mathrm{CN} \cap \kappa\}$, using Theorem 1.6.7, that

$$\kappa^\lambda \leq (\prod_{\xi<\mathrm{cf}(\kappa)} \kappa_\xi)^\lambda = \prod_{\xi<\mathrm{cf}(\kappa)} \kappa_\xi^\lambda \leq \prod_{\xi<\mathrm{cf}(\kappa)} \mu = \mu^{\mathrm{cf}(\kappa)} \leq (\kappa^\lambda)^{\mathrm{cf}(\kappa)} = \kappa^\lambda .$$

Next we prove c). If $f \in {}^\lambda\kappa$, then there is an $\alpha < \kappa$ such that $f \in {}^\lambda\alpha$, since $\lambda < \mathrm{cf}(\kappa)$. Furthermore we have $|\alpha|^\lambda = \nu^\lambda$ for all $\nu \in \kappa \cap \mathrm{CN}$ with $\omega \leq \nu \leq \alpha < \nu^+$, hence by Lemma 1.6.3

$$\kappa^\lambda \leq |\bigcup\{{}^\lambda\alpha : \alpha < \kappa\}| \leq \sum_{\alpha<\kappa} |\alpha|^\lambda \leq \sum_{\nu\in\kappa\cap\mathrm{CN}} \nu^+ \cdot \nu^\lambda$$

$$\leq \kappa \cdot \sum_{\nu\in\kappa\cap\mathrm{CN}} \nu^\lambda \leq \kappa \cdot \sup\{\nu^\lambda : \nu \in \kappa \cap \mathrm{CN}\} \leq \kappa^\lambda.$$

For d), we note that the assertion is easy to show for successor cardinals. So let us assume that κ is a limit cardinal. We can apply Lemma 1.6.11 and obtain a strictly increasing sequence $(\kappa_\xi : \xi < \mathrm{cf}(\kappa))$ of cardinals less than κ such that $\kappa = \sum_{\xi<\mathrm{cf}(\kappa)} \kappa_\xi$. Then

$$2^\kappa = (2^\kappa)^{\mathrm{cf}(\kappa)} \geq (2^{<\kappa})^{\mathrm{cf}(\kappa)} = \prod_{\xi<\mathrm{cf}(\kappa)} 2^{<\kappa} \geq \prod_{\xi<\mathrm{cf}(\kappa)} 2^{\kappa_\xi} = 2^{\sum \kappa_\xi} = 2^\kappa .$$

Corollary 1.6.16 *Let κ be an infinite cardinal number and $(\kappa_\xi : \xi < \mathrm{cf}(\kappa))$ be a sequence of cardinal numbers such that $\kappa = \sum_{\xi<\mathrm{cf}(\kappa)} \kappa_\xi$ and $2 \leq \kappa_\xi < \kappa$ for every $\xi < \mathrm{cf}(\kappa)$. Then*

a) $\prod_{\xi<\mathrm{cf}(\kappa)} \kappa_\xi = \kappa^{\mathrm{cf}(\kappa)}$.

b) If λ is a cardinal number with $\lambda \geq \mathrm{cf}(\kappa)$, then $\prod_{\xi<\mathrm{cf}(\kappa)} \kappa_\xi^\lambda = \kappa^\lambda$.

Proof If κ is regular, that means if $\mathrm{cf}(\kappa) = \kappa$, then $\kappa^{\mathrm{cf}(\kappa)} = \kappa^\kappa = 2^\kappa \leq \prod_{\xi<\kappa} \kappa_\xi \leq \kappa^\kappa = \kappa^{\mathrm{cf}(\kappa)}$. Now let κ be singular. Then from Lemma 1.6.3 and the assumption we can infer that the sequence $(\kappa_\xi : \xi < \mathrm{cf}(\kappa))$ is not bounded in κ. Therefore it has a strictly increasing subsequence $(\lambda_\xi : \xi < \mathrm{cf}(\kappa))$ having the same supremum κ. Thus, by Lemma 1.6.15 a), we have

$$\kappa^{\mathrm{cf}(\kappa)} \geq \prod_{\xi<\mathrm{cf}(\kappa)} \kappa_\xi \geq \prod_{\xi<\mathrm{cf}(\kappa)} \lambda_\xi = \kappa^{\mathrm{cf}(\kappa)}.$$

Part b) follows from part a) together with Corollary 1.5.12 and Lemma 1.6.2.

Note that in Lemma 1.6.15 a) we have already proved assertion b) of the following theorem for cardinal numbers β. A slight modification of this proof shows that this assertion is true for any limit ordinal β such that $\beta \leq \omega_1$. Tarski believed that it is true for arbitrary limit ordinals. Remark 1.6.18 shows that it is consistent with the axioms of ZFC if ZF is consistent, using the famous result of K. Gödel that ZFC + (GCH) is consistent only if ZF is consistent.

It was not until the results of P. J. Cohen that this hypothesis was refuted; Magidor [Ma] produced a model with a counterexample for $\beta = \omega_1 + \omega$, and Jech and Shelah [JS] showed that if there is any counterexample, then there is one with $\beta = \omega_1 + \omega$.

Theorem 1.6.17

a) For any ordinal $\beta > 0$, $\prod_{\xi\le\beta}\aleph_\xi = \aleph_\beta^{|\beta|}$. For limit ordinals β, also $\prod_{\xi<\beta}\aleph_\xi = \aleph_\beta^{|\beta|}$.

b) If β is a principal number of addition and $(\alpha_\xi : \xi < \beta)$ is a strictly increasing sequence of ordinals with supremum α, then

$$\prod_{\xi<\beta}\aleph_{\alpha_\xi} = \aleph_\alpha^{|\beta|}.$$

Proof We start with the proof of part b). So let $\beta > 1$ be a principal number of addition and $(\alpha_\xi : \xi < \beta)$ be strictly increasing such that $\sup\{\alpha_\xi : \xi < \beta\} = \alpha$. Then we show first:

$$(*) \qquad (\prod_{\xi<\beta}\aleph_{\alpha_\xi})^{|\beta|} = \prod_{\xi<\beta}\aleph_{\alpha_\xi}.$$

For the proof of $(*)$, we put $\kappa_\xi := \aleph_{\alpha_\xi}$ and $\tau_{\xi,\eta} := \xi$ for all $\eta, \xi < \beta$. Since, by assumption and Lemma 1.4.7 a), β is not a sum of smaller ordinals, the natural sum $\xi\#\eta$ is less than β for all $\xi, \eta < \beta$. Further we have $\xi \le \xi\#\eta$ and $\eta \le \xi\#\eta$. This gives $\{\tau_{\xi,\eta} : \xi, \eta < \beta\} = \{\tau_{\xi,\eta} : \zeta < \beta \wedge \xi\#\eta = \zeta\}$. By the definition of the natural sum, the equation $\xi\#\eta = \zeta$ has for a given ζ only finitely many solutions, hence $\prod_{\xi\#\eta=\zeta}\kappa_{\tau_{\xi,\eta}} = \kappa_{\tau_{\zeta,0}} = \kappa_\zeta$ by Corollary 1.5.12. From this we can conclude with a general associative law that

$$\begin{aligned}(\prod_{\xi<\beta}\kappa_\xi)^{|\beta|} &= \prod_{\xi<\beta}\kappa_\xi^{|\beta|} = \prod_{\xi<\beta}(\prod_{\eta<\beta}\kappa_\xi) \\ &= \prod_{\xi<\beta}(\prod_{\eta<\beta}\kappa_{\tau_{\xi,\eta}}) = \prod_{\zeta<\beta}(\prod_{\xi\#\eta=\zeta}\kappa_{\tau_{\xi,\eta}}) = \prod_{\zeta<\beta}\kappa_\zeta,\end{aligned}$$

which proves $(*)$.

Now the sequence $(\alpha_\xi : \xi < \beta)$ is strictly increasing and the aleph function is continuous, and thus $\sup\{\aleph_{\alpha_\xi} : \xi < \beta\} = \aleph_\alpha$ and $\beta = \sup\{\xi : \xi < \beta\} \le \sup\{\alpha_\xi : \xi < \beta\} = \alpha$. By Lemma 1.6.3 we can infer that $\sum_{\xi<\beta}\aleph_{\alpha_\xi} = \aleph_\alpha$. From $(*)$ and Corollary 1.6.8 it follows that $\aleph_\alpha^{|\beta|} = (\sum_{\xi<\beta}\aleph_{\alpha_\xi})^{|\beta|} = (\prod_{\xi<\beta}\aleph_{\alpha_\xi})^{|\beta|} = \prod_{\xi<\beta}\aleph_{\alpha_\xi}$. This completes the proof of b).

Next we turn to the proof of the second equation in part a). To get a contradiction, we assume that it is false and choose β as the smallest limit ordinal for which it fails. Then part b) tells us that β is not additively indecomposable. If we take the Cantor normal form of β for the base ω, $\beta = \sum_{i<n}\omega^{\beta_i}$, then $n > 1$, since ω^δ is additively indecomposable for every $\delta > 0$, $\beta_0 \ge \ldots \ge \beta_{n-1}$, and $\beta_{n-1} > 0$. Consequently, $\gamma := \omega^{\beta_{n-1}}$ is a principal number of addition, and β has the form $\beta = \sigma + \gamma$. We have $\gamma \leq \sigma < \beta$ and $|\beta| = |\sigma|$, since the cardinality of the ordinal sum of two infinite ordinals equals the maximum of the cardinalities of the summands. By the choice of β, we have $\prod_{\xi<\sigma} \aleph_\xi = \aleph_\sigma^{|\sigma|} = \aleph_\sigma^{|\beta|}$, and by part b), $\prod_{\xi<\gamma} \aleph_{\sigma+\xi} = \aleph_\beta^{|\gamma|}$. So we can conclude that $\prod_{\xi<\beta} \aleph_\xi = \prod_{\xi<\sigma} \aleph_\xi \cdot \prod_{\xi<\gamma} \aleph_{\sigma+\xi} = \aleph_\sigma^{|\beta|} \cdot \aleph_\beta^{|\gamma|}$. Furthermore, Tarski's formula in Lemma 1.6.12 yields

$$\aleph_\beta^{|\beta|} = \aleph_{\sigma+\gamma}^{|\beta|} = \aleph_\sigma^{|\beta|} \cdot \aleph_{\sigma+\gamma}^{|\gamma|} = \aleph_\sigma^{|\beta|} \cdot \aleph_\beta^{|\gamma|} = \prod_{\xi<\beta} \aleph_\xi.$$

This contradiction shows that the second equation in a) is true.

Now the first equation in a) follows by transfinite induction. For $\beta = 1$ it is true, and for limit ordinals β it follows from the second, since $\aleph_\beta^{|\beta|} \cdot \aleph_\beta = \aleph_\beta^{|\beta|}$. Finally we get, by the inductive hypothesis and Hausdorff's formula,

$$\prod_{\xi\leq\beta+1} \aleph_\xi = (\prod_{\xi\leq\beta} \aleph_\xi) \cdot \aleph_{\beta+1} = \aleph_\beta^{|\beta|} \cdot \aleph_{\beta+1} = \aleph_{\beta+1}^{|\beta|} = \aleph_{\beta+1}^{|\beta+1|}.$$

Remark 1.6.18 Assuming the generalized continuum hypothesis, we can conclude for limit ordinals β and strictly increasing sequences $(\alpha_\xi : \xi < \beta)$ with supremum α that

$$\aleph_{\alpha+1} = \prod_{\xi<\beta} \aleph_{\alpha_\xi} = \aleph_\alpha^{|\beta|}.$$

To see this we infer from Corollary 1.6.4, König's lemma, and its corollary, that $\aleph_\alpha = \sum_{\xi<\beta} \aleph_{\alpha_\xi} < \prod_{\xi<\beta} \aleph_{\alpha_\xi}$. From this we get

$$\aleph_\alpha < \prod_{\xi<\beta} \aleph_{\alpha_\xi} \leq \aleph_\alpha^{|\beta|} \leq \aleph_\alpha^{|\alpha|} \leq \aleph_\alpha^{\aleph_\alpha} = 2^{\aleph_\alpha} = \aleph_{\alpha+1},$$

and this immediately yields the desired claim.

Exercises

1) Prove Lemma 1.6.1.
2) Prove Lemma 1.6.2.
3) Assume that κ is an infinite cardinal and $n \in \omega\backslash\{0\}$. Show that $\sum_{\xi<\kappa} n = \kappa$.
4) Assume that $\kappa \in \mathrm{ICN}$, $\mu \in \mathrm{cf}(\kappa) \cap \mathrm{CN}$ and $(\kappa_\xi : \xi < \mu)$ is a sequence of cardinal numbers less than κ. Prove that $\sum_{\xi<\mu} \kappa_\xi < \kappa$.
5) Show that $\aleph_\omega^{\aleph_0} = \prod_{n\in\omega} \aleph_n$.
6) Assume that κ is an infinite cardinal. Show that $(\kappa^+)^\kappa = 2^\kappa$.

7) Assume that κ is an infinite cardinal. Show that

$$2^{\kappa} = \Big(\sum_{\nu\in\kappa\cap\mathrm{CN}} 2^{\nu}\Big)^{\mathrm{cf}(\kappa)}.$$

8) Prove that $\prod_{\alpha<\aleph_1+\omega}\aleph_\alpha = \aleph_{\aleph_1+\omega}^{\aleph_1}$.
Hint: Use Tarski's formula.

9) Assume that σ is a limit ordinal and $(\eta(\xi) : \xi < \sigma)$ is a normal function.

   a) Prove: $\sup\{\sum_{\xi<\tau}\aleph_{\eta(\xi)} : \tau < \sigma\} = \sum_{\xi<\sigma}\aleph_{\eta(\xi)}$.
   b) Refute: $\sup\{\prod_{\xi<\tau}\aleph_{\eta(\xi)} : \tau < \sigma\} = \prod_{\xi<\sigma}\aleph_{\eta(\xi)}$.

10) Prove that $\prod_{\xi<\tau} 2^{\aleph_\xi} = 2^{\aleph_\tau}$ for any limit ordinal τ.
Hint: Assign to $X \subseteq \aleph_\tau$ the function $h_X \in \prod_{\xi<\tau}\mathcal{P}(\aleph_\xi)$ given by $h_X(\xi) := X \cap \aleph_\xi$.

11) Assume that κ is an infinite cardinal such that $2^{\nu} < \kappa$ for all $\nu \in \kappa\cap\mathrm{CN}$.
Prove that $\mathrm{cf}\,(\kappa^{\mathrm{cf}(\kappa)}) > \kappa$.

12) Prove: If κ is regular and $2^{\mu} \le \kappa$ for all cardinals $\mu < \kappa$, then $\sum_{\mu\in\kappa\cap\mathrm{CN}}\kappa^{\mu} = \kappa$.

13) Prove: If $\kappa = \lambda^+$, then $\sum_{\mu\in\kappa\cap\mathrm{CN}}\kappa^{\mu} = 2^{\lambda}$.

14) Refute: If κ and λ are cardinal numbers such that $0 < \lambda < \mathrm{cf}\,(\kappa)$, then $\kappa^{\lambda} = \sum_{\nu\in\kappa\cap\mathrm{CN}}\nu^{\lambda}$.

15) Assume that α is a limit ordinal and γ is an ordinal satisfying $\aleph_\gamma < \mathrm{cf}\,(\alpha)$.
Prove that $\aleph_\alpha^{\aleph_\gamma} = \sum_{\xi<\alpha}\aleph_\xi^{\aleph_\gamma} = \sup\{\aleph_\xi^{\aleph_\gamma} : \xi < \alpha\}$.
Hint: Use a golden rule.

1.7 Further Properties of κ^λ – the Singular Cardinal Hypothesis

In this section we will characterize the cardinal power κ^λ by its values for smaller bases or exponents[17]. Furthermore it will be shown that this power is uniquely determined in the axiom system ZF+(SCH) by the function $(2^\lambda : \lambda \in \mathrm{ICN})$, and that it can be easily computed in the system ZFC+(GCH). (SCH) denotes the so-called "singular cardinal hypothesis", given by

$$(\mathrm{SCH})\quad \forall\kappa\in\mathrm{ICN}(2^{\mathrm{cf}(\kappa)} < \kappa \Longrightarrow \kappa^{\mathrm{cf}(\kappa)} = \kappa^+)\ .$$

[17]The results in 1.6.12, 1.6.13, and 1.6.15 have certainly already dealt with this subject. We needed them to obtain further findings on products of cardinal numbers.

Lemma 1.7.1 (Tarski's Recursion Formula) *If β is a limit ordinal, α is the supremum of the strictly increasing sequence $(\alpha_\xi : \xi < \beta)$ and $\aleph_\gamma < \mathrm{cf}(\beta)$, then*

$$\aleph_\alpha^{\aleph_\gamma} = \sum_{\xi<\beta} \aleph_{\alpha_\xi}^{\aleph_\gamma} = \sup\{\aleph_{\alpha_\xi}^{\aleph_\gamma} : \xi < \beta\} .$$

In particular, for limit ordinals α and ordinals γ satisfying $\aleph_\gamma < \mathrm{cf}(\alpha)$, we have

$$\aleph_\alpha^{\aleph_\gamma} = \sum_{\xi<\alpha} \aleph_\xi^{\aleph_\gamma} = \sup\{\aleph_\xi^{\aleph_\gamma} : \xi < \alpha\} .$$

Proof By assumption, we have $\aleph_\gamma < \mathrm{cf}(\beta) = \mathrm{cf}(\alpha) = \mathrm{cf}(\aleph_\alpha)$. Therefore we get

$${}^{\aleph_\gamma}\aleph_\alpha = \bigcup\{{}^{\aleph_\gamma}\aleph_\xi : \xi < \alpha\} = \bigcup\{{}^{\aleph_\gamma}\aleph_{\alpha_\xi} : \xi < \beta\}.$$

Since obviously the cardinality of the index set is at most the supremum of the summands, Lemma 1.6.3 says that

$$\aleph_\alpha^{\aleph_\gamma} \leq \sum_{\xi<\beta} \aleph_{\alpha_\xi}^{\aleph_\gamma} = \sup\{\aleph_{\alpha_\xi}^{\aleph_\gamma} : \xi < \beta\} \leq \aleph_\alpha^{\aleph_\gamma}.$$

Definition *Assume that κ is an infinite and λ is an uncountable cardinal number. We say that λ is* **κ-strong** *iff $\rho^\kappa < \lambda$ for every cardinal number ρ less than λ. λ is called a* **strong limit cardinal** *iff λ is ν-strong for every cardinal number ν less than λ, that means iff $\rho^\nu < \lambda$ for all cardinal numbers ρ and ν less than λ. It is easy to see that λ is a strong limit cardinal iff $2^\nu < \lambda$ for all cardinal numbers ν less than λ.*

We say that κ is a **(strongly) inaccessible cardinal number** *iff κ is regular, uncountable and a strong limit cardinal, and a* **weakly inaccessible cardinal number** *iff κ is regular, uncountable and a limit cardinal.*

Remark and Example Obviously we have $\kappa < \lambda$ for any κ-strong cardinal λ.

The existence of inaccessible cardinal numbers cannot be proved in ZFC. For strongly inaccessible cardinals κ this results from the fact that the von Neumann level V_κ with the $\in$-relation is a model of ZFC. Therefore from the existence of an inaccessible cardinal in ZFC the consistency of ZFC could be proved in ZFC, which would contradict a famous theorem of K. Gödel concerning the impossibility of certain consistency proofs. For details, the reader is referred to Drake's book [Dr], Chapter 4.

The strong limit cardinals form a proper class. To show this, consider a fixed infinite cardinal number μ. We define by recursion a strictly increasing sequence $(\kappa_n : n \in \omega)$ of cardinal numbers by $\kappa_0 := \mu$ and $\kappa_{n+1} := 2^{\kappa_n}$. Then $\kappa := \sup\{\kappa_n : n \in \omega\}$ is a strong limit cardinal. Namely, if ρ and ν are smaller than κ, then there is a natural number n such that $\max\{\rho, \nu\} \leq \kappa_n$. We can conclude that $\rho^\nu \leq \kappa_n^{\kappa_n} = 2^{\kappa_n} = \kappa_{n+1} < \kappa$. Thus the class of strong limit cardinals is unbounded in the class of cardinals. Furthermore the supremum of a set of strong limit cardinals is also strong limit. Consequently, the class of strong limit cardinals is unbounded and closed in the class of cardinal numbers, which makes the assertion that these cardinal numbers occur very often precise. The same holds, for any infinite cardinal λ, for the class of λ-strong cardinals.

Lemma 1.7.2 *If κ is a strong limit cardinal, then $2^{<\kappa} = \kappa$ and $2^\kappa = \kappa^{\mathrm{cf}(\kappa)}$.*

Proof The first assertion is easy to show. For the second we can conclude, using Lemma 1.6.15, that $\kappa^{\mathrm{cf}(\kappa)} \leq \kappa^\kappa = 2^\kappa = (2^{<\kappa})^{\mathrm{cf}(\kappa)} = \kappa^{\mathrm{cf}(\kappa)}$.

Lemma 1.7.3 *Let α and β be ordinals with $\beta < \alpha$.*[18] *Then*

a) If $\aleph_\alpha$ is $\aleph_\beta$-strong, then $\aleph_\alpha = \sup\{\aleph_\gamma^{\aleph_\beta} : \gamma < \alpha\}$ and

$$\aleph_\alpha^{\aleph_\beta} = \begin{cases} \aleph_\alpha & \text{if } \mathrm{cf}(\aleph_\alpha) > \aleph_\beta, \\ \aleph_\alpha^{\mathrm{cf}(\alpha)} = \aleph_\alpha^{\mathrm{cf}(\aleph_\alpha)} & \text{otherwise.} \end{cases}$$

b) If $\gamma < \alpha$ and $\aleph_\alpha \leq \aleph_\gamma^{\aleph_\beta}$, then $\aleph_\alpha^{\aleph_\beta} = \aleph_\gamma^{\aleph_\beta}$.

c) Assume that $\aleph_\alpha^{\aleph_\beta} > 2^{\aleph_\beta}$[19]*, $\aleph_\alpha$ is not $\aleph_\beta$-strong, and $\gamma < \alpha$ is the smallest ordinal such that $\aleph_\gamma^{\aleph_\beta} \geq \aleph_\alpha$. Then $\aleph_\gamma$ is $\aleph_\beta$-strong and singular, and satisfies $\mathrm{cf}(\aleph_\gamma) \leq \aleph_\beta < \aleph_\gamma$; furthermore, $\aleph_\alpha^{\aleph_\beta} = \aleph_\gamma^{\mathrm{cf}(\aleph_\gamma)}$.*

Proof First let, for the proof of a), α be a limit ordinal. From the assumption of a) we can conclude that $\aleph_\alpha = \sup\{\aleph_\gamma : \gamma < \alpha\} \leq \sup\{\aleph_\gamma^{\aleph_\beta} : \gamma < \alpha\} \leq \aleph_\alpha$, and thus

$$(*) \qquad \aleph_\alpha = \sup\{\aleph_\gamma^{\aleph_\beta} : \gamma < \alpha\}.$$

Let first $\mathrm{cf}(\aleph_\alpha) > \aleph_\beta$. Then ${}^{\aleph_\beta}\aleph_\alpha = \bigcup\{{}^{\aleph_\beta}\aleph_\gamma : \gamma < \alpha\}$, and Lemma 1.6.3 says that

$$\aleph_\alpha^{\aleph_\beta} \leq |\alpha| \cdot \sup\{\aleph_\gamma^{\aleph_\beta} : \gamma < \alpha\} = \aleph_\alpha.$$

Now assume that $\mathrm{cf}(\aleph_\alpha) \leq \aleph_\beta$. By assumption, we have $\aleph_\beta < \aleph_\alpha$, hence $\aleph_\alpha$ is singular and α is a limit ordinal. From Lemma 1.6.15 and (*) we can conclude that

$$\aleph_\alpha^{\aleph_\beta} = (\sup\{\aleph_\gamma^{\aleph_\beta} : \gamma < \alpha\})^{\mathrm{cf}(\aleph_\alpha)} = \aleph_\alpha^{\mathrm{cf}(\aleph_\alpha)} = \aleph_\alpha^{\mathrm{cf}(\alpha)}.$$

If α is the successor ordinal $\sigma + 1$, then $\aleph_{\sigma+1}^{\aleph_\beta} = \aleph_{\sigma+1} \cdot \aleph_\sigma^{\aleph_\beta} = \aleph_{\sigma+1}$, since $\aleph_\alpha$ is $\aleph_\beta$-strong.

---

[18] For $\alpha \leq \beta$ it follows from Theorem 1.6.12 that $\aleph_\alpha^{\aleph_\beta} = 2^{\aleph_\beta}$.

[19] Certainly this is the case if $\aleph_\alpha > 2^{\aleph_\beta}$.

Part b) follows from $\aleph_\gamma^{\aleph_\beta} \le \aleph_\alpha^{\aleph_\beta} \le (\aleph_\gamma^{\aleph_\beta})^{\aleph_\beta} = \aleph_\gamma^{\aleph_\beta}$.

For the proof of part c), we choose the smallest ordinal $\gamma < \alpha$ satisfying $\aleph_\gamma^{\aleph_\beta} \ge \aleph_\alpha$. Then $\aleph_\gamma^{\aleph_\beta} = \aleph_\alpha^{\aleph_\beta}$, and $\beta < \gamma$, since otherwise we would get $\aleph_\alpha^{\aleph_\beta} = \aleph_\gamma^{\aleph_\beta} = 2^{\aleph_\beta}$, contradicting the assumption $2^{\aleph_\beta} < \aleph_\alpha^{\aleph_\beta}$. Furthermore we have $\aleph_\sigma^{\aleph_\beta} < \aleph_\alpha$ and thus $\aleph_\sigma^{\aleph_\beta} < \aleph_\alpha^{\aleph_\beta} = \aleph_\gamma^{\aleph_\beta}$ for all $\sigma < \gamma$. The supposition $\aleph_\sigma^{\aleph_\beta} \ge \aleph_\gamma$ for some $\sigma < \gamma$ would give, together with part b), the contradiction $\aleph_\sigma^{\aleph_\beta} = \aleph_\gamma^{\aleph_\beta}$. Therefore we get $\aleph_\sigma^{\aleph_\beta} < \aleph_\gamma$ for all $\sigma < \gamma$, and $\aleph_\gamma$ is $\aleph_\beta$-strong. The supposition $\mathrm{cf}(\aleph_\gamma) > \aleph_\beta$ leads with part a) to $\aleph_\alpha^{\aleph_\beta} = \aleph_\gamma^{\aleph_\beta} = \aleph_\gamma < \aleph_\alpha$. So there only remains $\mathrm{cf}(\aleph_\gamma) \le \aleph_\beta < \aleph_\gamma$, and part a) yields $\aleph_\alpha^{\aleph_\beta} = \aleph_\gamma^{\aleph_\beta} = \aleph_\gamma^{\mathrm{cf}(\aleph_\gamma)}$.

Theorem 1.7.4 (Jech) *Let α and β be ordinals. If $\aleph_\alpha^{\aleph_\beta} \notin \{2^{\aleph_\beta}, \aleph_\alpha\}$, then there is an ordinal $\gamma \le \alpha$ with $\mathrm{cf}(\aleph_\gamma) \le \aleph_\beta < \aleph_\gamma$, such that $\aleph_\gamma$ is $\aleph_\beta$-strong and satisfies $\aleph_\alpha^{\aleph_\beta} = \aleph_\gamma^{\mathrm{cf}(\aleph_\gamma)}$. Therefore*

$$\aleph_\alpha^{\aleph_\beta} = \begin{cases} 2^{\aleph_\beta} & \textit{or} \\ \aleph_\alpha & \textit{or} \\ \aleph_\gamma^{\mathrm{cf}(\aleph_\gamma)} & \textit{for some } \gamma \le \alpha \textit{ with } \mathrm{cf}(\aleph_\gamma) \le \aleph_\beta < \aleph_\gamma. \end{cases}$$

Proof Let $\aleph_\alpha^{\aleph_\beta} \notin \{2^{\aleph_\beta}, \aleph_\alpha\}$. Then, by Theorem 1.6.12, $\beta < \alpha$. If $\aleph_\alpha$ is $\aleph_\beta$-strong, then the assertion follows from Lemma 1.7.3 a), with $\gamma := \alpha$, and from the assumption that $\aleph_\alpha^{\aleph_\beta} \ne \aleph_\alpha$. Otherwise, it follows from Lemma 1.7.3 c).

Now we turn to the discussion of the singular cardinal hypothesis. Let κ be an infinite cardinal number. By Corollary 1.6.10 we have $2^{\mathrm{cf}(\kappa)} \ne \kappa$. If $2^{\mathrm{cf}(\kappa)} > \kappa$, then $\kappa^{\mathrm{cf}(\kappa)} \le (2^{\mathrm{cf}(\kappa)})^{\mathrm{cf}(\kappa)} = 2^{\mathrm{cf}(\kappa)}$, and thus $2^{\mathrm{cf}(\kappa)} = \kappa^{\mathrm{cf}(\kappa)}$. In other words, in the case that $\kappa < 2^{\mathrm{cf}(\kappa)}$, the cardinal $\kappa^{\mathrm{cf}(\kappa)}$ equals the value of the continuum function at the regular argument $\mathrm{cf}(\kappa)$. Since, from Easton's result, the continuum function for regular arguments is not subject to any limitations except monotonicity and the condition $\kappa < \mathrm{cf}(2^\kappa)$, this holds in this case also for $\kappa^{\mathrm{cf}(\kappa)}$; so there are models of ZFC in which $\kappa^{\mathrm{cf}(\kappa)} > \kappa^+$ is valid. However, if $2^{\mathrm{cf}(\kappa)} < \kappa$, then $2^{\mathrm{cf}(\kappa)} < \kappa^{\mathrm{cf}(\kappa)}$, i.e., the continuum function and the continuum function for κ (s.b.) differ at the regular argument $\mathrm{cf}(\kappa)$.

Since $\kappa < \kappa^{\mathrm{cf}(\kappa)} \le \kappa^\kappa = 2^\kappa$, it clearly follows from $2^\kappa = \kappa^+$ that $\kappa^{\mathrm{cf}(\kappa)} = \kappa^+$. Therefore the singular cardinal hypothesis can be inferred from the generalized continuum hypothesis.

If we know the values of the continuum function $(2^\kappa : \kappa \in \mathrm{ICN})$, and if we know the values $\kappa^{\mathrm{cf}(\kappa)}$ for any singular κ, then Theorem 1.7.4 shows that we know the set of values of the function $(\kappa^\lambda : \kappa, \lambda \in \mathrm{ICN})$. The additional axiom (SCH) makes the course of values of the power function for cardinal numbers, which is a subject of actual research and the main topic of this book, easily surveyed (see Theorem 1.7.8). This applies all the more if we take the generalized continuum hypothesis as an axiom – see the next lemma.

Corollary 1.7.5 *Assume* (GCH). *Then for all cardinals* $\kappa \geq 2$ *and* $\lambda \geq \omega$,

$$\kappa^\lambda = \begin{cases} \kappa & \text{if } \lambda < \mathrm{cf}(\kappa), \\ \kappa^+ & \text{if } \mathrm{cf}(\kappa) \leq \lambda < \kappa, \\ \lambda^+ & \text{if } \kappa \leq \lambda. \end{cases}$$

Proof If $\kappa \leq \lambda$, then $\kappa^\lambda = 2^\lambda = \lambda^+$ by Theorem 1.6.12. So let $\lambda < \kappa$; then κ is infinite. If $\lambda < \mathrm{cf}(\kappa)$, then we can infer from $\nu^\lambda \leq \max\{2^\nu, 2^\lambda\} \leq \max\{\nu^+, \lambda^+\}$, using Lemma 1.6.15, that

$$\kappa \leq \kappa^\lambda = \kappa \cdot \sup\{\nu^\lambda : \nu \in \mathrm{CN} \cap \kappa\} \leq \kappa \cdot \sup\{\max\{\nu^+, \lambda^+\} : \nu \in \mathrm{CN} \cap \kappa\} \leq \kappa.$$

If $\mathrm{cf}(\kappa) \leq \lambda < \kappa$, then we get $\kappa < \kappa^{\mathrm{cf}(\kappa)} \leq \kappa^\lambda \leq \kappa^\kappa = 2^\kappa = \kappa^+$, hence $\kappa^\lambda = \kappa^+$.

Definition *The function* $(\kappa^{\mathrm{cf}(\kappa)} : \kappa \in \mathrm{ICN})$ *is called the* **gimel function**; *instead of* $\kappa^{\mathrm{cf}(\kappa)}$ *we also write* $\gimel(\kappa)$. *If* $\kappa = 2$ *or if* κ *is an infinite cardinal number, then we call the function* $(\kappa^\rho : \rho \in \mathrm{ICN})$ *the* **continuum function for** κ; *if* $\kappa = 2$, *we call it the* **continuum function**. *We say that the continuum function for* κ *is* **eventually constant below** λ *iff there is a cardinal number* ρ_0 *with* $\rho_0 < \lambda$, *such that* $\kappa^\rho = \kappa^{\rho_0}$ *for all cardinal numbers* ρ *with* $\rho_0 \leq \rho < \lambda$. *In this case we also say that the continuum function for* κ *is constant below* λ **from** ρ_0 **on**.

Lemma 1.7.6 *Assume that* κ *and* λ *are cardinal numbers such that* $\kappa \geq 2$ *and* $\lambda \geq \omega$. *Then*

a) Either the continuum function for κ *is eventually constant below* λ – *if this happens at* $\rho_0 < \lambda$, *then obviously* $\kappa^{<\lambda} = \kappa^\rho$ *for every cardinal* ρ *with* $\rho_0 \leq \rho < \lambda$ – *or there is a strictly increasing sequence* $(\lambda_\xi : \xi < \mathrm{cf}(\lambda))$ *which is cofinal in* λ, *such that the sequence* $(\kappa^{\lambda_\xi} : \xi < \mathrm{cf}(\lambda))$ *is strictly increasing and cofinal in* $\kappa^{<\lambda}$.

b) In a), it holds that $\mathrm{cf}(\kappa^{<\lambda}) \geq \lambda$ *in the first case and* $\mathrm{cf}(\kappa^{<\lambda}) = \mathrm{cf}(\lambda)$ *in the second case.*

c) If $\nu > 0$ *is a cardinal number, then*

$$(\kappa^{<\lambda})^\nu = \begin{cases} \kappa^{<\lambda} & \text{if } 0 < \nu < \mathrm{cf}(\lambda), \\ \kappa^\lambda & \text{if } \mathrm{cf}(\lambda) \leq \nu < \lambda, \\ \kappa^\nu & \text{if } \lambda \leq \nu. \end{cases}$$

d) Assume (GCH). Then

$$\kappa^{<\lambda} = \begin{cases} \kappa & \text{if } \lambda \leq \mathrm{cf}(\kappa), \\ \kappa^+ & \text{if } \mathrm{cf}(\kappa) < \lambda \leq \kappa^+, \\ \lambda & \text{if } \kappa^+ < \lambda. \end{cases}$$

Proof If λ is a successor cardinal μ^+, then $\kappa^{<\lambda} = \kappa^\mu$, and clearly the first assertion of the alternative is true. Now let λ be a limit cardinal and assume that the first assertion is not true for λ. Then, for each cardinal $\rho < \lambda$, there is a cardinal ν such that $\rho < \nu < \lambda$ and $\kappa^\rho < \kappa^\nu$. If ρ runs through the values of a sequence of length $\mathrm{cf}(\lambda)$ which is cofinal in λ, then, by the previous condition, we obtain by recursion a strictly increasing subsequence $(\lambda_\xi : \xi < \mathrm{cf}(\lambda))$ of this sequence which is cofinal in λ and therefore has also length $\mathrm{cf}(\lambda)$ and the further property that the sequence $(\kappa^{\lambda_\xi} : \xi < \mathrm{cf}(\lambda))$ is strictly increasing. By construction, the last sequence must be cofinal in $\kappa^{<\lambda}$.

Now we prove part b). Assume that the first assertion in a) is true. If $\lambda = \aleph_0$, then it is easy to see that $\kappa \geq \omega$ and $\kappa^{<\lambda} = \kappa$. So we get $\mathrm{cf}(\kappa^{<\lambda}) = \mathrm{cf}(\kappa) \geq \lambda$. Now let $\lambda > \aleph_0$. Theorem 1.6.9 tells us that $\mathrm{cf}(\kappa^{<\lambda}) = \mathrm{cf}(\kappa^\rho) > \rho$ for every infinite cardinal number ρ with $\rho_0 \leq \rho < \lambda$. Thus we get $\mathrm{cf}(\kappa^{<\lambda}) \geq \lambda$. If the second assertion holds, the cardinal $\kappa^{<\lambda}$ has cofinality $\mathrm{cf}(\lambda)$, since it is the supremum of the strictly increasing sequence $(\kappa^{\lambda_\xi} : \xi < \mathrm{cf}(\lambda))$ and since $\mathrm{cf}(\lambda)$ is regular.

Next we turn to the proof of part c). Let ν be a cardinal number, and assume first that the continuum function for κ is constant below λ from ρ_0 on. If $0 < \nu < \mathrm{cf}(\lambda)$, then we choose a cardinal number ρ_1 such that $\max\{\rho_0, \nu\} \leq \rho_1 < \lambda$. This gives $(\kappa^{<\lambda})^\nu = (\kappa^{\rho_1})^\nu = \kappa^{\rho_1 \cdot \nu} = \kappa^{\rho_1} = \kappa^{<\lambda}$. For the case $\lambda \leq \nu$ we obtain in the same way $(\kappa^{<\lambda})^\nu = \kappa^{\rho_0 \cdot \nu} = \kappa^\nu$. If $\mathrm{cf}(\lambda) \leq \nu < \lambda$, then we know that λ is singular. λ can be represented by Lemma 1.6.11 as a sum $\sum_{\xi<\mathrm{cf}(\lambda)} \lambda_\xi$, where $(\lambda_\xi : \xi < \mathrm{cf}(\lambda))$ is a strictly increasing sequence of cardinal numbers less than λ satisfying $\max\{\mathrm{cf}(\lambda), \rho_0\} \leq \lambda_0$. Lemma 1.6.2 says that

$$\kappa^{<\lambda} = \kappa^{\rho_0} = \kappa^{\rho_0 \cdot \mathrm{cf}(\lambda)} = \prod_{\xi<\mathrm{cf}(\lambda)} \kappa^{\rho_0} = \prod_{\xi<\mathrm{cf}(\lambda)} \kappa^{\lambda_\xi} = \kappa^{\sum \lambda_\xi} = \kappa^\lambda,$$

and thus $(\kappa^{<\lambda})^\nu = (\kappa^\lambda)^\nu = \kappa^\lambda$.

Now we consider the case that in the alternative in a) the second assertion is true. Let the sequences in this assertion be given. Since $\mathrm{cf}(\lambda) = \mathrm{cf}(\kappa^{<\lambda}) \leq \kappa^{<\lambda} = \sup\{\kappa^{\lambda_\xi} : \xi < \mathrm{cf}(\lambda)\}$, Lemma 1.6.3 says that $\kappa^{<\lambda} = \sum(\kappa^{\lambda_\xi} : \xi < \mathrm{cf}(\lambda))$. Let $\nu < \mathrm{cf}(\lambda) = \mathrm{cf}(\kappa^{<\lambda})$ and, without loss of generality, $\nu \leq \lambda_0$. Then, by Lemma 1.6.15 c),

$$(\kappa^{<\lambda})^\nu = \kappa^{<\lambda} \cdot \sum(\mu^\nu : \mu \in \mathrm{CN} \cap \kappa^{<\lambda}) = \sum_{\xi<\mathrm{cf}(\lambda)} (\kappa^{\lambda_\xi})^\nu = \sum_{\xi<\mathrm{cf}(\lambda)} \kappa^{\lambda_\xi} = \kappa^{<\lambda}.$$

This proves our claim for the case $\nu < \mathrm{cf}(\lambda)$. From Corollary 1.6.8 we can conclude, if $\nu \geq \omega$, that

$$(*) \quad (\kappa^{<\lambda})^\nu = (\sum_{\xi<\mathrm{cf}(\lambda)} \kappa^{\lambda_\xi})^\nu = (\prod_{\xi<\mathrm{cf}(\lambda)} \kappa^{\lambda_\xi})^\nu = \prod_{\xi<\mathrm{cf}(\lambda)} (\kappa^{\lambda_\xi})^\nu = \kappa^{\sum_{\xi<\mathrm{cf}(\lambda)} \lambda_\xi \cdot \nu}.$$

If $\mathrm{cf}(\lambda) \leq \nu < \lambda$, we can again assume without loss of generality that $\lambda_0 \geq \nu$, and we get $\sum_{\xi<\mathrm{cf}(\lambda)} \lambda_\xi \cdot \nu = \sum_{\xi<\mathrm{cf}(\lambda)} \lambda_\xi = \lambda$. If $\lambda \leq \nu$, then $\sum_{\xi<\mathrm{cf}(\lambda)} \lambda_\xi \cdot \nu = \sum_{\xi<\mathrm{cf}(\lambda)} \nu = \mathrm{cf}(\lambda) \cdot \nu = \nu$. Putting this in $(*)$ completes the proof of c).

Finally, we prove d). If κ is finite, then $\kappa < \lambda$, and from Theorem 1.6.12 a) we obtain

$$\kappa^{<\lambda} = \sup\{\kappa^\rho : \rho < \lambda\} = \sup\{2^\rho : \rho < \lambda\} = \sup\{\rho^+ : \rho < \lambda\} = \lambda.$$

So assume that $\kappa \geq \omega$. We continue to use Corollary 1.7.5.

If $\lambda \leq \mathrm{cf}(\kappa)$, then $\kappa^{<\lambda} = \sup\{\kappa^\rho : \rho < \lambda\} = \sup\{\kappa : \rho < \lambda\} = \kappa$. If $\mathrm{cf}(\kappa) < \lambda \leq \kappa^+$, then $\kappa^{<\lambda} = \sup\{\kappa^+ : \rho < \lambda\} = \kappa^+$. And if $\kappa^+ < \lambda$, then $\kappa^{<\lambda} = \sup\{\kappa^\rho : \kappa^+ \leq \rho < \lambda\} = \sup\{\rho^+ : \rho < \lambda\} = \lambda$.

Theorem 1.7.7 (Bukovsky, Hechler) *If λ is a singular cardinal number, and if the continuum function for κ is eventually constant below λ from ρ_0 on, then $\kappa^\lambda = \kappa^{<\lambda} = \kappa^{\rho_0}$.*

Proof Since $\mathrm{cf}(\lambda) < \lambda$, we can choose a cardinal number ρ_0 satisfying the assumption which is greater than $\mathrm{cf}(\lambda)$. Then $\kappa^{<\lambda} = \kappa^{\rho_0}$ and, by Lemma 1.7.6 c), $\kappa^\lambda = (\kappa^{<\lambda})^{\mathrm{cf}(\lambda)} = (\kappa^{\rho_0})^{\mathrm{cf}(\lambda)} = \kappa^{\rho_0}$.

Remark In contrast to Easton's result for regular cardinals, Theorem 1.7.7 makes it possible that the value of the function $(\kappa^\lambda : \lambda \in \mathrm{ICN})$ at singular arguments λ depends on its values at arguments below λ.

Theorem 1.7.8 *Assume* (SCH).

a) *If κ is a singular cardinal number, then*

$$2^\kappa = \begin{cases} 2^{<\kappa} & \text{if } \exists \nu_0 \in \mathrm{ICN} \cap \kappa\, \forall \nu \in \mathrm{CN}(\nu_0 \leq \nu < \kappa \Longrightarrow 2^\nu = 2^{\nu_0}) \\ & \text{(the continuum function is eventually} \\ & \text{constant below } \kappa\text{),} \\ (2^{<\kappa})^+ & \text{otherwise.} \end{cases}$$

b) *If κ and λ are infinite cardinal numbers, then*

$$\kappa^\lambda = \begin{cases} \kappa & \text{if } 2^\lambda < \kappa \text{ and } \lambda < \mathrm{cf}(\kappa), \\ \kappa^+ & \text{if } 2^\lambda < \kappa \text{ and } \lambda \geq \mathrm{cf}(\kappa), \\ 2^\lambda & \text{if } 2^\lambda \geq \kappa. \end{cases}$$

Proof For a), we note that Theorem 1.7.7 yields $2^\kappa = 2^{\nu_0} = 2^{<\kappa}$ if the continuum function is constant below κ from ν_0 on. So let us assume that the continuum function is not constant below κ. Then Lemma 1.7.6 b) tells us that $2^{<\kappa}$ has cofinality $\mathrm{cf}(\kappa)$. Our assumption gives $2^{\mathrm{cf}(2^{<\kappa})} = 2^{\mathrm{cf}(\kappa)} < 2^{<\kappa}$. From (SCH) and Lemma 1.6.15 we can conclude that $2^\kappa = (2^{<\kappa})^{\mathrm{cf}(\kappa)} = (2^{<\kappa})^{\mathrm{cf}(2^{<\kappa})} = (2^{<\kappa})^+$.

The third case in b) is easy to show. If $\kappa \le 2^\lambda$, then $2^\lambda \le \kappa^\lambda \le (2^\lambda)^\lambda = 2^\lambda$, since λ is infinite.

Now we prove part b), using transfinite induction on the cardinal number κ. By the previous remark, we can assume without loss of generality that $2^\lambda < \kappa$, hence $\lambda < \kappa$.

First let κ be a successor cardinal ν^+. Then $2^\lambda \le \nu$. If $2^\lambda < \nu$, then the inductive hypothesis gives $\nu^\lambda \in \{\nu, \nu^+\}$. If $2^\lambda = \nu$, then $\nu^\lambda = 2^\lambda < \kappa$. In any case we have $\nu^\lambda \le \kappa$, and from Theorem 1.6.12 we can conclude that $\kappa^\lambda = (\nu^+)^\lambda = \nu^\lambda \cdot \nu^+ = \nu^+ = \kappa$.

Now let κ be a limit cardinal. For $\nu < \kappa$, the inductive hypothesis gives $\nu^\lambda \in \{\nu, \nu^+, 2^\lambda\}$, in particular $\nu^\lambda < \kappa$. From Lemma 1.7.3 a) we can conclude that $\kappa^\lambda = \kappa$, if $\mathrm{cf}(\kappa) > \lambda$, and $\kappa^\lambda = \kappa^{\mathrm{cf}(\kappa)}$, if $\mathrm{cf}(\kappa) \le \lambda$. In this case we have $2^{\mathrm{cf}(\kappa)} \le 2^\lambda < \kappa$. With (SCH), we obtain $\kappa^\lambda = \kappa^{\mathrm{cf}(\kappa)} = \kappa^+$.

Lemma 1.7.9 *The continuum function $(2^\kappa : \kappa \in \mathrm{ICN})$ can be characterized by the gimel function. More precisely:*

a) If κ is a regular cardinal number, then $2^\kappa = \gimel(\kappa)$.

b) If κ is a singular cardinal number and if the continuum function is eventually constant below κ, then $2^\kappa = 2^{<\kappa}$.

c) If κ is a singular cardinal number and if the continuum function is not eventually constant below κ, then $2^\kappa = \gimel(2^{<\kappa})$.

Proof Part a) holds, since $\kappa = \mathrm{cf}(\kappa)$ gives $2^\kappa = 2^{\mathrm{cf}(\kappa)} \le \kappa^{\mathrm{cf}(\kappa)} \le \kappa^\kappa = 2^\kappa$.

For b) we note that $\gimel(\kappa) \le 2^\kappa$, since $\kappa^{\mathrm{cf}(\kappa)} \le \kappa^\kappa = 2^\kappa$. If κ is singular, and if the continuum function is eventually constant below κ, then $2^\kappa = 2^{<\kappa}$ follows from the theorem of Bukovsky and Hechler.

If the continuum function is not eventually constant below κ, then Lemma 1.7.6 b) says that $\mathrm{cf}(\kappa) = \mathrm{cf}(2^{<\kappa})$.
Part c) of this lemma tells us that $(2^{<\kappa})^{\mathrm{cf}(2^{<\kappa})} = 2^\kappa$. This proves part c).

Remark Lemma 1.7.9 illustrates the importance of the gimel function. An essential aim of this book is to give nontrivial upper bounds for the power $\kappa^{\mathrm{cf}(\kappa)}$ for singular cardinals κ.

Exercises

1) Assume that κ is a μ-strong successor cardinal. Show that κ^+ is also μ-strong.
2) Prove: If μ and ν are infinite cardinals, then there is a μ-strong regular cardinal greater than ν.
3) Let μ be an infinite cardinal. Find the first $\omega + 1$ μ-strong cardinals.
   Hint: Consider the cardinal $(2^\mu)^+$.

4) Prove: If κ is a weakly inaccessible cardinal, then $\kappa = \aleph_\kappa$. Show that on the other hand the singular fixed points of the aleph function form a proper class.
5) Let κ be an infinite cardinal. Prove that κ is strongly inaccessible iff $\sum_{\alpha<\lambda} \kappa_\alpha < \kappa$ and $\prod_{\alpha<\lambda} \kappa_\alpha < \kappa$ for all cardinals $\lambda < \kappa$ and all sequences $(\kappa_\alpha : \alpha < \lambda)$ of cardinals less than κ.
6) Assume that κ is a strongly inaccessible cardinal number. Prove that $|\mathrm{V}_\alpha| < \kappa$ for all $\alpha < \kappa$.
   Hint: Use transfinite induction on α according to the definition of the sets V_α.
7) The **beth function** $\beth : \mathrm{ON} \longrightarrow \mathrm{ICN}$ is defined by $\beth(0) = \aleph_0$, $\beth(\alpha+1) = 2^{\beth(\alpha)}$ and $\beth(\gamma) = \sup\{2^{\beth(\alpha)} : \alpha < \gamma\}$, if γ is a limit ordinal. As usual we also write $\beth_\alpha$ instead of $\beth(\alpha)$. Prove:
   a) $|\mathrm{V}_{\omega+\alpha}| = \beth_\alpha$.
   b) If κ is an uncountable regular cardinal, then κ is a fixed point of the beth function iff κ is strongly inaccessible.
8) Prove: If $2^{\aleph_1} < \aleph_\omega$ and $\gimel(\aleph_\omega) > \aleph_{\omega_1}$, then $\gimel(\aleph_{\omega_1}) = \gimel(\aleph_\omega)$.
9) Prove: If $2^{\aleph_0} \geq \aleph_{\omega_1}$, then $\gimel(\aleph_\omega) = 2^{\aleph_0}$ and $\gimel(\aleph_{\omega_1}) = 2^{\aleph_1}$.
10) (Jech) Assume that λ is a $\mathrm{cf}(\lambda)$-strong singular cardinal. Show that

$$\mathrm{cf}(\gimel(\lambda)) \geq \min\{\nu : \exists \mu \in \lambda \cap \mathrm{CN}\ \lambda \leq \mu^\nu\}.$$

11) a) Prove: If $2^{\aleph_0} = \aleph_{\omega+1}$ and $2^{\aleph_1} = \aleph_{\omega+2}$, then $\aleph_\omega^{\aleph_0} < 2^{\aleph_\omega}$.
   Hint: Use 1.6.15.
   b) Prove that the monotonicity of the gimel function is not provable in ZFC.
   Hint: By Easton's result, ZFC is consistent with the assertions $2^{\aleph_0} = \aleph_{\omega+1}$, $2^{\aleph_1} = \aleph_{\omega+2}$ and $2^{\aleph_2} = (2^{\aleph_\omega})^+$.
12) Assume that λ is a singular cardinal, $\kappa = \mathrm{cf}(\lambda)$ and $\Phi \in {}^\kappa\kappa$. Further let $(\mu_\xi : \xi < \kappa)$ be a sequence unbounded in λ such that the set $\{\xi < \kappa : 2^{\mu_\xi} \leq \mu_\xi^{+\Phi(\xi)}\}$ is unbounded in κ. Prove that λ is a strong limit cardinal.
13) Assume that κ and λ are cardinals with $\kappa \geq 2$ and $\lambda \geq \omega$, such that $\mathrm{cf}(\kappa^{<\lambda}) > \lambda$. Prove that the continuum function for κ is eventually constant below λ.

1.8 Clubs and Stationary Sets

Definition *If α is a limit ordinal, then we know that a subset C of α is* **unbounded in α** *iff* $\sup(C) = \alpha$; *we call C* **closed in α** *iff $\gamma \in C$ for every limit ordinal $\gamma < \alpha$ in which $C \cap \gamma$ is unbounded. Every closed unbounded subset of α is called a* **club** *in α. If λ is a singular cardinal of uncountable cofinality, then C is called a* **special club** *in λ iff C is a club in λ of order type* $\mathrm{cf}(\lambda)$ *all of whose members are singular cardinals greater than* $\mathrm{cf}(\lambda)$. *A* **club sequence** *is any sequence of the form $(C_\alpha : \alpha \in A)$, where A is a set of limit ordinals and, for every $\alpha \in A$, C_α is a club in α of order type* $\mathrm{cf}(\alpha)$.

For any ordinal α and any sequence $(X_\xi : \xi < \alpha)$ of subsets of α, we call the set

$$\begin{aligned}\Delta(X_\xi : \xi < \alpha) &:= \Delta_{\xi<\alpha} X_\xi \\ &:= \{\beta \in \alpha : \forall \xi < \beta(\beta \in X_\xi)\} = \{\beta < \alpha : \beta \in \bigcap_{\xi<\beta} X_\xi\}\end{aligned}$$

the **diagonal intersection** *and*

$$\begin{aligned}\nabla(X_\xi : \xi < \alpha) &:= \nabla_{\xi<\alpha} X_\xi \\ &:= \{\beta \in \alpha : \exists \xi < \beta(\beta \in X_\xi)\} = \{\beta < \alpha : \beta \in \bigcup_{\xi<\beta} X_\xi\}\end{aligned}$$

the **diagonal union** *of the sequence $(X_\xi : \xi < \alpha)$.*

If C is an arbitrary class of ordinals, then we say that $\gamma \in \mathrm{ON} \setminus \{0\}$ is a **limit point** *of C iff $\gamma = \bigcup(C \cap \gamma)$.*

Remarks and examples Assume that α is a limit ordinal.

a) The **order topology** on a nonempty subset A of ON is that topology on A having as a base the set

$$\{(\sigma,\tau)_{\mathrm{ON}} \cap A : \sigma, \tau \in A\} \cup \{\sigma \cap A : \sigma \in A\} \cup \{A \setminus (\sigma + 1) : \sigma \in A\} \cup \{A\}.$$

We remind the reader that a **topology** on a fixed nonempty set A is a set $\mathcal{T}$ of subsets of A which has A and $\emptyset$ amongst its members and is closed under finite intersections and arbitrary unions. The members of $\mathcal{T}$ are called **open sets**, their complements in A are called **closed sets**. We say that a set $\mathcal{B} \subseteq \mathcal{T}$ is a **base** of this topology iff every open set is a union of members of $\mathcal{B}$. If A is given without a topology, then we call a set $\mathcal{B} \subseteq \mathcal{P}(A)$ a **base of a topology on A** iff $\bigcup \mathcal{B} = A$ and $B_1 \cap B_2$ is a union of members of $\mathcal{B}$ for all sets $B_1, B_2 \in \mathcal{B}$. In this case, the set $\{\bigcup \mathcal{B}' : \mathcal{B}' \subseteq \mathcal{B}\}$ is a topology on A. It is called the topology on A **generated by the base $\mathcal{B}$**. It is easy to see that any subset C of a limit ordinal α is closed in the above sense iff it is closed relative to the order topology on α.

b) A subset C of α is closed in α iff the **limit**, i.e., the supremum, of any sequence of members of C which is bounded in α is an element of C.

c) For every ordinal $\gamma < \alpha$, the set $[\gamma, \alpha)$ is a club in α – every final segment of α is a club in α.

d) If $\mathrm{cf}(\alpha) > \omega$, then $\alpha \cap \mathrm{Lim}$ is a club in α (see the next lemma). Without the assumption on $\mathrm{cf}(\alpha)$, this assertion is obviously false, since the set of limit ordinals less than α needs not to be unbounded in α; consider ω or $\omega \cdot 2$ $(= \omega + \omega)$, where $+$ and $\cdot$ denote the ordinal operations.

e) The set of all successor ordinals in $\alpha > \omega$ is not a club in α – later on, we will see that it is the standard example of a "thin" set, since its intersection with the club of limit ordinals is empty.

f) If $\alpha > \aleph_0$ is regular and thus a regular cardinal, and if $f : \alpha \longrightarrow \alpha$ is a function, then the set $C := \{\beta < \alpha : f[\beta] \subseteq \beta\}$ is a club in α.

   To prove this, let $(\beta_\xi : \xi < \gamma)$ be a strictly increasing sequence in C with $\gamma < \mathrm{cf}(\alpha) = \alpha$, and $\beta := \sup\{\beta_\xi : \xi < \gamma\}$. Then $\beta \in \alpha$ and, since $f[\beta_\xi] \subseteq \beta_\xi \subseteq \beta$ for every $\xi < \beta$, $f[\beta] = f[\bigcup\{\beta_\xi : \xi < \gamma\}] = \bigcup\{f[\beta_\xi] : \xi < \gamma\} \subseteq \beta$. Consequently, C is closed. For unboundedness, we apply a typical construction. Fix $\sigma < \alpha$. Since α is regular, we have $\beta_0 := \max\{\sigma+1, \sup(f[\sigma])+1\} < \alpha$. Assume that the monotone sequence $(\beta_j : j \leq n)$ with $\beta_n < \alpha$ is already defined. Put, analogously to the basis of the induction, $\beta_{n+1} := \max\{\beta_n + 1, \sup(f[\beta_n]) + 1\}$. By assumption we have $\beta := \sup\{\beta_n : n \in \omega\} \in \alpha$, and we obtain that $f[\beta] = \bigcup\{f[\beta_n] : n < \omega\} \subseteq \bigcup\{\beta_{n+1} : n < \omega\} = \beta$. Thus $\beta \in C$ and $\beta > \sigma$, which shows that C is unbounded in α.

Lemma 1.8.1 *Let α be a limit ordinal with uncountable cofinality. Then the following holds.*

a) $\alpha \cap \mathrm{Lim}$ *is a club in* α.

b) α *contains a club of order type* $\mathrm{cf}(\alpha)$ *(of course this is also true if* $\mathrm{cf}(\alpha) = \omega$*). Furthermore, if A is a subset of* Lim, *then there exists, by* (AC), *a club sequence* $(C_\alpha : \alpha \in A)$.

c) If E is any unbounded subset of α, then the set E' of limit points of E less than α is a club in α; the same holds for $E \cup E'$.

d) If $g : \alpha \longrightarrow \mathrm{ON}$ *is a normal function, then* $\mathrm{ran}(g)$ *is a club in* $\sup(\mathrm{ran}(g))$*; if in addition* $g : \alpha \longrightarrow \alpha$*, then* $\mathrm{ran}(g)$ *is a club in* α*. Conversely, every club C in α is the range of a normal function f with* $\mathrm{dom}(f) = \alpha$ *if α is regular. If α is singular, then every club C in α is the range of a normal function* $f : \mathrm{otp}(C) \longrightarrow \alpha$.

e) If α is a singular cardinal, then α includes a special club.

Proof For a), it obviously suffices to prove unboundedness. Fix $\sigma \in \alpha \cap \mathrm{Lim}$ and put $\sigma_0 := \sigma$. If $\sigma_j < \alpha$ is already defined for all $j \leq n$ such that the sequence $(\sigma_j : j \leq n)$ is strictly monotone, then there is an ordinal σ_{n+1} such that $\sigma_n < \sigma_{n+1} < \alpha$. The ordinal $\tau := \sup\{\sigma_n : n \in \omega\}$ is a limit ordinal, has cofinality ω and does not dominate α. The assumption yields that it is a member of $\alpha \cap \mathrm{Lim}$.

For b), observe that we obtain from Lemma 1.5.15 a continuous, strictly monotone function $h : \mathrm{cf}(\alpha) \longrightarrow \alpha$ such that $\sup(\mathrm{ran}(h)) = \alpha$. The last property implies that $\mathrm{ran}(h)$ is unbounded in α, and from the continuity of h and Lemma 1.4.1 we can conclude that $\mathrm{ran}(h)$ is closed. Since $h : \mathrm{cf}(\alpha) \longrightarrow \mathrm{ran}(h)$ is an $\in$-isomorphism, $\mathrm{ran}(h)$ is a club in α of order type $\mathrm{cf}(\alpha)$, as desired.

For c), the unboundedness of E' is proved analogously to the proof of a), using the the unboundedness of E: If $\sigma \in \alpha$, then choose $\sigma_0 \in E$ with $\sigma_0 > \sigma$. If $\sigma_j \in E$ is already defined for all $j \leq n$ such that the sequence $(\sigma_j : j \leq n)$ is strictly increasing, then there is an ordinal $\sigma_{n+1} \in E$ such that $\sigma_n < \sigma_{n+1} < \alpha$. The ordinal $\tau := \sup\{\sigma_n : n \in \omega\}$ is greater than σ, is a limit point of E, has cofinality ω and does not dominate α. Therefore it is, by assumption, a member of $\alpha \cap E'$.

If γ is a limit ordinal and $E' \cap \gamma$ is unbounded in γ, then $E \cap \gamma$ is unbounded in γ. This is easy to see, since for every $\beta < \gamma$ there is a limit point σ of E such that $\beta + 1 < \sigma < \gamma$ and, by the definition of a limit point, there exists $\tau \in \sigma \cap E$ with $\beta < \tau$. Thus we get $\gamma = \sup(E \cap \gamma)$ and $\gamma \in E'$.

For d), we observe that C is unbounded in $\sup(\mathrm{ran}(g))$, if C is the range of a normal function g from α into ON; in the case that $g : \alpha \longrightarrow \alpha$, C is unbounded in α, since $\beta \leq g(\beta)$ for every $\beta < \alpha$. C is closed in $\sup(\mathrm{ran}(g))$, since g is continuous.

Assume that α is regular, C is a club in α, and, according to Corollary 1.3.11, $f : \gamma \longrightarrow C$ is the order isomorphism from $\gamma := \mathrm{otp}(C)$ onto C. f is strictly monotone and continuous, since C is closed, and $\mathrm{ran}(f) = C$ is unbounded in α. This gives $\alpha = \mathrm{cf}(\alpha) \leq \gamma$. Since $\xi \leq f(\xi)$ for all $\xi < \gamma$, we get $\gamma \leq \sup(\mathrm{ran}(f)) = \alpha$. In the same way one proves the corresponding assertion for singular α.

For e), Lemma 1.6.11 says that α is the supremum of a continuous, strictly increasing sequence $(\lambda_\xi : \xi < \mathrm{cf}(\alpha))$ of singular cardinals greater than $\mathrm{cf}(\alpha)$. The range of this sequence is a club in α by part d); by definition, it is a special club.

Lemma 1.8.2 *Let α be an ordinal with uncountable cofinality.*

a) The intersection of less than $\mathrm{cf}(\alpha)$ clubs in α is a club in α.

b) Assume that $(C_\gamma : \gamma < \alpha)$ is a sequence of clubs in α. If the set $\bigcap\{C_\gamma : \gamma < \beta\}$ is unbounded in α for all $\beta < \alpha$, then the diagonal intersection of

the sequence $(C_\gamma : \gamma < \alpha)$ *is a club in* α. *This holds in particular if the sequence* $(C_\gamma : \gamma < \alpha)$ *is decreasing under* $\subseteq$.
If $\alpha > \aleph_0$ *is regular, then* $\Delta(C_\gamma : \gamma < \alpha)$ *is a club in* α.

Proof For a), let $(C_\gamma : \gamma < \tau)$ with $0 < \tau < \mathrm{cf}(\alpha)$ be a sequence of clubs in α, and let $C := \bigcap\{C_\gamma : \gamma < \tau\}$. Certainly C is closed in α. We show that C is unbounded in α. For this we use transfinite induction and assume that every intersection of less than τ clubs in α is unbounded in α. Fix $\sigma < \alpha$. If $\tau = 2$, then choose $\alpha_n \in C_0$ and $\beta_n \in C_1$ for each $n \in \omega$ such that $\sigma < \alpha_1 < \beta_1 < \alpha_2 < \beta_2 < \dots$. We then have $\sigma < \beta := \sup\{\alpha_n : n \in \omega\} = \sup\{\beta_n : n \in \omega\}$ and $\beta \in C_0 \cap C_1 \cap \alpha$, since $\mathrm{cf}(\alpha) > \aleph_0$. If $\tau = \xi + 1$, then the same proof, together with the inductive hypothesis, yields that $C = C_\xi \cap \bigcap\{C_\gamma : \gamma < \xi\}$ is unbounded in α. Finally, if τ is a limit ordinal, choose for $i < \tau$ ordinals $\gamma_i < \alpha$ such that $\gamma_0 > \sigma$, $\gamma_i \in \bigcap\{C_j : j < i\}$, and $\gamma_i > \sup\{\gamma_j : j < i\}$, which is possible by the inductive hypothesis and because $i < \mathrm{cf}(\alpha)$, hence $\sup\{\gamma_j : j < i\} < \alpha$. Put $\beta := \sup\{\gamma_i : i < \tau\}$. We get $\sigma < \beta < \alpha$. Furthermore $\beta = \sup\{\gamma_i : i_0 < i < \tau\}$ for any $i_0 < \tau$, and thus β is a supremum of elements of C_{i_0}. This gives $\beta \in C_{i_0}$ for every $i_0 < \tau$, and we can conclude that $\beta \in C$.

For the proof of b), we show first that $C := \Delta(C_\gamma : \gamma < \alpha)$ is closed. So assume that $\tau < \alpha$ is a limit ordinal and $C \cap \tau$ is unbounded in τ, and fix $\sigma < \tau$. Then we have $\tau = \sup\{i \in C : \sigma < i < \tau\}$. If $\sigma < i \in C$, then $i \in C_\sigma$, and since C_σ is a club, we get $\tau \in C_\sigma$. This gives $\tau \in C$.

For unboundedness, fix $\sigma < \alpha$. Choose $\sigma_0 \in C_0 \backslash \sigma$ and $\sigma_{n+1} \in \bigcap\{C_\gamma : \gamma < \sigma_n\} \backslash (\sigma_n + 1)$. This is possible, since, by assumption, the set $\bigcap\{C_\gamma : \gamma < \sigma_n\}$ is unbounded in α. Then we have $\beta := \sup\{\sigma_n : n \in \omega\} < \alpha$, since $\mathrm{cf}(\alpha) > \aleph_0$. Let $i < \beta$. Then there is an $n_0 \in \omega$ with $i < \sigma_{n_0}$, and further $\beta = \sup\{\sigma_n : n > n_0\}$ is a supremum of elements of C_i, which gives $\beta \in C_i$. Since $\beta \in \bigcap\{C_i : i < \beta\}$, we can conclude that $\beta \in C$, and clearly $\beta > \sigma$.

If $\alpha > \aleph_0$ is regular, then, for every $\beta < \alpha$, the set $\bigcap\{C_\gamma : \gamma < \beta\}$ is a club in α by part a); in particular it is unbounded in α. Using the previous result, we can now conclude that $\Delta\{C_\gamma : \gamma < \alpha\}$ is a club in α.

Definition *If A is a nonempty set, then we say that a set F of subsets of A is a* **filter** *on A iff the following conditions are satisfied:*

(F1) $A \in F$ *and* $\emptyset \notin F$.
(F2) If $X, Y \in F$, *then* $X \cap Y \in F$.
(F3) If $X \in F$, Y *is a subset of* A *and* $X \subseteq Y$, *then* $Y \in F$.

Every $\subseteq$*-maximal filter on A is called an* **ultrafilter** *on A. If a nonempty set* $S \subseteq \mathcal{P}(A)$ *has the* **finite intersection property** (fip)*, i.e., if* $\bigcap S' \neq \emptyset$ *for all* $S' \in \mathcal{P}_{\mathrm{fin}}(S)$, *then (see Lemma 1.8.3)*

$$F := \langle S \rangle := \{X \subseteq A : \exists S' \in \mathcal{P}_{\mathrm{fin}}(S) \bigcap S' \subseteq X\}$$

is a filter on A such that $S \subseteq F$. *It is the* **filter generated by** $\boldsymbol{S}$. *If* $S = \{Y\}$, *we also call it the filter generated by* Y. *If* F *is an arbitrary filter on* A *and* X *is a subset of* A *such that* $A \setminus X \notin F$, *we denote, with regard to the corresponding denotations for ideals, by* $F \restriction X$ *the filter generated by* $F \cup \{X\}$.

For any $a \in A$, *the set* $\{X \subseteq A : a \in X\}$ *is an ultrafilter on* A; *we call it the* **principal ultrafilter on** $\boldsymbol{A}$ **generated by** $\{\boldsymbol{a}\}$. *Every ultrafilter which is not principal is called* **nonprincipal** *or* **free**.

We say that a set I *of subsets of* A *is an* **ideal** *on* A *iff it satisfies the following properties:*

(I1) $A \notin I$ *and* $\emptyset \in I$.

(I2) If $X \in I$ *and* $Y \in I$, *then* $X \cup Y \in I$.

(I3) If $X \in I$ *and* $Y \subseteq X$, *then* $Y \in I$.

The set $\mathrm{dom}(I) := \bigcup I$ *is called the* **domain of the ideal** $\boldsymbol{I}$. *If* F *is a set of subsets of* A, *one easily checks that* F *is a filter on* A *iff the set* $I := \{A \setminus X : X \in F\}$ *is an ideal on* A. *We call* I *the* **dual ideal of** $\boldsymbol{F}$. *If* F *is an ultrafilter on* A, *then the dual ideal of* F *is called a* **prime ideal** *or a* **maximal ideal**. *Conversely, if an ideal* I *on* A *is given, we can define a filter* F *in the same way and call it the* **dual filter of** $\boldsymbol{I}$.

Assume that α *is a limit ordinal. We say that a filter* F *on* α *is* **normal** *iff, for any sequence* $(X_\xi : \xi < \mathrm{cf}(\alpha))$ *of members of* F, *the diagonal intersection* $\Delta(X_\xi : \xi < \mathrm{cf}(\alpha))$ *is a member of* F. *If* $\mathrm{cf}(\alpha) > \aleph_0$, *then we call the filter*

$$F_C = F_C(\alpha) := \{X \subseteq \alpha : \exists C(C \text{ is a club in } \alpha \wedge C \subseteq X)\}$$

(see Lemma 1.8.2) the **clubfilter** *on* α.

Any subset S *of* α *is called* **stationary** *in* α *or a* **stationary subset of** $\boldsymbol{\alpha}$ *iff* $C \cap S \neq \emptyset$ *for every club* C *in* α. *Every subset of* α *which is not stationary in* α *is called* **thin** *or, as usual,* **nonstationary** *in* α.

Lemma 1.8.3 *Let* A *be a nonempty set.*

a) If a nonempty set $S \subseteq \mathcal{P}(A)$ *has the finite intersection property, then* $\langle S\rangle$ *is the smallest filter on* A *such that* $S \subseteq \langle S\rangle$.

b) If U *is a filter on* A, *then the following conditions are equivalent.*

(i) U *is an ultrafilter.*

(ii) $\forall X, Y \in \mathcal{P}(A)(X \cup Y \in U \implies X \in U \vee Y \in U)$.

(iii) $\forall X \in \mathcal{P}(A)(X \notin U \implies A \setminus X \in U)$.

c) If I *is an ideal and* F *a filter on* A *with* $I \cap F = \emptyset$[20], *then there exists an ultrafilter* D *on* A *such that* $F \subseteq D$ *and* $D \cap I = \emptyset$. *In particular we get:*

[20] Such a filter always exists; take for example the trival filter $\{A\}$.

If $d \notin I$, then there is an ultrafilter D on A such that $D \cap I = \emptyset$ and $d \in D$.

For every filter F on A there is an ultrafilter D on A such that $F \subseteq D$ ("every filter can be extended to an ultrafilter"[21]).

There is a free ultrafilter on any infinite set.

d) *If A is infinite and U an ultrafilter on A, then the following conditions are equivalent:*

(i) *U is nonprincipal.*

(ii) *$F_{\text{cof}} := \{X \subseteq A : A \setminus X \text{ is finite}\} \subseteq U$.*

(iii) *No member of U is a finite set.*

Proof The proof of a) and d) is left to the reader. We give an indirect proof for the implication "(i) $\Longrightarrow$ (iii)" in b) and assume that there is a subset X of A such that $X \notin U$ and $A \setminus X \notin U$. Then we get $Y \cap X \neq \emptyset$ for every $Y \in U$, since $Y \cap X = \emptyset$ would imply $Y \subseteq A \setminus X$ and thus $A \setminus X \in U$. Put $F := \{Z \subseteq A : \exists Y \in U(Y \cap X \subseteq Z)\}$. Obviously, $U \subseteq F$ and $X \in F$. Further it is easy to see with the filter properties of U that F is a filter. This gives $U = F$, since U is maximal, hence $X \in U$, a contradiction.

For the proof of "(iii) $\Longrightarrow$ (ii)" we assume, that there are sets $X, Y \in \mathcal{P}(A)$ such that $X \cup Y \in U$, $X \notin U$ and $Y \notin U$. From (iii) we get $A \setminus X \in U$ and $A \setminus Y \in U$. Thus the intersection of these sets, that is $Z := A \setminus (X \cup Y)$, is a member of U. Now we can conclude that $\emptyset = (X \cup Y) \cap Z \in U$, contradicting our assumption that U is a filter.

To prove "(ii) $\Longrightarrow$ (i)", assume that $F \supseteq U$ is a filter and $X \in F$. We will show that $X \in U$. Fix $B \in U$. Then $B = (B \cap X) \cup (B \setminus X)$. Assumption (ii) gives $X \in U$ or $B \setminus X \in U$. But from the second condition we could conclude with $U \subseteq F$ that $X \cap (B \setminus X) = \emptyset \in F$ which would contradict the fact that F is a filter.

For the proof of c), it is easy to check that the set

$$\mathcal{M} := \{G \subseteq \mathcal{P}(A) : G \text{ is a filter on } A \wedge F \subseteq G \wedge G \cap I = \emptyset\}$$

satisfies the assumptions of Zorn's lemma. So let D be a maximal element of $\mathcal{M}$. We will show that D is an ultrafilter on A and assume, contrary to our hopes, that there is a set $X \subseteq A$ such that $X \notin D$ and $A \setminus X \notin D$. If $Y \in D$, then $X \cap Y \neq \emptyset$, since otherwise $A \setminus X \in D$ would follow from $Y \subseteq A \setminus X$. In the same way we get $Y \cap (A \setminus X) \neq \emptyset$. This shows that the sets $D \cup \{X\}$ and $D \cup \{A \setminus X\}$

[21] Of course this assertion follows easily if we apply Zorn's lemma to the set of all filters on A containing F.

have the finite intersection property. Let F_1 and F_2 be the filters generated by them according to part a). Since D is maximal in $\mathcal{M}$, we can choose $Z_1 \in F_1 \cap I$ and $Z_2 \in F_2 \cap I$. Part a) gives $F_1 = \{Y \subseteq A : \exists d \in D(d \cap X \subseteq Y)\}$, since D is a filter, and an analogous representation holds for F_2. Thus we obtain sets $d_1, d_2 \in D$, such that $d_1 \cap X \in I$ and $d_2 \cap (A \setminus X) \in I$. Therefore the set $(d_1 \cap X) \cup (d_2 \cap (A \setminus X))$ is a member of I. But $d_1 \cap d_2$ is a subset of it, and so we get $d_1 \cap d_2 \in I \cap D$, contradicting $D \cap I = \emptyset$. This completes the proof of the first assertion in c).

If $d \subseteq A$ and $d \notin I$, then the filter $F_d := \{Y \subseteq A : d \subseteq Y\}$ generated by d is a filter on A with $I \cap F_d = \emptyset$, as desired. The second special consequence follows at once, taking $I = \{\emptyset\}$. Also the third consequence is easy to show, taking $I := \{X \subseteq A : X \text{ is finite}\}$. Then I is an ideal on A, and from d) we can conclude that any ultrafilter disjoint from I is nonprincipal.

Lemma 1.8.4 *Assume that α is a limit ordinal with uncountable cofinality.*

a) *Every superset of a club in α is stationary in α.*
b) *Every stationary subset of α is unbounded in α.*
c) *Every union of less than $\mathrm{cf}(\alpha)$ sets that are nonstationary in α is nonstationary in α.*
d) *For every regular cardinal $\lambda < \mathrm{cf}(\alpha)$, the set*

$$\Sigma_{\alpha,\lambda} := \{\beta < \alpha : \mathrm{cf}(\beta) = \lambda\}$$

is stationary in α. Furthermore, every club C in α has a member β with $\mathrm{cf}(\beta) = \lambda$ such that $C \cap \beta$ is a club in β.

Proof Part a) follows from Lemma 1.8.2 a). The same holds for part c): If $\gamma < \mathrm{cf}(\alpha)$ and $(T_\xi : \xi < \gamma)$ is a sequence of nonstationary sets, then choose for each $\xi < \gamma$ a club C_ξ in α with $C_\xi \cap T_\xi = \emptyset$; $C := \bigcap\{C_\xi : \xi < \gamma\}$ is a club which does not meet the set $\bigcup\{T_\xi : \xi < \gamma\}$. Part b) follows at once if we choose, for each $\gamma < \alpha$, a member of the intersection of the club $\{\beta \in \alpha : \gamma \leq \beta\}$ with the given stationary set.

For the proof of d), consider a fixed club C in α. We can define by recursion, since $\lambda < \mathrm{cf}(\alpha)$ and clubs are unbounded, a strictly increasing continuous sequence g of members of C having length λ. Since C is closed and λ is regular, the supremum β of this sequence is an element of C and has cofinality λ, hence $\beta \in \Sigma_{\alpha,\lambda}$. The second assertion follows, too, since $\mathrm{ran}(g)$ is a club in β by Lemma 1.8.1, and $\mathrm{ran}(g) \subseteq C$.

Definition *Assume that M is a subset of $\mathcal{P}(A)$ such that $\bigcup M' \neq A$ for every finite subset M' of M. Then*

$$\langle M \rangle := \{X \subseteq A : \exists M' \in \mathcal{P}_{\mathrm{fin}}(M)\ X \subseteq \bigcup M'\}$$

is an ideal on A called the **ideal generated by** $\boldsymbol{M}$. *It is the smallest ideal on A including M. If* $A \setminus B \notin I$, *then there exists the ideal* $I[B] := \langle I \cup \{B\} \rangle$. *We say that* $\boldsymbol{B}$ **generates the ideal** $\boldsymbol{I[B]}$ **over** $\boldsymbol{I}$. *If* $X \notin I$, *then the set*

$$I \restriction X := \{Z \subseteq A : X \cap Z \in I\}$$

is an ideal on A, the **restriction of the ideal** $\boldsymbol{I}$ **to** $\boldsymbol{X}$. *Note that* $I \restriction X = I[A \setminus X]$.

Let α *be a limit ordinal. We say that an ideal I on* α *is* **normal** *iff, for any sequence* $(X_\xi : \xi < \mathrm{cf}(\alpha))$ *of members of I, the diagonal union* $\nabla(X_\xi : \xi < \mathrm{cf}(\alpha))$ *is a member of I. If* κ *is a cardinal with* $\mathrm{cf}(\alpha) \geq \kappa$, *then an ideal I on* α *is called* $\boldsymbol{\kappa}$**-complete** *iff every union of less than* κ *members of I is a member of I. I.e., whenever* $(X_\alpha : \alpha < \gamma)$ *is a family of members of I and* $\gamma < \kappa$, *then* $\bigcup\{X_\alpha : \alpha < \gamma\} \in I$. *Any* $\aleph_1$*-complete ideal also is called* $\boldsymbol{\sigma}$**-complete**. *If* $\mathrm{cf}(\alpha) > \aleph_0$, *then we call the dual ideal of the clubfilter* F_C *on* α *the* **ideal of thin subsets** *of* α, *writing* $\mathrm{I_d}(\alpha)$; *i.e,*

$$\mathrm{I_d}(\alpha) = \{X \subseteq A : \exists C (C \textit{ is a club in } \alpha \wedge C \cap X = \emptyset)\}.$$

For any nonempty set A of ordinals with $\sup(A) \notin A$, *the ideal*

$$\mathrm{I_b}(A) := \{X \subseteq A : \exists \beta \in A\ X \subseteq \beta\}$$

on A is called the **ideal of bounded subsets** *of A. An ultrafilter D on A is called* **unbounded** *iff* $D \cap \mathrm{I_b}(A) = \emptyset$.

Lemma 1.8.5 *If* κ *is an uncountable regular cardinal, then* $\mathrm{I_d}(\kappa)$ *is a* κ*-complete normal ideal on* κ *and* $\mathrm{I_b}(\kappa) \subseteq \mathrm{I_d}(\kappa)$.

Proof By definition, $\mathrm{I_d}(\kappa)$ is closed under subsets. If $(X_\xi : \xi < \kappa)$ is a sequence of members of this ideal, then for each $\xi < \kappa$ there is a club C_ξ in κ such that $C_\xi \cap X_\xi = \emptyset$. Lemma 1.8.2 tells us that the diagonal intersection D of $(C_\xi : \xi < \kappa)$ and, for every $\gamma < \kappa$, $D_\gamma := \bigcap\{C_\xi : \xi < \gamma\}$ are clubs in κ. An easy calculation shows that $D \cap \nabla\{X_\xi : \xi < \kappa\} = \emptyset$ and $D_\gamma \cap \bigcup\{X_\xi : \xi < \gamma\} = \emptyset$. Thus $\mathrm{I_d}(\kappa)$ is κ-complete and normal, as desired.

Lemma 1.8.6 *Assume that* κ *is a regular uncountable cardinal and I is a normal ideal on* κ.

a) If I is κ*-complete and* $\mathrm{dom}(I) = \kappa$, *then* $\mathrm{I_b}(\kappa) \subseteq I$, *and* $\{\xi + 1 : \xi < \kappa\}$ *is a member of I.*

b) If $\mathrm{I_b}(\kappa) \subseteq I$, *then even* $\mathrm{I_d}(\kappa) \subseteq I$.
Thus $\mathrm{I_d}(\kappa)$ *is the smallest* κ*-complete normal ideal J on* κ *such that* $\mathrm{dom}(J) = \kappa$.

Proof For the proof of a) we observe that, for each $\alpha < \kappa$, the set $X_\alpha := \{\xi < \kappa : \xi \leq \alpha{+}1\}$ is a member of I, since I is κ-complete and any singleton $\{\xi\}$, $\xi < \kappa$, is a member of I on account of $\mathrm{dom}(I) = \kappa$. Now I is normal, therefore the set $\triangledown_{\alpha<\kappa} X_\alpha$ is a member of I, and we have $\triangledown_{\alpha<\kappa} X_\alpha = \{\xi + 1 : \xi < \kappa\}$.

Next we prove part b). Fix $T \in \mathrm{I_d}(\kappa)$. Then there is a club C in κ with $T \cap C = \emptyset$. C has order type κ, since κ is regular. So let $(\delta_\alpha : \alpha < \kappa)$ be the $\in$-isomorphism from κ onto C, and let $Y_\alpha := \{\xi < \kappa : \xi \leq \delta_\alpha\}$. By assumption, we have $\mathrm{I_b}(\kappa) \subseteq I$, and thus $Y_\alpha \in I$ and $X_\alpha \in I$ for every $\alpha < \kappa$. From this we can conclude that $\triangledown_{\alpha<\kappa} Y_\alpha \in I$ and $\triangledown_{\alpha<\kappa} X_\alpha \in I$, since I is normal. If we can prove the assertion

$$(*) \qquad T \subseteq \triangledown_{\alpha<\kappa} X_\alpha \cup \triangledown_{\alpha<\kappa} Y_\alpha,$$

then $T \in I$ follows, and b) will be shown. We have $\kappa \setminus (\triangledown_{\alpha<\kappa} X_\alpha \cup \triangledown_{\alpha<\kappa} Y_\alpha) = (\kappa \setminus \triangledown_{\alpha<\kappa} X_\alpha) \cap (\kappa \setminus \triangledown_{\alpha<\kappa} Y_\alpha) = \Delta_{\alpha<\kappa}(\kappa \setminus X_\alpha) \cap \Delta_{\alpha<\kappa}(\kappa \setminus Y_\alpha)$. If $\beta \in \kappa \setminus (\triangledown_{\alpha<\kappa} X_\alpha \cup \triangledown_{\alpha<\kappa} Y_\alpha)$, then β is a limit ordinal, since $\beta \in \Delta_{\alpha<\kappa}(\kappa \setminus X_\alpha)$, and $\delta_\alpha < \beta$ for all $\alpha < \beta$, since $\beta \in \Delta_{\alpha<\kappa}(\kappa \setminus Y_\alpha)$. Now, firstly the sequence $(\delta_\alpha : \alpha < \kappa)$ is continuous as a normal function, and we get $\delta_\beta = \sup\{\delta_\alpha : \alpha < \beta\} \leq \beta$, and secondly it is strictly increasing, which gives $\beta \leq \delta_\beta$. Altogether we get $\beta = \delta_\beta$ and thus $\beta \in C$. Since $T \cap C = \emptyset$, we can conclude that $\beta \notin T$, which proves $(*)$.

Lemma 1.8.7 *If κ is an uncountable regular cardinal and I is a normal ideal on κ such that $\mathrm{I_b}(\kappa) \subseteq I$, then I is κ-complete.*

Proof Consider a cardinal $\mu < \kappa$ and a sequence $(X_\alpha : \alpha < \mu)$ of members of I. We will show that $\bigcup_{\alpha<\mu} X_\alpha \in I$. Let the sequence $(Y_\alpha : \alpha < \kappa)$ be defined by

$$Y_\alpha := \begin{cases} X_\alpha \setminus \mu & \text{if } \alpha < \mu, \\ \emptyset & \text{otherwise.} \end{cases}$$

Since I is normal, we know that $\triangledown_{\alpha<\kappa} Y_\alpha \in I$. Furthermore it is easy to see that $\triangledown_{\alpha<\kappa} Y_\alpha = \bigcup\{Y_\alpha : \alpha < \mu\}$, and thus $\bigcup\{Y_\alpha : \alpha < \mu\} \in I$. Now $\mu \in \mathrm{I_b}(\kappa)$ gives $\mu \in I$, and from this we can conclude that $\bigcup_{\alpha<\mu} X_\alpha \subseteq \bigcup_{\alpha<\mu} Y_\alpha \cup \mu \in I$ and thus $\bigcup_{\alpha<\mu} X_\alpha \in I$, as desired.

Theorem 1.8.8 (Fodor) *Assume that α is a limit ordinal with $\mathrm{cf}(\alpha) > \aleph_0$ and $S \subseteq \alpha$ is stationary in α. Let further $f : S \longrightarrow \alpha$ be a* **regressive function** *(i.e., $f(\gamma) < \gamma$ for all $\gamma \in S \setminus \{0\}$). Then, for some $\eta < \alpha$, the set $f^{-1}[\eta]$ is stationary in α.*

If α is regular, then there even exists $\eta < \alpha$ such that $f^{-1}[\{\eta\}]$ is stationary in α, i.e., f is constant on a stationary subset of S.

Proof To get a contradiction we assume that, for each $\eta < \alpha$, there is a club D_η in α such that $D_\eta \cap f^{-1}[\eta] = \emptyset$. Choose, according to Lemma 1.8.1, a club $E \subseteq \alpha$ in α of order type $\operatorname{cf}(\alpha)$, and put $C_\eta := \bigcap\{D_\xi : \xi \in E \cap (\tau+1)\}$, where $\tau = \tau(\eta) := \min\{\xi \in E : \xi \geq \eta\}$. C_η is a club in α, being the intersection of less than $\operatorname{cf}(\alpha)$ clubs in α. Furthermore we have $C_\eta \cap f^{-1}[\eta] = \emptyset$, since $C_\eta \subseteq D_{\tau(\eta)}$, $\tau(\eta) \geq \eta$ and $D_{\tau(\eta)} \cap f^{-1}[\tau(\eta)] = \emptyset$ and thus $D_{\tau(\eta)} \cap f^{-1}[\eta] = \emptyset$. Now the sequence $(C_\eta : \eta < \alpha)$ is descending, and Lemma 1.8.2 says that $C := \Delta(C_\eta : \eta < \alpha)$ and thus $C \cap \mathrm{Lim}$ are clubs in α. Consequently, there exists a limit ordinal γ such that $\gamma \in C \cap S$, since S is stationary in α. Furthermore we have $f(\gamma) < \gamma$, since f is regressive, hence $\sigma := f(\gamma) + 1 < \gamma$ and $\gamma \in C_\sigma$. But on the other hand we have $\gamma \in f^{-1}[\sigma]$ by the definition of σ, contradicting $C_\sigma \cap f^{-1}[\sigma] = \emptyset$.

Now assume in addition that α is regular and $f^{-1}[\eta]$ is stationary in α for some $\eta < \alpha$. From $f^{-1}[\eta] = \bigcup\{f^{-1}[\{\sigma\}] : \sigma < \eta\}$ we can conclude from Lemma 1.8.4 that there is an ordinal $\sigma < \eta$ such that $f^{-1}[\{\sigma\}]$ is stationary in α.

Lemma 1.8.9 *Assume that κ is a regular cardinal and I is an ideal on κ. Then I is normal iff there exists, for every set $S \subseteq \kappa$ with $S \notin I$ and every function f regressive on S, a set $S_0 \subseteq S$ such that $S_0 \notin I$ and f is constant on S_0.*

Proof First we show that the previous condition is sufficient for the normality of I. So assume that $(X_\xi : \xi < \kappa)$ is a sequence of members of I. We must show that $S := \nabla_{\xi<\kappa} X_\xi$ is in I. If $\sigma \in S$, then the ordinal $f(\sigma) := \min\{\xi < \sigma : \sigma \in X_\xi\}$ exists. The function f defined by this equation is regressive on S. If S was not a member of I, then our assumption would yield a set $S_0 \subseteq S$ with $S_0 \notin I$ and an ordinal $\gamma < \kappa$ such that $\operatorname{ran}(f \restriction S_0) = \{\gamma\}$. We could infer that $S_0 \subseteq X_\gamma$ by the definition of f, and thus $S_0 \in I$, a contradiction. Consequently, we get $S \in I$, as desired.

Conversely assume that I is normal, $S \subseteq \kappa$ is a set with $S \notin I$, and $f : S \longrightarrow \kappa$ is a regressive function. If $0 \in S$ and $S \setminus \{0\} \in I$, then $\{0\} \notin I$, and f is constant on $S_0 := \{0\}$. Otherwise we have $S \setminus \{0\} \notin I$ or $0 \notin S$, and we can assume without loss of generality that $0 \notin S$. Put $X_\xi := f^{-1}[\{\xi\}]$ for every $\xi < \kappa$. Then there must exist an ordinal $\xi < \kappa$ with $X_\xi \notin I$. The supposition $X_\xi \in I$ for every $\xi < \kappa$ would lead to $\nabla_{\xi<\kappa} X_\xi \in I$, since I is normal. But on the other hand we have $S = \nabla_{\xi<\kappa} X_\xi$, since f is regressive, and $S \notin I$. Therefore there exists a set X_ξ such that $X_\xi \notin I$. Clearly we have $X_\xi \subseteq S$, and the function f is constant on X_ξ.

Corollary 1.8.10 *If κ is a regular cardinal, I is a normal ideal on κ and $X \in \mathcal{P}(\kappa) \setminus I$, then $I \restriction X$ is a normal ideal on κ.*

Proof Assume that $S \in \mathcal{P}(\kappa) \setminus (I \restriction X)$ and $f : S \longrightarrow \kappa$ is a regressive function. Then $X \cap S \notin I$ and, by assumption, there is a set $S_0 \subseteq S \cap X$ with $S_0 \notin I$ and an ordinal $\gamma < \kappa$ such that $\operatorname{ran}(f \restriction S_0) = \{\gamma\}$. It is obvious that $S_0 \notin I \restriction X$.

Next we prove a well-known theorem of Solovay which says for regular uncountable cardinals κ: If $S \subseteq \kappa$ and $S \notin \mathrm{I_d}(\kappa)$, then S can be partitioned into κ classes which are not members of $\mathrm{I_d}(\kappa)$. For the proof we need two preparatory lemmas.

Lemma 1.8.11 *Assume that κ is an uncountable regular cardinal and S is a stationary subset of κ. Then the set T is stationary in κ, where*

$$T := \{\alpha \in S : \mathrm{cf}\,(\alpha) = \omega \vee (\mathrm{cf}\,(\alpha) > \omega \wedge S \cap \alpha \textit{ is not stationary in } \alpha)\}.$$

Proof If C is a club in κ, then Lemma 1.8.1 tells us that the set $C' := \{\sigma < \kappa : \sigma \neq 0 \wedge \sup(C \cap \sigma) = \sigma\}$ of limit points of C is a club in κ. Let $\alpha := \min(C' \cap S)$. We show that $\alpha \in C \cap T$. If $\mathrm{cf}\,(\alpha) = \omega$, then this holds by definition. So let $\mathrm{cf}\,(\alpha) > \omega$. Then, by Lemma 1.8.1, $C' \cap \alpha$ is a club in α, since α is a limit point of C. Further we have $(C' \cap \alpha) \cap (S \cap \alpha) = \emptyset$ by the choice of α, which means that $S \cap \alpha$ is not stationary in α.

Lemma 1.8.12 *Let κ be an uncountable regular cardinal and T be a stationary subset of $\kappa \cap \mathrm{Lim}$. If f_α is, for all $\alpha \in T$, a normal sequence cofinal in α, then one of the following assertions is true.*

(1) $\exists \xi < \kappa\ \forall \eta < \kappa (T_\eta := \{\alpha \in T : \xi \in \mathrm{dom}(f_\alpha) \wedge f_\alpha(\xi) \geq \eta\}$ *is stationary in* $\kappa)$.

(2) $\exists D \subseteq \kappa(D$ *is a club in* $\kappa \wedge \forall \gamma, \alpha \in D \cap T(\gamma < \alpha \implies \gamma = f_\alpha(\gamma)))$.

Proof Assume that the first assertion is false, that means that for each $\xi < \kappa$ there is an ordinal $\eta(\xi) < \kappa$ and a club C_ξ in κ such that $C_\xi \cap T_{\eta(\xi)} = \emptyset$. This defines a function η from κ into κ. Put $C := \Delta_{\xi<\kappa} C_\xi$ and $D := \{\alpha \in C : \eta[\alpha] \subseteq \alpha \wedge \mathrm{Lim}(\alpha)\}$. From the example following the definition of a club and Lemma 1.8.2 we can conclude that D is a club in κ.

If we consider $\alpha, \gamma \in D \cap T$ with $\gamma < \alpha$, then $\alpha \in C$, and thus $\alpha \in C_\xi$ for all $\xi < \alpha$ and $\alpha \notin T_{\eta(\xi)}$ for all $\xi < \alpha$. From this we can infer for every $\xi \in \gamma \cap \mathrm{dom}(f_\alpha)$ that $f_\alpha(\xi) < \eta(\xi)$, since $\alpha \notin T_{\eta(\xi)}$, and $\eta(\xi) < \gamma$, since $\gamma \in D$. Now the supposition $\mathrm{dom}(f_\alpha) \leq \gamma$ would lead to $\mathrm{ran}(f_\alpha) \subseteq \gamma < \alpha$, contradicting the assumption on f_α. So we get $\gamma \in \mathrm{dom}(f_\alpha)$. Since γ is a limit ordinal and f_α is continuous, the previous considerations show that $f_\alpha(\gamma) = \sup\{f_\alpha(\xi) : \xi < \gamma\} \leq \gamma$. Furthermore we have $\gamma \leq f_\alpha(\gamma)$, since f_α is strictly increasing. This gives $f_\alpha(\gamma) = \gamma$, and the second assertion is true.

Theorem 1.8.13 (Solovay) *Assume that κ is an uncountable regular cardinal and S is a stationary subset of κ. Then S can be partitioned into κ stationary subsets of κ.*

Proof Since every superset of a stationary set is also stationary, it obviously suffices to show that the assertion is true for some stationary subset T of S. If f is a function from T into κ and $\gamma_1, \gamma_2 \in \kappa$ are different, then $f^{-1}[\{\gamma_1\}] \cap f^{-1}[\{\gamma_2\}] = \emptyset$. Thus f yields the desired partition if the set $\{\gamma \in \kappa : f^{-1}[\{\gamma\}]$ is stationary$\}$ is unbounded in κ. Now, if f is regressive and $T_\eta := \{\alpha \in T : f(\alpha) \geq \eta\}$ is stationary for all $\eta < \kappa$, then Fodor's theorem tells us that, for each $\eta < \kappa$, there is an ordinal $\gamma_\eta < \kappa$ such that $f^{-1}[\{\gamma_\eta\}] \cap T_\eta$ is stationary in κ. Since $\gamma_\eta \geq \eta$ by the definition of T_η, the set $\{\gamma_\eta : \eta < \kappa\}$ is unbounded in κ, and we are ready.

We choose the set $T \subseteq \mathrm{Lim}$ from Lemma 1.8.11 as the stationary subset T of S which will suit our purposes. For each $\alpha \in T$ we can find a sequence f_α which is cofinal in α and satisfies $\mathrm{ran}(f_\alpha) \cap T = \emptyset$. Namely, if $\mathrm{cf}(\alpha) > \aleph_0$, then $S \cap \alpha$ is nonstationary in α, and Lemma 1.8.1 d) says that there is a club C_α in α which is the range of a normal function f_α and does not meet $S \cap \alpha$. If $\mathrm{cf}(\alpha) = \aleph_0$, it is enough to choose a strictly increasing sequence $g_\alpha : \omega \longrightarrow \alpha$ with $\sup(\mathrm{ran}(g_\alpha)) = \alpha$ and to put $f_\alpha(n) := g_\alpha(n) + 1$. Now a club D satisfying the conditions in assertion (2) of Lemma 1.8.12 cannot exist. Otherwise $D \cap T$ would be stationary, in particular there would be ordinals $\alpha, \gamma \in D \cap T$ such that $\gamma < \alpha$ and $f_\alpha(\gamma) = \gamma$, that means with $\gamma \in \mathrm{ran}(f_\alpha) \cap T$, contradicting $\mathrm{ran}(f_\alpha) \cap T = \emptyset$. Therefore some $\xi < \kappa$ satisfies the condition in assertion (1) of this lemma, and we define for this ξ and for $\alpha \in T$

$$f(\alpha) := \begin{cases} f_\alpha(\xi) & \text{if } \xi \in \mathrm{dom}(f_\alpha), \\ 0 & \text{otherwise.} \end{cases}$$

If $\eta > 0$, then $\{\alpha \in T : f(\alpha) \geq \eta\} = \{\alpha \in T : \xi \in \mathrm{dom}(f_\alpha) \wedge f_\alpha(\xi) \geq \eta\}$, and thus the set $\{\alpha \in T : f(\alpha) \geq \eta\}$ is stationary in κ for every $\eta < \kappa$. Since f is regressive, our introductory remarks show that the proof of the theorem is completed.

Definition *Let α be a limit ordinal. We say that a stationary subset S of α is* **nonreflecting** *iff, for all limit ordinals $\beta < \alpha$, the set $S \cap \beta$ is not stationary in β.*

Lemma 1.8.14 *If λ is a regular cardinal, then the set*

$$\Sigma_{\lambda^+,\lambda} = \{\sigma < \lambda^+ : \mathrm{cf}(\sigma) = \lambda\}$$

is a nonreflecting stationary subset of λ^+.

Proof By Lemma 1.8.4, the set $\Sigma_{\lambda^+,\lambda}$ is stationary in λ^+.

Now let $\alpha < \lambda^+$ be a fixed limit ordinal. Then $\mu := \mathrm{cf}(\alpha) \leq \lambda$. Take $f : \mu \longrightarrow \alpha$ as a strictly monotone function such that $\sup(\mathrm{ran}(f)) = \alpha$. Then the function g with $g(\gamma) := f(\gamma) + 1$ has the same properties. Put $C' := \mathrm{ran}(g)$ and

$$C := C' \cup \{\gamma < \alpha : \sup(C' \cap \gamma) = \gamma\}.$$

For $\mu = \omega$, C is a club in α, and the same holds for $\mu > \omega$ by Lemma 1.8.1. By the definition of C, every limit ordinal $\gamma \in C$ is the supremum of the set $\{g(\sigma) : \sigma < \mu \wedge g(\sigma) < \gamma\}$. If we choose α_γ as the least ordinal in μ such that $\gamma \leq g(\alpha_\gamma)$, then $\gamma = \sup(\operatorname{ran}(g \restriction \alpha_\gamma))$, and we get $\operatorname{cf}(\gamma) \leq \alpha_\gamma < \mu \leq \lambda$. So we have shown that $\Sigma_{\lambda^+,\lambda} \cap C = \emptyset$, and $\Sigma_{\lambda^+,\lambda} \cap \alpha$ is not stationary in α, as desired.

Exercises[22] $A, B, \ldots, X, Y, \ldots$, etc. denote sets.

1) Prove: If $\mathcal{F}$ is a nonempty set of filters on A, then $\bigcap \mathcal{F}$ is a filter on A.
2) Prove: If $\mathcal{K}$ is a nonempty $\subseteq$-chain of filters on A, then $\bigcup \mathcal{K}$ is a filter on A.
3) Let X be a subset of A. Prove:

   a) If F is a filter (ultrafilter) on A and $X \in F$, then $F \cap \mathcal{P}(X)$ is a filter (ultrafilter) on X.

   b) If G is a filter on X, then G can be extended to an ultrafilter on A.

4) If F is a filter on A, we put $F^+ := \{X \subseteq A : A \setminus X \notin F\}$. Prove that $F \subseteq F^+$ and that F^+ is a filter iff F is an ultrafilter.
5) Let A be an infinite set. Prove that the set $\{X \subseteq A : A \setminus X \text{ is finite}\}$ is a filter on A. It is called the **Fréchet-Filter** on A.
6) Let A be an infinite set. Prove:

   a) There is an ultrafilter F on A such that all members of F have the same cardinality as A (a so-called **uniform ultrafilter** on A).
   Hint: The set $\{A \setminus X : |X| < |A|\}$ can be extended to an ultrafilter.

   b) The set of all uniform ultrafilters on A and the set of nonprincipal ultrafilters on A coincide iff A is countable (see also Exercise 7 and Exercise 8 below).
   Hint: If A is not countable, choose $X \subseteq A$ with $\aleph_0 \leq |X| < |A|$, and apply part a) to X.

7) Prove that there are no normal nonprincipal ultrafilters on ω.
   Hint: Assume, to get a contradiction, that F is such a filter. Then $\omega \setminus \{0\}$ is not an element of the dual ideal of F. Define $f : \omega \setminus \{0\} \longrightarrow \omega$ by $f(n+1) := n$. Now apply Lemma 1.8.9.
8) Let κ be an infinite cardinal. A filter F on A is called **κ-complete** iff $\bigcap M \in F$ for any nonempty subset M of F of cardinality less than κ; otherwise it is called **κ-incomplete**. An $\aleph_1$-complete filter on A is also called **σ-complete**, an $\aleph_1$-incomplete filter on A is also called **σ-incomplete**.
   Prove:

[22] Exercises on ideals can be found in Section 2.1.

a) If A is infinite, then every nonprincipal ultrafilter on A is $|A|^+$-incomplete. In particular, all nonprincipal ultrafilters on countable sets are σ-incomplete.

b) If U is a nonprincipal $|A|$-complete ultrafilter on A, then U is a uniform ultrafilter.

Hint: For any nonprincipal ultrafilter U on A, we have $A \setminus \{x\} \in U$ for every $x \in A$.

9) We call an uncountable cardinal κ a **measurable cardinal** iff there is a nonprincipal κ-complete ultrafilter on κ. Prove that every measurable cardinal is regular and strong limit, i.e., it is an inaccessible cardinal.

10) Prove for filters F on A: If

$$\mu := \sup\{\kappa : F \text{ is } \kappa\text{-complete}\}$$

exists, then μ is regular and F is μ-complete.

Hint: Assume that μ is singular and use the fact that F is μ^+-incomplete.

11) Assume that A is infinite and D is an ultrafilter on A. Prove that D is κ-complete iff, for every $\tau < \kappa$ and every partition $(X_\sigma : \sigma < \tau)$ of A (that means $\bigcup\{X_\sigma : \sigma < \tau\} = A$ and $X_\sigma \cap X_\xi = \emptyset$ whenever $\xi < \sigma < \tau$), there exists a set $X_{\sigma_0} \in D$.

12) (Shelah) If A is infinite, F is a filter on A, and κ is a cardinal, then F is called **κ-regular** iff there is a sequence $(X_\xi : \xi < \kappa)$ of members of F such that $\bigcap\{X_\xi : \xi \in N\} = \emptyset$ for every denumerable subset N of κ. Prove that

$$\operatorname{reg}(F) := \min\{\kappa : F \text{ is not } \kappa\text{-regular}\}$$

exists and is a regular cardinal.

13) We call a filter F on A **regular** iff F is $|A|$-regular. Prove: If F is a filter on A, then F is regular iff there is a function $f : A \longrightarrow [A]^{<\omega}$, such that $\{x \in A : a \in f(x)\} \in F$ for all $a \in A$.

14) a) If A is infinite, then there is a regular ultrafilter on A.
Hint: Consider a bijection $f : A \longrightarrow [A]^{<\omega}$ and let $X_a := \{a \in A : a \in f(x)\}$ for every $a \in A$.

b) If F is a regular ultrafilter, then F is σ-incomplete.

15) If A is denumerable, then every free ultrafilter on A is regular.

16) Assume that A and B are infinite sets and F is an ultrafilter on A. For any functions $f, g \in {}^A B$ we define, in analogy to the definition for ideals,

$$f =_F g \text{ iff } \{a \in A : f(a) = g(a)\} \in F.$$

Show that this gives an equivalence relation on ${}^{A}B$. If $[f] := \{g \in {}^{A}B : f =_F g\}$ is the equivalence class of F, then let $\prod_{a\in A} B/F := \{[f] : f \in {}^{A}B\}$ be the set of equivalence classes. Prove:
If C is a set with $|B| = |C|$, then $|\prod_{a\in A} B/F| = |\prod_{a\in A} C/F|$. If U is a regular ultrafilter on A, then $|\prod_{a\in A} B/U| = |B|^{|A|}$.
Hint: For the last assertion, use $\kappa := |A|$ instead of A and $C := \bigcup\{{}^{n}B : n \in \omega\}$ instead of B.

17) (Shelah) Assume that F is a filter on an infinite set A. F is called **weakly κ-regular** iff there is a sequence $(X_\xi : \xi < \kappa)$ of elements of F^+ such that $\bigcap\{X_\xi : \xi \in N\} = \emptyset$ for every $N \in [\kappa]^{\aleph_0}$. Let

$$\mathrm{reg}_*(F) := \min\{\kappa : F \text{ is not weakly } \kappa\text{-regular}\}.$$

Prove: The cardinal number $\mathrm{reg}_*(F)$ exists; furthermore, if $\nu < \mathrm{reg}_*(F)$ is a cardinal, then there exists a ν-regular filter D on A with $F \subseteq D$.
Hint: Let $\nu \in \mathrm{reg}_*(F) \cap \mathrm{ICN}$ and $(X_\xi : \xi < \nu)$ be a sequence of members of F^+ which witnesses that F is weakly ν-regular. Further let $(u_\xi : \xi < \nu)$ be an enumeration of $[\nu]^{<\aleph_0}$. Put

$$D := \{X \subseteq A : \exists u \in [\nu]^{<\aleph_0} \forall \xi < \nu (u \subseteq u_\xi \implies (A \setminus X_\xi) \cup X \in F)\}.$$

If $Y_\xi := \bigcup\{X_\zeta : \xi \in u_\zeta\}$, then the sequence $(Y_\xi : \xi < \nu)$ witnesses that D is ν-regular.

18) Let κ be an uncountable regular cardinal and $f : \kappa \longrightarrow \kappa$ be a normal function. Prove that the set C of all fixed points of f is a club in κ.
Hint: It is easy to verify that C is closed. For unboundedness, let $\sigma < \kappa$ and define $\alpha_0 := \sigma$, $\alpha_{n+1} := f(\alpha_n)$, and $\alpha := \sup\{\alpha_n : n \in \omega\}$.

19) Assume that α is a limit ordinal and C is a subset of α. Prove that C is closed in α iff, for every sequence of members of C which is bounded in α, its supremum is an element of C.

20) Assume that κ is a cardinal with uncountable cofinality, $S \subseteq \kappa$, and $(\alpha_\xi : \xi < \mathrm{cf}(\kappa))$ is a normal sequence cofinal in κ. Prove: S is stationary in κ iff $\{\xi < \mathrm{cf}(\kappa) : \alpha_\xi \in S\}$ is stationary in $\mathrm{cf}(\kappa)$.

21) Assume that κ is an uncountable regular cardinal and S is a subset of κ. Prove that the following assertions are equivalent:

   (i) S is stationary in κ.

   (ii) For every regressive function $f : S \longrightarrow \kappa$ there is an ordinal $\alpha < \kappa$ such that $f^{-1}[\alpha]$ is unbounded in κ.

   Hint: Assume that C is a club in κ such that $C \cap S = \emptyset$. Let $\sigma := \min(C)$ and define $f : S \longrightarrow \kappa$ by $f(\xi) = 0$, if $\xi \le \sigma$, $f(\xi) = \sup(C \cap \xi)$ otherwise.

22) Prove: If λ is a κ-strong cardinal with $\aleph_0 < \kappa = \mathrm{cf}(\lambda) < \lambda$, then the set

$$\{\mu < \lambda : \mu \text{ is } \kappa\text{-strong}\}$$

is a club in λ.

Hint: For unboundedness, consider a normal sequence $(\lambda_\xi : \xi < \mathrm{cf}(\kappa))$ of singular cardinals less than λ such that $\lambda = \sup\{\lambda_\xi : \xi < \mathrm{cf}(\lambda)\}$, according to 1.6.11.

23) a) If κ is an inaccessible cardinal, then the set

$$\{\mu < \kappa : \mu \text{ is strong limit}\}$$

is a club in κ_0.

b) If κ_0 is the smallest inaccessible cardinal, then the set

$$\{\lambda < \kappa_0 : \lambda \text{ is strong limit and singular}\}$$

is a club in κ.

24) Assume that κ is the α-th inaccessible cardinal, where $\alpha < \kappa$. Prove that the set $T := \{\lambda < \kappa : \lambda \text{ is regular}\}$ is not stationary in κ.

Hint: Observe that the set of inaccessible cardinals less than κ is bounded in κ, and use the previous question.

25) A cardinal κ is called a **Mahlo cardinal** iff κ is inaccessible and the set $\{\lambda < \kappa : \lambda \text{ is regular}\}$ is stationary in κ. Assume that κ is a Mahlo cardinal. Prove that $\{\lambda < \kappa : \lambda \text{ is inaccessible}\}$ is stationary in κ, and that κ is the κ-th inaccessible cardinal.

26) Assume that $\kappa := \min\{\lambda : \lambda \text{ is the } \lambda\text{-th inaccessible cardinal}\}$. Prove that κ is not a Mahlo cardinal.

Hint: Take the order isomorphism of the set of inaccessible cardinals less than κ and use Fodor's theorem to show that this set is not stationary in κ.

27) Assume that κ is a Mahlo cardinal. Prove that the set

$$\{\lambda < \kappa : \lambda \text{ is the } \lambda\text{-th inaccessible cardinal}\}$$

is unbounded in κ.

Hint: Take the order isomorphism f of the set of inaccessible cardinals less than κ – $f(\alpha)$ is the α-th inaccessible cardinal – and observe, that the set $\{\nu < \kappa : f^{-1}(\nu) \leq \nu\}$ is stationary and $\{\nu < \kappa : f^{-1}(\nu) < \nu\}$ is not stationary in κ.

28) Assume that κ is a regular uncountable cardinal, and let $X_\alpha := \kappa \setminus \alpha$ for every $\alpha < \kappa$. Prove that $\Delta_{\alpha<\kappa} X_\alpha = \kappa$ and $\bigcap_{\alpha<\kappa} X_\alpha = \emptyset$.

29) Assume that κ is a regular uncountable cardinal and $(X_\alpha : \alpha < \kappa)$ is a family of subsets of κ.

a) Let $Y_\alpha := X_\alpha \cap (\kappa \setminus (\alpha+1))$ for every $\alpha < \kappa$. Prove that $\Delta_{\alpha<\kappa} X_\alpha = \Delta_{\alpha<\kappa} Y_\alpha$.

b) Prove that $\Delta_{\alpha<\kappa} X_\alpha = \bigcap_{\alpha<\kappa}(X_\alpha \cup (\alpha+1))$.

30) Assume that κ is a regular uncountable cardinal, $f : \kappa \longrightarrow \kappa$ is a normal function and C is the set of fixed points of f which are limit ordinals. Prove that $C = \Delta_{\alpha<\kappa}(f(\alpha)+1, \kappa)_{\mathrm{ON}}$.

31) (Jech) Assume that κ and λ are cardinals such that $\omega < \kappa \le \lambda$ and $\mathrm{cf}(\kappa) = \kappa$. In analogy to the definitions for cardinals, we call a set $C \subseteq [\lambda]^{<\kappa}$ **closed** in $[\lambda]^{<\kappa}$ iff, for every chain K (in the partial ordering $([\lambda]^{<\kappa}, \subseteq)$) of members of C with $|K| < \kappa$, we have $\bigcup K \in C$. C is called **unbounded** in $[\lambda]^{<\kappa}$ iff, for every set $X \in [\lambda]^{<\kappa}$, there is a set $c \in C$, so that $X \subseteq c$. Again, every closed and unbounded subset of $[\lambda]^{<\kappa}$ is called a **club** in $[\lambda]^{<\kappa}$. A set $S \subseteq [\lambda]^{<\kappa}$ is called **stationary** in $[\lambda]^{<\kappa}$ iff $S \cap C \neq \emptyset$ for every club C in $[\lambda]^{<\kappa}$. Prove:

a) The intersection of less than κ clubs in $[\lambda]^{<\kappa}$ is a club in $[\lambda]^{<\kappa}$.

b) If $(C_\xi : \xi < \lambda)$ is a sequence of clubs in $[\lambda]^{<\kappa}$, then the diagonal intersection

$$\Delta_{\xi<\lambda} C_\xi := \{X \subseteq [\lambda]^{<\kappa} : X \in \bigcap\{C_\xi : \xi \in X\}\}$$

is a club in $[\lambda]^{<\kappa}$.

c) If $S \subseteq [\lambda]^{<\kappa}$ is stationary in $[\lambda]^{<\kappa}$ and $\Phi : S \longrightarrow \lambda$ is a choice function for S, then there is a stationary set $S_0 \subseteq S$ and an ordinal α, such that $\mathrm{ran}(\Phi \restriction S_0) = \{\alpha\}$.
Hint: The proof runs parallel to the proof of Lemma 1.8.2 and Theorem 1.8.8.

32) Assume that λ is a cardinal with uncountable cofinality and S is a stationary subset of λ. Prove that S can be partitioned into $\mathrm{cf}(\lambda)$ stationary subsets of λ.
Hint: Take a normal function $f : \mathrm{cf}(\lambda) \longrightarrow \lambda$ with $\sup(\mathrm{ran}(f)) = \lambda$ and apply Solovay's theorem.

1.9 The Erdös-Rado Partition Theorem

There are two versions of the pigeonhole principle, a finite one and an infinite one.

Pigeonhole Principle (finite version) If $t, k \in \omega \setminus \{0\}$, and if $t \cdot k+1$ pigeons fly into k holes, then one hole contains $t+1$ pigeons.

Pigeonhole Principle (infinite version) If κ is a cardinal and μ an infinite cardinal, and if we partition μ into κ sets, where $\kappa < \operatorname{cf}(\mu)$, then one set contains μ members.

The finite version of the pigeonhole principle has interesting generalizations. We illustrate this by the following example.

> At every party of six persons, there are three persons who know each other, or there are three persons who don't know each other.

For a proof let $\{0, 1, ..., 5\} = 6$ be the set of participants at the party. We partition the set $[6]^2$ of the subsets of 6 containing two elements into two classes by

$$\begin{aligned} P_0 &= \{\{i,j\} \in [6]^2 : i,j \text{ know each other}\}, \\ P_1 &= \{\{i,j\} \in [6]^2 : i,j \text{ do not know each other}\}. \end{aligned}$$

Now we fix some person, without loss of generality the person 0. The set $M := \{\{0,i\} : 1 \leq i < 6\}$ is partitioned by the partition (P_0, P_1) into two classes. Since $|M| = 5$, we can infer from the pigeonhole principle that there are three persons i_1, i_2, i_3, so that $\{0, i_1\}, \{0, i_2\}, \{0, i_3\}$ are in the same class, say P_0.

If $\{\{i_1, i_2\}, \{i_1, i_3\}, \{i_2, i_3\}\} \cap P_0 \neq \emptyset$ and, without loss of generality, $\{i_1, i_2\} \in P_0$, then 0, i_1 and i_2 know each other. On the other hand, if $\{\{i_1, i_2\}, \{i_1, i_3\}, \{i_2, i_3\}\} \cap P_0 = \emptyset$, then i_1, i_2 and i_3 do not know each other.

Definition *Let $g : A \longrightarrow B$ be a function. Then we know that the set $\{g^{-1}[\{b\}] : b \in \operatorname{ran}(g)\}$ is a partition of A; thus we will speak of the* **partition g**.

If κ, λ, μ and ν are cardinals such that $\nu \leq \mu$ and $\lambda \leq \mu$, then we write

$$\mu \longrightarrow (\lambda)^\nu_\kappa$$

for the following assertion: If K and M are sets with $|M| = \mu$ and $|K| = \kappa$, and if $f : [M]^\nu \longrightarrow K$ is a function, then there is a subset L of M with $|L| = \lambda$ and an element $k_0 \in K$ such that $f[[L]^\nu] = \{k_0\}$. L is called **homogeneous** *with respect to the partition f.*

In the new notation, our previous example can be written as $6 \longrightarrow (3)^2_2$. The pigeonhole principles can be formulated as follows.

Pigeonhole Principle (finite version) If $t, k \in \omega \setminus \{0\}$, then

$$t \cdot k + 1 \longrightarrow (t+1)^1_k .$$

Pigeonhole Principle (infinite version) If κ and μ are cardinal numbers with $1 \leq \kappa < \mathrm{cf}(\mu)$, then

$$\mu \longrightarrow (\mu)^1_\kappa .$$

In this section we will prove, for $n < \omega$ and $\mu \in \mathrm{ICN}$, a theorem of the form

$$(*) \qquad \mu \longrightarrow (\lambda)^n_\kappa \quad \Longrightarrow \quad \exists \nu \in \mathrm{ICN}(\nu \longrightarrow (\lambda)^{n+1}_\kappa) .$$

An essential step of the proof is the following lemma.

Lemma 1.9.1 *Assume that $n \in \omega$ and $\mu \in \mathrm{ICN}$ satisfy $\mu \longrightarrow (\lambda)^n_\kappa$. Further let $\nu \in \mathrm{ICN}$ and $f : [\nu]^{n+1} \longrightarrow \kappa$ be a function. If $(\tau_i : i < \mu)$ is an injective function from μ into ν satisfying*

$$f(\{\tau_{i_1}, \ldots, \tau_{i_n}, \tau_j\}) = f(\{\tau_{i_1}, \ldots, \tau_{i_n}, \tau_{i_n+1}\})$$

for all $i_1 < \ldots < i_n < j < \mu$, then there is a set $M_0 \subseteq \{\tau_i : i < \mu\}$ with $|M_0| = \lambda$, and $k < \kappa$ such that $f[[M_0]^{n+1}] = \{k\}$.

Proof Let us assume that a function $(\tau_i : i < \mu)$ with the above property exists. If $i_1 < \ldots < i_n < \mu$, then we put

$$g(\{\tau_{i_1}, \ldots, \tau_{i_n}\}) := f(\{\tau_{i_1}, \ldots, \tau_{i_n}, \tau_{i_n+1}\}).$$

From the assumption $\mu \longrightarrow (\lambda)^n_\kappa$ we can infer that there is a set $M_0 \subseteq \{\tau_i : i < \mu\}$ with $|M_0| = \lambda$ and $k \in \kappa$, such that $g[[M_0]^n] = \{k\}$. If $i_1 < \ldots < i_n < i_{n+1} < \mu$ and $\tau_{i_1}, \ldots, \tau_{i_n}, \tau_{i_{n+1}} \in M_0$, then the assumption about f gives us

$$f(\{\tau_{i_1}, \ldots, \tau_{i_{n+1}}\}) = g(\{\tau_{i_1}, \ldots, \tau_{i_n}\}) = k.$$

Lemma 1.9.1 shows that, for the solution of $(*)$, we must solve the following problem: If $f : [\nu]^{n+1} \longrightarrow \kappa$ is a partition, find an injective sequence $(\tau_i : i < \mu)$ of elements of ν, such that

$$f(\{\tau_{i_1}, \ldots, \tau_{i_n}, \tau_j\}) = f(\{\tau_{i_1}, \ldots, \tau_{i_n}, \tau_{i_n+1}\})$$

for all $i_1 < \ldots < i_n < j < \mu$. Note that we did not yet say anything about the size of ν.

Let us solve the problem with a tree argument.

Definition *A partial ordering* $(T,<)$ *is called a* **tree** *iff for every* $x \in T$ *the set* $\hat{x} := \{y \in T : y < x\}$ *is well-ordered by* $<$. *A* **branch** *in* $(T,<)$ *is a* $\subseteq$*-maximal chain in* $(T,<)$*; thus every branch is well-ordered by* $<$. *The order type of this well-ordering is the* **length of the branch**. *We define the* **rank function** $\rho : T \longrightarrow \mathrm{ON}$ *of* $(T,<)$ *by* $\rho(x) := \sup\{\rho(y)+1 : y \in T \wedge y < x\}$. *The ordinal* $\alpha := \sup\{\rho(x)+1 : x \in T\} =: l(T)$ *is called the* **length of the tree**. *An* **α-tree** *is a tree of length* α. *The set* $S_\beta := \{x \in T : \rho(x) = \beta\}$ *is the* **β-th level** *of* $(T,<)$. *Note that* $S_\beta = \emptyset$ *iff* $\beta \geq l(T)$.

To prove a theorem of the form $(*)$, let us assume that $f : [\nu]^{n+1} \longrightarrow \kappa$ is a partition and $\nu \in \mathrm{ICN}$ is chosen large enough to carry through the following construction. We define by recursion a tree T of length μ whose elements are nonempty subsets of ν, and define the relation $<$ by

$$A < B :\Longleftrightarrow A \supsetneqq B.$$

It is our aim that, for all $\alpha < \mu$, we can arrange the following for the levels S_α of the tree $(T,<)$.

(i) $\bigcup S_\alpha = \nu \setminus \{\min(A) : A \in \bigcup\{S_\beta : \beta < \alpha\}\}$.
(ii) Any two different elements of S_α are disjoint.
(iii) If $\xi < \alpha$ and $B \in S_\alpha$, then there is a unique set $A \in S_\xi$ with $A < B$.

For this, we define by recursion the levels S_α of the tree T. We put $S_0 = \{\nu\}$. For the successor step, let us assume that S_ξ is already defined and satisfies properties (i)–(iii) for all ξ with $\xi \leq \alpha$. If $B \in S_\alpha$, then, by (i)–(iii), there exists a unique chain of sets such that

$$\nu = A_0 \supsetneqq A_1 \supsetneqq \ldots \supsetneqq A_\xi \supsetneqq \ldots \supsetneqq A_\alpha = B$$

and $A_\xi \in S_\xi$ for all $\xi \leq \alpha$. We put $\tau_\xi := \min(A_\xi)$ for $\xi \leq \alpha$ and define on $B \setminus \{\tau_\alpha\}$ a relation $\sim_B$ by

$$\rho \sim_B \sigma :\Longleftrightarrow$$
$$f(\{\tau_{\xi_1}, \ldots, \tau_{\xi_n}, \rho\}) = f(\{\tau_{\xi_1}, \ldots, \tau_{\xi_n}, \sigma\}) \text{ for all } \xi_1 < \ldots < \xi_n \leq \alpha.$$

It is easy to check that $\sim_B$ is an equivalence relation. Let $\mathcal{E}_B$ be the set of its equivalence classes, and let

$$S_{\alpha+1} := \bigcup\{\mathcal{E}_B : B \in S_\alpha\}$$

be the $\alpha+1$-th level of T. By construction, $S_{\alpha+1}$ satisfies properties (i)–(iii).

We will give an estimate for $|S_{\alpha+1}|$. Obviously we have $|S_{\alpha+1}| = \sum(|\mathcal{E}_B| : B \in S_\alpha)$. For $B \in S_\alpha$ and $\rho \in B \setminus \{\tau_\alpha\}$ we consider the function $g_\rho : [\{\tau_\xi : \xi \leq \alpha\}]^n \longrightarrow \kappa$, defined by the equation $g_\rho(\{\tau_{\xi_1}, \ldots, \tau_{\xi_n}\}) := f(\{\tau_{\xi_1}, \ldots, \tau_{\xi_n}, \rho\})$. Since $\rho \sim_B \sigma$ iff $g_\rho = g_\sigma$, the number of equivalence classes is majorized by the number of functions from $[\{\tau_\xi : \xi \le \alpha\}]^n$ into κ, and thus, using Theorem 1.5.14, by $\kappa^{\max\{\aleph_0,|\alpha|\}}$. So we get

(iv) $|S_{\alpha+1}| \le |S_\alpha| \cdot \kappa^{|\alpha|+\aleph_0} \wedge (\kappa < \aleph_0 \wedge \alpha < \aleph_0 \Longrightarrow |S_{\alpha+1}| < \aleph_0)$.

Now let $\alpha < \mu$ be a limit ordinal. Take

$$S_\alpha := \{\bigcap_{\xi<\alpha} h(\xi) : h \in \prod_{\xi<\alpha} S_\xi\} \setminus \{\emptyset\}.$$

Putting $T_\alpha := \bigcup\{S_\xi : \xi < \alpha\}$, we can infer from property (ii) that S_α is the set of nonempty intersections of branches of length α in the tree T_α. Using transfinite induction, we show that properties (i)–(iii) hold for S_α. So assume that they are satisfied for every level S_ξ, $\xi < \alpha$. For (i), let $\rho \in \nu \setminus \{\min(A) : A \in \bigcup_{\xi<\alpha} S_\xi\}$. If $\xi < \alpha$, then $\rho \in \nu \setminus \{\min(A) : A \in \bigcup_{\beta<\xi} S_\beta\}$, and thus the inductive hypothesis and (i) give $\rho \in \bigcup S_\xi$. By (ii), there is a unique set A_ξ satisfying $A_\xi \in S_\xi$ and $\rho \in A_\xi$. Let the function $h \in \prod_{\xi<\alpha} S_\xi$ be defined by $h(\xi) := A_\xi$ for $\xi < \alpha$. Then $\rho \in \bigcap\{h(\xi) : \xi < \alpha\}$, and thus $\rho \in \bigcup S_\alpha$.

For the proof of the reverse inclusion, let $\rho \in \bigcup S_\alpha$. Then there is a function $h \in \prod_{\xi<\alpha} S_\xi$ such that $\rho \in \bigcap\{h(\xi) : \xi < \alpha\}$. Fix $\xi < \alpha$. If $A \in S_\xi$ and $A \ne h(\xi)$, then $\rho \notin A$ by (ii), and thus $\rho \ne \min(A)$. If $A \in S_\xi$ and $A = h(\xi)$, then $\min(A) = \min(h(\xi)) \notin h(\xi+1)$ by (i), and from $\rho \in h(\xi+1)$ we can conclude that $\rho \ne \min(A)$. This shows that $\rho \in \nu \setminus \{\min(A) : A \in \bigcup\{S_\xi : \xi < \alpha\}\}$, and (i) is proved for S_α.

Properties (ii) and (iii) follow from the inductive hypothesis and the construction. Altogether we have shown that $(T, <)$ is a tree with the desired properties. It is obvious that

(v) $\mathrm{Lim}(\alpha) \Longrightarrow |S_\alpha| \le \prod_{\xi<\alpha} |S_\xi|$.

Let us assume that T is a μ-tree. If we can show that T possesses a branch $(A_i : i < \mu)$ of length μ, we can complete the proof of $(*)$. In this case put $\tau_i := \min(A_i)$ for each $i < \mu$. Assume that $i_1 < \ldots < i_n < j < \mu$. Then $i_n + 1 \le j$, and from $A_j \subseteq A_{i_n}$ and $A_{i_n+1} \subseteq A_{i_n}$ we can infer that $\tau_j, \tau_{i_n+1} \in A_{i_n}$. Consequently, τ_j and τ_{i_n+1} are elements of the same equivalence class, that means we have

$$f(\{\tau_{i_1}, \ldots, \tau_{i_n}, \tau_j\}) = f(\{\tau_{i_1}, \ldots, \tau_{i_n}, \tau_{i_n+1}\}).$$

Now, assuming $\mu \longrightarrow (\lambda)^n_\kappa$, we can apply Lemma 1.9.1, and $(*)$ is proved.

There remains to show, under an appropriate choice of ν, that T has a branch of length μ.

First, we prove two lemmas, due to D. König and Ramsey.

Lemma 1.9.2 (D. König) *If T is an ω-tree with finite levels only, then T has a branch of length ω.*

Proof As usual let S_n, $n \in \omega$, be the levels of T. By induction, we define an ω-branch $(x_i : i < \omega)$ of T. Since T is infinite and S_0 is finite, we can choose $x_0 \in S_0$ such that $|\{x \in T : x_0 < x\}| = \aleph_0$, using the pigeonhole principle. Assume that $x_n \in S_n$ with $|\{x \in T : x_n < x\}| = \aleph_0$ is already defined. Applying the previous argument to the sets $\{x \in T : x_n < x\}$ and S_{n+1}, we obtain an element x_{n+1} of S_{n+1} satisfying $x_n < x_{n+1}$ and $|\{x \in T : x_{n+1} < x\}| = \aleph_0$. The sequence $(x_n : n < \omega)$ is a branch of T of length ω.

Theorem 1.9.3 (Ramsey) *If $k, n \in \omega \setminus \{0\}$, then*

$$\aleph_0 \longrightarrow (\aleph_0)^n_k.$$

Proof Consider the above construction of the tree T and take $\mu = \aleph_0$, $\nu = \aleph_0$ and $\kappa = k < \omega$. Property (iv) shows that every level of the tree T is finite, hence T is an ω-tree. Lemma 1.9.2 says that T has an ω-branch, and so we can conclude that

$$\aleph_0 \longrightarrow (\aleph_0)^n_k \quad \Longrightarrow \quad \aleph_0 \longrightarrow (\aleph_0)^{n+1}_k.$$

Since $\aleph_0 \longrightarrow (\aleph_0)^1_k$ is obviously true, an $(n-1)$-times application of this assertion proves the lemma.

Theorem 1.9.4 (Erdös and Rado) *Assume that μ and λ are infinite cardinals with $\lambda \leq \mu$. Let further κ be a cardinal with $0 < \kappa < \mu$ and $n \geq 1$ be a natural number. Then*

$$\mu \longrightarrow (\lambda)^n_\kappa \quad \Longrightarrow \quad (2^{<\mu})^+ \longrightarrow (\lambda)^{n+1}_\kappa.$$

Proof If $\mu = \aleph_0$, then $\lambda = \aleph_0$, since $\lambda \leq \mu$. The proof of Theorem 1.9.3 shows that

$$\aleph_0 \longrightarrow (\aleph_0)^n_\kappa \quad \Longrightarrow \quad \aleph_0 \longrightarrow (\aleph_0)^{n+1}_\kappa,$$

and thus all the more

$$\aleph_0 \longrightarrow (\aleph_0)^n_\kappa \quad \Longrightarrow \quad (2^{<\aleph_0})^+ \longrightarrow (\aleph_0)^{n+1}_\kappa.$$

So we can assume that $\mu > \aleph_0$. Let T be the μ-tree constructed above. We use induction on α to show that $|S_\alpha| \leq 2^{|\alpha|+\aleph_0+\kappa}$ for all $\alpha < \mu$. Clearly, $|S_0| = 1 < 2^{\aleph_0+\kappa}$. For the successor step we obtain, using property (iv):

$$|S_{\alpha+1}| \leq |S_\alpha| \cdot \kappa^{|\alpha|+\aleph_0} \leq 2^{|\alpha|+\aleph_0+\kappa} \cdot \kappa^{|\alpha|+\aleph_0+\kappa} = 2^{|\alpha|+\aleph_0+\kappa} = 2^{|\alpha+1|+\aleph_0+\kappa}.$$

If $\alpha < \mu$ is a limit ordinal, we can conlude with property (v) that $|S_\alpha| \leq \prod_{\xi<\alpha} |S_\xi| \leq \prod_{\xi<\alpha} 2^{|\xi|+\aleph_0+\kappa} \leq \prod_{\xi<\alpha} 2^{|\alpha|+\aleph_0+\kappa} = 2^{|\alpha|+\aleph_0+\kappa}$. So we get

$$\begin{aligned} |\bigcup\{S_\alpha : \alpha < \mu\}| &\leq \mu \cdot \sup\{|S_\alpha| : \alpha < \mu\} \\ &\leq \mu \cdot \sup\{2^{|\alpha|+\aleph_0+\kappa} : \alpha < \mu\} \leq \mu \cdot 2^{<\mu} = 2^{<\mu}, \end{aligned}$$

where the last equation follows from Corollary 1.6.5.

Now we put $\nu := (2^{<\mu})^+$ and define

$$S_\mu := \{\bigcap_{\xi<\mu} h(\xi) : h \in \prod_{\xi<\mu} S_\xi\} \setminus \{\emptyset\}.$$

As in the proof of (i) it can be shown that $\bigcup S_\mu = \nu \setminus \{\min(A) : A \in \bigcup\{S_\xi : \xi < \mu\}\ \}$. Clearly we have $|\{\min(A) : A \in \bigcup\{S_\xi : \xi < \mu\}\}| \leq |\bigcup\{S_\xi : \xi < \mu\}| \leq 2^{<\mu}$. So the choice of ν yields $\bigcup S_\mu \neq \emptyset$, hence $S_\mu \neq \emptyset$, and therefore T has a branch $(A_\xi : \xi < \mu)$ of length μ. The above considerations show that this completes the proof of the theorem.

Corollary 1.9.5 (Erdös and Rado) *If κ is an infinite cardinal number, then*

$$(2^\kappa)^+ \longrightarrow (\kappa^+)^2_\kappa.$$

Proof Obviously, we have $\kappa^+ \longrightarrow (\kappa^+)^1_\kappa$, and from Theorem 1.9.4 we can infer that $(2^{<\kappa^+})^+ \longrightarrow (\kappa^+)^2_\kappa$. Since $2^{<\kappa^+} = 2^\kappa$, the assertion is proved.

Exercises

1) Every infinite partial ordering has an infinite chain or an infinite antichain.

   Hint: Use Ramsey's theorem.
2) Show that $2^\kappa \not\longrightarrow (3)^2_\kappa$ for all cardinals $\kappa \geq 1$.
3) Assume that κ is an infinite cardinal, and let $\exp_n(\kappa)$ be defined by $\exp_0(\kappa) := \kappa$ and $\exp_{n+1}(\kappa) := 2^{\exp_n(\kappa)}$ for $n \in \omega$. Show that

   $$(\exp_n(\kappa))^+ \longrightarrow (\kappa^+)^{n+1}_\kappa .$$

   Hint: Use Theorem 1.9.4.
4) Assume that $\kappa \in \mathrm{ICN}$ and $f : [\kappa]^2 \longrightarrow 2$ is a partition. Prove: If there exists, for every set $A \in [\kappa]^\kappa$, a member a of A such that $|\{x \in A : f(\{a, x\}) = 1\}| = \kappa$, then there is a set $C \in [\kappa]^{\aleph_0}$ such that $\mathrm{ran}(f \upharpoonright [C]^2) = \{1\}$.
5) (Dushnik, Miller) Assume that $\kappa \in \mathrm{ICN}$ and $f : [\kappa]^2 \longrightarrow 2$ is a partition. Show that there is a subset X of κ such that $|X| = \kappa$ and $\mathrm{ran}(f \upharpoonright [X]^2) = \{0\}$ or $|X| = \aleph_0$ and $\mathrm{ran}(f \upharpoonright [X]^2) = \{1\}$.

   Hint: Assume that the assertion is true for all $\mu \in \kappa \cap \mathrm{ICN}$ and that there no set $C \in [\kappa]^{\aleph_0}$ such that $\mathrm{ran}(f \upharpoonright [C]^2) = \{1\}$. By Exercise 4 there is a set $A \in [\kappa]^\kappa$ such that $|\{\alpha \in A : f(\{a, \alpha\}) = 1\}| < \kappa$ for every $a \in A$. Now distinguish the cases that κ is regular and κ is singular and apply the inductive hypothesis in the second case.

6) Show that there is a partition $f : [\omega]^{<\omega} \longrightarrow \{0,1\}$ such that, for every infinite subset x of ω, $f^{-1}[\{0\}] \cap [x]^{<\omega} \neq \emptyset \neq f^{-1}[\{1\}] \cap [x]^{<\omega}$. So we get $\omega \not\longrightarrow (\omega)^{<\omega}_2$.
   Hint: If $x \in [\omega]^{<\omega}$, consider the property $|x| \in x$.
7) Show that there is a partition $f : [\omega]^{\omega} \longrightarrow \{0,1\}$ such that, for every infinite subset $x \subseteq \omega$, $f^{-1}[\{0\}] \cap [x]^{\omega} \neq \emptyset \neq f^{-1}[\{1\}] \cap [x]^{\omega}$. So we get $\omega \not\longrightarrow (\omega)^{\omega}_2$.
   Hint: Choose a well-ordering $\prec$ of $[\omega]^{\omega}$ and consider the property of a set $y \in [\omega]^{\omega}$ that there is a set $z \in [y]^{\omega}$ with $z \prec y$.
8) Show that there is a partition $f : [\mathbb{R}]^2 \longrightarrow \{0,1\}$ such that, for every subset x of $\mathbb{R}$ with $|x| = \aleph_1$,

$$f^{-1}[\{0\}] \cap [x]^2 \neq \emptyset \neq f^{-1}[\{1\}] \cap [x]^2.$$

   So we get $2^{\aleph_0} \not\longrightarrow (\aleph_1)^2_2$. In particular, the obvious generalization of Ramsey's theorem, namely $\aleph_1 \longrightarrow (\aleph_1)^2_2$, is false.
   Hint: Let $<$ be the usual ordering on $\mathbb{R}$ and $<^*$ be a well-ordering of $\mathbb{R}$. If $\{x, y\} \in [\mathbb{R}]^2$ and $x < y$, then $x <^* y$ or $y <^* x$.
9) (Kurepa, Sierpinski) For every infinite cardinal κ,

$$2^{\kappa} \not\longrightarrow (\kappa^+)^2_2.$$

   Hint: In the exercises of Section 1.5 we have seen that there is no strictly increasing or decreasing sequence of length κ^+ in $P := {}^{\kappa}\{0,1\}$ under the lexicographical ordering. Let $2^{\kappa} =: \lambda$ and $\{f_\alpha : \alpha < \lambda\}$ be an enumeration of ${}^{\kappa}\{0,1\}$, and take $\prec$ as the linear ordering on λ induced by the lexicographical ordering on P. Now define a partition by $F(\{\alpha, \beta\}) := 1$ iff $\alpha \in \beta$ and $\alpha \prec \beta$, $F(\{\alpha, \beta\} := 0$ otherwise.
10) A cardinal κ is called **weakly compact** iff it is uncountable and satisfies the partition property $\kappa \longrightarrow (\kappa)^2_2$. Prove that every weakly compact cardinal is inaccessible.
   Hint: To show that κ is regular, assume that κ is the disjoint union of nonempty sets A_ξ, $\xi < \lambda$, where $\lambda < \kappa$ and $|A_\xi| < \kappa$ for all $\xi < \lambda$, and consider the partition of $[\kappa]^2$ given by this partition of κ. To show that κ is strong limit, assume that $\kappa \leq 2^{\lambda}$ for some $\lambda < \kappa$, and use the previous question.

Remark An infinite cardinal κ is called a **Ramsey cardinal** iff it satisfies the partition property $\kappa \longrightarrow (\kappa)^{<\omega}_2$. Exercise 6 shows that ω is not a Ramsey cardinal.
One can prove (see for example Jech's book [Je2]): Every measurable cardinal is a Ramsey cardinal, and every Ramsey cardinal is weakly compact and hence inaccessible, see above.

Chapter 2

The Galvin-Hajnal Theorem

The so-called **singular cardinal problem** consists of the description of the possible size of the cardinal $\aleph_\eta^{\mathrm{cf}(\aleph_\eta)}$, that is $\gimel(\aleph_\eta)$, the value of the gimel function at the argument $\aleph_\eta$, for singular cardinals $\aleph_\eta$. An estimate for this cardinal power is given by the Galvin-Hajnal theorem if $\aleph_\eta$ is an $\aleph_0$-strong singular cardinal with uncountable cofinality. The centre of our investigations will be the Galvin-Hajnal formula, from which all other results on cardinals in this chapter will follow. For the first time it turns out that a profound cardinal property is a source of cardinal arithmetic.

2.1 Ideals and the Reduction of Relations

In the following let A be a nonempty set and I be an ideal on A, that is a nonempty set of subsets of A which does not contain the member A and is closed under finite unions and under inclusions. Any ideal on A provides us with a measure of the size of a set. Elements of I are subsets of A which we can regard as *small.* We also say that they are of measure zero modulo I. If κ is an uncountable regular cardinal, then $\mathrm{I_b}(\kappa)$ and $\mathrm{I_d}(\kappa)$, the set of bounded and the set of thin subsets of κ respectively, are two ideals on κ which will play an important role in this book.

We remind the reader of some denotations. If I is an ideal on the set A, then $\mathrm{dom}(I) := \bigcup I$ is the domain of I and $I^+ := \{X \subseteq A : X \notin I\}$ is the set of subsets of A which are positive modulo I. We say that the sets $X \in I$ are **null sets** or **of measure zero** modulo I. Any union of two positive sets or null sets is positive or a null set, respectively, any intersection of two null sets is a null set, subsets of null sets are null sets, and supersets of positive sets are positive. If X is positive and Y is a null set, then $X \setminus Y$ is positive. Since $I \neq \mathcal{P}(A)$, the set $(A \setminus X) \cap (A \setminus Y)$ is positive for any null sets X and Y. Note that X and $A \setminus X$ can both be positive.

Definition *We say that a property $\varphi(a)$* **holds modulo I for almost all $a \in A$** *iff the set $\{a \in A : \neg\varphi(a)\}$, that means the set of exceptions, is a member of I. The elements of the set*

$$I^+ := \{X \subseteq A : X \notin I\}$$

are called **I-positive sets** *or* **sets of positive measure** *modulo I.*

Furthermore we say that an ideal J **extends** *an ideal I iff $I \subseteq J$.*

Often we will consider ultrafilters instead of maximal ideals. If D is an ultrafilter on A and I is an ideal on A with $D \cap I = \emptyset$, then I is a subset of the maximal ideal dual to D. This justifies the phrase **D extends I** *for ultrafilters D on A with $D \cap I = \emptyset$.*

To give an example, let us consider the assertion $B \subseteq_I C$. It is to be read as "B is a subset of C modulo I", for B and C subsets of A. $B \subseteq_I C$ means that modulo I for almost all $a \in A$ the implication $a \in B \implies a \in C$ is true. We are mainly interested in such properties later on, when we will investigate products of structures and especially products of ordinals with the standard well-ordering $\in$.

Definition *Let A be a nonempty set. Remember that we have called functions from A into* ON **ordinal functions** *on A. We will omit the addition "on A" when the context makes it plain which basic set is meant. Usually,* **we denote ordinal functions by f, g, h,** *also with indices. In this chapter, families of almost disjoint functions $\mathcal{F}, \mathcal{G}, \ldots$, whose members we also denote by $f, g, \ldots$, will play an important role. Therefore* **we will denote in this chapter**, *and sometimes in later chapters,* **ordinal functions also by Φ** *and Ψ.*

If f and g are ordinal functions on A and I is an ideal on A, then we say that f is **modulo I less** *than g, and write $f <_I g$, iff $f(a) < g(a)$ for almost all $a \in A$; that means that the set*

$$\{a \in A : \neg(f(a) < g(a))\}, \text{ i.e., } \{a \in A : g(a) \leq f(a)\},$$

is a member of the ideal I.

In particular if $I = \{\emptyset\}$, we write $f <_A g$ instead of $f <_I g$ or simply $f < g$ if it is clear which basic set is meant; then f is **pointwise less** *than g on A, that means $f(a) < g(a)$ holds for* all $a \in A$. *If $B \subseteq A$, we write $f <_B g$ for $f \restriction B <_B g \restriction B$. Furthermore we put*

$$
\begin{array}{lll}
B \subseteq_I C & :\Longleftrightarrow & B \setminus C \in I, \\
B =_I C & :\Longleftrightarrow & (B \setminus C) \cup (C \setminus B) \in I, \\
f =_I g & :\Longleftrightarrow & \{a \in A : f(a) \neq g(a)\} \in I, \\
f \leq_I g & :\Longleftrightarrow & \{a \in A : g(a) < f(a)\} \in I, \\
f <_I g & :\Longleftrightarrow & \{a \in A : g(a) \leq f(a)\} \in I, \\
f \lneqq_I g & :\Longleftrightarrow & f \leq_I g \wedge \neg(f =_I g).
\end{array}
$$

Remark The definition of the relation $\lneqq_I$ is a bit out of the ordinary, since it is not a property that holds almost everywhere, but only a property that holds for all members of an I-positive set. We have included it for reasons which will be discussed later on.

Lemma 2.1.1 *Assume that A is a nonempty set and I is an ideal on A. Then for all subsets B, C and D of A, the following holds.*

1. $B =_I B$.
2. $B =_I C \Longrightarrow C =_I B$.
3. $B =_I C \wedge C =_I D \Longrightarrow B =_I D$.
4. $B = C \Longrightarrow B =_I C$.
5. $B \subseteq_I B$.
6. $B \subseteq_I C \wedge C \subseteq_I D \Longrightarrow B \subseteq_I D$.
7. $B \subseteq C \Longrightarrow B \subseteq_I C$.
8. $B =_I C \Longleftrightarrow B \subseteq_I C \wedge C \subseteq_I B$.
9. $B \subseteq_I C \Longleftrightarrow A \setminus C \subseteq_I A \setminus B$.
10. $B \subseteq_I C \wedge C \in I \Longrightarrow B \in I$.

Lemma 2.1.2 *Assume that A is a nonempty set and I is an ideal on A. Then for all ordinal functions f, g and h on A, the following holds.*

1. $f =_I f$.
2. $f =_I g \Longrightarrow g =_I f$.
3. $f =_I g \wedge g =_I h \Longrightarrow f =_I h$.
4. $f = g \Longrightarrow f =_I g$.
5. $f \leq_I f$.
6. $f \leq_I g \wedge g \leq_I f \Longleftrightarrow f =_I g$.
7. $f \leq_I g \wedge g \leq_I h \Longrightarrow f \leq_I h$.
8. $f \leq g \Longrightarrow f \leq_I g$.
9. $f \lneqq_I g \Longrightarrow \{a \in A : f(a) < g(a)\} \in I^+$.
10. $\neg(f <_I f)$.
11. $f <_I g \wedge g <_I h \Longrightarrow f <_I h$.
12. $f < g \Longrightarrow f <_I g$.
13. $f <_I g \Longrightarrow f \lneqq_I g$.
14. $f \lneqq_I g \Longrightarrow f \leq_I g$.
15. $f \leq_I g \wedge g \lneqq_I h \Longrightarrow f \lneqq_I h$.
16. $f \leq_I g \wedge g <_I h \Longrightarrow f <_I h$.

Remark Note that in general the relations $\subseteq_I$ and $\leq_I$ are not partial orderings; they are only antisymmetric modulo I in the sense of property 8. and property 6. from the preceding lemmas, respectively. In the same way the equivalence

$$f <_I g \Longleftrightarrow f \leq_I g \wedge \neg(f =_I g)$$

in general does *not* hold. This is the reason for the definition of the relation $\lneqq_I$.

One could avoid this problem by introducing reduced products, as is the common practice in Model Theory or Algebra. But our way of proceeding will be considerably more practical for our purposes.

Lemma 2.1.3 *Assume that A is a nonempty set, I is an ideal on A, and J is an ideal on A extending I. Furthermore let $\varphi(x)$ be a formula such that $\varphi(a)$ holds modulo I for almost all $a \in A$. Then $\varphi(a)$ holds modulo J for almost all $a \in A$. In particular this applies to maximal ideals J which are dual to ultrafilters extending I. Thus, if D is an ultrafilter on A extending I, then*

$$\{a \in A : \varphi(a)\} \in D.$$

Remark By the previous lemma, the relations $=_I$, $\subseteq_I$, $\leq_I$ and $<_I$ are "transmitted" from I to extensions J of I. But this is not true in general for the relation $\lneqq_I$.

If A is a nonempty set, I is an ideal on A, and B is a subset of A, then we have by Section 1.8: There is an ideal J on A such that $I \subseteq J$ and $(A \setminus B) \in J$ iff $B \notin I$ (that means iff $\neg(A \setminus B =_I A)$). We denoted the smallest ideal with this property by $I \upharpoonright B$. The following lemma shows the connection between this ideal and the ideal $I \cap \mathcal{P}(B)$.

Lemma 2.1.4 *Asssume that A is a nonempty set, I is an ideal on A, and $B \in I^+$. If f and g are ordinal functions on A, then*

$$f <_{I \upharpoonright B} g \iff f \upharpoonright B <_{I \cap \mathcal{P}(B)} g \upharpoonright B.$$

An analogous assertion is true for the relations $=_{I \upharpoonright B}$ and $\leq_{I \upharpoonright B}$.

If we replace the ideal I by an ideal of the form $I \upharpoonright B$, then we continue to speak about functions which are defined on A but are only interested in their values at elements of B.

Definition *We define the* **pointwise supremum** *of a set $\mathcal{S}$ of ordinal functions on A as the function* $\sup(\mathcal{S}) : A \longrightarrow \mathrm{ON}$ *with*

$$\sup(\mathcal{S})(a) := \sup\{f(a) : f \in \mathcal{S}\}$$

for all $a \in A$. Analogously the functions $\max(\mathcal{S})$ *(for finite sets $\mathcal{S}$) and* $\min(\mathcal{S})$ *are defined. Finally let $f+1$ be the ordinal function on A given by*

$$(f+1)(a) := f(a) + 1.$$

All notions defined in this section for ideals can be defined for filters F if one takes the respective definition for the ideal dual to F. If for example F is a filter on A, then

$$f <_F g \iff \{a \in A : f(a) < g(a)\} \in F.$$

If I is maximal, then the relations $\leq_I$ and $<_I$ are linear orderings modulo I. Reformulating this by means of filters, we obtain the following lemma.

Lemma 2.1.5 *Let A be a nonempty set and D be an ultrafilter on A. If f and g are ordinal functions on A, then*

a) $f \leq_D g \vee g \leq_D f$.

b) $f \leq_D g \Longleftrightarrow f <_D g \vee f =_D g$.

c) $f <_D g \vee f =_D g \vee g <_D f$.

d) $f <_D g \Longleftrightarrow f \lneqq_D g$.

Exercises

1) Show: If I is an ideal on A and $X \in I$, then $X \subseteq_I Y$ for every subset Y of A.
2) Show: If I is an ideal on A, $X \in \mathcal{P}(A) \setminus I$, and $Z \in I$, then $I \restriction (X \cup Z) = I \restriction X$.
3) Assume that I is an ideal on A and $X_1, X_2 \in \mathcal{P}(A) \setminus I$. Show that
$$I \restriction (X_1 \cup X_2) = I \restriction X_1 \cap I \restriction X_2.$$
4) a) Give examples of a set A, an ideal I on A, and functions $f, g \in {}^{A}\mathrm{ON}$ such that
   a1) $f \lneqq_I g$ and not $f <_I g$.
   a2) $f \leq_I g$ and not $f \lneqq_I g$.
   b) Prove that the relations $\lneqq_I$ and $<_I$ coincide for maximal ideals I on a set A.
5) Assume that I is an ideal on A, $X \in \mathcal{P}(A) \setminus I$, and D is an ultrafilter on A. Prove that $X \in D$ and $D \cap I = \emptyset$ iff $D \cap I \restriction X = \emptyset$.
6) The dual version of a principal filter is a principal ideal. We call an ideal on A **principal** iff there is a subset X_0 of A such that $I = \{Y \subseteq A : Y \cap X_0 = \emptyset\}$. Prove:
   a) If κ is an infinite cardinal and $|A| \geq \kappa$, then the set $I := \{Y \subseteq A : |Y| < \kappa\}$ is a nonprincipal ideal on A.
   b) Every principal ideal on an infinite cardinal is normal.
7) Assume that I is an ideal on A. Show that I is principal iff $\bigcup I \in I$.
8) Assume that A is an infinite set and B is a nonempty subset of A. Further let I be an ideal on B and, for each $b \in B$, I_b be an ideal on A. Prove that
$$J := \{X \subseteq A : \{b \in B : X \notin I_b\} \in I\}$$
is an ideal on A.

9) If I is an ideal on the cardinal κ, then 0 is not an element of any diagonal union.
10) Assume that I is an ideal on A, γ is an ordinal, and $(d_\alpha : \alpha < \gamma)$ is a sequence of I-positive subsets of A such that $d_\beta \subseteq_I d_\alpha$ for all $\alpha, \beta \in \gamma$ with $\alpha < \beta$. Prove that there is an ultrafilter D on A such that $D \cap I = \emptyset$ and $d_\alpha \in D$ for all $\alpha < \gamma$.
Hint: Prove that $\{d_\alpha : \alpha < \gamma\} \cup \{Y \subseteq A : A \setminus Y \in I\}$ has the finite intersection property.
11) If κ is a singular cardinal, then there is no normal ideal on κ which contains all bounded subsets of κ.
Hint: As usual, κ is the supremum of a strictly increasing sequence $(\kappa_\alpha : \alpha < \mathrm{cf}(\kappa))$ of cardinals with $\kappa_0 \geq \mathrm{cf}(\kappa)$. Assume that there is such an ideal. Consider the sequence $(X_\alpha : \alpha < \kappa)$, where $X_\alpha := \kappa_\alpha$ for $\alpha < \mathrm{cf}(\kappa)$, and show that $\kappa \setminus \mathrm{cf}(\kappa) \subseteq \nabla_{\alpha < \mathrm{cf}(\kappa)} X_\alpha$.
12) (Ulam) Assume that κ is an infinite cardinal, $\kappa = \lambda^+$, and I is a κ-complete ideal on κ with $\mathrm{dom}(I) = \kappa$. Prove that there is a sequence $(X_\xi : \xi < \kappa)$ of pairwise disjoint I-positive subsets of κ.
Hint: Choose an injective function $f_\eta : \eta \longrightarrow \lambda$ for each $\eta < \kappa$, and define the $\kappa \times \lambda$-matrix $(X_{\alpha,\xi})$ by $X_{\alpha,\xi} := \{\eta > \alpha : f_\eta(\alpha) = \xi\}$. The coefficients in the columns of this matrix are pairwise disjoint subsets of κ. For every $\alpha < \kappa$ we have $\bigcup\{X_{\alpha,\xi} : \xi < \lambda\} = (\alpha, \kappa)_{\mathrm{ON}}$. Since $(\alpha, \kappa)_{\mathrm{ON}} \notin I$, there is $\xi_\alpha < \lambda$ such that $X_{\alpha,\xi_\alpha} \notin I$.
13) Assume that κ is a singular cardinal. Show that every κ-complete ideal on κ is principal.
Hint: If I is a κ-complete ideal on κ, then note that $\alpha \cap \bigcup I \in I$ for every $\alpha < \kappa$ and show that $\bigcup I \in I$.

2.2 The Galvin-Hajnal Formula

Definition *Every function $K : A \longrightarrow \mathcal{P}(A)$ is called a* **set function.** *We call a subset B of A* **independent** *with respect to the set function K iff $x \notin K(y)$ for all $x, y \in B$ with $x \neq y$.*

We can interpret a set function as a function which assigns to each $x \in A$ the set of people which are known to x, where this relation needs not be symmetric. An independent set can then be interpreted as a subset of A all of whose members do not know each other. Note that $\emptyset$ is independent and that the set of independent subsets of A is closed under union of chains. By Zorn's lemma, there exists an $\subseteq$-maximal independent subset of A; by the same reasons, we obtain, for any subset Y of A, an independent subset of A which is $\subseteq$-maximal w.r.t. the property of being disjoint to Y.

Lemma 2.2.1 (Lázár) *Assume that κ is a regular and λ is a – finite or infinite – cardinal with $\lambda < \kappa$, A is a set with $|A| = \kappa$, and $K : A \longrightarrow \mathcal{P}(A)$ is a set function such that $|K(x)| < \lambda$ for all $x \in A$. Then there is a subset B of A of cardinality κ which is independent w.r.t. K.*[1]

Proof For any subset X of A let $X_u := \{a \in A : K(a) \cap X = \emptyset\}$ be the set of members a of A who do not know any member of X. We call X popular iff $|X_u| < \kappa$ and unpopular iff $|X_u| = \kappa$. Further let $k_X := \bigcup\{K(x) : x \in X\}$ be the set of members of A which are known to members of X. If $|X| < \kappa$, then by the assumption on K we have $|k_X| < \kappa$ since κ is regular. First we notice the following fact:
If $Z \subseteq A$ is independent and unpopular, $|Z| < \kappa$, $Y \subseteq A$, $|Y| < \kappa$, and $Z \cap Y = \emptyset$, then

$$Z_u \setminus (k_Z \cup Y \cup Z) \neq \emptyset.$$

If we take a member z of this set, we have $K(z) \cap Z = \emptyset$ and $z \notin k_Z \cup Y$. Thus $Z^* := Z \cup \{z\}$ is independent and satisfies $Z^* \cap Y = \emptyset$. If we can choose Y in such a way that $|Y| < \kappa$ and $Z \cap Y = \emptyset$ implies that Z is unpopular, then any $\subseteq$-maximal independent set Z disjoint from Y, which exists by Zorn's lemma, must be of size κ.

It is clear that the condition $Z \cap Y = \emptyset$ implies that Z is unpopular if Y is the union of an $\subseteq$-maximal set of pairwise disjoint popular sets which again exists by Zorn's lemma. If we can show that $|Y| < \kappa$ for such a set Y, we get the desired result. So assume that $\mathcal{P}$ is a set of pairwise disjoint popular sets such that $|\mathcal{P}| = \lambda$. The set $W := \bigcup\{X_u : X \in \mathcal{P}\}$ is of size less than κ since κ is regular. Choose $y \in A \setminus W$. Then $K(y) \cap X \neq \emptyset$ for all $X \in \mathcal{P}$, and consequently $K(y)$ includes a set of representatives for $\mathcal{P}$. So we get $|\mathcal{P}| \leq |K(y)| < \lambda$. This contradiction shows that $|\mathcal{P}| < \lambda$ for every $\subseteq$-maximal set $\mathcal{P}$ of pairwise disjoint popular sets. Thus we can choose such a set $\mathcal{P}$, and $Y := \bigcup \mathcal{P}$ has the desired properties.

Lemma 2.2.2 *If I is a σ-complete ideal on A and α is an ordinal, then the relation $<_I$ is well-founded on ${}^A\alpha$.*

Proof We give an indirect proof and apply Lemma 1.3.8. Let $(f_n : n \in \omega)$ be a sequence of members of ${}^A\alpha$ such that $f_{n+1} <_I f_n$ for all $n \in \omega$. On account of the definition of $<_I$ and the σ-completeness of the ideal I, the set $\bigcup\{\{a \in A : f_n(a) \leq f_{n+1}(a)\} : n \in \omega\}$ is a member of I, and so its complement $\bigcap\{\{a \in A : f_{n+1}(a) < f_n(a)\} : n \in \omega\}$ is I-positive. Choose a member a of this complement; then $(f_n(a) : n \in \omega)$ is a strictly decreasing sequence of ordinals. This contradicts the well-foundedness of the $\in$-relation on the ordinals.

[1] For singular cardinals this was proved by A. Hajnal.

Definition *Let α be an ordinal and I be a σ-complete ideal on the set A. Lemma 2.2.2 says that $<_I$ is a well-founded relation on the set ${}^A\alpha$. By transfinite recursion we define for every function $\Phi \in {}^A\alpha$ its* **rank** *or* **norm** $\|\Phi\|_{I,\alpha}$ *with respect to the ideal I by*

$$\|\Phi\|_{I,\alpha} := \sup\{\|\Psi\|_{I,\alpha} + 1 : \Psi \in {}^A\alpha \text{ and } \Psi <_I \Phi\}.$$

In view of the following remark we write $\|\Phi\|_I$ instead of $\|\Phi\|_{I,\alpha}$ and will no longer refer to the set ${}^A\alpha$ of functions. It is always possible to choose α large enough to fulfil all necessessary implications.

Remark For all $\Phi \in {}^A\alpha$ and all $\beta > \alpha$,

$$\|\Phi\|_{I,\alpha} = \|\Phi\|_{I,\beta}.$$

Proof We prove the assertion by transfinite induction on the well-founded relation $<_I$ on ${}^A\alpha$. Clearly we have $\|\Phi\|_{I,\alpha} \leq \|\Phi\|_{I,\beta}$, since ${}^A\alpha \subseteq {}^A\beta$. For the reverse inequality we apply the definition of $\|\Phi\|_{I,\alpha}$ and want to find, for each function $\Psi \in {}^A\beta$ with $\Psi <_I \Phi$, a function $\Psi' \in {}^A\alpha$ satisfying $\|\Psi\|_{I,\beta} = \|\Psi'\|_{I,\beta}$ and $\Psi' <_I \Phi$. Then the inductive hypothesis tells us that

$$\|\Psi\|_{I,\beta} + 1 = \|\Psi'\|_{I,\beta} + 1 = \|\Psi'\|_{I,\alpha} + 1 \leq \|\Phi\|_{I,\alpha},$$

from which we can conclude that $\|\Phi\|_{I,\beta} \leq \|\Phi\|_{I,\alpha}$. So let Ψ be given as above. Then the set $\{a \in A : \Phi(a) \leq \Psi(a)\}$ is a member of I. For its subset $B := \{a \in A : \alpha \leq \Psi(a)\}$, we get $B \in I$. Let the function Ψ' result from Ψ by changing the values of Ψ on B to 0. Then $\Psi' \in {}^A\alpha$ and $\Psi' <_I \Phi$. Furthermore, Lemma 2.2.3 a) says that $\|\Psi\|_{I,\beta} = \|\Psi'\|_{I,\beta}$, which remained to be shown.

Lemma 2.2.3 *Assume that $\Phi, \Phi_1, \Phi_2 \in {}^A\alpha$. With the denotations of the preceding definition, we have:*

a) If $\Phi_1 =_I \Phi_2$, then $\|\Phi_1\|_{I,\alpha} = \|\Phi_2\|_{I,\alpha}$.

b) If $\Phi_1 <_I \Phi_2$, then $\|\Phi_1\|_I < \|\Phi_2\|_I$.

c) If $\Phi_1 \leq_I \Phi_2$, then $\|\Phi_1\|_I \leq \|\Phi_2\|_I$.

d) If $(\Phi_\xi : \xi \leq \gamma)$ is a sequence of members of ${}^A\alpha$, strictly increasing under $<_I$, then $\|\Phi_\gamma\|_I \geq \gamma$.

e) If $\tau < \|\Phi\|_I$, then there is a function $\Psi \in {}^A\alpha$ satisfying $\Psi <_I \Phi$ and $\|\Psi\|_I = \tau$.

Proof For c), assume that $\Phi_1 \leq_I \Phi_2$. Then, by Lemma 2.1.2, $\|\Phi_1\|_I = \sup\{\|\Psi\|_I + 1 : \Psi <_I \Phi_1\} \leq \sup\{\|\Psi\|_I + 1 : \Psi <_I \Phi_2\} = \|\Phi_2\|_I$.

For the proof of e), we use transfinite induction on $\|\Phi\|_I$. Put $\|\Phi\| := \|\Phi\|_I$ and assume that $\tau < \|\Phi\|$. If $\|\Phi\| = 0$, the assertion is trivial. Let $\|\Phi\| = \sigma + 1$.

By the definition of $\|\Phi\|$, there is a function $\Psi_0 \in {}^A\alpha$ such that $\Psi_0 <_I \Phi$ and $\|\Phi\| = \|\Psi_0\| + 1$. Then $\tau \leq \|\Psi_0\|$. If $\tau = \|\Psi_0\|$, take $\Psi := \Psi_0$. Otherwise we can use the inductive hypothesis to get a function $\Psi \in {}^A\alpha$ such that $\Psi < \Psi_0$ and $\|\Psi\| = \tau$. Since the relation $<_I$ is transitive, the assertion is proved for the successor step. Finally, if $\|\Phi\|$ is a limit ordinal, then the definition of $\|\Phi\|$ yields a function Ψ_0 such that $\Psi_0 <_I \Phi$ and $\tau < \|\Psi_0\| + 1$. Then $\tau \leq \|\Psi_0\|$, and we can proceed as in the successor step.

Lemma 2.2.4 *Assume that α is an ordinal and I is a σ-complete ideal on the set A. If $\Phi \in {}^A\alpha$, $Y \in I^+$ and $X := \{a \in Y : \Phi(a)$ is not a successor ordinal $\} \in I$, then $\|\Phi\|_{I \restriction Y}$ is a successor ordinal.*

Proof Let $\|\Phi\| := \|\Phi\|_{I \restriction Y}$. We define a function $\Psi \in {}^A\alpha$ by the equations $\Phi(a) =: \Psi(a) + 1$ for $a \in Y \setminus X$ and $\Psi(a) := 0$ for $a \in A \setminus (Y \setminus X)$. Then we have $\{a \in Y : \Phi(a) \leq \Psi(a)\} \in I$, since this set is a subset of X. By the definition of the ideal $I \restriction Y$, we get $\Psi <_{I \restriction Y} \Phi$, hence $\|\Psi\| < \|\Phi\|$.

We claim that $\|\Psi\| + 1 = \|\Phi\|$. Let $\Phi_1 <_{I \restriction Y} \Phi$. Then the set $Z := \{a \in Y : \Phi(a) \leq \Phi_1(a)\}$ is a member of I. But, for any $a \in Y \setminus (X \cup Z)$, we have $\Phi_1(a) < \Phi(a)$ and $\Phi(a) = \Psi(a) + 1$, hence $\Phi_1(a) \leq \Psi(a)$. Consequently, the set $\{a \in Y : \Psi(a) < \Phi_1(a)\}$ is also a member of I, since it is a subset of $X \cup Z$. This gives $\Phi_1 \leq_{I \restriction Y} \Psi$, and thus, using Lemma 2.2.3 c), $\|\Phi_1\| \leq \|\Psi\|$. By the definition of $\|\Phi\|$, we get $\|\Phi\| = \|\Psi\| + 1$.

Lemma 2.2.5 *Assume that $\kappa > \omega$ is a regular cardinal, $\Phi \in {}^\kappa\mathrm{ON}$, and I is a κ-complete normal ideal on κ with* $\mathrm{dom}(I) = \kappa$. *Then, for every ordinal $\sigma < \kappa$,*

$$\|\Phi\|_I \leq \sigma \iff S_\sigma := \{\xi < \kappa : \Phi(\xi) \leq \sigma\} \in I^+.$$

Proof Let $\sigma < \kappa$ and first $\|\Phi\|_I \leq \sigma$. Assume, to get a contradiction, that $S_\sigma \in I$. We put $\Psi_\eta := \kappa \times \{\eta\}$ for all $\eta < \kappa$. Then $\Psi_\eta <_I \Phi$ for every $\eta \leq \sigma$. Further we have $\Psi_\rho <_I \Psi_\eta$ for all ρ and η with $\rho < \eta \leq \sigma$, and from this we conclude that $\|\Phi\|_I \geq \sigma + 1$, using Lemma 2.2.3. This contradiction proves one direction of the equivalence.

We prove the other one by transfinite induction on σ. For this, consider a fixed function $\Psi \in {}^\kappa\mathrm{ON}$ satisfying $\Psi <_I \Phi$. Together with Lemma 1.8.6 we know that the sets $\{\xi < \kappa : \xi < \sigma\}$ and $\{\xi < \kappa : \Phi(\xi) \leq \Psi(\xi)\}$ are members of I, and thus the set

$$S := \{\xi < \kappa : \Psi(\xi) < \Phi(\xi) \leq \sigma \leq \xi\}$$

is a member of I^+. Otherwise S_σ, as a subset of the union of S with these sets, would be a member of I. Futhermore, Ψ is regressive on S. Since I is a normal ideal, Lemma 1.8.9 shows that there is a set $S_0 \subseteq S$ with $S_0 \in I^+$ and an ordinal $\alpha < \sigma$, such that $\mathrm{ran}(\Psi \restriction S_0) = \{\alpha\}$. With $\alpha < \sigma$ we get $\|\Psi\|_I \leq \alpha < \sigma$

by the inductive hypothesis. These considerations hold for all Ψ with $\Psi <_I \Phi$, and so we get $\|\Phi\|_I \leq \sigma$.

Definition *Assume that A is a nonempty set of ordinals and* $\sup(A) \notin A$. *Any two functions $f, g \in {}^A\mathrm{V}$ are called* **almost disjoint** *if the set $\{\xi \in A : f(\xi) = g(\xi)\}$ is a member of $\mathrm{I_b}(A)$. We say that a set $\mathcal{F} \subseteq {}^A\mathrm{V}$ is a* **set of almost disjoint functions** *iff every two distinct members of $\mathcal{F}$ are almost disjoint.*

If κ is an infinite cardinal and $\Psi \in {}^\kappa\mathrm{ICN}$ is a function, then let

$$T(\Psi) := \sup\{|\mathcal{F}| : \mathcal{F} \subseteq \prod_{\xi<\kappa} \Psi(\xi) \text{ is a set of almost disjoint functions}\}.$$

In the case that $\sigma \in \mathrm{ON}$ and $\Psi \in {}^\kappa\mathrm{ON}$ is defined by $\Psi(\xi) = \sigma$ for all $\xi < \kappa$, we write $T(\kappa, \aleph_\sigma)$ instead of $T(\aleph \circ \Psi)$.

Theorem 2.2.6 (Galvin-Hajnal Formula) *Assume that $\kappa > \omega$ is a regular cardinal and I is a κ-complete normal ideal on κ with* $\mathrm{dom}(I) = \kappa$. *Further let $(\eta(\xi) : \xi < \kappa)$ be an increasing continuous sequence of ordinals and*

$$\aleph_\Theta := 2^\kappa \cdot \sup\{T(\kappa, \aleph_{\eta(\xi)}) : \xi < \kappa\}.$$

If $\Phi \in {}^\kappa\mathrm{ON}$ is a function and $\Psi \in {}^\kappa\mathrm{ON}$ is defined by $\Psi(\xi) := \eta(\xi) + \Phi(\xi)$, then

$$T(\aleph \circ \Psi) \leq \aleph_{\Theta + \|\Phi\|_I}.$$

Proof[2] Let $\mathcal{F} \subseteq \prod_{\xi<\kappa} \aleph_{\eta(\xi)+\Phi(\xi)}$ be a set of almost disjoint functions. To show that $T(\aleph \circ \Psi) \leq \aleph_\Theta$, we use transfinite induction on the ordinal $\|\Phi\|_I$ and will distinguish the cases $\|\Phi\|_I = 0$ and $\|\Phi\|_I > 0$. In the first case, we will represent $\mathcal{F}$, using the fundamental property of normal ideals given by Lemma 1.8.9, as a union of sets $\mathcal{F}_{X,\beta}$ for certain sets $X \subseteq \kappa$ and ordinals $\beta < \kappa$. $\mathcal{F}_{X,\beta}$ can be mapped one-to-one onto a set $\mathcal{F}''_{X,\beta} \subseteq \prod_{\xi\in\kappa} \aleph_{\eta(\beta)}$ of almost disjoint functions. This will already motivate the choice of Θ: There are at most 2^κ such pairs (X, β), and the definition of $T(\kappa, \aleph_{\eta(\beta)})$ will yield $|\mathcal{F}_{X,\beta}| \leq T(\kappa, \aleph_{\eta(\beta)})$, hence $|\mathcal{F}| \leq 2^\kappa \cdot \sup\{T(\kappa, \aleph_{\eta(\xi)}) : \xi < \kappa\} = \aleph_\Theta$. The proof of the second case is more complicated and will apply two combinatorial results, Lemma 2.2.1 and the Erdös-Rado theorem.

Case 1 $\|\Phi\|_I = 0$.

Observe first that Lemma 1.8.6 says that $\mathrm{I_b}(\kappa) \subseteq I$, hence every set $X \in I^+$ has cardinality κ. Together with Lemma 2.2.5, Case 1 gives $\{\xi < \kappa : \Phi(\xi) = 0\} \in I^+$. By Lemma 1.8.6, the set $\{\xi + 1 : \xi < \kappa\}$ is a member of I. Furthermore $\mathrm{dom}(I) = \kappa$ yields $\{0\} \in I$, and consequently we have

$$X_0 := \{\xi < \kappa : \Phi(\xi) = 0 \wedge \mathrm{Lim}(\xi)\} \in I^+.$$

[2]We follow the proof in [GH].

Fix $f \in \mathcal{F}$. Since $\mathcal{F} \subseteq \prod_{\xi<\kappa} \aleph_{\eta(\xi)+\Phi(\xi)}$, we get $f(\xi) < \aleph_{\eta(\xi)}$ for any $\xi \in X_0$. By assumption, the function $(\eta(\xi) : \xi < \kappa)$ is continuous, and so there is an ordinal $\alpha_f(\xi) < \xi$ such that $f(\xi) < \aleph_{\eta(\alpha_f(\xi))}$. Thus we obtain the regressive function $(\alpha_f(\xi) : \xi \in X_0)$, and from Lemma 1.8.9 and the fact that I is normal we can conclude that there is a set $X_f \in \mathcal{P}(X_0) \setminus I$ and an ordinal $\beta_f < \kappa$ such that $\alpha_f(\xi) = \{\beta_f\}$ for all $\xi \in X_f$, hence $f \restriction X_f \in \prod_{\xi \in X_f} \aleph_{\eta(\beta_f)}$.

Fix $X \in \mathcal{P}(X_0) \setminus I$ and $\beta < \kappa$ and let

$$\mathcal{F}_{X,\beta} := \{f \in \mathcal{F} : f \restriction X \in \prod_{\xi \in X} \aleph_{\eta(\beta)}\}.$$

Every member of $\mathcal{P}(X_0) \setminus I$ is unbounded in κ, hence we obtain via $\mathcal{F}'_{X,\beta} := \{f \restriction X : f \in \mathcal{F}_{X,\beta}\}$ a set of almost disjoint functions with $|\mathcal{F}_{X,\beta}| = |\mathcal{F}'_{X,\beta}|$, included in $\prod_{\xi \in X} \aleph_{\eta(\beta)}$. Any bijection from X onto κ induces a bijection from $\prod_{\xi \in X} \aleph_{\eta(\beta)}$ onto $\prod_{\xi<\kappa} \aleph_{\eta(\beta)}$ which preserves the property of two functions of being almost disjoint since κ is regular. Therefore we get $|\mathcal{F}_{X,\beta}| \leq T(\kappa, \aleph_{\eta(\beta)}) \leq \aleph_\Theta$.

Since $\mathcal{F} = \bigcup\{\mathcal{F}_{X,\beta} : X \in \mathcal{P}(X_0) \setminus I \text{ and } \beta < \kappa\}$ and there are only 2^κ such pairs (X, β), our assertion is proved for the case $\|\Phi\|_I = 0$.

Case 2 $\|\Phi\|_I =: \alpha > 0$.
Let us assume that the assertion is true for every function $\Phi' \in {}^\kappa\mathrm{ON}$ with $\|\Phi'\|_I < \alpha$. Again we want to represent our given set $\mathcal{F} \subseteq \prod_{\xi<\kappa} \aleph_{\eta(\xi)+\Phi(\xi)}$ of almost disjoint functions as a union of sets which are not too large; more precisely, as the union of $|\alpha|$ sets, since $|\alpha| \leq \alpha \leq \aleph_{\Theta+\|\Phi\|_I}$.

Fix $f \in \mathcal{F}$. If $\Phi(\xi) \neq 0$, then there is a least $\delta < \Phi(\xi)$ such that $|f(\xi)| \leq \aleph_{\eta(\xi)+\delta}$. But the assumption $\|\Phi\|_I > 0$ implies $\{\xi < \kappa : \Phi(\xi) = 0\} \in I$ by Lemma 2.2.5, and thus $\Phi(\xi) > 0$ holds modulo I for almost all $\xi < \kappa$. So we can conclude, if we define functions $\Phi_f \in {}^\kappa\mathrm{ON}$ and $\Psi_f \in {}^\kappa\mathrm{ON}$ by

$$\Phi_f(\xi) := \min\{\delta : |f(\xi)| \leq \aleph_{\eta(\xi)+\delta}\} \quad \text{and} \quad \Psi_f(\xi) := \eta(\xi) + \Phi_f(\xi),$$

that $\Phi_f(\xi) < \Phi(\xi)$ holds modulo I for almost all $\xi < \kappa$. Thus we get $\Phi_f <_I \Phi$, hence $\|\Phi_f\|_I < \|\Phi\|_I = \alpha$, and clearly we have $\mathcal{F} = \bigcup\{\mathcal{F}_\sigma : \sigma < \alpha\}$, where

$$\mathcal{F}_\sigma := \{f \in \mathcal{F} : \|\Phi_f\|_I = \sigma\}$$

for $\sigma < \alpha$. Now we have found our candidates: For the proof of $|\mathcal{F}| \leq \aleph_{\Theta+\|\Phi\|_I}$ it suffices to show that $|\mathcal{F}_\sigma| \leq \aleph_{\Theta+\|\Phi\|_I}$. We will prove a slightly stronger version, namely $|\mathcal{F}_\sigma| \leq \aleph_{\Theta+\sigma+1}$ for every $\sigma < \alpha$.

So let $\sigma < \alpha$ be fixed. First we will state an important observation the proof of which will use the inductive hypothesis, namely that the set of $\leq_\kappa$-predecessors in $\mathcal{F}_\sigma$ of any function $f \in \mathcal{F}_\sigma$, that means the set $Pr(f) := \{g \in \mathcal{F}_\sigma : g \leq_\kappa f\}$, has at most the size $\aleph_{\Theta+\|\Phi_f\|_I}$. Consider a fixed $f \in \mathcal{F}_\sigma$. For $g \leq_\kappa$

f, we have $g \in \prod_{\xi<\kappa}(f(\xi)+1)$. If $\xi < \kappa$, then we know that $|f(\xi)| \leq \aleph_{\eta(\xi)+\Phi_f(\xi)}$ and can choose an injection i_ξ from $f(\xi)+1$ into $\aleph_{\eta(\xi)+\Phi_f(\xi)}$. For each $g \in Pr(f)$ we can define a function $g^* \in \prod_{\xi<\kappa} \aleph_{\eta(\xi)+\Phi_f(\xi)}$ by $g^*(\xi) := i_\xi(g(\xi))$. If we put $Pr^*(f) := \{g^* : g \in Pr(f)\}$, then $|Pr(f)| = |Pr^*(f)|$, and $Pr^*(f)$ is a set of almost disjoint functions included in $\prod_{\xi<\kappa} \aleph_{\eta(\xi)+\Phi_f(\xi)}$. From $\|\Phi_f\|_I < \alpha$ and the inductive hypothesis we infer that $|Pr(f)| = |Pr^*(f)| \leq T(\aleph \circ \Psi_f) \leq \aleph_{\Theta+\|\Phi_f\|_I} = \aleph_{\Theta+\sigma}$.

Now we can apply our combinatorial results. Let us assume, to get a contradiction, that $|\mathcal{F}_\sigma| \geq \aleph_{\Theta+\sigma+2}$ for some $\sigma < \alpha$. Without loss of generality let $|\mathcal{F}_\sigma| = \aleph_{\Theta+\sigma+2}$, since otherwise we could take a suitable subset of $\mathcal{F}_\sigma$. Then $|\mathcal{F}_\sigma|$ is a regular cardinal, and we have shown that the assumptions of Lemma 2.2.1 are satisfied if we choose the set function $Pr : \mathcal{F}_\sigma \longrightarrow \mathcal{P}(\mathcal{F}_\sigma)$ which we have defined by

$$Pr(f) = \{g \in \mathcal{F}_\sigma : g \leq_\kappa f\},$$

since $|Pr(f)| \leq \aleph_{\Theta+\sigma}$ for all $f \in \mathcal{F}_\sigma$. Consequently there exists a subset of $\mathcal{F}_\sigma$ with cardinality $\aleph_{\Theta+\sigma+2}$ which is independent with respect to Pr. By the definition of Θ, we have $\aleph_{\Theta+\sigma+2} > 2^\kappa$, and so there is an injective sequence $(f_\xi : \xi < (2^\kappa)^+)$ of elements of $\mathcal{F}_\sigma$ such that $f_\zeta \notin Pr(f_\xi)$ for any ζ and ξ with $\zeta < \xi < (2^\kappa)^+$. The last assertion implies that there is a smallest ordinal $l(\zeta,\xi) < \kappa$ such that $f_\zeta(l(\zeta,\xi)) > f_\xi(l(\zeta,\xi))$. By this property, a function $l : [(2^\kappa)^+]^2 \longrightarrow \kappa$ is defined. Now the Erdös-Rado theorem, Corollary 1.9.5, says that

$$(2^\kappa)^+ \longrightarrow (\kappa^+)^2_\kappa. \tag{2.1}$$

Consequently there is a set $A \subseteq (2^\kappa)^+$ with $|A| = \kappa^+$ and an ordinal $\gamma < \kappa$ satisfying $\operatorname{ran}(l \restriction [A]^2) = \{\gamma\}$. In particular, we can choose a strictly increasing sequence $(\xi_n : n < \omega)$ of members of A. Then, by construction, $(f_{\xi_n}(\gamma) : n \in \omega)$ is a strictly decreasing sequence of ordinals. This contradiction completes the proof of the case $\|\Phi\|_I > 0$ and the proof of the theorem.

Lemma 2.2.7 *Let κ be an infinite cardinal and $(\gamma(\xi) : \xi < \kappa)$ be a sequence of ordinals. Then there is a set $\mathcal{F} \subseteq \prod_{\alpha<\kappa}(\prod_{\xi<\alpha} \aleph_{\gamma(\xi)})$ of almost disjoint functions such that*

$$|\mathcal{F}| = \prod_{\xi<\kappa} \aleph_{\gamma(\xi)}.$$

Proof We put $\mu := |\prod_{\xi<\kappa} \aleph_{\gamma(\xi)}|$ and choose an enumeration $(g_\xi : \xi < \mu)$ of the cartesian product $\prod_{\xi<\kappa} \aleph_{\gamma(\xi)}$. For $\xi < \mu$ let the function $f_\xi \in \prod_{\alpha<\kappa}(\prod_{\xi<\alpha} \aleph_{\gamma(\xi)})$ be defined by $f_\xi(\alpha) = g_\xi \restriction \alpha$. Then $\mathcal{F} = \{f_\xi : \xi < \mu\}$ is a set of almost disjoint functions, and in particular $|\mathcal{F}| = \mu = \prod_{\xi<\kappa} \aleph_{\gamma(\xi)}$.

Theorem 2.2.8 *Assume that κ is an uncountable regular cardinal, I is a normal κ-complete ideal on κ with* $\mathrm{dom}(I) = \kappa$, $(\eta(\xi) : \xi < \kappa)$ *is an increasing continuous sequence of ordinals, and*

$$\aleph_\Theta := 2^\kappa \cdot \sup\{T(\kappa, \aleph_{\eta(\xi)}) : \xi < \kappa\}.$$

If $\Phi \in {}^\kappa\mathrm{ON}$ and $(\aleph_{\gamma(\xi)} : \xi < \kappa)$ is a sequence of cardinals satisfying

$$\prod_{\xi<\alpha} \aleph_{\gamma(\xi)} \leq \aleph_{\eta(\alpha)+\Phi(\alpha)}$$

for all $\alpha < \kappa$, then

$$\prod_{\xi<\kappa} \aleph_{\gamma(\xi)} \leq \aleph_{\Theta+\|\Phi\|_I}.$$

Proof By Lemma 2.2.7, there is a set $\mathcal{F}$ of almost disjoint functions satisfying

$$\mathcal{F} \subseteq \prod_{\alpha<\kappa}\Big(\prod_{\xi<\alpha} \aleph_{\gamma(\xi)}\Big) \quad \text{and} \quad |\mathcal{F}| = \prod_{\xi<\kappa} \aleph_{\gamma(\xi)}.$$

Since $\prod_{\xi<\alpha} \aleph_{\gamma(\xi)} \leq \aleph_{\eta(\alpha)+\Phi(\alpha)}$, there is a set $\mathcal{F}'$ of almost disjoint functions such that $\mathcal{F}' \subseteq \prod_{\alpha<\kappa} \aleph_{\eta(\alpha)+\Phi(\alpha)}$ and $|\mathcal{F}| = |\mathcal{F}'|$. If $\Psi \in {}^\kappa\mathrm{ON}$ is defined by $\Psi(\xi) := \eta(\xi) + \Phi(\xi)$, then $T(\aleph \circ \Psi) \leq \aleph_{\Theta+\|\Phi\|_I}$ by Theorem 2.2.6, and with $|\mathcal{F}'| \leq T(\aleph \circ \Psi)$ the assertion of the theorem follows.

It is obvious that an estimate for $\|\Phi\|_I$ would improve the preceding theorem. We will obtain such an estimate in Lemma 2.2.10. For its proof we need the following technical lemma.

Lemma 2.2.9 *Assume that A is a nonempty set, I is a σ-complete ideal on A, $\Phi \in {}^A\mathrm{ON}$ is a function, and $X, Y \in I^+$. Then*

a) If $X \subseteq Y$, then $\|\Phi\|_{I \restriction Y} \leq \|\Phi\|_{I \restriction X}$.

b) If $Z \in I$, then $\|\Phi\|_{I \restriction (X \cup Z)} = \|\Phi\|_{I \restriction X}$.

c) $\|\Phi\|_{I \restriction (X \cup Y)} = \min\{\|\Phi\|_{I \restriction X}, \|\Phi\|_{I \restriction Y}\}$.

Proof For part a), we have $I \restriction Y = I[A \setminus Y] \subseteq I[A \setminus X] = I \restriction X$, and therefore $\Psi <_{I \restriction Y} \Phi$ implies $\Psi <_{I \restriction X} \Phi$. Using transfinite induction on the relation $<_{I \restriction X}$, which is well-founded by Lemma 2.2.2, we can conclude that $\|\Phi\|_{I \restriction Y} = \sup\{\|\Psi\|_{I \restriction Y} + 1 : \Psi <_{I \restriction Y} \Phi\} \leq \sup\{\|\Psi\|_{I \restriction Y} + 1 : \Psi <_{I \restriction X} \Phi\} \leq \sup\{\|\Psi\|_{I \restriction X} + 1 : \Psi <_{I \restriction X} \Phi\} = \|\Phi\|_{I \restriction X}$.

For b), we observe that the above assertion gives $I \restriction (X \cup Z) \subseteq I \restriction X$. Since $Z \in I$, we get $I \restriction (X \cup Z) = I \restriction X$.

For c), we note that part a) implies

(1) $\|\Phi\|_{I\restriction(X\cup Y)} \leq \min\{\|\Phi\|_{I\restriction X}, \|\Phi\|_{I\restriction Y}\}$.

We prove the remaining inequality

(2) $\min\{\|\Phi\|_{I\restriction X}, \|\Phi\|_{I\restriction Y}\} \leq \|\Phi\|_{I\restriction(X\cup Y)}$

by transfinite induction on the well-founded relation $<_{I\restriction(X\cup Y)}$. Without loss of generality let $\|\Phi\|_{I\restriction X} \leq \|\Phi\|_{I\restriction Y}$. Then it remains to be shown that

(3) $\|\Phi\|_{I\restriction X} \leq \|\Phi\|_{I\restriction(X\cup Y)}$.

On account of $\|\Phi\|_{I\restriction X} = \sup\{\|\Psi\|_{I\restriction X}+1 : \Psi <_{I\restriction X} \Phi\}$, it suffices to show that

(4) $\Psi <_{I\restriction X} \Phi$ implies $\|\Psi\|_{I\restriction X} < \|\Phi\|_{I\restriction(X\cup Y)}$.

So assume that $\Psi <_{I\restriction X} \Phi$. Then $\|\Psi\|_{I\restriction X} < \|\Phi\|_{I\restriction X} \leq \|\Phi\|_{I\restriction Y}$, and consequently $\|\Psi\|_{I\restriction X} < \|\Phi\|_{I\restriction Y}$. Now Lemma 2.2.3 says that there is a function $\Psi' <_{I\restriction Y} \Phi$ such that $\|\Psi'\|_{I\restriction Y} = \|\Psi\|_{I\restriction X}$. We distinguish two cases.

Case 1 $X \cap Y = \emptyset$.

Let the function $\Psi'' \in {}^{A}\mathrm{ON}$ be defined by

$$\Psi''(a) = \begin{cases} \Psi(a) & \text{if } a \in A\setminus Y, \\ \Psi'(a) & \text{otherwise.} \end{cases}$$

With Lemma 2.1.4 we get $\{a \in X : \Psi(a) \geq \Phi(a)\} \in I$ from $\Psi <_{I\restriction X} \Phi$, and $\{a \in Y : \Psi'(a) \geq \Phi(a)\} \in I$ from $\Psi' <_{I\restriction Y} \Phi$. Therefore the set $\{a \in X\cup Y : \Phi(a) \leq \Psi''(a)\}$ is a member of I as a subset of the union of these two sets, and so we can infer that

(5) $\Psi'' <_{I\restriction(X\cup Y)} \Phi$.

By the inductive hypothesis, we have $\|\Psi''\|_{I\restriction(X\cup Y)} = \min\{\|\Psi''\|_{I\restriction X}, \|\Psi''\|_{I\restriction Y}\}$. Furthermore Lemma 2.2.3 implies $\|\Psi''\|_{I\restriction Y} = \|\Psi'\|_{I\restriction Y} = \|\Psi\|_{I\restriction X} = \|\Psi''\|_{I\restriction X}$, since $\Psi'' =_{I\restriction Y} \Psi'$ and $\Psi'' =_{I\restriction X} \Psi$. So we get $\|\Psi''\|_{I\restriction(X\cup Y)} = \|\Psi''\|_{I\restriction X} = \|\Psi''\|_{I\restriction Y}$. Now (5) yields $\|\Psi\|_{I\restriction X} = \|\Psi''\|_{I\restriction X} = \|\Psi''\|_{I\restriction(X\cup Y)} < \|\Phi\|_{I\restriction(X\cup Y)}$, hence (4) and hereby (3) is proved.

Case 2 $X \cap Y \neq \emptyset$.

If $X\setminus Y \in I$, then $Y \cup (X\setminus Y) = X \cup Y$, and from b) we can conclude that $\|\Phi\|_{I\restriction(X\cup Y)} = \|\Phi\|_{I\restriction Y} \geq \|\Phi\|_{I\restriction X}$. If $X\setminus Y \notin I$, we can apply Case 1 to the sets $X\setminus Y$ and Y. Therefore a) yields the estimate $\|\Phi\|_{I\restriction X} \leq \min\{\|\Phi\|_{I\restriction(X\setminus Y)}, \|\Phi\|_{I\restriction Y}\} = \|\Phi\|_{I\restriction(X\cup Y)}$. This completes the proof of (3).

Lemma 2.2.10 *Let κ be an uncountable regular cardinal and I be a κ-complete normal ideal on κ with* $\mathrm{dom}(I) = \kappa$. *Then* *a)* *For every function* $\Phi \in {}^\kappa\mathrm{ON}$, $\|\Phi\|_I < (\prod_{\xi<\kappa} |\Phi(\xi)+1|)^+$.

b) *For every function* $\Phi \in {}^\kappa\mathrm{ON}$, $I' := I \cup \{Y \subseteq \kappa : Y \in I^+$ *and* $\|\Phi\|_I < \|\Phi\|_{I \restriction Y}\}$ *is a* κ*-complete ideal on* κ *with* $\mathrm{dom}(I') = \kappa$.

c) **(Shelah)** *If* $\lambda > \kappa$ *is a regular cardinal with* $\mu^\kappa < \lambda$ *for every cardinal* $\mu < \lambda$ *with* $\mathrm{cf}(\mu) = \kappa$, *then there is, for every function* $\Phi \in {}^\kappa\lambda$ *with* $\|\Phi\|_I \in \mathrm{Lim}$, *a set* $Y \in \mathcal{P}(\kappa)\setminus I'$ *and a set* $\mathcal{F} \subseteq {}^\kappa\lambda$ *with* $|\mathcal{F}| < \lambda$, *such that* $\mathcal{F}$ *is cofinal in* Φ *modulo* $I \restriction Y$; *that means that* $f <_{I \restriction Y} \Phi$ *for all* $f \in \mathcal{F}$ *and, for every function* $g \in {}^\kappa\mathrm{ON}$ *with* $g <_{I \restriction Y} \Phi$, *there is a function* $h \in \mathcal{F}$ *such that* $g <_{I \restriction Y} h$. *In particular we have* $\mathrm{cf}(\prod \Phi, <_{I \restriction Y}) < \lambda$.

d) **(Shelah)** *If* $\lambda > \kappa$ *is a regular cardinal satisfying* $\mu^\kappa < \lambda$ *for all cardinals* $\mu < \lambda$ *with* $\mathrm{cf}(\mu) = \kappa$, *and if* $\Phi \in {}^\kappa\lambda$, *then* $\|\Phi\|_I < \lambda$.

Proof For part a), Lemma 2.2.3 says that there is a sequence $(\Psi_\alpha : \alpha < \|\Phi\|_I)$ of elements of ${}^\kappa\mathrm{ON}$, so that $\|\Psi_\alpha\|_I = \alpha$ and $\Psi_\alpha <_I \Phi$ for all ordinals $\alpha < \|\Phi\|_I$. These properties are preserved, by Lemma 2.2.3 a), if we change for all $\alpha < \|\Phi\|_I$ the value $\Psi_\alpha(\xi)$ to 0 for each member ξ of the null set $\{\xi < \kappa : \Phi(\xi) \le \Psi_\alpha(\xi)\}$. So we can assume without loss of generality that $\Psi_\alpha \in \prod_{\xi<\kappa}(\Phi(\xi)+1)$. Then we have $|\|\Phi\|_I| \le \prod_{\xi<\kappa} |\Phi(\xi)+1|$, hence $\|\Phi\|_I < (\prod_{\xi<\kappa} |\Phi(\xi)+1|)^+$.

For b), note that $\kappa \notin I'$. Lemma 2.2.9 tells us that I' is an ideal on κ. To show that I' is κ-complete, we consider a sequence $(Y_\alpha : \alpha < \beta)$ of elements of I' which has length $\beta < \kappa$. Lemma 2.2.9 says for $Z \in I$ and $Y \in I' \setminus I$ that $\|\Phi\|_I < \|\Phi\|_{I \restriction Y} = \|\Phi\|_{I \restriction (Y \cup Z)}$, and thus $Y \cup Z \in I'$. Therefore we can assume without loss of generality that $Y_\alpha \in I' \setminus I$ for all $\alpha < \beta$, because I is κ-complete. So we get $\|\Phi\|_I < \|\Phi\|_{I \restriction Y_\alpha}$ for all $\alpha < \beta$. Further we can assume without loss of generality that $Y_\alpha \cap Y_{\alpha'} = \emptyset$ for all α and α' with $\alpha < \alpha' < \beta$. To arrange this, we can make the members of the given sequence pairwise disjoint, putting as usual $Y'_\alpha := Y_\alpha \setminus \bigcup\{Y_\xi : \xi < \alpha\}$ and using $\bigcup\{Y_\alpha : \alpha < \beta\} = \bigcup\{Y'_\alpha : \alpha < \beta\}$, and removing again the resulting members of I; the remaining ones are members of $I' \setminus I$.

We put $Y := \bigcup\{Y_\alpha : \alpha < \beta\}$ and want to show that $\|\Phi\|_I < \|\Phi\|_{I \restriction Y}$. This can be achieved by proving, using transfinite induction on γ, the following slightly more general assertion:

$(*)$ If $\Psi \in {}^\kappa\mathrm{ON}$, $\gamma \in \mathrm{ON}$ and, for all $\alpha < \beta, \gamma \le \|\Psi\|_{I \restriction Y_\alpha}$, then $\gamma \le \|\Psi\|_{I \restriction Y}$.

We assume that this assertion is true for all $\Psi \in {}^\kappa\mathrm{ON}$ and all $\gamma' < \gamma$ and that, for a given function Ψ, we have $\gamma \le \|\Psi\|_{I \restriction Y_\alpha}$ for all $\alpha < \beta$. Now let $\gamma' < \gamma$ be fixed. Using Lemma 2.2.3 we can find, for each $\alpha < \beta$, a function $\Psi'_\alpha \in {}^\kappa\mathrm{ON}$, so that $\Psi'_\alpha <_{I \restriction Y_\alpha} \Psi$ and $\gamma' \le \|\Psi'_\alpha\|_{I \restriction Y_\alpha}$. Let $\Psi' \in {}^\kappa\mathrm{ON}$ be defined by

$$\Psi' := \bigcup\{\Psi'_\alpha \restriction Y_\alpha : \alpha < \beta\} \cup ((\kappa \setminus Y) \times \{0\}).$$

Obviously $\Psi' <_{I \restriction Y} \Psi$, because I is κ-complete and the sets Y_α, $\alpha < \kappa$, are pairwise disjoint. Furthermore, we have $\gamma' \leq \|\Psi'\|_{I \restriction Y_\alpha}$ for all $\alpha < \beta$. Therefore the inductive hypothesis gives $\gamma' \leq \|\Psi'\|_{I \restriction Y}$. From $\|\Psi'\|_{I \restriction Y} < \|\Psi\|_{I \restriction Y}$ we can conclude that $\gamma \leq \|\Psi\|_{I \restriction Y}$. This proves $(*)$. If we put $\gamma := \|\Phi\|_I + 1$ and $\Psi := \Phi$, then γ and Ψ satisfy the premisses of $(*)$ since $Y_\alpha \in I' \backslash I$ for all $\alpha < \beta$. So we can infer that $\|\Phi\|_I + 1 \leq \|\Phi\|_{I \restriction Y}$, hence $Y \in I'$.

For c), let λ and $\Phi \in {}^\kappa\lambda$ with the given properties be fixed. We distinguish two cases.

Case 1 cf $\circ\, \Phi$ is constant on a set $Y \in \mathcal{P}(\kappa) \setminus I'$.
Then there is a cardinal μ such that $(\mathrm{cf} \circ \Phi) \restriction Y = Y \times \{\mu\}$. Since $Y \notin I'$, we have $\|\Phi\|_I = \|\Phi\|_{I \restriction Y}$. From Lemma 2.2.5 we can conclude that $\mu \neq 0$, since $\mathrm{ran}(\mathrm{cf} \circ \Phi \restriction Y) = \{0\}$ would give $Y = \{\xi \in Y : \Phi(\xi) = 0\} \in I^+$, hence $\|\Phi\| = 0$. Furthermore we have $\mu \neq 1$. Otherwise we would have $Y = \{\xi \in Y : \Phi(\xi)$ is a successor ordinal$\}$, and $\|\Phi\|_{I \restriction Y}$ would be a successor ordinal by Lemma 2.2.4, contradicting our assumption that $\|\Phi\|_I$ is a limit ordinal. So we have shown that μ is a regular cardinal.

For each $\xi \in Y$ let $(g_\alpha(\xi) : \alpha < \mu)$ be a normal sequence which is cofinal in $\Phi(\xi)$, for each $\xi \in \kappa \setminus Y$ and $\alpha < \mu$ put $g_\alpha(\xi) := 0$. This defines a set $\mathcal{G} = \{g_\alpha : \alpha < \mu\}$ of functions from κ into λ. If $\mu > \kappa$, then let $\mathcal{G}$ be our required set $\mathcal{F}$ of functions. If $\xi \in Y$, then $\mathrm{cf}\,(\Phi(\xi)) = \mu \leq \Phi(\xi) < \lambda$, and so we get $|\mathcal{F}| \leq \mu < \lambda$. If $\mu \leq \kappa$, then let $\mathcal{F}$ be the set of all functions $f \in {}^\kappa\mathrm{ON}$ satisfying $f(\xi) \in \{g_\alpha(\xi) : \alpha < \mu\}$ for all $\xi < \kappa$. By assumption, we can conclude that $|\mathcal{F}| \leq \mu^\kappa \leq \kappa^\kappa < \lambda$. Now in both cases one can easily verify that $\mathcal{F}$ is cofinal in Φ modulo $I \restriction Y$.

Case 2 cf $\circ\, \Phi$ is not constant on any set $Z \in \mathcal{P}(\kappa) \setminus I'$.
For each cardinal $\nu < \lambda$, let $X_\nu := \{\xi < \kappa : \mathrm{cf}\,(\Phi(\xi)) = \nu\}$ (for $\nu \geq \lambda$, this set is empty). By Case 2, we have $X_\nu \in I'$ for all $\nu < \lambda$. Let μ be the smallest cardinal such that

$$Y := \bigcup\{X_\nu : \nu < \mu\} \setminus (X_0 \cup X_1) \notin I'.$$

Since $\kappa \setminus (X_0 \cup X_1) = \bigcup\{X_\nu : \nu \in \lambda\} \setminus (X_0 \cup X_1) \notin I'$, such a cardinal exists, and we have $\mu \leq \lambda$. Next we observe that μ is a limit cardinal. Namely, if $\mu = \rho^+$ and $T := \{\nu < \mu : \nu \notin \{0,1\} \wedge X_\nu \neq \emptyset\}$, then $Y = \bigcup\{X_\nu : \nu \in T \cap \rho\} \cup X_\rho$, and $X_\rho \in I'$ gives $\bigcup\{X_\nu : \nu \in T \cap \rho\} \notin I'$, contradicting the choice of μ. Now the set T is unbounded in μ, and we have $Y = \bigcup\{X_\nu : \nu \in T\}$. From the fact that I' is κ-complete and $Y \notin I'$, we can conclude that $|T| \geq \kappa$. Since the sets X_ν, $\nu \in T$, are pairwise disjoint nonempty subsets of κ, we have $|T| = \kappa$, hence $\mathrm{cf}\,(\mu) \leq \kappa$. Since T is unbounded in μ, we can easily find a subset T' of T which is unbounded in μ and has the cardinality $\mathrm{cf}\,(\mu)$. For $\nu \in T'$, put $Z_\nu := \bigcup\{X_\rho : \rho \in T \cap \nu\}$; then $Z_\nu \in I'$ by the choice of μ, and

$Y = \bigcup\{Z_\nu : \nu \in T'\}$, since T' is unbounded in T. So the assumption $\mathrm{cf}(\mu) < \kappa$ leads to $Y \in I'$, contradicting again the choice of μ. Therefore we get $\mathrm{cf}(\mu) = \kappa$. Since λ is regular, $\kappa < \lambda$ and $\mu \leq \lambda$, we get $\mu < \lambda$ and thus, by assumption, $\mu^\kappa < \lambda$. Now we choose, for each ordinal $\xi \in Y$ with $\Phi(\xi) \neq 0$, a cofinal subset C_ξ of $\Phi(\xi)$ of order type $\mathrm{cf}(\Phi(\xi))$; for the remaining $\xi \in \kappa$ we put $C_\xi := \{0\}$. We define $\mathcal{F} := \prod_{\xi<\kappa} C_\xi$. Then $|\mathcal{F}| \leq \mu^\kappa$, hence $|\mathcal{F}| < \lambda$. By construction, $\mathcal{F}$ is cofinal in Φ modulo $I \restriction Y$.

For the proof of part d), we assume, to get a contradiction, that there is a function $\Phi \in {}^\kappa\lambda$ such that $\|\Phi\|_I \geq \lambda$. If $\|\Phi\|_I > \lambda$, Lemma 2.2.3 yields a function $\Psi \in {}^\kappa\lambda$ such that $\|\Psi\|_I = \lambda$. So we can assume without loss of generality that $\|\Phi\|_I = \lambda$. By part c), there is a set $\mathcal{F} \subseteq {}^\kappa\lambda$ with $|\mathcal{F}| < \lambda$ and a set $Y \in \mathcal{P}(\kappa) \setminus I'$, such that $\mathcal{F}$ is cofinal in Φ modulo $I \restriction Y$. Since $Y \notin I'$, we have $\|\Phi\|_I = \lambda = \|\Phi\|_{I \restriction Y}$. We will show that $\|\Phi\|_{I \restriction Y} = \sup\{\|f\|_{I \restriction Y} : f \in \mathcal{F}\}$ which will contradict the assumption that λ is a regular cardinal.

By definition, we have $\|\Phi\|_{I \restriction Y} = \sup\{\|\Psi\|_{I \restriction Y} + 1 : \Psi <_{I \restriction Y} \Phi\}$. If $f \in \mathcal{F}$, then $f <_{I \restriction Y} \Phi$, hence $\|f\|_{I \restriction Y} < \|\Phi\|_{I \restriction Y}$. On the other hand, if $\Psi <_{I \restriction Y} \Phi$, then c) yields a function $h \in \mathcal{F}$ such that $\Psi <_{I \restriction Y} h$ and thus $\|\Psi\|_{I \restriction Y} + 1 < \|h\|_{I \restriction Y} + 1$. This shows that $\|\Phi\|_{I \restriction Y} = \sup\{\|f\|_{I \restriction Y} : f \in \mathcal{F}\}$, as desired.

Exercises

1) Assume that κ is a regular cardinal and I is an ideal on κ. We call a function $g \in {}^\kappa\kappa$ **minimal for I** iff it satisfies the following conditions:

   (1) $g \leq_I \mathrm{id}_\kappa$.

   (2) For every $\eta < \kappa$, we have $\{\xi < \kappa : g(\xi) = \eta\} \in I$.

   (3) If $f \in {}^\kappa\kappa$ is regressive on κ, then $f \circ g$ is constant on an I-positive set.

   Prove: If I is a σ-complete ideal on κ with $\bigcup I = \kappa$, then there is a function $g \in {}^\kappa\kappa$ which is minimal for I.

   Hint: The set of all functions $h \in {}^\kappa\kappa$ which satisfy (1) and (2) possesses a $<_I$-minimal element.

2) Prove Lemma 2.2.3.

3) Assume that I is a σ-complete ideal on the nonempty set A and $\Phi \in {}^A\mathrm{ON}$ is a function. Prove:

   a) If $\|\Phi\|_I$ is a successor ordinal, then the set $\{a \in A : \exists \alpha \in \mathrm{ON}\ \Phi(a) = \alpha + 1\}$ is I-positive. Can one show that $\Phi(a)$ is a successor ordinal for almost all $a \in A$ modulo I?

   b) If $\Phi(a)$ is a successor ordinal for almost all $a \in A$ modulo I, then $\|\Phi\|_I$ is a successor ordinal.

c) If $\|\Phi\|_I$ is a limit ordinal, then the set $\{a \in A : \Phi(a) \text{ is a limit ordinal}\}$ is I-positive. Can one show that $\Phi(a)$ is a limit ordinal for almost all $a \in A$ modulo I?

d) If $\Phi(a)$ is a limit ordinal for almost all $a \in A$ modulo I, then $\|\Phi\|_I$ is a limit ordinal.

4) Assume that $\kappa > \aleph_0$ is a regular cardinal, $\Phi \in {}^{\kappa}\mathrm{ON}$, $I = \mathrm{I_d}(\kappa)$, $\sigma < \kappa$ is an ordinal, and $\{\xi < \kappa : \Phi(\xi) = \sigma\}$ is a club in κ. Prove that $\|\Phi(\xi)\|_I = \sigma$.

5) Assume that $\kappa > \omega$ is a regular cardinal. Prove that there are functions $\Phi, \Psi \in {}^{\kappa}\mathrm{ON}$ such that $\Phi \restriction X = \Psi \restriction X$ for some $X \in \mathrm{I_d}(\kappa)^+$ and $\|\Phi\|_{\mathrm{I_d}(\kappa)} \neq \|\Psi\|_{\mathrm{I_d}(\kappa)}$.

Hint: Use Lemma 2.2.5.

6) Assume that κ is an uncountable regular cardinal and $\Phi \in {}^{\kappa}\mathrm{ON}$ is a function. Prove: If $\sigma < \kappa$ and $S := \{\xi < \kappa : \Phi(\xi) < \sigma\}$ is stationary in κ, then $\|\Phi\|_{\mathrm{I_d}(\kappa)} < \sigma$.

7) Assume that $\kappa > \omega$ is a regular cardinal, $I := \mathrm{I_d}(\kappa)$, and $\alpha < \kappa$. Further let $\Phi \in {}^{\kappa}\kappa$, $+$ be the ordinal sum, and $S := \{\xi < \kappa : \Phi(\xi) \leq \xi + \alpha\}$ be stationary in κ. Prove that $\|\Phi\|_{\mathrm{I_d}(\kappa)} \leq \kappa + \alpha$.

Hint: Use Fodor's theorem, Lemma 1.8.4 and Lemma 2.2.5.

8) Assume that $\kappa > \omega$ is a regular cardinal, $I := \mathrm{I_d}(\kappa)$, and $\alpha < \kappa$. Further let $\|\Phi\| := \|\Phi\|_{\mathrm{I_d}(\kappa)}$ for any $\Phi \in {}^{\kappa}\kappa$ and $+$ be the ordinal sum. Prove: If $\Psi_\alpha(\xi) := \xi + \alpha$ for all $\xi < \kappa$, then $\|\Psi_\alpha\| = \kappa + \alpha$.

9) Assume that $\kappa > \omega$ is a regular cardinal, $I := \mathrm{I_d}(\kappa)$, $\Phi \in {}^{\kappa}\kappa$, and $+$ is the ordinal sum. Prove:

a) If $S := \{\xi < \kappa : \Phi(\xi) \leq \xi + \xi\}$ is stationary in κ, then $\|\Phi\|_{\mathrm{I_d}(\kappa)} \leq \kappa + \kappa$.

b) If $\Phi(\xi) := \xi + \xi$ for all $\xi < \kappa$, then $\|\Phi\|_{\mathrm{I_d}(\kappa)} = \kappa + \kappa$.

Hint: Use transfinite induction, Fodor's theorem and Lemma 1.8.4.

10) (Galvin, Hajnal) Assume that κ is an uncountable regular cardinal and I is a κ-complete normal ideal on κ with $\mathrm{dom}(I) = \kappa$. Prove without using Lemma 2.2.10:

a) If μ is an infinite cardinal and $\Phi \in {}^{\kappa}\mu$, then $\|\Phi\|_I < (\mu^\kappa)^+$.

b) If λ is an uncountable, regular, κ-strong cardinal and $\Phi \in {}^{\kappa}\lambda$, then $\|\Phi\|_I < \lambda$.

11) Assume that κ is an uncountable regular cardinal, $\alpha \geq 2$ is an ordinal, and I is an ideal on κ.

a) Prove that $=_I$ is an equivalence relation on ${}^{\kappa}\alpha$.

b) For the case that $I = \mathrm{I_b}(\kappa)$ find two equivalence classes A and B w.r.t. $=_I$ such that, for any $f \in A$ and $g \in B$, f and g are not almost disjoint.

12) Assume that κ and λ are cardinals satisfying $\omega \leq \lambda \leq \kappa$. Prove that there is a set $\mathcal{F} \subseteq {}^{\lambda}\kappa$ of almost disjoint functions such that $|\mathcal{F}| > \lambda$.

Hint: Every subset of ${}^{\lambda}\kappa$ of almost disjoint functions which is maximal under inclusion has cardinality greater than λ.

2.3 Applications of the Galvin-Hajnal Formula

Corollary 2.3.1 *Assume that λ and κ are regular cardinals satisfying $\aleph_0 < \kappa < \lambda$ and $\nu^\kappa < \lambda$ for every cardinal $\nu < \lambda$ with $\mathrm{cf}(\nu) = \kappa$. Further let μ be an infinite cardinal with $\mu^\kappa < \mu^{+\lambda}$, and $(\aleph_{\gamma(\xi)} : \xi < \kappa)$ be a sequence of cardinals satisfying*

$$\prod_{\xi<\alpha} \aleph_{\gamma(\xi)} < \mu^{+\lambda}$$

for all $\alpha < \kappa$. Then

$$\prod_{\xi<\kappa} \aleph_{\gamma(\xi)} < \mu^{+\lambda}.$$

Proof To apply Theorem 2.2.8, let $\Phi \in {}^{\kappa}\mathrm{ON}$ be defined by

$$\Phi(\alpha) := \min\{\beta \in \mathrm{ON} : \prod_{\xi<\alpha} \aleph_{\gamma(\xi)} \leq \mu^{+\beta}\}.$$

By assumption, we have $\Phi \in {}^{\kappa}\lambda$, and Lemma 2.2.10 d) tells us that $\|\Phi\|_I < \lambda$ for every κ-complete normal ideal I on κ with $\mathrm{dom}(I) = \kappa$, in particular for $I := \mathrm{I_d}(\kappa)$ by Lemma 1.8.5. If $\mu =: \aleph_\beta$ and $\eta(\xi) := \beta$ for all $\xi < \kappa$, then we have $\prod_{\xi<\alpha} \aleph_{\gamma(\xi)} \leq \aleph_{\beta+\Phi(\alpha)}$, and the cardinal $\aleph_\Theta$ in Theorem 2.2.8 can be estimated by $\aleph_\Theta \leq 2^\kappa \cdot \mu^\kappa = \mu^\kappa < \mu^{+\lambda} = \aleph_{\beta+\lambda}$. Therefore there is $\sigma < \lambda$ such that $\aleph_\Theta = \aleph_{\beta+\sigma}$. From this together with $\|\Phi\|_I < \lambda$ and Lemma 1.5.9 we can conclude that $\aleph_{\Theta+\|\Phi\|_I} = \aleph_{\beta+\sigma+\|\Phi\|_I} < \aleph_{\beta+\lambda}$. Moreover, the definitions of the functions η and Φ satisfy the assumptions of Theorem 2.2.8. So we get $\prod_{\xi<\kappa} \aleph_{\gamma(\xi)} \leq \aleph_{\Theta+\|\Phi\|_I} < \mu^{+\lambda}$, as desired.

Corollary 2.3.2 *Assume that $\aleph_\eta$ is a κ-strong singular cardinal, where $\kappa := \mathrm{cf}(\aleph_\eta) > \omega$, $\Phi \in {}^{\kappa}\mathrm{ON}$ is a function, $(\eta(\xi) : \xi < \kappa)$ is a normal function cofinal in η, and*

$$\mathcal{F} \subseteq \prod_{\xi<\kappa} \aleph_{\eta(\xi)+\Phi(\xi)}$$

is a set of almost disjoint functions. Then

$$|\mathcal{F}| \leq \aleph_{\eta+\|\Phi\|_{\mathrm{I_d}(\kappa)}}.$$

In other words, if $\Psi \in {}^{\kappa}\mathrm{ON}$ *is defined by* $\Psi(\xi) := \eta(\xi) + \Phi(\xi)$, *then*

$$T(\aleph \circ \Psi) \leq \aleph_{\eta+\|\Phi\|_{\mathrm{I_d}(\kappa)}}.$$

Proof We get an estimate for the cardinal $\aleph_\Theta$ defined in Theorem 2.2.6 by

$$\aleph_\Theta = 2^\kappa \cdot \sup\{T(\kappa, \aleph_{\eta(\xi)}) : \xi < \kappa\} \leq 2^\kappa \cdot \sup\{\aleph_{\eta(\xi)}^\kappa : \xi < \kappa\} \leq 2^\kappa \cdot \aleph_\eta = \aleph_\eta.$$

From Theorem 2.2.6 we can conclude, putting $\Psi := (\eta(\xi) + \Phi(\xi) : \xi < \kappa)$, that $|\mathcal{F}| \leq T(\aleph \circ \Psi) \leq \aleph_{\Theta+\|\Phi\|_{\mathrm{I_d}(\kappa)}} \leq \aleph_{\eta+\|\Phi\|_{\mathrm{I_d}(\kappa)}}$.

Lemma 2.3.3 *Assume that* $\aleph_\eta$ *is a limit cardinal,* $\kappa = \mathrm{cf}(\aleph_\eta)$, $(\eta(\xi) : \xi < \kappa)$ *is an increasing sequence cofinal in* η, *and* $\lambda \geq 1$ *is a cardinal. Then there exists a set* $\mathcal{F} \subseteq \prod_{\xi<\kappa} {}^{\lambda}\aleph_{\eta(\xi)}$ *of almost disjoint functions such that* $|\mathcal{F}| = \aleph_\eta^\lambda$.

Proof For each function $\Phi \in {}^{\lambda}\aleph_\eta$, we define a function $f_\Phi \in \prod_{\xi<\kappa} {}^{\lambda}\aleph_{\eta(\xi)}$ by

$$f_\Phi(\xi)(\alpha) = \begin{cases} \Phi(\alpha) & \text{if } \Phi(\alpha) < \aleph_{\eta(\xi)}, \\ 0 & \text{otherwise.} \end{cases}$$

If $\Phi, \Phi' \in {}^{\lambda}\aleph_\eta$ and $\Phi \neq \Phi'$, then there is an ordinal $\alpha_0 < \lambda$ with $\Phi(\alpha_0) \neq \Phi'(\alpha_0)$. Choose an ordinal $\xi_0 < \kappa$ such that $\aleph_{\eta(\xi_0)}$ is greater than both $\Phi(\alpha_0)$ and $\Phi'(\alpha_0)$. Then for every ξ with $\xi_0 \leq \xi < \kappa$ we can conclude that $f_\Phi(\xi)(\alpha_0) = \Phi(\alpha_0) \neq \Phi'(\alpha_0) = f_{\Phi'}(\xi)(\alpha_0)$, and thus $f_\Phi(\xi) \neq f_{\Phi'}(\xi)$ for all $\xi \in [\xi_0, \kappa)_{\mathrm{ON}}$. Consequently, the functions f_Φ and $f_{\Phi'}$ are almost disjoint, and the set $\mathcal{F} := \{f_\Phi : \Phi \in {}^{\lambda}\aleph_\eta\}$ has the desired properties.

Corollary 2.3.4 (Galvin-Hajnal Lemma) *Assume that* $\aleph_\eta$ *is a* κ*-strong singular cardinal, where* $\kappa := \mathrm{cf}(\aleph_\eta) > \omega$, $(\eta(\xi) : \xi < \kappa)$ *is a normal sequence cofinal in* η, *and* $\lambda \geq 1$ *is a cardinal. Further let* $\Phi \in {}^{\kappa}\mathrm{ON}$ *be a function satisfying*

$$\aleph_{\eta(\xi)}^\lambda = \aleph_{\eta(\xi)+\Phi(\xi)}$$

for all $\xi < \kappa$. *Then*

$$\aleph_\eta^\lambda \leq \aleph_{\eta+\|\Phi\|_{\mathrm{I_d}(\kappa)}}.$$

Proof By Lemma 2.3.3, there is a set $\mathcal{F} \subseteq \prod_{\xi<\kappa} {}^{\lambda}\aleph_{\eta(\xi)}$ of almost disjoint functions such that $|\mathcal{F}| = \aleph_\eta^\lambda$. Corollary 2.3.2 tells us that $|\mathcal{F}| \leq \aleph_{\eta+\|\Phi\|_{\mathrm{I_d}(\kappa)}}$.

Corollary 2.3.5 (Galvin-Hajnal Theorem) *Let $\aleph_\eta$ be a singular cardinal such that $\kappa := \mathrm{cf}(\aleph_\eta) > \omega$. If $\aleph_\eta$ is $\aleph_0$-strong, then, for every cardinal λ with $\kappa \le \lambda < \aleph_\eta$,*

$$\aleph_\eta^\lambda < \aleph_{(|\eta|^\lambda)^+}.$$

If $\aleph_\eta$ is even a strong limit cardinal, then

$$2^{\aleph_\eta} < \aleph_{(2^{|\eta|})^+},$$

and if in addition $\eta < \aleph_\eta$, then

$$2^{\aleph_\eta} < \aleph_{\aleph_\eta}.$$

Proof First assume in addition to our assumption that $\aleph_\eta$ is λ-strong. Then clearly $\aleph_\eta$ is κ-strong. We take a normal sequence $(\eta(\xi) : \xi < \kappa)$ cofinal in η. If $\Phi \in {}^\kappa\mathrm{ON}$ is defined by the equation $\aleph_{\eta(\xi)}^\lambda = \aleph_{\eta(\xi)+\Phi(\xi)}$, then Corollary 2.3.4 tells us that $\aleph_\eta^\lambda \le \aleph_{\eta+\|\Phi\|_{\mathrm{I_d}(\kappa)}}$. Since $\aleph_\eta$ is λ-strong, we can conclude that $\aleph_{\eta(\xi)}^\lambda < \aleph_\eta$ for all $\xi < \kappa$. Thus $\Phi \in {}^\kappa\eta$ and hereby $\|\Phi\|_{\mathrm{I_d}(\kappa)} < (|{}^\kappa\eta|)^+ \le (|\eta|^\lambda)^+$. This implies $\eta + \|\Phi\|_{\mathrm{I_d}(\kappa)} < (|\eta|^\lambda)^+$, which gives $\aleph_\eta^\lambda < \aleph_{(|\eta|^\lambda)^+}$ for this case.

Now let, according to our assumption, $\aleph_\eta$ be ω-strong but not λ-strong. We distinguish two cases.

Case 1 $2^\lambda \ge \aleph_\eta$.

Then we have $\aleph_\eta^\lambda \le (2^\lambda)^\lambda = 2^\lambda \le \aleph_{2^\lambda} < \aleph_{(2^\lambda)^+} \le \aleph_{(|\eta|^\lambda)^+}$.

Case 2 $2^\lambda < \aleph_\eta$.

Since $\aleph_\eta$ is not λ-strong, there is $\beta < \eta$ such that $\aleph_\beta^\lambda \ge \aleph_\eta$. Case 2 yields that the assumptions of Theorem 1.7.4 are fulfilled. So there is $\alpha \le \beta$ such that $\aleph_\alpha$ is singular and λ-strong, has cofinality at most λ and satisfies $\aleph_\beta^\lambda = \aleph_\alpha^{\mathrm{cf}(\aleph_\alpha)}$. Now we can conclude that $\aleph_\eta^\lambda = \aleph_\beta^\lambda$ and

$$(*) \qquad \aleph_\eta^\lambda = \aleph_\alpha^{\mathrm{cf}(\aleph_\alpha)}.$$

We want to apply the first part of the proof to $\aleph_\alpha$ instead of $\aleph_\eta$ and assume, contrary to our hopes, that $\mathrm{cf}(\aleph_\alpha) = \omega$. Since $\aleph_\eta$ is ω-strong, we get $\aleph_\alpha^{\aleph_0} < \aleph_\eta$, hence $\aleph_\alpha^{\aleph_0} < \aleph_\eta \le \aleph_\eta^\lambda$, contradicting $(*)$. Thus $\mathrm{cf}(\aleph_\alpha) > \omega$. By the above considerations we get $\aleph_\eta^\lambda = \aleph_\alpha^{\mathrm{cf}(\aleph_\alpha)} < \aleph_{(|\alpha|^{\mathrm{cf}(\aleph_\alpha)})^+} \le \aleph_{(|\eta|^\lambda)^+}$.

If in addition $\aleph_\eta$ is a strong limit cardinal, then Lemma 1.7.2 says that $2^{\aleph_\eta} = \aleph_\eta^{\mathrm{cf}(\aleph_\eta)}$. Together with $\kappa = \mathrm{cf}(\aleph_\eta) = \mathrm{cf}(\eta)$ and $|\eta|^\kappa \le |\eta|^{|\eta|} = 2^{|\eta|}$, our claim follows from the inequality $\aleph_\eta^\kappa < \aleph_{(|\eta|^\kappa)^+}$ which we have just shown.

Corollary 2.3.6 *If σ is an ordinal with $\sigma < \omega_1$ such that the set $\{\xi < \omega_1 : \aleph_\xi^{\aleph_1} \le \aleph_{\xi+\sigma}\}$ is stationary in $\aleph_1$, then*

$$\aleph_{\omega_1}^{\aleph_1} \le \aleph_{\omega_1+\sigma}.$$

Proof If $\beta < \omega_1$, then our assumption yields $\xi < \omega_1$ such that $\beta < \xi$ and $\aleph_\xi^{\aleph_1} \leq \aleph_{\xi+\sigma} < \aleph_{\omega_1}$, hence $\aleph_\beta^{\aleph_1} \leq \aleph_\xi^{\aleph_1} \leq \aleph_{\xi+\sigma}$. Therefore $\aleph_{\omega_1}$ is $\aleph_1$-strong, and the function Φ, defined by the equation $\aleph_\xi^{\aleph_1} = \aleph_{\xi+\Phi(\xi)}$, is a member of ${}^{\omega_1}\omega_1$. Now the set $\{\xi < \omega_1 : \Phi(\xi) \leq \sigma\}$ is stationary in $\aleph_1$, i.e., it is an element of $\mathrm{I_d}(\aleph_1)^+$. Therefore we know from Lemma 2.2.5 that $\|\Phi\|_{\mathrm{I_d}(\aleph_1)} \leq \sigma$. Furthermore we have $\mathrm{cf}(\aleph_{\omega_1}) = \aleph_1$. Putting $\eta(\xi) := \xi$ for all $\xi < \omega_1$, we can conclude, using Corollary 2.3.4, that $\aleph_{\omega_1}^{\aleph_1} \leq \aleph_{\omega_1+\sigma}$.

Corollary 2.3.7 *Assume that* $2^{\aleph_1} < \aleph_{\omega_1}$ *and* σ *is an ordinal with* $\sigma < \omega_1$ *such that the set* $\{\xi < \aleph_1 : \aleph_\xi^{\aleph_0} \leq \aleph_{\xi+\sigma}\}$ *is stationary in* $\aleph_1$. *Then*

$$\aleph_{\omega_1}^{\aleph_1} \leq \aleph_{\omega_1+\sigma}.$$

Proof We have $2^{\aleph_1} = \aleph_\rho$ for some $\rho < \omega_1$. Thus $\aleph_\xi^{\aleph_0} \geq \aleph_\rho^{\aleph_0} = 2^{\aleph_1}$ for every ξ with $\max\{\rho, \aleph_0\} < \xi < \aleph_1$, and from Corollary 1.6.13 we can conclude that $\aleph_\xi^{\aleph_1} = 2^{\aleph_1} \cdot \aleph_\xi^{\aleph_0} = \aleph_\xi^{\aleph_0}$. It follows that

$$\{\xi < \aleph_1 : \rho < \xi \wedge \omega < \xi \wedge \aleph_\xi^{\aleph_0} \leq \aleph_{\xi+\sigma}\} = \{\xi < \aleph_1 : \rho < \xi \wedge \omega < \xi \wedge \aleph_\xi^{\aleph_1} \leq \aleph_{\xi+\sigma}\}.$$

Now the desired assertion can be inferred from Corollary 2.3.6.

Lemma 2.3.8 *If* $\kappa > \omega$ *is a regular cardinal and* $\Phi \in {}^\kappa\mathrm{ON}$ *is a function, then* $\|\Phi\|_{\mathrm{I_d}(\kappa)} < \kappa$ *iff the set* $\{\xi < \kappa : \Phi(\xi) < \xi\}$ *is a member of* $\mathrm{I_d}(\kappa)^+$, *i.e., is stationary in* κ.

Proof If the set $S := \{\xi < \kappa : \Phi(\xi) < \xi\}$ is stationary in κ, then, by Fodor's Theorem, there is an ordinal $\alpha < \kappa$ and a stationary set $S^* \subseteq S$ such that $\Phi \restriction S^* = S^* \times \{\alpha\}$. Together with Lemma 2.2.5 we can conclude that $\|\Phi\|_{\mathrm{I_d}(\kappa)} \leq \alpha < \kappa$.

If conversely $\|\Phi\|_{\mathrm{I_d}(\kappa)} = \alpha < \kappa$, then we know from Lemma 2.2.5 that the set $S_\alpha := \{\xi < \kappa : \Phi(\xi) \leq \alpha\}$ is stationary in κ. Therefore $S_\alpha \setminus (\alpha + 1)$ is also stationary in κ; since $S_\alpha \setminus (\alpha + 1) \subseteq S$, this also holds for S.

Corollary 2.3.9 *If the set* $\{\xi < \omega_1 : \aleph_\xi^{\aleph_1} < \aleph_{\xi+\xi}\}$ *is stationary in* $\aleph_1$, *then*

$$\aleph_{\omega_1}^{\aleph_1} < \aleph_{\omega_1+\omega_1}.$$

Proof The equation $\aleph_\xi^{\aleph_1} = \aleph_{\xi+\Phi(\xi)}$ defines a function $\Phi \in {}^{\omega_1}\mathrm{ON}$. By assumption, we have $\{\xi < \omega_1 : \Phi(\xi) < \xi\} \in \mathrm{I_d}(\aleph_1)^+$, hence $\|\Phi\|_{\mathrm{I_d}(\omega_1)} < \aleph_1$ by Lemma 2.3.8, and furthermore $\aleph_{\omega_1}$ is $\aleph_1$-strong. Now the assumptions of Corollary 2.3.4 are fulfilled, taking $\eta(\xi) := \xi$ for all $\xi < \kappa$. So we get $\aleph_{\omega_1}^{\aleph_1} \leq \aleph_{\omega_1+\|\Phi\|_{\mathrm{I_d}(\omega_1)}} < \aleph_{\omega_1+\omega_1}$.

Lemma 2.3.10 *Let $\aleph_\eta$ be a limit cardinal, $\kappa := \mathrm{cf}(\aleph_\eta)$, and $(\eta(\xi) : \xi < \kappa)$ be an increasing sequence that is cofinal in η. Then there is a set $\mathcal{F}$ of almost disjoint functions such that $\mathcal{F} \subseteq \prod_{\xi<\kappa} {}^{\aleph_{\eta(\xi)}}2$ and*

$$2^{\aleph_\eta} = |\mathcal{F}| = |\prod_{\xi<\kappa} {}^{\aleph_{\eta(\xi)}}2|.$$

Proof For each $X \in \mathcal{P}(\aleph_\eta)$, we define a function $f_X \in \prod_{\xi<\kappa} \mathcal{P}(\aleph_{\eta(\xi)})$ by $f_X(\xi) = X \cap \aleph_{\eta(\xi)}$. If $X, Y \in \mathcal{P}(\aleph_\eta)$ are different, then there is an ordinal $\xi_0 < \kappa$ such that $f_X(\xi_0) \neq f_Y(\xi_0)$. Since $(\eta(\xi) : \xi < \kappa)$ is nondecreasing, we get $f_X(\xi) \neq f_Y(\xi)$ for all $\xi \geq \xi_0$, and thus f_X and f_Y are almost disjoint. On account of $2^{\aleph_{\eta(\xi)}} = |\mathcal{P}(\aleph_{\eta(\xi)})|$ we can identify f_X with a function in $\prod_{\xi<\kappa} {}^{\aleph_{\eta(\xi)}}2$ and put

$$\mathcal{F} := \{f_X : X \in \mathcal{P}(\aleph_\eta)\}.$$

Then we get $2^{\aleph_\eta} = |\mathcal{P}(\aleph_\eta)| = |\mathcal{F}| \leq |\prod_{\xi<\kappa} {}^{\aleph_{\eta(\xi)}}2| \leq (2^{\aleph_\eta})^\kappa = 2^{\aleph_\eta}$.

Corollary 2.3.11 *Assume that $\aleph_\eta$ is a singular cardinal with $\kappa := \mathrm{cf}(\aleph_\eta) > \omega$, $(\eta(\xi) : \xi < \kappa)$ is a normal function cofinal in η, and $\sigma < \kappa$ is an ordinal, such that $\{\xi < \kappa : 2^{\aleph_{\eta(\xi)}} \leq \aleph_{\eta(\xi)+\sigma}\} \in \mathrm{I_d}(\kappa)^+$. Then*

$$2^{\aleph_\eta} \leq \aleph_{\eta+\sigma}.$$

Proof As in the proof of Corollary 2.3.6 it can be verified that the equation $2^{\aleph_{\eta(\xi)}} = \aleph_{\eta(\xi)+\Phi(\xi)}$ defines a function $\Phi \in {}^\kappa\kappa$, and that $\aleph_\eta$ is a strong limit cardinal. Since $\{\xi < \kappa : \Phi(\xi) \leq \sigma\} \in \mathrm{I_d}(\kappa)^+$, Lemma 2.2.5 says that $\|\Phi\|_{\mathrm{I_d}(\kappa)} \leq \sigma$. Lemma 2.3.10 tells us that there is a set $\mathcal{F} \subseteq \prod_{\xi<\kappa} \aleph_{\eta(\xi)+\Phi(\xi)}$ of almost disjoint functions such that $|\mathcal{F}| = 2^{\aleph_\eta}$. By Corollary 2.3.2, we get $|\mathcal{F}| \leq \aleph_{\eta+\|\Phi\|_{\mathrm{I_d}(\kappa)}} \leq \aleph_{\eta+\sigma}$.

Corollary 2.3.12 (J. Silver) *If $\aleph_\eta$ is a singular cardinal of uncountable cofinality, and if the set $\{\kappa < \aleph_\eta : \kappa \in \mathrm{ICN} \wedge 2^\kappa = \kappa^+\}$ is stationary in $\aleph_\eta$, then $2^{\aleph_\eta} = \aleph_{\eta+1}$.*

Corollary 2.3.13 *Assume that κ and λ are regular cardinals such that $\aleph_0 < \kappa < \lambda$ and $\mu^\kappa < \lambda$ for every cardinal $\mu < \lambda$ with $\mathrm{cf}(\mu) = \kappa$. Further let $\Phi \in {}^\kappa\lambda$ be a function. Then, for every set $\mathcal{F} \subseteq \prod_{\xi<\kappa} \aleph_{\Phi(\xi)}$ of almost disjoint functions, we have $|\mathcal{F}| < \aleph_\lambda$.*

Proof Let the function $(\eta(\xi) : \xi < \kappa)$ be defined by $\eta(\xi) = 0$ for all $\xi < \kappa$. Then we obtain an estimate for the cardinal $\aleph_\Theta$ in Theorem 2.2.6 by $\aleph_\Theta \leq 2^\kappa \cdot \aleph_0^\kappa \leq \kappa^\kappa \cdot \kappa^\kappa < \lambda \leq \aleph_\lambda$, and this gives $\Theta < \lambda$ for the ordinal Θ. Now Lemma 2.2.10 says that $\|\Phi\|_I < \lambda$, where $I := \mathrm{I_d}(\kappa)$, and from Theorem 2.2.6 we can conclude that $|\mathcal{F}| \leq \aleph_{\Theta+\|\Phi\|_I} < \aleph_\lambda$.

Corollary 2.3.14 *Assume that κ and λ are regular cardinals with $\omega < \kappa < \lambda$, satisfying $\mu^\kappa < \lambda$ for all cardinals $\mu < \lambda$ with $\mathrm{cf}(\mu) = \kappa$. Further let $(\mu_\xi : \xi < \kappa)$ be a sequence of cardinals such that $\prod_{\xi<\alpha} \mu_\xi < \aleph_\lambda$ for all $\alpha < \kappa$. Then $\prod_{\xi<\kappa} \mu_\xi < \aleph_\lambda$.*

Proof Let the equation $\aleph_{\Phi(\alpha)} = \prod_{\xi<\alpha} \mu_\xi$, for $\alpha < \kappa$, define the function Φ on κ. Then $\Phi \in {}^\kappa\lambda$, and by Lemma 2.2.7 there is a set $\mathcal{F}$ of almost disjoint functions such that $\mathcal{F} \subseteq \prod_{\alpha<\kappa}(\prod_{\xi<\alpha} \mu_\xi) = \prod_{\xi<\kappa} \aleph_{\Phi(\xi)}$ and $|\mathcal{F}| = \prod_{\xi<\kappa} \mu_\xi$. Now Corollary 2.3.13 says that $|\mathcal{F}| < \aleph_\lambda$.

Corollary 2.3.15 *Assume that κ and λ are regular cardinals with $\omega < \kappa < \lambda$, λ is κ-strong, and μ is an arbitrary cardinal. If $\mu^\nu < \aleph_\lambda$ for all $\nu \in \kappa \cap \mathrm{CN}$, then $\mu^\kappa < \aleph_\lambda$.*

Corollary 2.3.16 *Assume that κ and λ are regular cardinals, $\omega < \kappa$, λ is κ-strong, and ν is a cardinal with $\mathrm{cf}(\nu) = \kappa$. If $2^\mu < \aleph_\lambda$ for all $\mu \in \mathrm{CN} \cap \nu$, then $2^\nu < \aleph_\lambda$.*

Proof We choose a sequence $(\nu_\xi : \xi < \kappa)$ of cardinals satisfing $\nu_\xi < \nu$ for all $\xi < \kappa$ and $\nu = \sum_{\xi<\kappa} \nu_\xi$. For each $\xi < \kappa$, we put $\mu_\xi := 2^{\nu_\xi}$. Then $\prod_{\xi<\alpha} \mu_\xi = 2^{\sum_{\xi<\alpha} \nu_\xi} < \aleph_\lambda$ for every ordinal $\alpha < \kappa$, and this together with Corollary 2.3.14 gives $2^{\sum_{\xi<\kappa} \nu_\xi} = 2^\nu < \aleph_\lambda$.

Corollary 2.3.17 *Let α be a limit ordinal with $\mathrm{cf}(\alpha) > \omega$. If $2^{\aleph_\xi} < \aleph_{(|\alpha|^{\mathrm{cf}(\alpha)})^+}$ for all ordinals $\xi < \alpha$, then*

$$2^{\aleph_\alpha} < \aleph_{(|\alpha|^{\mathrm{cf}(\alpha)})^+}.$$

Proof Let $\kappa := \mathrm{cf}(\alpha)$, $\lambda := (|\alpha|^\kappa)^+$, and $\nu := \aleph_\alpha$. Then κ and λ are regular cardinals with $\kappa < \lambda$, λ is κ-strong, and $\mathrm{cf}(\nu) = \mathrm{cf}(\alpha) = \kappa$. Thus all assumptions of Corollary 2.3.16 are satisfied, and we obtain $2^{\aleph_\alpha} < \aleph_{(|\alpha|^\kappa)^+}$.

Corollary 2.3.18 *If $\aleph_\xi^{\aleph_0} < \aleph_{(2^{\aleph_1})^+}$ for all $\xi < \omega_1$, then $\aleph_{\omega_1}^{\aleph_1} < \aleph_{(2^{\aleph_1})^+}$.*

Proof The assumption and Lemma 1.6.15 imply that

$$\aleph_{\omega_1}^{\aleph_0} = \aleph_{\omega_1} \cdot \sum_{\xi<\omega_1} \aleph_\xi^{\aleph_0} < \aleph_{(2^{\aleph_1})^+}, \quad \text{since} \quad \mathrm{cf}(\aleph_{(2^{\aleph_1})^+}) = (2^{\aleph_1})^+ > \aleph_1.$$

If we put $\kappa := \aleph_1$, $\lambda := (2^{\aleph_1})^+$, and $\mu := \aleph_{\omega_1}$, then κ and λ are regular cardinals such that $\kappa < \lambda$. Furthermore, λ is κ-strong and $\mu^\nu < \aleph_\lambda$ for all $\nu \in \kappa \cap \mathrm{CN}$. From Corollary 2.3.15 we can conclude that $\aleph_{\omega_1}^{\aleph_1} < \aleph_{(2^{\aleph_1})^+}$.

Exercises

1) Assume that $\aleph_\eta$ is a κ-strong singular cardinal, where $\kappa := \mathrm{cf}(\aleph_\eta) > \aleph_0$. Further let $(\eta(\xi) : \xi < \kappa)$ be a normal sequence cofinal in η such that the set $S := \{\xi < \kappa : \gimel(\aleph_{\eta(\xi)}) = \aleph^+_{\eta(\xi)}\}$ is stationary in κ. Prove that $\gimel(\aleph_\eta) = \aleph^+_\eta$.
Hint: Exercise 1.8.22 says that $\{\mu < \aleph_\eta : \mu \text{ is } \kappa\text{-strong}\}$ is a club in $\aleph_\eta$, and thus
$$S^* := \{\xi < \kappa : \aleph_{\eta(\xi)} \text{ is } \kappa\text{-strong and } \gimel(\aleph_{\eta(\xi)}) = \aleph^+_{\eta(\xi)}\}$$
is stationary in κ. If $\xi \in S^*$ is a limit ordinal, use a golden rule to prove that $\aleph^\kappa_{\eta(\xi)} = \gimel(\aleph_{\eta(\xi)})$ and apply the Galvin-Hajnal lemma.
2) (Silver) If the singular cardinal hypothesis holds for all singular cardinals with countable cofinality, then it holds for all singular cardinals.
Hint: Use transfinite induction on singular cardinals to show that from $\omega < \kappa := \mathrm{cf}(\aleph_\eta) < \aleph_\eta$ and $2^{\mathrm{cf}(\aleph_\eta)} < \aleph_\eta$ it follows that $\gimel(\aleph_\eta) = \aleph^+_\eta$. Prove first that $\aleph_\eta$ is κ-strong, and apply the previous question.
3) Assume that $\aleph_\eta$ is a κ-strong singular cardinal, where $\kappa := \mathrm{cf}(\aleph_\eta) > \aleph_0$. Further let $(\eta(\xi) : \xi < \kappa)$ be a normal sequence cofinal in η, and $\Phi \in {}^\kappa\mathrm{ON}$ be a function, such that $2^{\aleph_{\eta(\xi)}} \leq \aleph_{\eta(\xi)+\Phi(\xi)}$ for all $\xi < \kappa$. Then
$$2^{\aleph_\eta} \leq \aleph_{\eta+\|\Phi\|_{I_d(\kappa)}}.$$
Hint: Apply Lemma 2.3.10 and Corollary 2.3.2.
4) Let $\beta < \aleph_1$ be an ordinal. If $2^{\aleph_\eta} < \aleph^{+\beta}_\eta$ holds for all singular cardinals with countable cofinality, then this assertion holds for all singular cardinals.
Hint: Apply Exercise 6 from Section 2.2.
5) Assume that $2^{\aleph_1} < \aleph_{\omega_1}$ and $\Phi \in {}^{\omega_1}\omega_1$ is a function such that the set $\{\xi < \omega_1 : \aleph^{\aleph_0}_\xi \leq \aleph_{\xi+\Phi(\xi)}\}$ is a club in ω_1. Prove that
$$\gimel(\aleph_{\omega_1}) \leq \aleph_{\omega_1+\|\Phi\|_{I_d(\omega_1)}}.$$
Hint: Remember that $\aleph^{\aleph_1}_\alpha = \max\{2^{\aleph_1}, \aleph^{\aleph_0}_\alpha\}$ for all α with $\aleph_0 \leq \alpha < \aleph_1$ (Corollary 1.6.13), and apply the Galvin-Hajnal lemma.
6) If the set $\{\xi < \omega_1 : 2^{\aleph_\xi} \leq \aleph_{\xi+\xi+2}\}$ is stationary in ω_1, then
$$2^{\aleph_{\omega_1}} \leq \aleph_{\omega_1+\omega_1+2}.$$
Hint: Prove that $\aleph_{\omega_1}$ is a strong limit cardinal, and use Lemma 2.3.10 and Corollary 2.3.2.
7) If $\aleph_{\omega_1}$ is a strong limit cardinal, and if $\{\xi < \omega_1 : \aleph^{\aleph_0}_\xi \leq \aleph_{\xi+\xi}\}$ is a club in ω_1, then
$$2^{\aleph_{\omega_1}} < \aleph_{\omega_1+\omega_1}.$$
Hint: Use the previous questions.

Chapter 3

Ordinal Functions

In the Galvin-Hajnal theorem, the question of whether it holds for singular cardinals with countable cofinality was left open. Let $\aleph_\delta$ be singular and ν be a cardinal. If $\delta = \aleph_\delta$, then $\aleph_\delta^\nu = |\delta|^\nu < \aleph_{(|\delta|^\nu)^+}$, and thus the Galvin-Hajnal theorem holds for the case that δ is a fixed point of the aleph function. Let us now turn to the case that $\delta < \aleph_\delta$. Then $|\delta| < \aleph_\delta$, and for every regular cardinal λ with $|\delta| < \lambda < \aleph_\delta$, the set $\mathsf{a} := [\lambda, \aleph_\delta)_{\text{reg}}$ is an interval of regular cardinals satisfying $|\mathsf{a}| < \min(\mathsf{a})$, since $|\mathsf{a}| \leq |\delta|$. Shelah defines an operator pcf which assigns to each set a of regular cardinals and each cardinal μ a set $\text{pcf}_\mu(\mathsf{a})$ of regular cardinals satisfying the following properties for $\mu \geq 1$:

a) $\mathsf{a} \subseteq \text{pcf}_\mu(\mathsf{a})$.
b) $\min(\mathsf{a}) = \min \text{pcf}_\mu(\mathsf{a})$.
c) If $|\mathsf{a}| < \min(\mathsf{a})$, then $|\text{pcf}_\mu(\mathsf{a})| \leq |\mathsf{a}|^\mu$.
d) If a is an *interval* of regular cardinals with $|\mathsf{a}| < \min(\mathsf{a})$, then $\text{pcf}_\mu(\mathsf{a})$ is an interval of regular cardinals.

Now we continue the discussion of the special case that $\mathsf{a} = [\lambda, \aleph_\delta)_{\text{reg}}$. We have $\lambda = \aleph_\alpha$ for some ordinal $\alpha < \delta$, and therefore $\sup \text{pcf}_\mu(\mathsf{a}) < \aleph_{\alpha+|\text{pcf}_\mu(\mathsf{a})|^+}$ by a), b) and d). If λ is a μ-strong cardinal, then Theorem 6.2.4, which is of course due to Shelah and represents a main result of this book, tells us that $\sup \text{pcf}_\mu(\mathsf{a}) = \aleph_\delta^\mu$. So we obtain $\aleph_\delta^\mu < \aleph_{\alpha+|\text{pcf}_\mu(\mathsf{a})|^+}$. By c) we have $|\text{pcf}_\mu(\mathsf{a})| \leq |\mathsf{a}|^\mu \leq |\delta|^\mu$, and $\alpha < \delta < (|\delta|^\mu)^+$ implies

$$(*) \qquad \aleph_\delta^\mu < \aleph_{(|\delta|^\mu)^+}.$$

With Lemma 1.7.3 one can show, roughly speaking, that a μ-strong cardinal λ with $|\delta| < \lambda < \aleph_\delta$ can always be found (for details see Section 8.1). Therefore $(*)$ is a powerful generalization of the Galvin-Hajnal theorem. If in particular $\mu = |\delta|$, then $|\text{pcf}_\mu(\mathsf{a})| < |\mathsf{a}|^{+4}$ by Theorem 7.2.6. In the case that $\aleph_\delta \leq 2^{|\delta|}$ we have $\aleph_\delta^{|\delta|} \leq 2^{|\delta|}$. If $2^{|\delta|} < \aleph_\delta$, we can conclude, using the fact that $\lambda := (2^{|\delta|})^+$ is a $|\delta|$-strong cardinal and the estimate $\aleph_\delta^{|\delta|} < \aleph_{\alpha+|\mathrm{pcf}_\mu(\mathsf{a})|^+}$ from above, that $\aleph_\delta^{|\delta|} < \aleph_{|\delta|^{+4}}$, since $\alpha < |\delta|^{+4}$. Altogether we obtain

$$(**) \qquad \aleph_\delta^{|\delta|} < \max\{\aleph_{|\delta|^{+4}}, (2^{|\delta|})^+\}.$$

Since $\aleph_\delta^{\mathrm{cf}(\delta)} \leq \aleph_\delta^{|\delta|}$, $(**)$ yields an estimate for the gimel function at singular arguments.

The proofs of the assertions $(*)$ and $(**)$ are complicated, and the methods and techniques used for it are manifold. The theory developed solely for these proofs, the so-called pcf-theory, originates with Saharon Shelah. It is self-contained and allows further interesting applications which will be treated in Chapter 8.

In this chapter, we define first in Section 3.1 the fundamental notion of the true cofinality of a set $\mathcal{S} \subseteq {}^A\mathrm{ON}$ of ordinal functions modulo I, written as $\mathrm{tcf}(\mathcal{S}/I)$, where I is an ideal on their common domain A. The special case that $\mathcal{S} = \prod f := \prod_{a\in A} f(a)$ for some ordinal function $f \in {}^A\mathrm{ON}$, will play the decisive role in our investigations. If $\mathrm{tcf}(\prod f/I)$ exists, then it is a cardinal, namely 0 in the trivial case that $\prod f = \emptyset$, 1 if $\prod f$ has a maximum under $\leq_I$, and a regular cardinal in all other cases. If I is a maximal ideal on A and D is the dual ultrafilter of I, then $\mathrm{tcf}(\prod f/I)$ exists, and we also write $\mathrm{cf}(\prod f/I)$ and $\mathrm{cf}(\prod f/D)$ instead of $\mathrm{tcf}(\prod f/I)$.

If the cardinal $\lambda = \mathrm{tcf}(\prod f/I)$ exists, then it turns out that the partial ordering $(\prod f, <_I)$ is λ-directed, i.e., that every subset S of $\prod f$ whose cardinality is less than λ has an upper bound in $\prod f$ under the relation $<_I$. Theorem 3.1.13 will tell us that it suffices for the investigation of true cofinalities of sets $\prod f$ to consider functions which have regular cardinals for all of their values. Therefore in Section 3.3 we concentrate our attention on the special case $f = \mathrm{id}_\mathsf{a}$, where a is a set of regular cardinals satisfying $|\mathsf{a}| < \min(\mathsf{a})$. Such a set a will be called progressive.

In Section 3.2, special sequences of ordinal functions are introduced, namely sequences satisfying the κ-bound property, κ-rapid sequences and small sequences. These notions lead to a comparatively short proof of the main lemma of pcf-theory whose importance we mentioned above: If a is a progressive interval of regular cardinals, then the set

$$\mathrm{pcf}(\mathsf{a}) := \{\mathrm{cf}(\prod \mathsf{a}/D) : D \text{ is an ultrafilter on } \mathsf{a}\}$$

of possible cofinalities of a is also an interval.

In Section 3.4, the important ideal $\mathcal{J}_{<\lambda}(\mathsf{a})$ is introduced and its essential properties are developed. We will see that, for progressive sets a, the set $\mathrm{pcf}(\mathsf{a})$ has a maximum which is exactly the cofinality of the partial ordering $(\prod \mathsf{a}, \leq_\mathsf{a})$.

3.1 Suprema and Cofinalities

For the rest of this chapter, A denotes a nonempty set and I an ideal on A. The domain of the considered ordinal functions is A if it is not explicitly specified otherwise. We remind the reader of the fact that we often omit the addition "modulo I" when we are dealing with assertions on ordinal functions.

Definition *Let $\mathcal{S}$ be a set (or a family) of ordinal functions and h be an ordinal function. h is called an* **upper bound of $\mathcal{S}$ modulo I** *iff $f \leq_I h$ for all members f of $\mathcal{S}$. We call h a* **supremum of $\mathcal{S}$ modulo I** *iff h is an upper bound of $\mathcal{S}$ modulo I and $h \leq_I g$ holds for every upper bound g of $\mathcal{S}$. If $\mathcal{T}$ is a subset (subfamily) of $\mathcal{S}$, then $\mathcal{T}$ is called* **unbounded in $\mathcal{S}$ modulo I** *iff no member of $\mathcal{S}$ is an upper bound of $\mathcal{T}$ modulo I.*

For ordinal functions f and g, we use the following denotations:

$$\begin{aligned} \mathrm{B}(f,g) &:= \{a \in A : f(a) < g(a)\}, \\ \mathrm{B}[f,g] &:= \{a \in A : f(a) \leq g(a)\}. \end{aligned}$$

Remark In general, the relation $\leq_I$ is not antisymmetric, and therefore suprema are not uniquely determined. On the other hand, *modulo I* all suprema of $\mathcal{S}$ are equal. Further we note:

$$g \leq_I f \iff \mathrm{B}(f,g) \in I \quad \text{and} \quad g <_I f \iff \mathrm{B}[f,g] \in I.$$

A simple and practical necessary and sufficient condition for the existence of suprema is given by the following lemma.

Lemma 3.1.1 *If $\mathcal{S}$ is a set of ordinal functions, then an ordinal function g is a supremum of $\mathcal{S}$ iff g is an upper bound of $\mathcal{S}$ and there is no upper bound h of $\mathcal{S}$ satisfying $h \lneqq_I g$.*

Proof Let g be an upper bound of $\mathcal{S}$ with the property that there is no upper bound h of $\mathcal{S}$ such that $h \lneqq_I g$. We assume, to get a contradiction, that g is not a supremum of $\mathcal{S}$. Then there is an upper bound h of $\mathcal{S}$ such that $\neg(g \leq_I h)$. Put $p := \min\{g, h\}$. Obviously p is an upper bound of $\mathcal{S}$, and we get $p \leq g$. Since $g \leq_I h$ is not the case and $p \leq h$, $g \leq_I p$ cannot hold, and therefore we have in particular *not* $p =_I g$. So we get $p \lneqq_I g$, contradicting the assumption. Consequently, g is a supremum of $\mathcal{S}$. The proof of the reverse direction of the equivalence is trivial.

Definition *If $\mathcal{S}$ is a set of ordinal functions and $\mathcal{T} \subseteq \mathcal{S}$, then $\mathcal{T}$ is called* **cofinal in $\mathcal{S}$ modulo I** *iff $\mathcal{T}$ is cofinal in $\mathcal{S}$ with respect to the relation $\leq_I$ on $\mathcal{S}$, that means iff for all $f \in \mathcal{S}$ there is $g \in \mathcal{T}$ such that $f \leq_I g$. The cardinal number*

$\mathrm{cf}(\mathcal{S}, \leq_I)$ *from Section 1.5 is also called the* **cofinality of $\mathcal{S}$ modulo I**; *remember that*

$$\mathrm{cf}(\mathcal{S}, \leq_I) = \min\{|T| : T \text{ is a cofinal subset of } \mathcal{S} \text{ modulo } I\}.$$

A sequence $(f_\xi : \xi < \sigma)$ *of ordinal functions is called* **increasing modulo I** *(***strictly increasing modulo I***) iff* $f_\alpha \leq_I f_\beta$ $(f_\alpha <_I f_\beta)$ *for all* α *and* β *with* $\alpha < \beta < \sigma$. *If* $\mathcal{S}$ *is a set of ordinal functions and* $\bar{f} = (f_\xi : \xi < \sigma)$ *is a sequence of members of* $\mathcal{S}$*, then we say that* $\bar{f}$ *is* **cofinal in $\mathcal{S}$ modulo I** *iff it is strictly increasing modulo* I *and its range* $\{f_\xi : \xi < \sigma\}$ *is cofinal in* $\mathcal{S}$ *modulo* I. *If there is such a sequence, we define the* **true cofinality of $\mathcal{S}$ modulo I** *(*true cofinality*) by*

$$\mathrm{tcf}(\mathcal{S}/I) := \min\left\{\sigma : \begin{array}{c}\sigma \text{ is the length of a sequence}\\ \text{which is cofinal in } \mathcal{S} \text{ modulo } I\end{array}\right\}.$$

$f \in \mathcal{S}$ *is called a* **maximal element of $\mathcal{S}$ modulo I** *iff there is no* $h \in \mathcal{S}$ *such that* $f \lneqq_I h$.

Remark The true cofinality of $\mathcal{S}$ does not always exist (for example, if $\mathcal{S}$ has two maximal elements modulo I which take different values for each member of a positive subset of A). *If we use the denotation* $\mathrm{tcf}(\mathcal{S}/I) = \lambda$, *this will always imply that there is a sequence of elements of* $\mathcal{S}$ *which is cofinal in* $\mathcal{S}$ *modulo* I, *i.e., that the true cofinality of* $\mathcal{S}$ *modulo* I *is defined.* On account of the duality of filters and ideals, the notions defined above can be transferred canonically to filters.

Definition *If* f *is an ordinal function on* A, *we put*

$$\prod f := \prod_{a \in A} f(a).$$

Obviously we have $g < f$ *iff* g *is a member of* $\prod f$. *If a sequence of members of* $\prod f$ *is cofinal in the set* $\prod f$ *we also say that it is* **cofinal in f**.

The following lemma is a direct consequence of Lemma 2.1.2. It says that the values of members of sequences which are cofinal in f (or more general, which are cofinal in arbitrary sets $\mathcal{S}$ of ordinal functions) may be changed appropriately on sets which are members of I. It is a simple standard tool in the following chapters. For those readers who are familiar with reduced products, it is trivial.

Lemma 3.1.2 *Assume that* f *is an ordinal function on* A *and* $(f_\xi : \xi < \sigma)$ *and* $(g_\xi : \xi < \sigma)$ *are sequences of members of* $\prod f$ *satisfying* $f_\xi =_I g_\xi$ *for all* $\xi < \sigma$. *Further let* $(f_\xi : \xi < \sigma)$ *be cofinal in* f. *Then the sequence* $(g_\xi : \xi < \sigma)$ *is also cofinal in* f. *Consequently, if one changes the values of* f_ξ *on a set* $X_\xi \in I$ *for every* $\xi < \sigma$ *in such a way that the resulting function* f'_ξ *is a member of* $\prod f$, *then the sequence* $(f'_\xi : \xi < \sigma)$ *is also cofinal in* f. *Corresponding assertions hold for arbitrary sets* $\mathcal{S}$ *of ordinal functions and for other notions such as strict monotonicity, supremum, upper bound, and so on.*

The next lemma tells us that the true cofinality is 0, 1 or a regular cardinal and that we already know it if we have found *one* cofinal sequence of regular length.

Lemma 3.1.3 *Let* $\mathcal{S}$ *be a set of ordinal functions which posesses a sequence* $(f_\xi : \xi < \sigma)$ *which is cofinal in* $\mathcal{S}$ *modulo* I. *Then we have* $\mathrm{tcf}(\mathcal{S}/I) = \mathrm{cf}(\sigma)$. *Furthermore the following assertions hold for* $\lambda := \mathrm{tcf}(\mathcal{S}/I)$.

a) $\lambda \in \{0,1\} \cup \mathrm{ICN}$ *and* $\mathrm{cf}(\lambda) = \lambda$.
b) $\lambda = 0$ *iff* $\mathcal{S} = \emptyset$.
c) $\lambda = 1$ *iff* $\mathcal{S}$ *has a maximum under* $\leq_I$.

Proof By assumption, the ordinal $\lambda := \mathrm{tcf}(\mathcal{S}/I)$ exists. b) and c) are obviously true. Now consider an arbitrary sequence $(f_\xi : \xi < \sigma)$ which is cofinal in $\mathcal{S}$ modulo I. If $\sigma \in \{0,1\}$, it is obvious that $\lambda = \mathrm{cf}(\sigma) = \sigma$. If $1 < \sigma < \aleph_0$ or σ is a successor ordinal, then $\mathrm{cf}(\sigma) = 1 = \mathrm{tcf}(\mathcal{S}/I)$, since clearly the one-element sequence $(f_{\sigma-1})$ is also cofinal in $\mathcal{S}$ modulo I. In particular we get $\lambda \notin \omega \setminus \{0,1\}$, and λ is not a successor ordinal. So we can assume that σ is a limit ordinal. Then we can choose a sequence $(\alpha_i : i < \mathrm{cf}(\sigma))$ of ordinal numbers which is cofinal in σ. It is clear that the sequence $(f_{\alpha_i} : i < \mathrm{cf}(\sigma))$ is also cofinal in $\mathcal{S}$, and we obtain $\lambda \leq \mathrm{cf}(\sigma)$. Furthermore it is easy to see that λ is infinite.

By definition there is a cofinal sequence $(g_\xi : \xi < \lambda)$ of members of $\mathcal{S}$. Assume that $\lambda < \mathrm{cf}(\sigma)$. For each $\xi < \lambda$, we can choose an ordinal $\beta_\xi < \mathrm{cf}(\sigma)$ such that $g_\xi \leq_I f_{\beta_\xi}$. Since $\mathrm{cf}(\sigma)$ is regular, the ordinal $\beta := \sup\{\beta_\xi + 1 : \xi < \lambda\}$ is smaller than $\mathrm{cf}(\sigma)$, and so we have by construction

$$g_\xi <_I g_{\xi+1} \leq_I f_{\beta_{\xi+1}} \leq_I f_\beta$$

for all $\xi < \lambda$. This contradicts the fact that the sequence $(g_\xi : \xi < \lambda)$ is cofinal in $\mathcal{S}$. Thus we have shown that $\mathrm{cf}(\sigma) = \lambda = \mathrm{tcf}(\mathcal{S}/I)$, and that $\mathrm{tcf}(\mathcal{S}/I) \notin \{0,1\}$ implies that $\mathrm{tcf}(\mathcal{S}/I)$ is a regular cardinal.

Lemma 3.1.4 *Assume that* τ *is a limit ordinal and* $(f_\xi : \xi < \tau)$ *is a sequence of ordinal functions which is cofinal in the ordinal function* f. *Then* f *is a supremum of* $(f_\xi : \xi < \tau)$.

Proof We want to apply Lemma 3.1.1. Clearly f is an upper bound of the sequence, since all members of the sequence are by definition elements of $\prod f$. For an indirect proof we assume that, modulo I, there is an upper bound

f' of the sequence such that $f' \lneqq_I f$. In particular, the set $X := \mathrm{B}(f', f) = \{a \in A : f'(a) < f(a)\}$ is positive. Since $\prod f \neq \emptyset$, we have $f(a) \neq 0$ for all $a \in A$. So we can define the ordinal function $f^* \in \prod f$ by

$$f^*(a) := \begin{cases} f'(a) & \text{if } a \in X, \\ 0 & \text{if } a \in A \setminus X. \end{cases}$$

By assumption, there is an ordinal $\alpha < \tau$ such that $f^* \leq_I f_\alpha$. Since $f_\alpha <_I f_{\alpha+1} \leq_I f'$, we get $f^* <_I f'$, hence $\mathrm{B}[f', f^*] \in I$. But clearly we have $X \subseteq \mathrm{B}[f', f^*]$ which gives $X \in I$, contradicting the fact that X is positive.

A simple example for the existence of a true cofinality which will be used later is given by the following lemma.

Lemma 3.1.5 *If X is a nonempty set and α is an ordinal with* $\mathrm{cf}(\alpha) > |X|$, *then* $\mathrm{tcf}({}^X\alpha/\{\emptyset\}) = \mathrm{cf}(\alpha)$.

Proof Since X is nonempty and $|X| < \mathrm{cf}(\alpha)$, we have $\mathrm{cf}(\alpha) \geq \omega$. We can choose a normal sequence $(\alpha_i : i < \mathrm{cf}(\alpha))$ which is cofinal in α. For each $i < \mathrm{cf}(\alpha)$ let f_i be the ordinal function with domain X that has the constant value α_i. If $f \in {}^X\alpha$ is arbitrary, then by assumption $\gamma := \sup \mathrm{ran}(f) < \alpha$. So there is an $i < \mathrm{cf}(\alpha)$ such that $\gamma \leq \alpha_i$, hence $f \leq_X f_i$. Therefore the sequence $(f_i : i < \mathrm{cf}(\alpha))$ is cofinal modulo $I := \{\emptyset\}$ in ${}^X\alpha$, and together with Lemma 3.1.3 we are done.

A further class of examples is given by the maximal ideals or the ultrafilters on A.

Definition *Let us assume that $\mathcal{S}$ is a set of ordinal functions with the common domain A, D is an ultrafilter on A, and $I := \mathcal{P}(A) \setminus D$ is the maximal ideal dual to D. Then we put*

$$\mathrm{cf}(\mathcal{S}/D) := \mathrm{cf}(\mathcal{S}, \leq_I) \quad \textit{and} \quad \mathrm{tcf}(\mathcal{S}/D) := \mathrm{tcf}(\mathcal{S}/I).$$

When we are working with ultrafilters, we will use, in view of the following lemma, in most cases the denotation "cf" instead of "tcf".

Lemma 3.1.6 *If D is an ultrafilter on A and $\mathcal{S}$ is a set of ordinal functions, then there is a sequence of elements of $\mathcal{S}$ which is cofinal in $\mathcal{S}$ modulo D, and* $\mathrm{cf}(\mathcal{S}/D) = \mathrm{tcf}(\mathcal{S}/D)$.

Proof Let T be a cofinal subset of $\mathcal{S}$ of cardinality $\mathrm{cf}(\mathcal{S}/D)$. We define by recursion a sequence of elements of T which is strictly increasing modulo D. Assume that $\sigma \in \mathrm{ON}$, and that $(f_\xi : \xi < \sigma)$ is already defined. If this sequence is not cofinal in $\mathcal{S}$, we can find a function $g \in \mathcal{S}$ such that $\neg(g \leq_D f_\xi)$ for all

$\xi < \sigma$. By Lemma 2.1.5, we have $f_\xi <_D g$ for all $\xi < \sigma$. We can choose $f_\sigma \in T$ as a function satisfying $g \leq_D f_\sigma$.

Since strictly increasing sequences are injective, the definition must stop for some $\sigma \in \mathrm{ON}$. By construction, the corresponding sequence of length σ is cofinal in $\mathcal{S}$ modulo D. Now Lemma 3.1.3 says that $\mathrm{tcf}(\mathcal{S}/D) = \mathrm{cf}(\sigma) \leq \mathrm{cf}(\mathcal{S}/D)$, since $\mathrm{cf}(\sigma) \leq |\sigma| \leq \mathrm{cf}(\mathcal{S}/D)$. Conversely, $\mathrm{cf}(\mathcal{S}/D) \leq \mathrm{tcf}(\mathcal{S}/D)$ is true because the range of each cofinal sequence is a cofinal subset of $\mathcal{S}$.

The following important properties will be used in later chapters, often without quoting them.

Lemma 3.1.7 *Let $\mathcal{S}$ be a set of ordinal functions for which the cardinal $\mathrm{tcf}(\mathcal{S}/I)$ exists. If J is an ideal on A and $I \subseteq J$, then $\mathrm{tcf}(\mathcal{S}/J)$ exists, and the cardinals are equal. In particular we have $\mathrm{tcf}(\mathcal{S}/I \restriction X) = \mathrm{tcf}(\mathcal{S}/I)$ for any positive subset X of A.*

Lemma 3.1.8 *Assume that $\mathcal{S}$ is a set of ordinal functions, X is an I-positive subset of A, and*

$$\mathcal{S}' := \{f \restriction X : f \in \mathcal{S}\}.$$

If one of the cardinals $\mathrm{tcf}(\mathcal{S}/(I \restriction X))$ and $\mathrm{tcf}(\mathcal{S}'/(I \cap \mathcal{P}(X)))$ exists, then so does the other one, and they are equal.

In particular we have: If D is an ultrafilter on A and $X \in D$, then $D \restriction X = D$, $D \cap \mathcal{P}(X)$ is an ultrafilter on X, and

$$\mathrm{cf}(\mathcal{S}/D) = \mathrm{cf}(\mathcal{S}'/(D \cap \mathcal{P}(X))).$$

Furthermore, if D is an ultrafilter on A extending $I \restriction X$ (such an ultrafilter exists by Lemma 1.8.3 c)), and if $\mathrm{tcf}(\mathcal{S}/I \restriction X)$ exists, then

$$\mathrm{cf}(\mathcal{S}/D) = \mathrm{tcf}(\mathcal{S}/I \restriction X) = \mathrm{tcf}(\mathcal{S}'/I \cap \mathcal{P}(X)) = \mathrm{cf}(\mathcal{S}'/D \cap \mathcal{P}(X)).$$

Proof Lemma 2.1.4 says that, for every sequence which is cofinal in $\mathcal{S}$ modulo $I \restriction X$, we obtain, by restricting its members to X, a sequence in $\mathcal{S}'$ which is cofinal in $\mathcal{S}'$ modulo $I \cap \mathcal{P}(X)$. Conversely, we obtain for any sequence of this kind, by extending its members $f \restriction X$ to f, a sequence in $\mathcal{S}$ which is cofinal modulo $I \restriction X$. Now the first assertion follows from Lemma 3.1.3.

Let D be an ultrafilter on A and $X \in D$. The filter generated by $D \cup \{X\}$, that is $D \restriction X$, clearly is D itself, and with the properties of D it is easy to show that $D \cap \mathcal{P}(X)$ is also an ultrafilter. If I is the maximal ideal dual to D, then correspondingly we have $I \restriction X = I$, and $I \cap \mathcal{P}(X)$ is a maximal ideal on X. Now the second assertion follows from the first one.

The last asssertion follows from the first part of the lemma and Lemma 3.1.7.

Closely related to the notion of true cofinality is the following notion.

Definition *If $\mathcal{S}$ is a set of ordinal functions with domain A and λ is a cardinal, then we call the partial ordering $(\mathcal{S}, <_I)$* **λ-directed** *iff every subset of $\mathcal{S}$ of cardinality less than λ has an upper bound in $\mathcal{S}$ under the relation $<_I$. That means that $(\mathcal{S}, <_I)$ is λ-directed iff, for all $T \subseteq \mathcal{S}$ with $|T| < \lambda$, there is a function $f \in \mathcal{S}$ satisfying $g <_I f$ for all $g \in T$.*

Lemma 3.1.9 *Assume that f is an ordinal function such that the cardinal $\lambda := \operatorname{tcf}(\prod f/I)$ is infinite. Then the partial ordering $(\prod f, <_I)$ is λ-directed and not λ^+-directed.*

Proof Fix a sequence $(f_\xi : \xi < \lambda)$ which is cofinal in $\prod f$ modulo I. If $T \subseteq \prod f$ and $|T| < \lambda$, we choose, for each function $g \in T$, an ordinal ξ_g such that $\xi_g < \lambda$ and $g \leq_I f_{\xi_g}$. Since λ is regular by Lemma 3.1.3, the ordinal

$$\alpha := \sup\{\xi_g + 1 : g \in T\}$$

is smaller than λ. By construction we have $g \leq_I f_{\xi_g} <_I f_\alpha$ for all $g \in T$. Therefore $(\prod f, <_I)$ is λ-directed.

Of course $(\prod f, <_I)$ is not λ^+-directed, since otherwise the set $\{f_\xi : \xi < \lambda\}$ would have an upper bound modulo I in $\prod f$, contradicting the fact that the sequence is cofinal in $\prod f$ modulo I.

Now we present a practical tool for inferring the existence of the true cofinality of a partial ordering from the existence of the true cofinality of another partial ordering.

Lemma 3.1.10 *Assume that A and B are nonempty sets, I is an ideal on A, J is an ideal on B, $\mathcal{S}$ is a set of ordinal functions with domain A, and $\mathcal{T}$ is a set of ordinal functions with domain B. Further let G be a function from $\mathcal{S}$ into $\mathcal{T}$ satisfying*

$$\forall h \in \mathcal{T}\ \exists p \in \mathcal{S}\ \forall q \in \mathcal{S}\ (p \leq_I q \Longrightarrow h <_J G(q)). \tag{3.1}$$

If $\operatorname{tcf}(\mathcal{S}/I)$ exists, then $\operatorname{tcf}(\mathcal{T}/J)$ exists, and the two cardinals are equal.

Proof Let $\lambda := \operatorname{tcf}(\mathcal{S}/I)$ and $(f_\xi : \xi < \lambda)$ be a sequence cofinal in $\mathcal{S}$ modulo I. We define by recursion a sequence $(\alpha_i : i < \lambda)$ in λ such that $(G(f_{\alpha_i}) : i < \lambda)$ is cofinal in $\mathcal{T}$ modulo J.

Let $i < \lambda$ and let the sequence $(\alpha_j : j < i)$ be already defined. For each $j < i$ there is, by (3.1), a function $p_j \in \mathcal{S}$, so that $G(f_{\alpha_j}) <_J G(q)$ for all $q \in \mathcal{S}$ with $p_j \leq_I q$. For each p_j we can find an ordinal $\alpha_{i,j} < \lambda$ satisfying $p_j \leq_I f_{\alpha_{i,j}}$. We put

$$\alpha_i := \max\{i, \sup\{\alpha_{i,j} : j < i\}\}$$

and show that sequence has the desired properties. If $i < \lambda$ and $j < i$, then $p_j \leq_I f_{\alpha_{i,j}} \leq_I f_{\alpha_i}$, and consequently we have $G(f_{\alpha_j}) <_J G(f_{\alpha_i})$ by the choice of p_j. Therefore the sequence $(G(f_{\alpha_i}) : i < \lambda)$ is strictly increasing. If $h \in \mathcal{T}$, then there is $p \in \mathcal{S}$ satisfying the property from (3.1) and an ordinal $\beta < \lambda$ with $p \leq_I f_\beta$. Now, by the definition of α_β, we have $\beta \leq \alpha_\beta$, and this gives $h <_J G(f_{\alpha_\beta})$.

An essential application of the previous lemma will be the proof of the following assertion. Its importance will be pointed out in the remark after the proof of Lemma 3.1.12.

Lemma 3.1.11 *If f is an ordinal function all of whose values are limit ordinals, and if one of the cardinals* $\operatorname{tcf}(\prod f/I)$ *and* $\operatorname{tcf}(\prod(\operatorname{cf}\circ f)/I)$ *exists, then so does the other one, and we have*

$$\operatorname{tcf}\Big(\prod(\operatorname{cf}\circ f)/I\Big) = \operatorname{tcf}\Big(\prod f/I\Big).$$

Proof We will define functions $F : \prod(\operatorname{cf}\circ f) \longrightarrow \prod f$ and $G : \prod f \longrightarrow \prod(\operatorname{cf}\circ f)$ which satisfy the assumption (3.1) in Lemma 3.1.10. For each $a \in A$, we can take a normal sequence $(\alpha_i^a : i < \operatorname{cf}(f(a)))$ which is cofinal in the limit ordinal $f(a)$, and put, for any $q \in \prod(\operatorname{cf}\circ f)$, $p \in \prod f$, and $a \in A$

$$F(q)(a) := \alpha^a_{q(a)} \text{ and } G(p)(a) := \min\{i < \operatorname{cf}(f(a)) : \alpha_i^a > p(a)\}.$$

It is clear that $F(q) \in \prod f$ and $G(p) \in \prod(\operatorname{cf}\circ f)$.

To show (3.1) for F, I and $J = I$, let $h \in \prod f$. Define $p \in \prod(\operatorname{cf}\circ f)$ by $p(a) := \min\{i < \operatorname{cf}(f(a)) : \alpha_i^a > h(a)\}$. If $q \in \prod(\operatorname{cf}\circ f)$ and $p \leq_I q$, then $p(a) \leq q(a)$ for almost all $a \in A$. By the definition of p we have $\alpha^a_{p(a)} > h(a)$, hence $F(q)(a) = \alpha^a_{q(a)} \geq \alpha^a_{p(a)} > h(a)$ for almost all $a \in A$. This yields $h <_I F(q)$.

Next we will prove the corresponding property for the function G. Consider $h \in \prod(\operatorname{cf}\circ f)$. If the function $p \in \prod f$ is defined by $p(a) := \alpha^a_{h(a)}$ for all $a \in A$, then obviously we have $h <_A G(p)$ since the sequence $(\alpha_i^a : i < \operatorname{cf}(f(a)))$ is strictly increasing. If $q \in \prod f$ and $p \leq_I q$, then $G(p) \leq_I G(q)$, and so we get $h <_I G(q)$. Therefore G has the desired property.

Now Lemma 3.1.10 yields the assertion of the lemma.

Lemma 3.1.12 *Assume that τ is a limit ordinal, $(f_\xi : \xi < \tau)$ is a strictly increasing sequence of ordinal functions, and f is a supremum of this sequence. Then $f_\xi <_I f$ for all $\xi < \tau$. Furthermore $f(a)$ is a limit ordinal for almost all $a \in A$. In particular, the last assertion is true if* $\operatorname{tcf}(\prod f/I)$ *exists and is infinite.*

Proof If $\xi < \tau$, then we have $f_\xi <_I f_{\xi+1} \leq_I f$ This gives $f_\xi <_I f$, hence $f_\xi(a) < f(a)$ for almost all $a \in A$ and in particular $f(a) \neq 0$ for almost all $a \in A$.

Now we assume, to get a contradiction, that on an I-positive set Z all values of f are successor ordinals. Then we can define an ordinal function f' with $f' \lneqq_I f$ by

$$f'(a) := \begin{cases} f(a) - 1 & \text{if } a \in Z, \\ f(a) & \text{if } a \in A \setminus Z. \end{cases}$$

By Lemma 3.1.1, f' is not an upper bound of the given sequence, and therefore there is $\beta < \tau$ such that $\neg (f_\beta \leq_I f')$, hence $\mathrm{B}(f', f_\beta)$ is positive. Now $f'(a) < f_\beta(a)$ gives $f(a) \leq f_\beta(a)$ for any $a \in A$, hence $\mathrm{B}(f', f_\beta) \subseteq \mathrm{B}[f, f_\beta]$. Thus $\mathrm{B}[f, f_\beta]$ is positive, contradicting $f_\beta <_I f$. Consequently, $f(a)$ is a limit ordinal for almost all $a \in A$.

For the proof of the last assertion of the lemma, assume that $\lambda = \mathrm{tcf}(\prod f/I)$ exists and is infinite. Then λ is a regular cardinal, and there is a sequence $(g_\xi : \xi < \lambda)$ of ordinal functions which is cofinal in $\prod f$ modulo I. Now Lemma 3.1.4 says that f is a supremum of this sequence, and the desired result follows from the first part of the proof.

Remark *The preceding lemmas say that it suffices to confine the investigation of true cofinalities of $\prod f$ to ordinal functions f all of whose values are regular cardinals.*

To see this, let λ be a regular cardinal and $(f_\xi : \xi < \lambda)$ be a sequence which is cofinal in $\prod f$. By Lemma 3.1.12, the set $X := \{a \in A : f(a) \text{ is not a limit ordinal}\}$ is a member of I. If we change the values of f on X to $\aleph_0$, and if we change correspondingly the values of the functions f_ξ $(\xi < \lambda)$ on X in such a way that the resulting functions f'_ξ are elements of $\prod f'$, where f' is the function resulting from f, then Lemma 3.1.2 says that the sequence $(f'_\xi : \xi < \lambda)$ is cofinal in $\prod f'$ modulo I. Thus we can infer from Lemma 3.1.3 that $\mathrm{tcf}(\prod f/I) = \mathrm{tcf}(\prod f'/I)$. Now Lemma 3.1.11 says that $\mathrm{tcf}(\prod f'/I) = \mathrm{tcf}((\mathrm{cf} \circ f')/I)$. Conversely, if $\mathrm{tcf}(\prod(\mathrm{cf} \circ f)/I)$ exists and is infinite, then, again by Lemma 3.1.12, $\mathrm{cf}(f(a))$ is a limit ordinal, hence a regular cardinal, for almost all $a \in A$. This means that $f(a)$ is a limit ordinal for almost all $a \in A$. By the same procedure as above we obtain a function f' such that $\mathrm{tcf}(\prod(\mathrm{cf} \circ f)/I) = \mathrm{tcf}(\prod(\mathrm{cf} \circ f')/I) = \mathrm{tcf}(\prod f'/I) = \mathrm{tcf}(\prod f/I)$. We summarize these results in Theorem 3.1.13.

Theorem 3.1.13 *Let f be an ordinal function for which one of the cardinals $\mathrm{tcf}(\prod f/I)$ and $\mathrm{tcf}(\prod(\mathrm{cf} \circ f)/I)$ exists and is infinite. Then also the other one exists, and we have*

$$\mathrm{tcf}\Big(\prod(\mathrm{cf} \circ f)/I\Big) = \mathrm{tcf}\Big(\prod f/I\Big).$$

Furthermore, there is an ordinal function g with domain A satisfying $\mathrm{cf} \circ f =_I g$ and $\mathrm{tcf}(\prod f/I) = \mathrm{tcf}(\prod g/I)$ all of whose values are regular cardinals.

Subsequent to a preparatory lemma we now turn to the proof of a theorem which bridges the gap between the notion of almost disjoint functions and Shelah's idea of the true cofinality of a function. Further it motivates an application of a very important theorem of pcf-theory, namely Theorem 3.3.6, which will play an outstanding role in this book. We already pointed out its importance in the introduction to this chapter.

Definition *Let I be an ideal on the set A and $\mathcal{F} \subseteq {}^A\mathrm{ON}$. We call $\mathcal{F}$* ***I-disjoint*** *iff $\{a \in A : f(a) = g(a)\} \in I$ holds for all $f, g \in \mathcal{F}$ with $f \neq g$.*

If μ is a cardinal, then we call $f \in {}^A\mathrm{ON}$ ***μ-large modulo I*** *iff there is an I-disjoint set $\mathcal{F} \subseteq \prod f$ with $|\mathcal{F}| = \mu$.*

Lemma 3.1.14 *Assume that λ is a κ-strong singular cardinal, where $\kappa := \mathrm{cf}(\lambda)$. Then there is an increasing function $f \in {}^\kappa\lambda$ satisfying $\mathrm{ran}(f) \subseteq \mathrm{Reg}$ and $\sup(\mathrm{ran}(f)) = \lambda$ which is λ^κ-large modulo $\mathrm{I_b}(\kappa)$. In particular we have $\lambda^\kappa = T(f)$, where $T(f)$ is the cardinal defined in Section 2.2.*

Proof Let $\aleph_\eta := \lambda$ and choose an increasing sequence $(\eta(\xi) : \xi < \kappa)$ of ordinals which is cofinal in η. By Lemma 2.3.3, there is a set $\mathcal{F}^* \subseteq \prod_{\xi<\kappa} {}^\kappa\aleph_{\eta(\xi)}$ of almost disjoint functions such that $|\mathcal{F}^*| = \lambda^\kappa$. For $\xi < \kappa$ we put

$$f(\xi) := (\aleph_{\eta(\xi)}^\kappa)^+.$$

By assumption, we have $f(\xi) < \lambda$ for every $\xi < \kappa$. Note that $\mathrm{ran}(f) \subseteq \mathrm{Reg}$ and $\sup(\mathrm{ran}(f)) = \lambda$. Now we choose, for each $\xi < \kappa$, an injective function $\tau_\xi : {}^\kappa\aleph_{\eta(\xi)} \longrightarrow f(\xi)$, and define, for each $h^* \in \mathcal{F}^*$, a function $h \in \prod f$ by $h(\xi) := \tau_\xi(h^*(\xi))$. Then $\mathcal{F} := \{h : h^* \in \mathcal{F}^*\}$ is an $\mathrm{I_b}(\kappa)$-disjoint set of members of $\prod f$ satisfying $|\mathcal{F}| = \lambda^\kappa$. So the function f has the desired properties. Clearly we have $T(f) = \lambda^\kappa$ since $|\prod f| \leq |\prod_{\xi<\kappa} \lambda| = \lambda^\kappa$.

Definition *For every ordinal function f with $\mathrm{dom}(f) = A$, we define the* **limit of f modulo I** *by*

$$\lim_I(f) := \min\{\mu \in \mathrm{CN} : \{a \in A : f(a) \leq \mu\} \in I^+\}.$$

Lemma 3.1.15 (Jech, Shelah) *Let λ be a κ-strong cardinal and $\aleph_0 < \kappa := \mathrm{cf}(\lambda) < \lambda$. If $\mu \in (\lambda, \lambda^\kappa]_{\mathrm{CN}}$ is regular, then there is a function $f \in {}^\kappa\lambda$ and an ideal I on κ such that $\lim_I(f) = \lambda$ and $\mathrm{tcf}(\prod f/I) = \mu$.*

Proof Let $\mu \in (\lambda, \lambda^\kappa]_{\mathrm{CN}}$ be a regular cardinal and $I := \mathrm{I_b}(\kappa)$. I is σ-complete, since κ is uncountable and regular. Therefore, by Lemma 2.2.2, the relation $<_I$ is well-founded on ${}^\kappa\lambda$. From Lemma 3.1.14 we obtain a function $f \in {}^\kappa\lambda$ with $\mathrm{ran}(f) \subseteq \mathrm{Reg}$ which is μ-large modulo I and $<_I$-minimal in the set ${}^\kappa\lambda$ with respect to this property. Now we put

$$I' := I \cup \{X \in I^+ : \exists g \in {}^\kappa\lambda(g <_{I \restriction X} f \wedge g \text{ is } \mu\text{-large modulo } I \restriction X)\}.$$

Claim 1 I' is an ideal on κ.

Proof We have $\emptyset \in I \subseteq I'$ and $\kappa \notin I'$ by the minimality of f, since $I \restriction \kappa = I$.

Consider $X \subseteq Y \in I'$; with no loss of generality we can assume that $X \notin I$ and thus $Y \notin I$. Then there is a function $g \in {}^{\kappa}\lambda$ with $g <_{I \restriction Y} f$ and an $I \restriction Y$-disjoint set $\mathcal{F} \subseteq \prod g$ with $|\mathcal{F}| = \mu$. $X \subseteq Y$ implies $I \restriction Y \subseteq I \restriction X$. So we get $g <_{I \restriction X} f$, and $\mathcal{F}$ is an $I \restriction X$-disjoint set of functions. Therefore X is a member of I'.

To show that I' is closed under finite unions, let $X, Y \in I'$. Without loss of generality let $X \cap Y = \emptyset$, since $X \cup Y = (X \setminus Y) \cup Y$ and we have just shown that $X \setminus Y \in I'$. Further we can assume without loss of generality that $X, Y \notin I$, because for $X \in I$ and $Y \notin I$ we have $I \restriction (X \cup Y) = I \restriction Y$. Then there are functions g and h satisfying $g <_{I \restriction X} f$ and $h <_{I \restriction Y} f$ which are μ-large modulo $I \restriction X$ and modulo $I \restriction Y$, respectively. We put

$$f'(\xi) := \begin{cases} g(\xi) & \text{if } \xi \in X, \\ h(\xi) & \text{if } \xi \in Y, \\ 1 & \text{if } \xi \in \kappa \setminus (X \cup Y). \end{cases}$$

Then we have $f' <_{I \restriction (X \cup Y)} f$, and the definition of f' demonstrates how to define, using the corresponding properties of g and h, an $I \restriction (X \cup Y)$-disjoint subset of $\prod f'$ which witnesses that f' is μ-large modulo $I \restriction (X \cup Y)$. This proves Claim 1.

Now let $\mathcal{F} \subseteq \prod f$ be an I-disjoint set of functions with $|\mathcal{F}| = \mu$.

Claim 2 If $g \in \prod f$, then $|\{h \in \mathcal{F} : \neg g <_{I'} h\}| < \mu$.

Proof Assume that there is a function $g \in \prod f$ such that $|H| = \mu$ holds for $H := \{h \in \mathcal{F} : \neg g <_{I'} h\}$. Then, for every $h \in H$, the set $X_h := \{\xi < \kappa : h(\xi) \leq g(\xi)\}$ is not a member of I'. Since $2^\kappa < \lambda < \mu$, the pigeonhole principle for the regular cardinal μ says that there is a set $X^* \in \mathcal{P}(\kappa) \setminus I'$ such that $|H^*| = \mu$, where $H^* := \{h \in H : X_h = X^*\}$. Then $h \restriction X^* \leq g \restriction X^*$ for all $h \in H^*$. Analogously we obtain sets $X^{**} \subseteq X^*$ and $H^{**} \subseteq H^*$ satisfying $|H^{**}| = \mu$ and $X^{**} = \{\xi < \kappa : h(\xi) < g(\xi)\}$ for all $h \in H^{**}$. Now we have $X^{**} \notin I'$ since $X^* \setminus X^{**} \in I'$. To see this, let $h, h' \in H^{**}$ and $h \neq h'$. Then $X^* \setminus X^{**} = \{\xi < \kappa : h(\xi) = g(\xi)\} = \{\xi < \kappa : h'(\xi) = g(\xi)\} \subseteq \{\xi < \kappa : h(\xi) = h'(\xi)\} \in I$, since $H^{**} \subseteq \mathcal{F}$ and $\mathcal{F}$ is I-disjoint.

Next we define $g^* := f \restriction (\kappa \setminus X^{**}) \cup g \restriction X^{**}$. g^* is μ-large modulo $I \restriction X^{**}$. Namely, if we put $h' := h \restriction X^{**} \cup (\kappa \setminus X^{**}) \times \{0\}$ for every $h \in H^{**}$, then $\{h' : h \in H^{**}\}$ is a subset of $\prod g^*$ which has cardinality μ and is $I \restriction X^{**}$-disjoint. Furthermore we have $g^* <_{I \restriction X^{**}} f$, contradicting $X^{**} \notin I'$. This proves Claim 2.

Now we can easily construct a sequence $(f_\alpha : \alpha < \mu)$ of members of $\mathcal{F}$ which is strictly increasing modulo I'. If $(f_\beta : \beta < \alpha)$ is already defined, then we infer from Claim 2 that $|\bigcup_{\beta<\alpha}\{h \in \mathcal{F} : \neg f_\beta <_{I'} h\}| < \mu$. Consequently, we can choose $f_\alpha \in \mathcal{F} \setminus \bigcup_{\beta<\alpha}\{h \in \mathcal{F} : \neg f_\beta <_{I'} h\}$.

It also follows that the sequence $(f_\alpha : \alpha < \mu)$ is cofinal in $\prod f$ modulo I' since $g \in \prod f$ implies, again by Claim 2, that $|\{\alpha < \mu : \neg g <_{I'} f_\alpha\}| < \mu$, and so there is $\alpha < \mu$ such that $g <_{I'} f_\alpha$. This gives $\operatorname{tcf}(\prod f/I') = \mu$.

Finally, to prove $\lim_{I'}(f) = \lambda$, let us assume that $\lim_{I'}(f) < \lambda$. Then we can find a cardinal $\nu < \lambda$ such that $X := \{\xi < \kappa : f(\xi) < \nu\} \notin I'$, and conclude, using Lemma 3.1.8 and the assumption that λ is κ-strong, that $\mu = \operatorname{tcf}(\prod f/I') = \operatorname{tcf}(\prod_{\xi \in X} f(\xi)/I' \cap \mathcal{P}(X)) \leq \nu^{|X|} \leq \nu^\kappa < \lambda < \mu$. This contradiction proves $\lim_{I'}(f) = \lambda$.

Exercises

1) Assume that $f \in {}^A\mathrm{ON}$ is a function, $\alpha \in \mathrm{ON} \setminus \{0\}$, I is an ideal on A, and $(f_\xi : \xi < \alpha)$ is a sequence of members of $\prod f$ which is increasing modulo I such that $\{f_\xi : \xi < \alpha\}$ is cofinal in $\prod f$ modulo I. Further let $f(a)$ be a limit ordinal for almost all $a \in A$. Prove:
   a) α is a limit ordinal.
   b) There is a club C in α such that $(f_\xi : \xi \in C)$ is strictly increasing modulo I.
2) Let $A := \{0, 1\}$ and the function $f \in {}^A\mathrm{ON}$ be defined by $f(0) = \omega$ and $f(1) = \omega_1$. Further let the ideals I_0, I_1 and I_2 be given by $I_0 := \{\emptyset\}$, $I_1 := \{\emptyset, \{0\}\}$ and $I_2 := \{\emptyset, \{1\}\}$. For which $j \in 3$ exists $\operatorname{tcf}(\prod f/I_j)$? Calculate it, if it exists!
3) (S. Neumann) Let $A := \aleph_1 \times \aleph_3$ and assume that $(A, \prec)$ is a linear extension of $(A, <)$, where $(\alpha, \beta) < (\sigma, \tau) \iff \alpha < \sigma \wedge \beta < \tau$; that means that $(A, \prec)$ is a linear ordering and $<$ is a subset of $\prec$. Prove that $\operatorname{cf}(A, \prec) \in \{\aleph_1, \aleph_3\}$. Show further that for every $\kappa \in \{\aleph_1, \aleph_3\}$ there exists a linear extension $(A, \prec)$ such that $\kappa = \operatorname{cf}(A, \prec)$.
4) Assume that I is an ideal on A, $\lambda \in \mathrm{ICN}$, $(f_\xi : \xi < \lambda)$ is a sequence of members of ${}^A\mathrm{ON}$, $f \in {}^A\mathrm{ON}$ is a supremum of $\{f_\xi : \xi < \lambda\}$ modulo I, and $X \in I^+$. Prove that $f \restriction X$ is a supremum of $\{f_\xi \restriction X : \xi < \lambda\}$ modulo $I \cap \mathcal{P}(X)$.
5) Assume that $f \in {}^A\mathrm{ON}$ is an ordinal function with $\prod f \neq \emptyset$. Prove that $\prod f$ has a maximum under $\leq_I$ iff, for almost all $a \in A$, $f(a)$ is a successor ordinal.
6) Assume that $\mathcal{S}$ is a subset of ${}^A\mathrm{ON}$, I is an ideal on A, and $\lambda = \operatorname{tcf}(\mathcal{S}/I)$. Prove: If $\mathcal{S}$ has a maximal element g modulo I, then g is a maximum of $\mathcal{S}$ under $\leq_I$.

7) Assume that I is an ideal on A, $f \in {}^A\mathrm{ON}$ is a function, and $\mathrm{tcf}(\prod f/I)$ is infinite. Prove that $X := \{a \in A : \exists\xi\ f(a) = \xi + 1\}$ is a member of I.
8) Assume that A is an infinite set, $f \in {}^A\mathrm{ON}$ is an ordinal function, I is an ideal on A, and $f(a)$ is a limit ordinal for almost all $a \in A$. Further let $\mu := \lim_I(\mathrm{cf} \circ f)$. Prove that $(\prod f, <_I)$ is μ-directed.
9) Let $f \in {}^A\mathrm{ON}$ be an ordinal function. Prove that $\lim_I(f)$ is the supremum of all cardinals ν^+ which satisfy $f(a) > \nu$ for almost all $a \in A$.
10) Assume that I is an ideal on the set A and $f \in {}^A\mathrm{ON}$ is an ordinal function. Then the cardinal number $\mathrm{tcf}(\prod f/I)$ exists and is equal to the infinite cardinal λ iff the following conditions are satisfied:

    (i) λ is regular.

    (ii) $(\prod f, <_I)$ is λ-directed.

    (iii) $\mathrm{cf}(\prod f, \leq_I) = \lambda$.

11) Assume that λ is a κ-strong cardinal, where $\aleph_0 < \kappa = \mathrm{cf}(\lambda) < \lambda$, and $\mu \in (\lambda, \lambda^\kappa]_{\mathrm{CN}}$ is a regular cardinal. Prove that there is a function $g \in {}^\kappa(\lambda \cap \mathrm{Reg})$ and an ideal I on κ such that $\lim_I(g) = \lambda$, $\mathrm{cf}(g(\xi)) > \kappa$ for all $\xi < \kappa$, and $\mu = \mathrm{tcf}(\prod g/I)$.
12) (Jech, Shelah) Assume that λ is a cardinal satisfying $\aleph_0 < \kappa := \mathrm{cf}(\lambda) < \lambda$ and $2^\kappa < \lambda$, and I is a σ-complete ideal on κ. Further let μ be a regular cardinal with $\lambda < \mu$ and $f \in {}^\kappa\lambda$ be a function which is μ-large modulo I and $<_I$-minimal with respect to this property. Prove:

    a) $I' := I \cup \{X \in I^+ : \exists g \in {}^\kappa\lambda(g <_{I\restriction X} f \wedge g \text{ is } \mu\text{-large modulo } I \restriction X)\}$ is a σ-complete ideal on κ.

    b) If I is normal, then so is I'.
    Hint: For any sequence $(Y_\alpha : \alpha < \kappa)$ of members of I' construct a sequence $(Y^1_\alpha : \alpha < \kappa)$ of members of I and a sequence $(Y^2_\alpha : \alpha < \kappa)$ of members of $I' \setminus I$ such that $\nabla_{\alpha<\kappa} Y_\alpha \subseteq \nabla_{\alpha<\kappa} Y^1_\alpha \cup \nabla_{\alpha<\kappa} Y^2_\alpha$.

13) If λ is a singular cardinal and $\kappa = \mathrm{cf}(\lambda)$, let

$$\mathrm{pp}(\lambda) := \sup\{\mathrm{tcf}(\prod g/I) : g \in {}^\kappa(\lambda \cap \mathrm{Reg}) \wedge \lim_I(g) = \lambda\}.$$

Prove: If $\aleph_0 < \kappa$ and λ is κ-strong, then $\mathrm{pp}(\lambda) = \lambda^\kappa$.

3.2 κ-rapid Sequences and the Main Lemma of pcf-Theory

For the rest of this chapter, A denotes an infinite set and I an ideal on A. The addition "modulo I" is often omitted.

Definition *Let κ be a cardinal and $\bar{f} := (f_\xi : \xi < \lambda)$ be a sequence of members of ${}^A\mathrm{ON}$. We say that $\bar{f}$ is a* **κ-ub[1] sequence modulo I** *iff $\bar{f}$ is strictly increasing modulo I and, for every family $(S_a : a \in A)$ of subsets of ON satisfying $|S_a| < \kappa$ for all $a \in A$, there exists an ordinal $\alpha < \lambda$ such that every function $h \in \prod_{a\in A} S_a$ with $f_\alpha <_I h$ is an upper bound of $\bar{f}$.*

This property yields a sufficient condition for the existence of suprema.

Theorem 3.2.1 *Let κ and λ be regular cardinals such that $|A| < \kappa < \lambda$. If $(f_\xi : \xi < \lambda)$ is modulo I a κ-ub sequence of ordinal functions which are members of ${}^A\mathrm{ON}$, then this sequence has a supremum modulo I.*

Proof Let κ and λ be regular cardinals such that $|A| < \kappa < \lambda$, and let $\bar{f} := (f_\xi : \xi < \lambda)$ be a κ-ub sequence of ordinal functions. We give an indirect proof and assume that $\bar{f}$ has no supremum. Lemma 3.1.1 says that, for every upper bound h of $\bar{f}$, there is an upper bound g of this sequence with $g \lneqq_I h$. We will construct by recursion a sequence $(h_i : i < |A|^+)$ of upper bounds of $\bar{f}$ satisfying $h_j \lneqq_I h_i$ for all i and j with $i < j < |A|^+$. Let $h_0 := \sup\{f_\xi + 1 : \xi < \lambda\}$. Using Lemma 3.1.1 it is clear how to choose the function h_{i+1} such that $h_{i+1} \lneqq_I h_i$ if h_i is already defined as an an upper bound of $\bar{f}$. Now let $i < |A|^+$ be a limit ordinal. For each $a \in A$, we define

$$S_a^i := \{h_j(a) : j < i\}.$$

Clearly we have $|S_a^i| \le |i| < |A|^+ \le \kappa$. The κ-ub property of $\bar{f}$ guarantees the existence of an ordinal $\alpha_i < \lambda$ such that every function $h \in \prod_{a\in A} S_a^i$ satisfying $f_{\alpha_i} <_I h$ is an upper bound of $\bar{f}$. For all $a \in A$ we put

$$h_i(a) := \min\{\beta \in S_a^i : \beta > f_{\alpha_i}(a)\}.$$

This minimum exists for every $a \in A$ by the choice of h_0, and certainly we have $f_{\alpha_i} <_A h_i$. Consequently, by the choice of α_i, the function h_i is an upper bound of the sequence $\bar{f}$. If $j < i$, then $f_{\alpha_i} <_I f_{\alpha_i+1} \le_I h_{j+1}$. This implies $h_i \le_I h_{j+1}$ by the definition of h_i, and $h_{j+1} \lneqq_I h_j$ holds by the definition of h_{j+1}, from which we infer that $h_i \lneqq_I h_j$. So the sequence $(h_i : i < |A|^+)$ has the desired properties.

[1] Upper bound.

Now let $\alpha := \sup\{\alpha_i : i \in \mathrm{Lim} \cap |A|^+\}$. Then we have $\alpha < \lambda$ since λ is regular. Further we put for every $a \in A$

$$S_a := \{h_i(a) : i < |A|^+\} = \bigcup\{S_a^i : i \in \mathrm{Lim} \cap |A|^+\}.$$

Analogously to the definition of h_i in the limit case, we define for $a \in A$ the ordinal

$$h(a) := \min\{\beta \in S_a : \beta > f_\alpha(a)\}.$$

If $i \in |A|^+$, then $f_\alpha <_I h_i$ and $h_i \in \prod_{a \in A} S_a$, hence $h \leq_I h_i$ by the definition of h. For each $a \in A$ there exists a limit ordinal $i(a) < |A|^+$ satisfying $h(a) \in S_a^{i(a)}$. If we put

$$i^* := \sup\{i(a) : a \in A\},$$

then $i^* < |A|^+$ and $h \in \prod_{a \in A} S_a^{i^*}$.

Now we have $f_{\alpha_{i^*}} \leq_I f_\alpha$ and $f_\alpha <_A h$ by the definition of h, hence $f_{\alpha_{i^*}}(a) < h(a)$ for almost all $a \in A$. The definition of h_{i^*} yields $h_{i^*} \leq_I h$. All in all we obtain the contradiction $h \leq_I h_{i^*+1} \lneqq_I h_{i^*} \leq_I h$.

Lemma 3.2.2 *Assume that κ and λ are regular cardinals with $|A| < \kappa < \lambda$, the sequence $\bar{f} = (f_\xi : \xi < \lambda)$ of ordinal functions is a κ-ub sequence modulo I, and f is a supremum of this sequence. Then* $\mathrm{cf}(f(a)) \geq \kappa$ *for almost all $a \in A$, and consequently $\kappa \leq \lim_I(\mathrm{cf} \circ f)$.*

Proof By Lemma 3.1.12 and Lemma 3.1.2, we can assume without loss of generality that all members of the range of the function f are limit ordinals. We assume, to get a contradiction, that the set $B := \{a \in A : \mathrm{cf}(f(a)) < \kappa\}$ is positive. For each $a \in B$ we can choose a cofinal subset S_a of $f(a)$ with $|S_a| < \kappa$, for $a \in A \setminus B$ we put $S_a := \{f(a)\}$. Since $\bar{f}$ is a κ-ub sequence, there is an $\alpha < \lambda$ such that every $h \in \prod_{a \in A} S_a$ with $f_\alpha <_I h$ is an upper bound of $\bar{f}$.

We have $f_\alpha <_I f_{\alpha+1} \leq_I f$, and thus the set $B' := B \setminus \mathrm{B}[f, f_\alpha]$ is positive. For each $a \in B'$ we can find, by the choice of S_a, a member $h(a)$ of S_a satisfying $f_\alpha(a) < h(a) < f(a)$, for $a \in A \setminus B'$ we choose $h(a) \in S_a$ arbitrarily; in particular, we have $h(a) = f(a)$ for all $a \in A \setminus B$. So we obtain an ordinal function $h \in \prod_{a \in A} S_a$ which satisfies

$$\mathrm{B}[h, f_\alpha] \subseteq \mathrm{B}[f, f_\alpha] \in I,$$

and consequently $f_\alpha <_I h$. Therefore, by the choice of α, h is an upper bound of $\bar{f}$. Further we have $h \lneqq_I f$ by construction, and from this we can conclude with Lemma 3.1.1 that f is not a supremum of $\bar{f}$, contradicting the assumption on f.

Definition *Let λ be an infinite cardinal, $(f_\xi : \xi < \lambda)$ be a sequence of ordinal functions and $\kappa > \omega$ be a regular cardinal. Then we call this sequence* **κ-rapid modulo I** *iff it is strictly increasing modulo I and, for all $\beta < \lambda$ with* $\mathrm{cf}(\beta) = \kappa$, *there is a club C in β of limit ordinals such that*

$$\sup\{f_\xi : \xi \in C \cap \gamma\} <_I f_\gamma$$

for all $\gamma \in C$.

Lemma 3.2.3 *If κ and λ are regular cardinals with $|A| < \kappa < \lambda$, then every κ-rapid sequence $(f_\xi : \xi < \lambda)$ of ordinal functions is a κ-ub sequence.*

Proof We assume that the assertion is false. Then there is a κ-rapid sequence $(f_\xi : \xi < \lambda)$ and a family $(S_a : a \in A)$ of subsets of ON satisfying $|S_a| < \kappa$ for every $a \in A$, for which the property in the definition of a κ-ub sequence is not fulfilled. By recursion, we define a normal sequence $(\alpha_i : i < \kappa)$ of ordinals less than λ and a sequence $(h_i : i < \kappa)$ of members of $\prod_{a \in A} S_a$.

Let $\alpha_0 < \lambda$ be arbitrary and $\alpha_i := \sup\{\alpha_j : j < i\}$ for limit ordinals $i < \kappa$. If $i < \kappa$ and and α_i is already defined, then we can find, by the choice of the family $(S_a : a \in A)$, a function $h_i \in \prod_{a \in A} S_a$ with $f_{\alpha_i} <_I h_i$ which is not an upper bound of the sequence $(f_\xi : \xi < \lambda)$. Therefore there is a limit ordinal $\alpha < \lambda$ such that $f_\alpha \leq_I h_i$ does *not* hold. Let α_{i+1} be chosen as such an ordinal; then it is greater than α_i, since otherwise we would have $f_{\alpha_{i+1}} \leq_I f_{\alpha_i} <_I h_i$.

The construction shows that $\beta := \sup\{\alpha_i : i < \kappa\} < \lambda$ and $\mathrm{cf}(\beta) = \kappa$. By assumption there is a club C in β, all of whose elements are limit ordinals, such that

$$s_\gamma := \sup\{f_\xi : \xi \in C \cap \gamma\} <_I f_\gamma$$

for all $\gamma \in C$. Without loss of generality we can assume that $\{\alpha_i : i < \kappa\} \subseteq C$, since otherwise we could take the intersection of C and the club $\{\alpha_i : i < \kappa\}$.

If $i < \kappa$, then we have $s_{\alpha_i} <_I f_{\alpha_i} <_I h_i$ and not $f_{\alpha_{i+1}} \leq_I h_i$. This means that the set

$$\begin{aligned} B_i &:= \mathrm{B}(h_i, f_{\alpha_{i+1}}) \setminus (\mathrm{B}[f_{\alpha_i}, s_{\alpha_i}] \cup \mathrm{B}[h_i, f_{\alpha_i}]) \\ &= \{a \in A : s_{\alpha_i}(a) < f_{\alpha_i}(a) < h_i(a) < f_{\alpha_{i+1}}(a)\} \end{aligned}$$

is positive; in particular it has at least one member, say a_i. Since $\kappa > |A|$ is regular, there is a set $Z \subseteq \kappa$ with $|Z| = \kappa$, so that the function $(a_i : i \in \kappa)$ has on Z the constant value a.

If $i, j \in Z$ with $i+1 < j$, then $\alpha_{i+1} \in C \cap \alpha_j$, and consequently $f_{\alpha_{i+1}}(a) \leq s_{\alpha_j}(a)$ by the definition of the function s_{α_j}. Furthermore we have $a \in B_i$ and $a \in B_j$, which implies $h_i(a) < f_{\alpha_{i+1}}(a)$ and $s_{\alpha_j}(a) < h_j(a)$. Altogether we get $h_i(a) < h_j(a)$, and thus the set $\{h_i(a) : i \in Z\}$ is a subset of S_a of cardinality κ, contradicting $|S_a| < \kappa$.

Definition *Let λ be an infinite cardinal and $(f_\xi : \xi < \lambda)$ be a sequence of ordinal functions in ${}^A\mathrm{ON}$. We say that an ordinal function $g \in {}^A\mathrm{ON}$* **cuts the sequence $(f_\xi : \xi < \lambda)$ diagonally modulo I** *iff, for each $\alpha < \lambda$, there exists $\beta < \lambda$ such that the set $\{a \in A : f_\alpha(a) \leq g(a) < f_\beta(a)\}$ is not a member of I. The sequence $(f_\xi : \xi < \lambda)$ is called* **small modulo I** *iff there is no ordinal function $g \in {}^A\mathrm{ON}$ cutting the sequence diagonally modulo I.*

Next we turn to the proof of the fact that, under certain assumptions, there even exist κ-rapid sequences which are small.

Lemma 3.2.4 *Assume that $f \in {}^A\mathrm{ON}$ is an ordinal function, $\prod f$ is nonempty, λ is a cardinal, I is an ideal on A, $(\prod f, <_I)$ is λ-directed and $\mu := \lim_I(\mathrm{cf} \circ f)$. Further let λ' be a regular cardinal satisfying $\mu \leq \lambda' < \lambda$. Then there is a sequence $(f_\xi : \xi < \lambda')$ of members of $\prod f$ which is κ-rapid for all $\kappa \in (\aleph_0, \mu)_{\mathrm{reg}}$. If in addition $|A|^+ < \mu$, then there will even exist such a sequence which is small modulo I.*

Proof If $(\aleph_0, \mu)_{\mathrm{reg}} = \emptyset$, then the assertion is trivially true. So we can assume that $\mu > \aleph_1$. Let $S := \{\beta < \lambda' : \mathrm{cf}(\beta) \in (\aleph_0, \mu)_{\mathrm{reg}}\}$ and, for each $\beta \in S$,

$$B_\beta := \{a \in A : \mathrm{cf}(f(a)) \leq \mathrm{cf}(\beta)\} \quad \text{and} \quad X_\beta := A \setminus B_\beta \quad \text{for } \beta \in S.$$

By the definition of μ, we have $B_\beta \in I$ and consequently $X_\beta \in I^+$ for all $\beta \in S$. For each $\beta \in S$, let C_β be a club in β of order type $\mathrm{cf}(\beta)$, all of whose members are limit ordinals. As well as the requested sequence $(f_\xi : \xi < \lambda')$ we define by recursion, for each $\beta \in S$, a sequence $(g^\beta_\xi : \xi \in C_\beta)$ of members of $\prod f$.

Let $f_0 \in \prod f$ be arbitrary and $\xi < \lambda'$ be fixed. If the function f_ξ is already defined, we can choose a function $f_{\xi+1} \in \prod f$ such that $f_\xi <_I f_{\xi+1}$. Such a function exists since, by assumption, $\prod f$ is in particular 2-directed. Now let $\xi < \lambda'$ be a limit ordinal and assume that the sequence $(f_\zeta : \zeta < \xi)$ is already defined; further let g^β_ζ be defined for each $\beta \in S$ and each $\zeta \in C_\beta \cap \xi$. For $a \in X_\beta$ we have $\mathrm{cf}(f(a)) > \mathrm{cf}(\beta) \geq |C_\beta \cap \xi|$. Therefore, if $\xi \in C_\beta$, then there is a function $g^\beta_\xi \in \prod f$, so that for all $a \in X_\beta$

$$\sup(\bigcup\{\{f_\zeta(a), g^\beta_\zeta(a)\} : \zeta \in C_\beta \cap \xi\}) < g^\beta_\xi(a).$$

Since $(\prod f, <_I)$ is λ-directed and $|S| < \lambda$, there is a function $f_\xi \in \prod f$ satisfying $f_\zeta <_I f_\xi$ for all $\zeta < \xi$ and $g^\beta_\xi <_I f_\xi$ for all $\beta \in S$ with $\xi \in C_\beta$.

The sequence $(f_\xi : \xi < \lambda')$ constructed in this way is obviously κ-rapid for all $\kappa \in (\aleph_0, \mu)_{\mathrm{reg}}$. Furthermore, for every $\beta \in S$, the sequence $(g^\beta_\xi \restriction X_\beta : \xi \in C_\beta)$ is strictly increasing under the relation $<_{X_\beta}$.

Now in addition let $|A|^+ < \mu$. To get a contradiction we assume that there is a function $g \in {}^A\mathrm{ON}$ which cuts the sequence $(f_\xi : \xi < \lambda')$ diagonally. By the definition of this property, we can choose by recursion a strictly increasing sequence $(\alpha_i : i < |A|^+)$ of ordinals less than λ', so that for all $i < |A|^+$

$$\{a \in A : f_{\alpha_i}(a) \leq g(a) < f_{\alpha_{i+1}}(a)\} \in I^+.$$

We put $\beta := \sup\{\alpha_i : i < |A|^+\}$. Then we have $\mathrm{cf}(\beta) = |C_\beta| = |A|^+$, $\beta \in S$, and $X_\beta \in I^+$.

Fix an ordinal $\xi \in C_\beta$. There is $i < |A|^+$ and limit ordinals $\delta, \gamma \in C_\beta$ such that $\xi < \alpha_i$ and $\alpha_{i+1} < \gamma < \delta$. We have $g_\xi^\beta <_I f_\xi <_I f_{\alpha_i}$ and $f_{\alpha_{i+1}} <_I f_\gamma <_I g_\delta^\beta$. From this we can conclude that the set

$$Z := B_\beta \cup \{a \in A : f_{\alpha_i}(a) \leq g_\xi^\beta(a)\} \cup \{a \in A : g_\delta^\beta(a) \leq f_{\alpha_{i+1}}(a)\}$$

is a member of I. Therefore we can infer that, modulo I,

$$\begin{aligned} U &:= \{a \in A : f_{\alpha_i}(a) \leq g(a) < f_{\alpha_{i+1}}(a)\} \\ &\subseteq \{a \in X_\beta : g_\xi^\beta(a) \leq g(a) < g_\delta^\beta(a)\} =: W. \end{aligned}$$

Namely, if $a \in U \setminus W$, then $f_{\alpha_i}(a) \leq g(a) < f_{\alpha_{i+1}}(a)$ and $a \notin X_\beta$ or $g(a) < g_\xi^\beta(a)$ or $g_\delta^\beta(a) \leq g(a)$. Each of the last three assertions implies $a \in Z$. So it is shown that $U \setminus W \in I$, that means $U \subseteq_I W$. Since U is not a member of I, this holds also for W. Thus we have proved: For every $\xi \in C_\beta$ there is a $\delta \in C_\beta$ with $\delta > \xi$ such that the set $\{a \in X_\beta : g_\xi^\beta(a) \leq g(a) < g_\delta^\beta(a)\}$ is not a member of I and is in particular nonempty.

Therefore there exists a strictly increasing sequence $(\delta_i : i < |A|^+)$ of ordinals in C_β such that the set $Y_i := \{a \in X_\beta : g_{\delta_i}^\beta(a) \leq g(a) < g_{\delta_{i+1}}^\beta(a)\}$ is not a member of I for all $i \in |A|^+$. Thus $(Y_i : i < |A|^+)$ is a sequence of nonempty, pairwise disjoint subsets of A. This contradiction shows that in the case $|A|^+ < \mu$ the sequence $(f_\xi : \xi < \lambda')$ is already small.

Definition *Let $(f_\xi : \xi < \lambda)$ be a sequence of elements of ${}^A\mathrm{ON}$. We call a function $f \in {}^A\mathrm{ON}$ an* **exact upper bound of $(f_\xi : \xi < \lambda)$ modulo I** *iff f is an upper bound of the sequence modulo I and, for every function $g \in {}^A\mathrm{ON}$ with $g <_I f$, there is an ordinal $\alpha < \lambda$ satisfying $g <_I f_\alpha$.*

Lemma 3.2.5 *Assume that the sequence $\bar{f} = (f_\xi : \xi < \lambda)$ is small and f is a supremum of $\bar{f}$. Then f is an exact upper bound of $\bar{f}$.*

Proof Let $g <_I f$. Since $\bar{f}$ is small, there is $\alpha < \lambda$ such that $Y_\beta := \{a \in A : f_\alpha(a) \leq g(a) < f_\beta(a)\} \in I$ for all $\beta < \lambda$. We will show that $g <_I f_\alpha$, which will prove the lemma. If $X := \{a \in A : f_\alpha(a) \leq g(a)\}$, then we put for $a \in A$

$$g'(a) := \begin{cases} g(a) & \text{if } a \in X, \\ f(a) & \text{if } a \in A \setminus X. \end{cases}$$

Now g' is also an upper bound of $\bar{f}$. Namely, if $\beta < \lambda$, then

$$\begin{aligned} & \{a \in A : g'(a) < f_\beta(a)\} \\ = \ & \{a \in X : g(a) < f_\beta(a)\} \cup \{a \in A \setminus X : f(a) < f_\beta(a)\} \\ \subseteq \ & \{a \in A : f_\alpha(a) \leq g(a) < f_\beta(a)\} \cup \{a \in A : f(a) < f_\beta(a)\} \in I, \end{aligned}$$

since $Y_\beta \in I$ and f is a supremum of $\bar{f}$. This implies $f \leq_I g'$, that means $\{a \in A : g'(a) < f(a)\} \in I$. Consequently, with $g <_I f$, we obtain

$$X \subseteq \{a \in A : g'(a) < f(a)\} \cup \{a \in A : f(a) \leq g(a)\} \in I,$$

since for $a \in X$ we have $g(a) = g'(a)$. Thus $X \in I$ and $g <_I f_\alpha$, as desired.

With the help of the preceding results we are now able to prove the so-called main lemma of pcf-theory. The importance of this theorem has been described in the introduction of this chapter.

Theorem 3.2.6 (Main Lemma, S. Shelah) *Assume that I is an ideal on A and $f \in {}^A\mathrm{Reg}$ is an ordinal function such that $\mu := \lim_I(f)$ is a limit cardinal with $|A| < \mu$. Further let λ' be a regular cardinal with $\lambda' > \mu$, for which the partial ordering $(\prod f, <_I)$ is $(\lambda')^+$-directed. Then there is an ordinal function $g \in {}^A\mathrm{ON}$, such that*

$$g <_I f, \quad \mu = \lim_I(\mathrm{cf} \circ g), \quad \textit{and } \lambda' = \mathrm{tcf}\left(\prod g/I\right).$$

Proof Let f and λ' with the above properties be given. From $\mathrm{ran}(f) \subseteq \mathrm{Reg}$ we get $\prod f \neq \emptyset$ and $f = \mathrm{cf} \circ f$. Now, with $\lambda := (\lambda')^+$, the assumptions of Lemma 3.2.4 are fulfilled. Furthermore we have $|A|^+ < \mu$, since $|A| < \mu$ and μ is a limit cardinal. Therefore there is a small sequence $(f_\xi : \xi < \lambda')$ of elements of $\prod f$ which is κ-rapid for all $\kappa \in (\aleph_0, \mu)_{\mathrm{reg}}$. By Lemma 3.2.3, the sequence $(f_\xi : \xi < \lambda')$ is a κ-ub sequence for every $\kappa \in (|A|, \mu)_{\mathrm{reg}}$, hence it has a supremum g by Theorem 3.2.1. Lemma 3.2.5 tells us that g is even an exact upper bound of $(f_\xi : \xi < \lambda')$. We now apply Lemma 3.1.2 several times and change the values of the given functions on appropriate sets in I. Since f is an upper bound of the sequence under $<_A$ and since $g \leq_I f$, we can assume that $g \leq_A f$ and, with $f_0 <_I g$, $\prod g \neq \emptyset$. Further, in view of $f_\xi <_I g$ for $\xi < \lambda'$, let without loss of generality every function f_ξ be a member of $\prod g$. From the property that g is an exact upper bound of the sequence we can now infer immediately that the set $\{f_\xi : \xi < \lambda'\}$ is cofinal in $\prod g$. Together with the strict monotonicity of the sequence we can conclude that $\mathrm{tcf}(\prod g/I) = \lambda'$.

Next we want to show that $g <_I f$, and for this we assume, to get a contradiction, that

$$X := \{a \in A : g(a) \geq f(a)\} = \{a \in A : g(a) = f(a)\} \notin I.$$

Then $\mathrm{tcf}(\prod g/I \restriction X) = \lambda'$ by Lemma 3.1.7. By assumption, $(\prod f, <_I)$ is $(\lambda')^+$-directed, and so this also holds for the partial ordering $(\prod f \restriction X, <_{I \cap \mathcal{P}(X)})$. Since $\prod f \restriction X = \prod g \restriction X$, $(\prod g \restriction X, <_{I \cap \mathcal{P}(X)})$ is $(\lambda')^+$-directed, and consequently this also holds for $(\prod g, <_{I \restriction X})$, contradicting $\mathrm{tcf}(\prod g/I \restriction X) = \lambda'$.

From Lemmas 3.2.3 and 3.2.2 we can conclude that $\kappa \le \lim_I(\mathrm{cf} \circ g)$ for all $\kappa \in (|A|, \mu)_{\mathrm{reg}}$. Since, by assumption, μ is a limit cardinal, we get $\mu \le \lim_I(\mathrm{cf} \circ g)$. Conversely the assertion $\lim_I(\mathrm{cf} \circ g) \le \lim_I(f) = \mu$ follows from $\mathrm{cf} \circ g \le_A g <_I f$. Altogether we obtain $\mu = \lim_I(\mathrm{cf} \circ g)$.

Exercises

1) Assume that A is a nonempty set, I is an ideal on A, λ is an infinite cardinal, and $(f_\xi : \xi < \lambda)$ is a sequence of members of ${}^A\mathrm{ON}$ strictly increasing modulo I. Further, let $g \in {}^A\mathrm{ON}$ be an exact upper bound of $(f_\xi : \xi < \lambda)$. Prove that there is a sequence $(f'_\xi : \xi < \lambda)$ of members of ${}^A\mathrm{ON}$ and an exact upper bound g' of $(f'_\xi : \xi < \lambda)$ modulo I such that $g =_I g'$, $f_\xi =_I f'_\xi$, $f'_\xi \in \prod g'$ for all $\xi < \lambda$, and $\mathrm{tcf}(\prod g'/I) = \lambda$.

2) Assume that A is an infinite set, I is an ideal on A, λ is a regular cardinal with $2^{|A|} < \lambda$, and $(f_\xi : \xi < \lambda)$ is a sequence of members of ${}^A\mathrm{ON}$ strictly increasing modulo I. Further let $g \in {}^A\mathrm{ON}$ be a supremum of $(f_\xi : \xi < \lambda)$ modulo I. Prove that g is an exact upper bound of $(f_\xi : \xi < \lambda)$ modulo I.

   Hint: For any $h <_I g$ and $\xi < \lambda$, consider the set $X_\xi := \{a \in A : f_\xi(a) \le h(a)\}$. Then $|\{\xi < \lambda : X_\xi = X\}| = \lambda$ for some $X \subseteq A$. Show that $X \in I$.

3) Assume that I is an ideal on the infinite set A, λ is an infinite cardinal, and $(f_\xi : \xi < \lambda)$ is a sequence of members of ${}^A\mathrm{ON}$ strictly increasing modulo I. Further let $g \in {}^A\mathrm{ON}$ be an exact upper bound of $(f_\xi : \xi < \lambda)$. Prove that g is modulo I a supremum of the sequence.

   Hint: Use Lemma 3.1.1.

4) Assume that λ is an infinite cardinal and $(f_\xi : \xi < \lambda)$ is a sequence of members of ${}^A\mathrm{ON}$. Further let $(g_i : i < \sigma)$ be a sequence of upper bounds modulo I of $\{f_\xi : \xi < \lambda\}$ satisfying $g_{i+1} \lneq_I g_i$ for all i with $i + 1 < \sigma$. We call the sequence $(g_i : i < \sigma)$ **normal** iff the following conditions are satisfied:

   (i) $\sup\{f_\xi : \xi < \lambda\} <_A g_0$.

   (ii) If $\gamma < \sigma$ is a limit ordinal, $S^\gamma_a := \{g_i(a) : i < \gamma\}$ for all $a \in A$, and $H^\gamma := \prod_{a \in A} S^\gamma_a$, then there is an ordinal $\alpha_\gamma < \lambda$ such that every function $h \in H^\gamma$ with $f_{\alpha_\gamma} <_I h$ is an upper $\le_I$-bound of $\{f_\xi : \xi < \lambda\}$ and $g_\gamma(a) = \min\{\beta \in S^\gamma_a : \beta > f_{\alpha_\gamma}(a)\}$ for all $a \in A$.

Now assume that the sequence $(g_i : i < \sigma)$ is normal and $\gamma < \sigma$ is a limit ordinal. Prove that g_γ is an upper $\leq_I$-bound of $(f_\xi : \xi < \lambda)$ and that $g_\gamma \lneq_I g_i$ for all $i < \gamma$.

5) (Jech) Assume that λ is a regular cardinal, the sequence $(f_\xi : \xi < \lambda)$ of members of ${}^A\mathrm{ON}$ is strictly increasing modulo I, and $|A|^+ < \lambda$. Further let $(g_i : i < \sigma)$ be a normal sequence of upper $\leq_I$-bounds of $(f_\xi : \xi < \lambda)$. Prove that $\sigma < |A|^+$.

   Hint: Assume to get a contradiction that $|A|^+ \leq \sigma$. Then also $(g_i : i < |A|^+)$ is a normal sequence of upper bounds of $\{f_\xi : \xi < \lambda\}$ modulo I. If $\gamma < |A|^+$ is a limit ordinal, let $\alpha_\gamma < \lambda$ be the ordinal according to property (ii) of a normal sequence. Put $\alpha := \sup\{\alpha_\gamma : \mathrm{Lim}(\gamma) \wedge \gamma < |A|^+\}$. For $a \in A$ let $S_a := \{g_i(a) : i < |A|^+\}$ and $g(a) := \min\{\beta \in S_a : \beta > f_\alpha(a)\}$. Prove that $g \leq_I g_i$ for all $i < |A|^+$. Then choose a limit ordinal $\gamma < |A|^+$ with $g \in \prod_{a\in A}\{g_i(a) : i < \gamma\}$ and show that $g_\gamma \leq_I g$. The assertion $g \leq_I g_{\gamma+1} \lneq_I g_\gamma \leq_I g$ yields a contradiction.

6) Assume that A is an infinite set and λ is a regular cardinal with $2^{|A|} < \lambda$. Prove that every sequence $(f_\xi : \xi < \lambda)$ of members of ${}^A\mathrm{ON}$ which is strictly increasing modulo I has a supremum modulo I.

   Hint: Let $(g_i : i < \sigma)$ be a normal sequence of $\leq_I$-upper bounds of $\{f_\xi : \xi < \lambda\}$, such that there is no $\leq_I$-upper bound g of $\{f_\xi : \xi < \lambda\}$ with $g \lneq_I g_i$ for all $i < \sigma$. If $\sigma = i + 1$, then Lemma 3.1.1 tells us that g_i is a supremum of $\{f_\xi : \xi < \lambda\}$ modulo I. Now show, using exercises 4 and 5, that σ cannot be a limit ordinal.

7) (Shelah) Assume that κ is an infinite cardinal, I is an ideal on κ, and η is an ordinal greater than 1. Prove that there is an ordinal $\alpha < (|\eta|^\kappa)^+$ such that there is no sequence of members of $\prod_{i<\kappa}\aleph_\eta$ of length $\aleph_{\alpha+1}$ which is strictly increasing modulo I. In particular we have $\mathrm{tcf}(\prod_{i<\kappa}\aleph_\eta/I) < \aleph_{(|\eta|^\kappa)^+}$ if this cardinal exists.

   Hint: Suppose to get a contradiction that for each ordinal $\alpha < (|\eta|^\kappa)^+$ there is a sequence $(f^\alpha_\xi : \xi < \aleph_{\alpha+1})$ of members of $\prod_{i<\kappa}\aleph_\eta$ which is strictly increasing modulo I. If $2^\kappa < \alpha < (|\eta|^\kappa)^+$, then $2^\kappa < \aleph_{\alpha+1}$, and Exercise 6 tells us that $(f^\alpha_\xi : \xi < \aleph_{\alpha+1})$ has a supremum g^α modulo I. By Exercise 2 we get $\mathrm{tcf}(\prod g^\alpha/I) = \aleph_{\alpha+1}$. Therefore we have $\mathrm{tcf}(\prod \mathrm{cf} \circ g^\alpha/I) = \aleph_{\alpha+1}$ for every $\alpha \in (2^\kappa, (|\eta|^\kappa)^+)_{\mathrm{ON}}$. Now consider the function $(\mathrm{cf} \circ g^\alpha : \alpha \in (2^\kappa, (|\eta|^\kappa)^+)_{\mathrm{ON}})$.

8) If $\aleph_\eta$ is a singular cardinal and $\kappa = \mathrm{cf}(\aleph_\eta)$, then $\mathrm{pp}(\aleph_\eta) < \aleph_{(|\eta|^\kappa)^+}$, where $\mathrm{pp}(\lambda) = \sup\{\mathrm{tcf}(\prod f/I) : f \in {}^\kappa\lambda \wedge \mathrm{ran}(f) \subseteq \mathrm{Reg} \wedge \lim_I(f) = \lambda\}$.

9) If $\aleph_\eta$ is a κ-strong cardinal, where $\aleph_0 < \kappa = \mathrm{cf}(\aleph_\eta) < \aleph_\eta$, then $\aleph_\eta^\kappa < \aleph_{(|\eta|^\kappa)^+}$.

   Hint: Use Exercise 3.13 and the previous question.

3.3 The Definition and Simple Properties of pcf(a)

Definition *If* a *is an arbitrary set of ordinal numbers and* I *is an ideal on* a, *then we define* $\operatorname{tcf}(\prod \mathsf{a}/I) := \operatorname{tcf}(\prod \mathrm{id}_{\mathsf{a}}/I)$ *where we know that* $\mathrm{id}_{\mathsf{a}} = \{(\xi,\xi) : \xi \in \mathsf{a}\}$ *is the identity on* a.

For any ordinal function $f \in {}^A\mathrm{ON}$, *we call* f **progressive** *iff* $\operatorname{cf}(f(a)) > |A|$ *for all* $a \in A$.

In this section we investigate the set

$$\mathrm{pcf}(f) = \{\operatorname{tcf}(\prod f/I) : I \text{ is an ideal on } \mathrm{dom}(f)\}$$

of true cofinalities of *progressive* ordinal functions f. Note that $\operatorname{cf}(f(a)) > |A|$ for all $a \in A$ implies that $f(a)$ is a limit ordinal for all $a \in A$. Thus we can conclude from Lemma 3.1.11 that $\mathrm{pcf}(f) = \mathrm{pcf}(\mathrm{cf} \circ f)$. Clearly also $\mathrm{cf} \circ f$ is progressive. The next lemma says for progressive functions, that we know $\mathrm{pcf}(f)$ if we can characterize the true cofinalities of the set $\prod \mathrm{ran}(f)$ of ordinal functions. Altogether we will have shown that it suffices for a characterization of $\mathrm{pcf}(f)$ to investigate the set of true cofinalities of sets $\prod \mathsf{a}$ where a is a set of regular cardinals.

Definition *If* I *is an ideal on* A *and* f *is a function with* $\mathrm{dom}(f) = A$, *we put*

$$f[I] := \{Y \subseteq \mathrm{ran}(f) : f^{-1}[Y] \in I\}.$$

It is easy to verify that this set is an ideal on $\mathrm{ran}(f)$. *We call it the* **image ideal of I under the function f**.

Lemma 3.3.1 *Assume that* $A \neq \emptyset$, I *is an ideal on* A, *and* f *is a progressive ordinal function on* A. *If there exists one of the cardinals* $\operatorname{tcf}(\prod f/I)$ *and* $\operatorname{tcf}(\prod \mathrm{ran}(f)/f[I])$, *then there exists the other one, and they are equal.*

Proof Let $\operatorname{tcf}(\prod f/I)$ exist. We want to apply Lemma 3.1.10 and hence define a function $G : \prod f \longrightarrow \prod \mathrm{ran}(f)$ which satisfies the condition (3.1) of this lemma. For $r \in \prod f$ and $\alpha \in \mathrm{ran}(f)$, we put

$$G(r)(\alpha) := \sup\{r(a) + 1 : a \in A \wedge f(a) = \alpha\}.$$

Since f is progressive, we have $|A| < \operatorname{cf}(f(a))$ and hereby $G(r)(\alpha) < \alpha$ for any $\alpha \in \mathrm{ran}(f)$. Consequently we get $G(r) \in \prod \mathrm{ran}(f)$. If $h \in \prod \mathrm{ran}(f)$, then $h \circ f$ is a member of $\prod f$. This function will play the role of the function p in Lemma 3.1.10. So consider $q \in \prod f$ with $h \circ f \leq_I q$, i.e. with $\mathrm{B}(q, h \circ f) \in I$. We will show that $h <_{f[I]} G(q)$, i.e. $\mathrm{B}[G(q), h] \in f[I]$ or, equivalently, $f^{-1}[\mathrm{B}[G(q), h]] \in I$.

If $\alpha \in \mathrm{ran}(f)$ and $a \in A$ have the properties $G(q)(\alpha) \leq h(\alpha)$ and $f(a) = \alpha$, then $q(a) < h(\alpha) = h(f(a))$ by the definition of G. So we have

$$f^{-1}[\mathrm{B}[G(q), h]] \subseteq \mathrm{B}(q, h \circ f) \in I,$$

and therefore $h <_{f[I]} G(q)$ is shown.

Now conversely let $\mathrm{tcf}\left(\prod \mathrm{ran}(f)/f[I]\right)$ exist. To apply Lemma 3.1.10, let the function $F : \prod \mathrm{ran}(f) \longrightarrow \prod f$ be defined by $F(q)(a) := q(f(a))$. For $h \in \prod f$ we can define the function $p \in \prod \mathrm{ran}(f)$ by

$$p(f(a)) := \sup\{h(a') + 1 : a' \in A \wedge f(a) = f(a')\}.$$

Since f is progressive, we get $p \in \prod \mathrm{ran}(f)$. Now let $q \in \prod \mathrm{ran}(f)$ be given such that $p \leq_{f[I]} q$. Then the set $\{a \in A : q(f(a)) < p(f(a))\}$ is an element of I, that means, $p(f(a)) \leq q(f(a)) = F(q)(a)$ holds for almost all $a \in A$. Because of $h(a) < p(f(a))$ for all $a \in A$, we get $h <_I F(q)$.

In the following sections of this chapter, a denotes a set of regular cardinals.

Definition *Let $g \in {}^A\mathrm{ON}$ be an ordinal function. The* **set of possible cofinalities**[2] **of g** *is defined by*

$$\mathrm{pcf}(g) := \{\mathrm{tcf}\left(\prod g/I\right) : I \text{ is an ideal on } A\}.$$

If a is a set of regular cardinals, then we put $\mathrm{pcf}(\mathsf{a}) := \mathrm{pcf}(\mathrm{id}_{\mathsf{a}})$ where we know that $\mathrm{id}_{\mathsf{a}} = \{(\mu, \mu) : \mu \in \mathsf{a}\}$ is the identity on a. That means that

$$\mathrm{pcf}(\mathsf{a}) = \{\mathrm{tcf}\left(\prod \mathsf{a}/I\right) : I \text{ is an ideal on } \mathsf{a}\}.$$

More generally – we will get $\mathrm{pcf}_{|\mathsf{a}|}(\mathsf{a}) = \mathrm{pcf}(\mathsf{a})$ in Lemma 3.3.4 – we define for every cardinal μ

$$\mathrm{pcf}_\mu(\mathsf{a}) := \bigcup\{\mathrm{pcf}(\mathsf{b}) : \mathsf{b} \in [\mathsf{a}]^{\leq\mu}\}.$$

We call the set a **progressive** *iff the function id_{a} is progressive, that is, iff $\mathsf{a} = \emptyset$ or $|\mathsf{a}| < \min(\mathsf{a})$. Further let*

$$\lim_I(\mathsf{a}) := \lim_I(\mathrm{id}_{\mathsf{a}}) = \min\{\mu \in \mathrm{CN} : \{\nu \in \mathsf{a} : \nu \leq \mu\} \notin I\}$$

be the **limit of a modulo I** *if I is an ideal on a. If D is an ultrafilter on a, then*

$$\lim_D(\mathsf{a}) = \min\{\mu \in \mathrm{CN} : \{\nu \in \mathsf{a} : \nu \leq \mu\} \in D\}$$

is the **limit of a modulo D**.

---

[2] *possible cofinalities*

Remark Obviously every subset of a progressive set is also progressive. Further it is clear by our convention that in the definition of pcf(a) only those ideals I can appear for which $\mathrm{tcf}(\prod \mathsf{a}/I)$ exists. We have $\mathrm{pcf}(\emptyset) = \emptyset = \mathrm{pcf}_0(\mathsf{a})$ since there are no (proper) ideals on the empty set. If $\mathsf{a} \neq \emptyset$, the cardinal $\mathrm{tcf}(\prod \mathsf{a}/I)$ is always infinite by Lemma 3.1.3, since $(\prod \mathsf{a}, \leq_I)$ has no maximum.

Lemma 3.3.2 *Assume that A is a nonempty set and $f \in {}^A\mathrm{ON}$ is a progressive ordinal function, and let $\mathsf{a} := \mathrm{ran}(\mathrm{cf} \circ f)$. Then*

$$\mathrm{pcf}(f) = \mathrm{pcf}(\mathrm{cf} \circ f) = \mathrm{pcf}(\mathsf{a}) = \{\mathrm{cf}(\prod \mathsf{a}/D) : D \text{ is an ultrafilter on } \mathsf{a}\}.$$

In particular, the last equation is true for any set a of regular cardinals.

Proof The first equation holds by Lemma 3.1.11 since $\mathrm{cf}(f(a)) > |A| \geq 1$ for all $a \in A$ and thus all values of f must be limit ordinals. The second equation follows from Lemma 3.3.1. For the last equation, let first D be an ultrafilter on a. Then $\mathrm{cf}(\prod \mathsf{a}/D)$ exists by Lemma 3.1.6, and for the dual ideal I of D we have $\mathrm{tcf}(\prod \mathsf{a}/I) = \mathrm{cf}(\prod \mathsf{a}/D)$. Conversely, if I is an ideal on a such that $\mathrm{tcf}(\prod \mathsf{a}/I)$ exists, then there is an ultrafilter D on a such that $D \cap I = \emptyset$. Now Lemma 3.1.7 says that $\mathrm{cf}(\prod \mathsf{a}/D) = \mathrm{tcf}(\prod \mathsf{a}/I)$.

Lemma 3.3.3 *If a is a nonempty progressive interval of regular cardinals which has no maximum, then all members of the set $\mathsf{a} \setminus \{\min(\mathsf{a})\}$ are successor cardinals. If $\aleph_\delta := \sup(\mathsf{a})$, then we have $\delta < \aleph_\delta$, and consequently $\aleph_\delta$ is singular.*

Proof Let ν be a limit cardinal, $\nu \in (\min(\mathsf{a}), \sup(\mathsf{a}))_{\mathrm{CN}}$. Then obviously ν is the supremum of $\mathsf{a} \cap \nu$. Thus we can conclude by assumption that $\mathrm{cf}(\nu) \leq |\mathsf{a}| < \min(\mathsf{a}) \leq \nu$, and ν is singular. For the proof of the second assertion note first that δ is a limit ordinal and assume, to get a contradiction, that $\delta = \aleph_\delta = \sup(\mathsf{a})$. For $\mathsf{b} := [\aleph_0, \min(\mathsf{a}))_{\mathrm{reg}}$ we have $\mathsf{b} \cup \mathsf{a} = [\aleph_0, \aleph_\delta)_{\mathrm{reg}}$, and therefore our supposition gives $|\mathsf{b}| + |\mathsf{a}| = |\delta| = \aleph_\delta$. Since $|\mathsf{b}| < \aleph_\delta$, we get $|\mathsf{a}| = \aleph_\delta$ which contradicts the fact that a is progressive. Now $\mathrm{cf}(\aleph_\delta) = \mathrm{cf}(\delta) \leq |\delta| \leq \delta < \aleph_\delta$ yields that the cardinal $\aleph_\delta$ is singular.

Now we want to list some simple properties of pcf(a) and $\mathrm{pcf}_\mu(\mathsf{a})$, respectively.

Lemma 3.3.4 *Let μ be a cardinal with $\mu > 0$. Then the following holds.*

a) $\mathrm{pcf}_\mu(\mathsf{a})$ *is a set of regular cardinals.*

b) $\sup \mathrm{pcf}_\mu(\mathsf{a}) \leq (\sup(\mathsf{a}))^\mu$.

c) If D is the principal ultrafilter on a generated by $\{\lambda\}$, then $\mathrm{cf}(\prod \mathsf{a}/D) = \lambda$ and $\lim_D(\mathsf{a}) = \lambda$. Conversely, if D is an ultrafilter on a, $\mathrm{cf}(\prod \mathsf{a}/D) = \lambda$ and $\mathsf{a} \cap \lambda$ is finite, then D is principal and generated by $\{\lambda\}$.

d) $\mathsf{a} \subseteq \mathrm{pcf}_\mu(\mathsf{a})$.

e) If a *is finite, then* $\mathrm{pcf}_\mu(\mathsf{a}) = \mathsf{a}$.

f) $\mathrm{pcf}_\mu(\mathsf{a}) \cap \min(\mathsf{a}) = \emptyset$.

g) If a *has no maximum and* D *is an unbounded ultrafilter on* a*, that is an ultrafilter on* a *which does not contain a bounded subset of* a*, then*

$$\mathrm{cf}\,(\prod \mathsf{a}/D) \geq \sup(\mathsf{a}).$$

h) If $\mathsf{b} \subseteq \mathsf{a}$*, then* $\mathrm{pcf}_\mu(\mathsf{b}) \subseteq \mathrm{pcf}_\mu(\mathsf{a})$ *and* $\mathrm{pcf}(\mathsf{b}) \subseteq \mathrm{pcf}(\mathsf{a})$*. So we get* $\mathrm{pcf}_\mu(\mathsf{a}) \subseteq \mathrm{pcf}(\mathsf{a})$ *and* $\mathrm{pcf}_{|\mathsf{a}|}(\mathsf{a}) = \mathrm{pcf}(\mathsf{a})$*. In particular, if* $\lambda \in \mathrm{pcf}_\mu(\mathsf{a})$*, then there exists an ultrafilter* D *on* a *such that* $\lambda = \mathrm{cf}\,(\prod \mathsf{a}/D)$*.*

i) If $\mathsf{b} \subseteq \mathrm{Reg}$*, then* $\mathrm{pcf}_\mu(\mathsf{a} \cup \mathsf{b}) = \mathrm{pcf}_\mu(\mathsf{a}) \cup \mathrm{pcf}_\mu(\mathsf{b})$ *and* $\mathrm{pcf}(\mathsf{a} \cup \mathsf{b}) = \mathrm{pcf}(\mathsf{a}) \cup \mathrm{pcf}(\mathsf{b})$.

j) If D *is an ultrafilter on* a *such that* $\lim_D(\mathsf{a}) = \kappa^+$ *or* $\lim_D(\mathsf{a}) = \min(\mathsf{a})$*, then* D *is principal and generated by* $\{\kappa^+\}$ *or* $\{\min(\mathsf{a})\}$*, respectively.*

k) Assume that D *is an ultrafilter and* I *is an ideal on* a *such that* $D \cap I = \emptyset$*. Then* $\lim_I(\mathsf{a}) \leq \lim_D(\mathsf{a})$*. If* I *is the dual ideal of* D*, then equality holds. Furthermore, if* $\sup(\mathsf{a}) \notin \mathsf{a}$ *and* $\mathrm{I_b}(\mathsf{a}) \subseteq I$*, then* $\lim_I(\mathsf{a}) = \lim_D(\mathsf{a}) = \sup(\mathsf{a})$.

l) Assume that $\sup(\mathsf{a}) \notin \mathsf{a}$ *and* D *is an ultrafilter on* a *such that* $\lim_D(\mathsf{a}) = \sup(\mathsf{a})$*. Then* $D \cap \mathrm{I_b}(\mathsf{a}) = \emptyset$*. Furthermore, if* $\mathrm{cf}\,(\prod \mathsf{a}/D) = \kappa$*,* $\mu < \sup(\mathsf{a})$ *and* $\mathsf{a}' = \mathsf{a} \setminus \mu$*, then* $\mathrm{cf}\,(\prod \mathsf{a}'/D \cap \mathcal{P}(\mathsf{a}')) = \kappa$*.*

Proof If $\mathsf{b} \subseteq \mathsf{a}$ and $f \in \prod \mathsf{b}$, then we have $f+1 \in \prod \mathsf{b}$ and $f < f+1$. Consequently, for any ultrafilter D on b, $\prod \mathsf{b}$ does not possess a maximum under $\leq_D$. Now the assertion in a) follows from Lemma 3.1.3.

To prove b), let $\lambda \in \mathrm{pcf}_\mu(\mathsf{a})$. Then there is a subset b of a with $|\mathsf{b}| \leq \mu$, an ultrafilter D on b and a sequence $(f_\xi : \xi < \lambda)$ cofinal in $\prod \mathsf{b}$ modulo D. This sequence is certainly injective, and thus we obtain

$$\lambda \leq |\prod \mathsf{b}| \leq (\sup(\mathsf{a}))^\mu.$$

For the proof of part c), consider the principal ultrafilter D on a generated by $\{\lambda\}$ for some $\lambda \in \mathsf{a}$. Then we have $\mathsf{a} \cap \lambda^+ \in D$ and $\mathsf{a} \cap \nu^+ \notin D$ for all $\nu < \lambda$, hence $\lim_D(\mathsf{a}) = \lambda$. If we put

$$f_\xi(\nu) := \begin{cases} 0 & \text{if } \nu \neq \lambda, \\ \xi & \text{if } \nu = \lambda \end{cases}$$

for $\xi < \lambda$ and $\nu \in \mathsf{a}$, then the sequence $(f_\xi : \xi < \lambda)$ is cofinal in $\prod \mathsf{a}$ modulo D. Now $\mathrm{cf}\,(\prod \mathsf{a}/D) = \lambda$ and part d) follow, and we also get e) since each ultrafilter on a finite set is principal. Next we prove the second assertion in c). If a is finite, then D is principal, and the first assertion implies that D is generated by $\{\lambda\}$ since $\mathrm{cf}\,(\prod \mathsf{a}/D) = \lambda$. So we can assume that a is infinite. If $\mathsf{a} \cap \lambda^+ \in D$, then D is principal since it has the finite member $\mathsf{a} \cap \lambda^+$, and we can argue as before. Suppose that $\mathsf{a} \cap \lambda^+ \notin D$. Then we have $\mathsf{a} \setminus \lambda^+ \in D$, and Lemma 3.1.8 says that $\lambda = \mathrm{cf}\,(\prod \mathsf{a}/D) = \mathrm{cf}\,(\prod(\mathsf{a} \setminus \lambda^+)/D \cap \mathcal{P}(\mathsf{a} \setminus \lambda^+))$, hence $\lambda \in \mathrm{pcf}(\mathsf{a} \setminus \lambda^+)$. From f) we can conclude that $\lambda \geq \min(\mathsf{a} \setminus \lambda^+) > \lambda$. This contradiction shows that $\mathsf{a} \cap \lambda^+ \in D$, as desired.

For the proof of part f) let $\lambda < \min(\mathsf{a})$. Assume that D is an ultrafilter on a nonempty subset b of a such that $|\mathsf{b}| \leq \mu$, and $(f_\xi : \xi < \lambda)$ is a sequence which is cofinal in $\prod \mathsf{b}$ modulo D. By the regularity of the members of b, the equation $g(\nu) := \sup\{f_\xi(\nu) + 1 : \xi < \lambda\}$, for $\nu \in \mathsf{b}$, defines a function $g \in \prod \mathsf{b}$ satisfying $f_\xi < g$ for all $\xi < \lambda$. Therefore, the sequence $(f_\xi : \xi < \lambda)$ is not cofinal in $\prod \mathsf{b}$ modulo D, which proves f).

For g), assume that the set a has no maximum, and let D be an unbounded ultrafilter on a. Further let λ be a regular cardinal with $\lambda < \sup(\mathsf{a})$ and $(f_\xi : \xi < \lambda)$ be a sequence of members of $\prod \mathsf{a}$. If we define

$$g(\nu) := \begin{cases} 0 & \text{if } \nu \leq \lambda, \\ \sup\{f_\xi(\nu) + 1 : \xi < \lambda\} & \text{if } \nu > \lambda \end{cases}$$

for every $\nu \in \mathsf{a}$, then

$$\{\nu \in \mathsf{a} : f_\xi(\nu) < g(\nu)\} = \mathsf{a} \cap (\lambda, \sup(\mathsf{a}))_{\mathrm{reg}} \in D,$$

i.e., we have $f_\xi <_D g$ for all $\xi < \lambda$. Consequently the sequence $(f_\xi : \xi < \lambda)$ cannot be cofinal in $\prod \mathsf{a}$ modulo D.

For the proof of h) we note first that the first inclusion holds by the definition of the set $\mathrm{pcf}_\mu(\mathsf{a})$. For the proof of the second inclusion let $\lambda \in \mathrm{pcf}(\mathsf{b})$, and let D be an ultrafilter on b satisfying $\mathrm{cf}\,(\prod \mathsf{b}/D) = \lambda$. We can choose an ultrafilter D' on a extending D. Then $\lambda = \mathrm{cf}\,(\prod \mathsf{a}/D') \in \mathrm{pcf}(\mathsf{a})$ by Lemma 3.1.7. So we get $\mathrm{pcf}(\mathsf{b}) \subseteq \mathrm{pcf}(\mathsf{a})$ and thus also $\mathrm{pcf}_\mu(\mathsf{a}) \subseteq \mathrm{pcf}(\mathsf{a})$ by the definition of $\mathrm{pcf}_\mu(\mathsf{a})$. On the other hand we have $\mathrm{pcf}(\mathsf{a}) \subseteq \mathrm{pcf}_{|\mathsf{a}|}(\mathsf{a})$ and consequently $\mathrm{pcf}_{|\mathsf{a}|}(\mathsf{a}) = \mathrm{pcf}(\mathsf{a})$.

For part i), we state first that the inclusion $\mathrm{pcf}(\mathsf{a}) \cup \mathrm{pcf}(\mathsf{b}) \subseteq \mathrm{pcf}(\mathsf{a} \cup \mathsf{b})$ follows from h). For the proof of the reverse inclusion let $\lambda \in \mathrm{pcf}(\mathsf{a} \cup \mathsf{b})$ and let D be an ultrafilter on $\mathsf{a} \cup \mathsf{b}$ such that $\lambda = \mathrm{cf}\,(\prod(\mathsf{a} \cup \mathsf{b})/D)$. On account of $\mathsf{a} \cup \mathsf{b} \in D$ we can assume without loss of generality, using Lemma 1.8.3, that $\mathsf{a} \in D$, and Lemma 3.1.8 again yields

$$\lambda = \mathrm{cf}\,(\prod(\mathsf{a} \cup \mathsf{b})/D) = \mathrm{cf}\,(\prod \mathsf{a}/(D \cap \mathcal{P}(\mathsf{a}))) \in \mathrm{pcf}(\mathsf{a}).$$

The first equation in i) now also follows.

For j), let D be an ultrafilter on a. If $\lim_D(\mathsf{a}) = \kappa^+$, then $\mathsf{a} \cap \kappa^{++} \in D$ and $\mathsf{a} \cap \kappa^+ \notin D$, hence $(\mathsf{a} \cap \kappa^{++}) \setminus (\mathsf{a} \cap \kappa^+) = \{\kappa^+\} \in D$, and D is the principal ultrafilter generated by $\{\kappa^+\}$. If $\lim_D(\mathsf{a}) = \min(\mathsf{a})$, then the definition of a limit shows that $\{\min(\mathsf{a})\} \in D$.

The proof of the last assertions is left as an exercise.

Definition *For any set B of cardinals, let*

$$\sup^+(B) := \min\{\kappa \in \mathrm{CN} : \forall \nu \in B \ \nu < \kappa\}.$$

Assume that λ is a singular cardinal such that $\mathrm{cf}(\lambda) \leq \kappa < \lambda$, *and let A be the set*

$$\{\mathrm{cf}(\textstyle\prod \mathsf{a}/D) : \mathsf{a} \subseteq \mathrm{Reg} \ \wedge \ \sup(\mathsf{a}) = \lambda \ \wedge \ |\mathsf{a}| \leq \kappa \ \wedge \ D \textit{ ultrafilter on } \mathsf{a} \ \wedge \ D \cap \mathrm{I_b}(\mathsf{a}) = \emptyset\}.$$

We define the cardinals $\mathrm{pp}_\kappa(\lambda)$ and $\mathrm{pp}^+_\kappa(\lambda)$, which will play an important role in the last chapters of this book, as

$$\mathrm{pp}_\kappa(\lambda) := \sup(A) \qquad \textit{and} \qquad \mathrm{pp}^+_\kappa(\lambda) := \sup^+(A).$$

If $\kappa = \mathrm{cf}(\lambda)$, then we put $\mathrm{pp}(\lambda) := \mathrm{pp}_\kappa(\lambda)$ and $\mathrm{pp}^+(\lambda) := \mathrm{pp}^+_\kappa(\lambda)$.

Lemma 3.3.5 *Assume that λ and κ are infinite cardinals and $\mathrm{cf}(\lambda) \leq \kappa < \lambda$. If μ is a regular cardinal such that $\lambda < \mu < \mathrm{pp}^+_\kappa(\lambda)$, then there is a subset a of λ satisfying $\mathsf{a} \subseteq \mathrm{Reg}$, $\sup(\mathsf{a}) = \lambda$ and $|\mathsf{a}| \leq \kappa$, and an ultrafilter D on a such that $D \cap \mathrm{I_b}(\mathsf{a}) = \emptyset$ and $\mathrm{cf}(\prod \mathsf{a}/D) = \mu$. In particular we have $\lim_D(\mathsf{a}) = \lambda$, and if $\kappa_0 \in (\kappa, \lambda)_{\mathrm{CN}}$, we can assume without loss of generality that $\kappa_0 < \min(\mathsf{a})$.*

Proof Let $\kappa_0 \in (\kappa, \lambda)_{\mathrm{CN}}$ and μ be a regular cardinal such that $\lambda < \mu < \mathrm{pp}^+_\kappa(\lambda)$. By the definition of $\mathrm{pp}^+_\kappa(\lambda)$, there is a set $\mathsf{b} \subseteq \lambda \cap \mathrm{Reg}$ such that $\sup(\mathsf{b}) = \lambda$ and $|\mathsf{b}| \leq \kappa$ and a maximal ideal I on b such that $\mathrm{I_b}(\mathsf{b}) \subseteq I$ and $\mathrm{tcf}(\prod \mathsf{b}/I) \geq \mu$. Without loss of generality we can assume that $\mathsf{b} \subseteq (\lambda \cap \mathrm{Reg}) \setminus \kappa_0$, since $\mathrm{I_b}(\mathsf{b}) \subseteq I$.

If $\mu = \mathrm{pp}_\kappa(\lambda)$, then $\mathrm{tcf}(\prod \mathsf{b}/I) = \mu$, and we can take $\mathsf{a} := \mathsf{b}$ and D as the dual ultrafilter of I, and get $\mu = \mathrm{cf}(\prod \mathsf{a}/D)$.

Now we discuss the case that $\mu < \mathrm{pp}_\kappa(\lambda)$. Without loss of generality let $\mu < \mathrm{tcf}(\prod \mathsf{b}/I)$. Then $(\prod \mathsf{b}, <_I)$ is μ^+-directed by Lemma 3.1.9. We can apply the main lemma 3.2.6 and Theorem 3.1.13 to the function $f := \mathrm{id}_\mathsf{b}$ and obtain a function $g \in {}^\mathsf{b}\mathrm{Reg}$ such that $g <_I f$, $\lim_I(g) = \lambda$, and $\mathrm{tcf}(\prod g/I) = \mu$. Note that $\{\nu \in \mathsf{b} : g(\nu) \leq \kappa_0\} \in I$. So we can assume that $\kappa_0 < \min(g[\mathsf{b}])$. We put $\mathsf{a} := g[\mathsf{b}]$. Then $\mathrm{I_b}(\mathsf{a}) \subseteq g[I]$. Choose an ultrafilter D on a such that $D \cap g[I] = \emptyset$. Lemma 3.3.1 and Lemma 3.1.7 say that $\mu = \mathrm{tcf}(\prod g/I) = \mathrm{cf}(\prod \mathsf{a}/D)$. It is easy to verify that a and D have the desired properties.

Theorem 3.3.6 *If* a *is a progressive interval of regular cardinals and* μ *is a cardinal, then* $\mathrm{pcf}_\mu(\mathsf{a})$ *is also an interval.*

Proof If a is a finite set, then $\mathrm{pcf}_\mu(\mathsf{a})$ is the empty set or the set a, hence in each case an interval. So let a be an infinite interval of regular cardinals. We have to show for arbitrary cardinals $\lambda \in \mathrm{pcf}_\mu(\mathsf{a})$ and $\lambda' \in [\min(\mathsf{a}), \lambda)_{\mathrm{reg}}$ that λ' is a member of $\mathrm{pcf}_\mu(\mathsf{a})$. By Lemma 3.3.4, this is certainly true for $\lambda' < \sup(\mathsf{a})$ since in this case $\lambda' \in \mathsf{a}$, and, if $\sup(\mathsf{a}) \in \mathsf{a}$, for $\lambda = \sup(\mathsf{a})$ since then $\sup(\mathsf{a}) = \max(\mathsf{a})$. If $\sup(\mathsf{a}) \notin \mathsf{a}$, then $\mathrm{cf}(\sup(\mathsf{a})) \leq |\mathsf{a}| < \min(\mathsf{a}) \leq \sup(\mathsf{a})$, and thus $\sup(\mathsf{a})$ is a singular cardinal.

So let $\sup(\mathsf{a}) < \lambda'$ and let b be a subset of a satisfying $|\mathsf{b}| \leq \mu$ and $\lambda \in \mathrm{pcf}(\mathsf{b})$. Then there is an ultrafilter D on b such that $\lambda = \mathrm{cf}(\prod \mathsf{b}/D)$. D is nonprincipal since $\lambda > \lambda' > \sup(\mathsf{a})$. If $\mu^* := \lim_D(\mathsf{b})$, then Lemma 3.3.4 says that $|\mathsf{b}| < \min(\mathsf{b}) < \mu^*$ and that μ^* is a limit cardinal, since D is nonprincipal.

Since $\lambda = \mathrm{cf}(\prod \mathsf{b}/D)$ and $(\lambda')^+ \leq \lambda$, the linear ordering $(\prod \mathsf{b}, <_D)$ is $(\lambda')^+$-directed. By the main lemma, Theorem 3.2.6, there is an ordinal function $g \in {}^{\mathsf{b}}\mathrm{ON}$ satisfying $g <_D \mathrm{id}_{\mathsf{b}}$, $\mu^* = \lim_D(\mathrm{cf} \circ g)$, and $\lambda' = \mathrm{cf}(\prod g/D)$. If we put $\mathsf{d} := \{\nu \in \mathsf{b} : g(\nu) < \nu \wedge \mathrm{cf}(g(\nu)) > \min(\mathsf{b})\}$, then $\mathsf{d} \in D$, $|\mathrm{ran}((\mathrm{cf} \circ g) \restriction \mathsf{d})| \leq \mu$, and $\mathrm{ran}((\mathrm{cf} \circ g) \restriction \mathsf{d})$ is a subset of a. The last assertion can be seen as follows. If $\nu \in \mathsf{d}$, then $\min(\mathsf{a}) \leq \min(\mathsf{b}) < \mathrm{cf}(g(\nu)) \leq g(\nu) < \nu$; since $\nu \in \mathsf{a}$, $\mathrm{cf}(g(\nu)) \in$ Reg and a is an interval, we obtain $\mathrm{cf}(g(\nu)) \in \mathsf{a}$. From Theorem 3.1.13 and Lemma 3.1.8 we can conclude that

$$\lambda' = \mathrm{cf}(\prod g/D) = \mathrm{cf}(\prod (\mathrm{cf} \circ g)/D) = \mathrm{cf}(\prod (\mathrm{cf} \circ g) \restriction \mathsf{d}/D \cap \mathcal{P}(\mathsf{d})),$$

and consequently, by Lemma 3.3.1, we get $\lambda' \in \mathrm{pcf}(\mathrm{ran}((\mathrm{cf} \circ g) \restriction \mathsf{d})) \subseteq \mathrm{pcf}_\mu(\mathsf{a})$. Therefore we get the desired assertion $\lambda' \in \mathrm{pcf}_\mu(\mathsf{a})$.

Lemma 3.3.7 *Assume that* $f \in {}^{A}\mathrm{Reg}$ *is a function,* B *is a set with* $|B| < \min(f[A])$, I *is an ideal on* B, *and for each* $b \in B$ *there is an ideal* I_b *on* A *such that* $\mathrm{tcf}(\prod f/I_b)$ *exists. Let* $g := (\mathrm{tcf}(\prod f/I_b) : b \in B)$ *and*

$$I^* := \{X \subseteq A : \{b \in B : X \notin I_b\} \in I\}.$$

Then I^* *is an ideal on* A. *If one of the cardinals* $\mathrm{tcf}(\prod f/I^*)$ *and* $\mathrm{tcf}(\prod g/I)$ *exists, then so does the other one, and they are equal.*

Proof The easy proof that I^* is an ideal on A is left to the reader. For each $b \in B$, we can choose a sequence $(f^b_\xi : \xi < g(b))$ which is cofinal in $\prod f$ modulo I_b. To apply Lemma 3.1.10, we can define functions $F : \prod g \longrightarrow \prod f$ and $G : \prod f \longrightarrow \prod g$ by

$$F(q)(a) := \sup\{f^b_{q(b)}(a) + 1 : b \in B\} \quad \text{and} \quad G(h)(b) := \min\{\xi < g(b) : h \leq_{I_b} f^b_\xi\}.$$

Since $|B| < \min(f[A])$, we have $F(q)(a) \in f(a)$ for all $a \in A$. Furthermore $h \leq_{I_b} f^b_{G(h)(b)}$ holds for every $b \in B$.

First let $\mathrm{tcf}(\prod g/I)$ exist. According to Lemma 3.1.10, we want to find for each function $h \in \prod f$ a function $p \in \prod g$ such that $h <_{I^*} F(q)$ for all $q \in \prod g$ with $p \leq_I q$. Define $p := G(h)$. Let $q \in \prod B$ be given such that $p \leq_I q$, that means that $Y := \{b \in B : q(b) < p(b)\} \in I$. We have to show that $h <_{I^*} F(q)$, that means that $X := \{b \in B : \{a \in A : F(q)(a) \leq h(a)\} \notin I_b\} \in I$. We will show $X \subseteq Y$ to complete this part of the proof. So let $b \in X$, that means $\{a \in A : F(q)(a) \leq h(a)\} \notin I_b$. Assume that $b \notin Y$, i.e., $p(b) \leq q(b)$. Then we would get, by the definition of F and G,

$$h \leq_{I_b} f^b_{G(h)(b)} = f^b_{p(b)} \leq_{I_b} f^b_{q(b)} <_A F(q),$$

consequently $h <_{I_b} F(q)$ and hereby $\{a \in A : F(q)(a) \leq h(a)\} \in I_b$, which contradicts $b \in X$. Therefore we have $b \in Y$, $X \subseteq Y$, and altogether $X \in I$.

For the converse let $\mathrm{tcf}(\prod f/I^*)$ exist and the function $q \in \prod g$ be given. We are looking for a function $p \in \prod f$ such that for all $h \in \prod f$ with $p \leq_{I^*} h$ we have $q <_I G(h)$. Define $p := F(q)$ and let $\{a \in A : h(a) < p(a)\} \in I^*$. We have to show that $X := \{b \in B : G(h)(b) \leq q(b)\} \in I$ and know that $Y := \{b \in B : \{a \in A : h(a) < p(a)\} \notin I_b\} \in I$. Again we prove $X \subseteq Y$. Let $b \in X$. This implies

$$h \leq_{I_b} f^b_{G(h)(b)} \leq_{I_b} f^b_{q(b)} <_A F(q) = p.$$

So we get $\{a \in A : p(a) \leq h(a)\} \in I_b$, and consequently $\{a \in A : h(a) < p(a)\} \notin I_b$. Thus we have $b \in Y$, as desired.

Corollary 3.3.8 *Assume that* b *is a nonempty subset of* $\mathrm{pcf}(\mathsf{a})$ *such that* $|\mathsf{b}| < \min(\mathsf{a})$ *and* I *is an ideal on* b. *For each* $\nu \in \mathsf{b}$, *let* I_ν *be an ideal on* a *satisfying* $\mathrm{tcf}(\prod \mathsf{a}/I_\nu) = \nu$. *Further let*

$$I^* := \{X \subseteq \mathsf{a} : \{\nu \in \mathsf{b} : X \notin I_\nu\} \in I\}.$$

Then I^* *is an ideal on* a. *If there exists one of the cardinals* $\mathrm{tcf}(\prod \mathsf{a}/I^*)$ *and* $\mathrm{tcf}(\prod \mathsf{b}/I)$, *then there exists the other one, and they are equal.*

Exercises

1) Assume that $\alpha \in \mathrm{ON}$, A is a nonempty set and $g := A \times \{\alpha\}$ is progressive. Determine $\mathrm{pcf}(g)$.
2) Let $g \in {}^A\mathrm{ON}$ be an ordinal function and $\mathsf{a} := \mathrm{ran}(\mathrm{cf} \circ g)$. Show that $\mathrm{pcf}(\mathsf{a}) \subseteq \mathrm{pcf}(g)$.

3) Let $g := \omega \times \{\omega\}$. Prove:

   a) $(\prod g, <_{\mathrm{I_b}(\omega)})$ is $\aleph_1$-directed.
   Hint: If $(f_n : n \in \omega)$ is a sequence of members of $\prod g$, define $f \in \prod g$ by $f(n) := \max\{f_m(n) + 1 : m \leq n\}$.

   b) If $2^{\aleph_0} = \aleph_1$ and $\mathsf{a} := \mathrm{ran}(\mathrm{cf} \circ g)$, then $\mathrm{tcf}(\prod g/\mathrm{I_b}(\omega)) = \aleph_1$, which gives $\aleph_1 \in \mathrm{pcf}(g) \setminus \mathrm{pcf}(\mathsf{a})$.

4) Let a be a set of regular cardinals and b be an initial segment of a, that means b is a subset of a and $\nu \geq \sup(\mathsf{b})$ for each $\nu \in \mathsf{a} \setminus \mathsf{b}$. If $\lambda \in \mathrm{pcf}(\mathsf{a})$ is a cardinal with $\lambda < \sup(\mathsf{b})$, then $\lambda \in \mathrm{pcf}(\mathsf{b})$.

5) Assume that D is an ultrafilter and I is an ideal on a such that $D \cap I = \emptyset$. Then $\lim_I(\mathsf{a}) \leq \lim_D(\mathsf{a})$. If I is the dual ideal of D, then equality holds. Furthermore, if $\sup(\mathsf{a}) \notin \mathsf{a}$ and $\mathrm{I_b}(\mathsf{a}) \subseteq I$, then $\lim_I(\mathsf{a}) = \lim_D(\mathsf{a}) = \sup(\mathsf{a})$.

6) Assume that a is an infinite set of regular cardinals, I is an ideal on a, $\lambda = \mathrm{tcf}(\prod \mathsf{a}/I)$, $\mu = \lim_I(\mathsf{a})$, and $\mu < \lambda$. Prove that μ is a limit cardinal.

7) Let a be a progressive set of regular cardinals. Characterize $\mathrm{pcf}(\mathsf{a})$ under the assumption that (GCH) holds.
   Hint: If $\lambda \in \mathrm{pcf}(\mathsf{a}) \setminus \mathsf{a}$, then $\lambda \in \mathrm{pcf}(\mathsf{a} \cap \lambda)$. Put $\nu := \sup(\mathsf{a} \cap \lambda)$ and show that $\lambda = \nu^+$.

8) Assume that a is a nonempty set of regular cardinals, μ is a cardinal, and D is an ultrafilter on a. Prove that $\mu = \lim_D(\mathsf{a})$ iff, for every ordinal $\alpha < \mu$, $(\alpha, \mu]_{\mathrm{reg}} \cap \mathsf{a} \in D$.

9) Assume that a is an infinite progressive set of regular cardinals. Prove:

   a) If $\mathsf{b} \subseteq \mathrm{pcf}(\mathsf{a})$ and $|\mathsf{b}| < \min(\mathsf{a})$, then $\mathrm{pcf}(\mathsf{b}) \subseteq \mathrm{pcf}(\mathsf{a})$.

   b) If $|\mathrm{pcf}(\mathsf{a})| < \min(\mathsf{a})$, then $\mathrm{pcf}(\mathrm{pcf}(\mathsf{a})) = \mathrm{pcf}(\mathsf{a})$.

   c) If $|\mathrm{pcf}(\mathsf{a})| < \min(\mathsf{a})$ and $\mathsf{b} \subseteq \mathrm{pcf}(\mathsf{a})$, then $\mathrm{pcf}(\mathrm{pcf}(\mathsf{b})) = \mathrm{pcf}(\mathsf{b})$.

   Hint: Use Corollary 3.3.8.

3.4 The Ideal $\mathcal{J}_{<\lambda}(\mathsf{a})$

Lemma 3.1.9 presents a necessary condition for $\lambda = \mathrm{tcf}(\prod \mathsf{a}/I)$: $(\prod \mathsf{a}, <_I)$ must be λ-directed.

Remember that $(\prod \mathsf{a}, <_I)$ is λ-directed iff, for every subset S of $\prod \mathsf{a}$ with $|S| < \lambda$, there is a function $f \in \prod \mathsf{a}$ such that $g <_I f$ for all $g \in S$. By our convention on a, we have $h + 1 \in \prod \mathsf{a}$ whenever $h \in \prod \mathsf{a}$, and thus $(\prod \mathsf{a}, <_I)$ is λ-directed iff every subset S of $\prod \mathsf{a}$ with $|S| < \lambda$ has a $\leq_I$-upper bound in $\prod \mathsf{a}$.

We want to investigate λ-directed partial orderings of the form $(\prod \mathsf{a}, <_I)$ in more detail and therefore introduce the following concept.

Definition *Let λ be a cardinal. We call an ideal I on the set* a **λ-mighty** *iff, for each I-positive set $\mathsf{b} \subseteq \mathsf{a}$, there is an ultrafilter D on* a *satisfying $\mathsf{b} \in D$, $D \cap I = \emptyset$, and* $\mathrm{cf}(\prod \mathsf{a}/D) \geq \lambda$.

Lemma 3.4.1 *Assume that I is a λ-mighty ideal on* a. *Then* $\mathsf{a} \cap \lambda \subseteq \mathrm{dom}(I)$.

Proof Let us assume, to get a contradiction, that there is a cardinal $\nu \in \mathsf{a} \cap \lambda$ such that $\{\nu\} \in I^+$. Since I is λ-mighty, there is an ultrafilter D on a satisfying $\{\nu\} \in D$, $D \cap I = \emptyset$ and $\mathrm{cf}(\prod \mathsf{a}/D) \geq \lambda$. But now Lemma 3.3.4 c) says that $\mathrm{cf}(\prod \mathsf{a}/D) = \nu < \lambda$. This contradiction proves $\mathsf{a} \cap \lambda \subseteq \mathrm{dom}(I)$.

Lemma 3.4.2 (Jech) *Assume that f is a progressive ordinal function on the nonempty set A, λ is a regular cardinal with $\lambda > |A|^+$, and $(f_\xi : \xi < \lambda)$ is a sequence of ordinal functions on A. Then there is an ordinal function $g \in {}^A\mathrm{ON}$ such that $g < f$ and, for all $h \in {}^A\mathrm{ON}$ with $g \leq h < f$, the set*

$$\{\xi < \lambda : \mathrm{B}(g, f_\xi) = \mathrm{B}(h, f_\xi)\}$$

is unbounded in λ and so has cardinality λ.

Proof First we consider the case that A is infinite and assume, to get a contradiction, that the assertion of the lemma is false. We will construct by recursion a sequence $(g_i : i < |A|^+)$ of members of $\prod f$ such that $g_i \leq g_j$ whenever $i < j < |A|^+$, and a sequence $(\alpha_i : i < |A|^+)$ of ordinals less than λ. $g_0 \in \prod f$ is chosen arbitrarily. If $i < |A|^+$ is a limit ordinal, we define $g_i := \sup\{g_j : j < i\}$. Since f is progressive and $|i| \leq |A|$, we have $g_i < f$. If $i < |A|^+$ and g_i is already defined, then by our supposition there is a function g_{i+1} with $g_i \leq g_{i+1} < f$ and an ordinal $\alpha_i < \lambda$ such that $\mathrm{B}(g_{i+1}, f_\xi) \neq \mathrm{B}(g_i, f_\xi)$ for all $\xi \in [\alpha_i, \lambda)_{\mathrm{ON}}$. Since $g_i \leq g_{i+1}$, we get $\mathrm{B}(g_{i+1}, f_\xi) \subsetneqq \mathrm{B}(g_i, f_\xi)$ for all $\xi \in [\alpha_i, \lambda)_{\mathrm{ON}}$.

Since λ is regular and greater than $|A|^+$, there is an ordinal $\alpha < \lambda$ such that $\alpha \geq \alpha_i$ for all $i < |A|^+$. Now, by construction, $(\mathrm{B}(g_i, f_\alpha) : i < |A|^+)$ is a strictly decreasing sequence of subsets of A with length $|A|^+$. This obviously leads to a contradiction.

If A is finite, we can proceed in the same way (the limit case does not occur), but will construct a sequence of length $|A|^{++} = |A| + 2$.

Corollary 3.4.3 *Assume that λ is a regular cardinal with $\lambda > |A|^+$, f is a progressive ordinal function on A, and $\bar{f} = (f_\xi : \xi < \lambda)$ is a sequence of ordinal functions in $\prod f$ which is strictly increasing modulo I. If $\bar{f}$ is unbounded in $\prod f$ modulo I, then there is a function $g \in \prod f$ and, for each $\xi < \lambda$, an I-positive set $B_\xi \subseteq A$ with the following properties:*

a) If $\zeta < \xi < \lambda$, then $B_\zeta \subseteq_I B_\xi$.

b) For all $\zeta < \lambda$, $\bar{f}$ is cofinal in f modulo $I \restriction B_\zeta$. In particular we get $\operatorname{tcf}(\prod f / I \restriction B_\zeta) = \lambda$.

c) If J is an ideal on A satisfying $I \subseteq J$ and $B_\xi \in J$ for all $\xi < \lambda$, then g is an upper bound of $\bar{f}$ modulo J.

Proof We can choose an ordinal function g such that $g < f$ and g satisfies the assertion from Lemma 3.4.2. By assumption, g is not an upper bound of the sequence $\bar{f} = (f_\xi : \xi < \lambda)$; hence there is an ordinal $\alpha < \lambda$ such that $\mathrm{B}(g, f_\alpha)$ is positive. We define

$$B_\xi := \mathrm{B}(g, f_{\alpha+\xi})$$

for each $\xi < \lambda$. Since $\bar{f}$ is increasing, we get

$$B_\zeta \setminus B_\xi \subseteq \mathrm{B}(f_{\alpha+\xi}, f_{\alpha+\zeta}) \in I$$

for all ζ and ξ with $\zeta < \xi < \lambda$. This proves part a), and certainly all sets B_ξ are positive, since B_0 is positive.

For the proof of part b), fix $\zeta < \lambda$. We note first that the given sequence is strictly increasing modulo $I \restriction B_\zeta$, since $I \subseteq I \restriction B_\zeta$. Next we consider an arbitrary function $h \in \prod f$ and will show that h is majorized under $<_{I \restriction B_\zeta}$ by some member of $\bar{f}$. Without loss of generality we can assume that $g \leq h$, since otherwise we could replace h by the function $\max\{g, h\}$. Then, by the choice of g, there is an ordinal $\beta \in [\alpha + \zeta, \lambda)_{\mathrm{ON}}$ such that $\mathrm{B}(g, f_\beta) = \mathrm{B}(h, f_\beta)$. $f_{\alpha+\zeta} \leq_I f_\beta$ yields

$$B_\zeta \cap \mathrm{B}[f_\beta, h] = B_\zeta \setminus \mathrm{B}(h, f_\beta) = B_\zeta \setminus \mathrm{B}(g, f_\beta) \subseteq \mathrm{B}(f_\beta, f_{\alpha+\zeta}) \in I.$$

So we get $h <_{I \restriction B_\zeta} f_\beta$.

If J is given as in part c), then we have $\mathrm{B}(g, f_\xi) \in J$, that means $f_\xi \leq_J g$, for all $\xi \in [\alpha, \lambda)_{\mathrm{ON}}$ by the definition of the sets B_ξ. From $I \subseteq J$ and the monotonicity of the sequence we can infer that $f_\xi \leq_J g$ for all $\xi < \lambda$, i.e., g is an upper bound of $\bar{f}$ modulo J.

Theorem 3.4.4 *Let λ be a cardinal, let the set a be progressive, and let I be a λ-mighty ideal on a. Then the partial ordering $(\prod \mathsf{a}, <_I)$ is λ-directed.*

Proof We have to show that every subset S of $\prod \mathsf{a}$ with $|S| < \lambda$ has an upper bound in $\prod \mathsf{a}$, i.e., that there is a function $g \in \prod \mathsf{a}$ satisfying $f \leq_I g$ for all $f \in S$, and will prove this by transfinite induction on the cardinal $|S|$. If $|S| \leq |\mathsf{a}|^+$, we put $g := \sup(S)$. Then we have $f \leq g$ for all $f \in S$. Since a is progressive, the function g will be in general a member of $\prod \mathsf{a}$. There is only a problem if

$$\nu := \min(\mathsf{a}) = |\mathsf{a}|^+ = |S|.$$

But in this case we have $\nu = |S| < \lambda$ and therefore $\{\nu\} \in I$ by Lemma 3.4.1. So we can change appropriately the value of g at ν.

For the inductive proof let $\mu := |S| > |\mathsf{a}|^+$ and $\bar{f} = (f_\xi : \xi < \mu)$ be an enumeration of S. Without loss of generality we can assume that this sequence is strictly increasing modulo I, since otherwise we can choose by recursion, using the inductive hypothesis, for each $\xi < \mu$ a function $f'_\xi \in \prod \mathsf{a}$ such that

$$f_\xi <_I f'_\xi \ \wedge \ \forall \zeta < \xi \ f'_\zeta <_I f'_\xi.$$

Then obviously it is enough to find an upper bound of the set $\{f'_\xi : \xi < \mu\}$. Furthermore, if $(\alpha_i : i < \mathrm{cf}(\mu))$ is a sequence cofinal in μ, it suffices to give an upper bound of the set $\{f_{\alpha_i} : i < \mathrm{cf}(\mu)\}$. So, by the inductive hypothesis, we can assume in addition that μ is regular.

Now all assumptions of Corollary 3.4.3 with μ and id_{a} instead of λ and f are fulfilled. Let us assume, to get a contradiction, that $\bar{f}$ is unbounded in $\prod \mathsf{a}$. Then we can conclude from the corollary that there is an I-positive set B such that $\bar{f}$ is cofinal in $\prod \mathsf{a}$ modulo $I \restriction B$. Thus we get $\mathrm{tcf}(\prod \mathsf{a}/I \restriction B) = \mu$. Since I is λ-mighty, there is an ultrafilter D on a extending the ideal $I \restriction B$ such that $\mathrm{cf}(\prod \mathsf{a}/D) \geq \lambda$. On the other hand, Lemma 3.1.7 says that $\mathrm{cf}(\prod \mathsf{a}/D) = \mathrm{tcf}(\prod \mathsf{a}/I \restriction B) = \mu < \lambda$. This contradiction shows that there is a function $g \in \prod \mathsf{a}$ which is modulo I an upper bound of the set S, as desired.

Now we want to know whether there are any λ-mighty ideals at all and how to characterize them if possible. Clearly any intersection of a family of λ-mighty ideals is again λ-mighty. So, if there exists a λ-mighty ideal on a, the following definition makes sense.

Definition *If there exists a λ-mighty ideal on a, then we define*

$$\mathcal{J}_{<\lambda}(\mathsf{a}) := \bigcap \{I : I \text{ is a } \lambda\text{-mighty ideal on } \mathsf{a}\};$$

then $\mathcal{J}_{<\lambda}(\mathsf{a})$ is the smallest λ-mighty ideal on a. Otherwise we put $\mathcal{J}_{<\lambda}(\mathsf{a}) := \mathcal{P}(\mathsf{a})$.

Lemma 3.4.5 *For any cardinal λ, we have*

$$\mathcal{J}_{<\lambda}(\mathsf{a}) = \{\mathsf{b} \subseteq \mathsf{a} : \forall D (D \text{ is an ultrafilter on } \mathsf{a} \wedge \mathsf{b} \in D \Longrightarrow \mathrm{cf}(\prod \mathsf{a}/D) < \lambda)\}.$$

Proof Let b be a subset of a such that $\mathsf{b} \notin \mathcal{J}_{<\lambda}(\mathsf{a})$. Then there is an ultrafilter D on a satisfying $\mathsf{b} \in D$, $D \cap \mathcal{J}_{<\lambda}(\mathsf{a}) = \emptyset$ and $\mathrm{cf}(\prod \mathsf{a}/D) \geq \lambda$, since $\mathcal{J}_{<\lambda}(\mathsf{a})$ is λ-mighty.

Conversely, if b is a subset of a and D is an ultrafilter on a such that $\mathsf{b} \in D$ and $\mathrm{cf}(\prod \mathsf{a}/D) \geq \lambda$, then the maximal ideal I dual to D is certainly λ-mighty and satisfies $\mathsf{b} \notin I$. By the definition of $\mathcal{J}_{<\lambda}(\mathsf{a})$, we get $\mathsf{b} \notin \mathcal{J}_{<\lambda}(\mathsf{a})$.

Corollary 3.4.6 *If λ is a cardinal, then*

$$\mathcal{J}_{<\lambda}(\mathsf{a}) = \{\mathsf{b} \subseteq \mathsf{a} : \mathrm{pcf}(\mathsf{b}) \subseteq \lambda\}.$$

In particular we have

$$\mathrm{pcf}(\mathsf{a}) \subseteq \lambda \iff \mathcal{J}_{<\lambda}(\mathsf{a}) = \mathcal{P}(\mathsf{a}).$$

Proof Let b be a subset of a. Any ultrafilter on b can be extended to an ultrafilter on a. Conversely, $D \cap \mathcal{P}(\mathsf{b})$ is an ultrafilter on b if D is an ultrafilter on a such that $\mathsf{b} \in D$. Using the fact that b is also progressive, we can infer from Lemma 3.3.2 that

$$\mathrm{pcf}(\mathsf{b}) = \{\mathrm{cf}\left(\prod \mathsf{a}/D\right) : D \text{ is an ultrafilter on } \mathsf{a} \wedge \mathsf{b} \in D\}.$$

Hereby our assertion follows immediately from the preceding lemma.

In particular, $\mathrm{pcf}(\mathsf{a}) \subseteq \lambda$ holds iff $\mathsf{a} \in \mathcal{J}_{<\lambda}(\mathsf{a})$. The last assertion is equivalent to $\mathcal{J}_{<\lambda}(\mathsf{a}) = \mathcal{P}(\mathsf{a})$.

Corollary 3.4.7 *If λ is a cardinal and $\mathsf{b} \subseteq \mathsf{a}$, then $\mathcal{J}_{<\lambda}(\mathsf{b}) = \mathcal{J}_{<\lambda}(\mathsf{a}) \cap \mathcal{P}(\mathsf{b})$.*

Corollary 3.4.8 *If κ and λ are cardinals with $\kappa < \lambda$, then $\mathcal{J}_{<\kappa}(\mathsf{a}) \subseteq \mathcal{J}_{<\lambda}(\mathsf{a})$.*

Lemma 3.4.9 *If a has no maximum and $\lambda = \sup(\mathsf{a})$, then every member of $\mathcal{J}_{<\lambda}(\mathsf{a})$ is a bounded subset of a, and thus $\mathcal{J}_{<\lambda}(\mathsf{a}) \subseteq \mathrm{I_b}(\mathsf{a})$.*

Proof Let b be an unbounded subset of a, i.e., $\mathsf{b} \notin \mathrm{I_b}(\mathsf{a})$. Then $\{\mathsf{b}\} \cup \{\mathsf{a} \setminus \nu : \nu < \lambda\}$ has the finite intersection property and can be extended to an ultrafilter D on a. Clearly D is an unbounded ultrafilter. By Lemma 3.3.4 we have $\mathrm{cf}\left(\prod \mathsf{a}/D\right) \geq \lambda$, hence $\mathsf{b} \notin \mathcal{J}_{<\lambda}(\mathsf{a})$.

A further characterization of $\mathcal{J}_{<\lambda}(\mathsf{a})$ is the following.

Lemma 3.4.10 *Let a be progressive, λ be an infinite cardinal, and I be an ideal on a. Then the following assertions are equivalent:*

(i) $(\prod \mathsf{a}, <_I)$ *is λ-directed.*

(ii) *I is λ-mighty.*

(iii) $\mathcal{J}_{<\lambda}(\mathsf{a}) \subseteq I$.

Thus, if $\mathcal{J}_{<\lambda}(\mathsf{a}) \neq \mathcal{P}(\mathsf{a})$, then $\mathcal{J}_{<\lambda}(\mathsf{a})$ is the smallest ideal J on a such that $(\prod \mathsf{a}, <_J)$ is λ-directed.

Proof If $(\prod \mathsf{a}, <_I)$ is λ-directed and D is an ultrafilter on a extending I, then $(\prod \mathsf{a}, <_D)$ also is λ-directed. By Lemma 3.1.9, we get $\mathrm{cf}\left(\prod \mathsf{a}/D\right) \geq \lambda$. So we can infer that I is λ-mighty.

If I is any λ-mighty ideal on a, then $\mathcal{J}_{<\lambda}(\mathsf{a}) \subseteq I$ by the definition of $\mathcal{J}_{<\lambda}(\mathsf{a})$.

To prove that (iii) implies (i), let us assume that $\mathcal{J}_{<\lambda}(\mathsf{a}) \subseteq I$. Then $\mathcal{J}_{<\lambda}(\mathsf{a}) \neq \mathcal{P}(\mathsf{a})$ and thus $\mathcal{J}_{<\lambda}(\mathsf{a})$ is a λ-mighty ideal on a. Theorem 3.4.4 says that the partial ordering $(\prod \mathsf{a}, <_{\mathcal{J}_{<\lambda}(\mathsf{a})})$ is λ-directed and thus the same holds for $(\prod \mathsf{a}, <_I)$.

Remark Together with Corollary 3.4.6 this Lemma completely answers for progressive sets a our question of the existence of ideals I on a such that $(\prod \mathsf{a}, <_I)$ is λ-directed.

There are ideals I on a such that $(\prod \mathsf{a}, <_I)$ is λ-directed iff $\mathrm{pcf}(\mathsf{a}) \not\subseteq \lambda$.

Corollary 3.4.11 *If* a *is progressive,* λ *is a cardinal, and* D *is an ultrafilter on* a*, then*

$$\mathrm{cf}\,(\prod \mathsf{a}/D) < \lambda \iff D \cap \mathcal{J}_{<\lambda}(\mathsf{a}) \neq \emptyset.$$

Proof Assume that $\mathrm{cf}\,(\prod \mathsf{a}/D) < \lambda$. If $\mathcal{J}_{<\lambda}(\mathsf{a}) = \mathcal{P}(\mathsf{a})$, the clearly we have $D \cap \mathcal{J}_{<\lambda}(\mathsf{a}) \neq \emptyset$. Otherwise $\mathcal{J}_{<\lambda}(\mathsf{a})$ is λ-directed by the previous lemma, and Lemma 3.1.9 says that $(\prod \mathsf{a}, <_D)$ is not λ-directed. Therefore D is no extension of $\mathcal{J}_{<\lambda}(\mathsf{a})$, i.e. we have $D \cap \mathcal{J}_{<\lambda}(\mathsf{a}) \neq \emptyset$. Conversely, this assertion yields $\mathrm{cf}\,(\prod \mathsf{a}/D) < \lambda$ by Lemma 3.4.5.

Corollary 3.4.12 *If* a *is progressive and* λ *is a cardinal, then*

$$\lambda \in \mathrm{pcf}(\mathsf{a}) \iff \mathcal{J}_{<\lambda}(\mathsf{a}) \subsetneqq \mathcal{J}_{<\lambda^+}(\mathsf{a}).$$

Proof Corollary 3.4.8 says that $\mathcal{J}_{<\lambda}(\mathsf{a}) \subseteq \mathcal{J}_{<\lambda^+}(\mathsf{a})$. Assume first that $\lambda \in \mathrm{pcf}(\mathsf{a})$. Then there is an ultrafilter D on a such that $\lambda = \mathrm{cf}\,(\prod \mathsf{a}/D)$. By Corollary 3.4.11 there is a set $\mathsf{b} \in D \cap \mathcal{J}_{<\lambda^+}(\mathsf{a})$, and from Lemma 3.4.5 we can conclude that $\mathsf{b} \notin \mathcal{J}_{<\lambda}(\mathsf{a})$. If conversely $\mathsf{b} \in \mathcal{J}_{<\lambda^+}(\mathsf{a}) \setminus \mathcal{J}_{<\lambda}(\mathsf{a})$, then there is, again by this lemma, an ultrafilter D on a such that $\mathrm{cf}\,(\prod \mathsf{a}/D) \geq \lambda$ and $\mathsf{b} \in D$, and we have $\mathrm{cf}\,(\prod \mathsf{a}/D) < \lambda^+$, hence $\mathrm{cf}\,(\prod \mathsf{a}/D) = \lambda$.

Theorem 3.4.13 *If* a *is progressive and* μ *is a cardinal, then*

$$|\mathrm{pcf}_\mu(\mathsf{a})| \leq |\mathsf{a}|^\mu.$$

Proof Consider a fixed cardinal $\lambda \in \mathrm{pcf}_\mu(\mathsf{a})$. We can choose a subset b of a such that $|\mathsf{b}| \leq \mu$ and $\lambda \in \mathrm{pcf}(\mathsf{b})$. By Corollary 3.4.12, there is a set $\mathsf{c}_\lambda \in \mathcal{J}_{<\lambda^+}(\mathsf{b}) \setminus \mathcal{J}_{<\lambda}(\mathsf{b})$. In particular, using Corollary 3.4.7, we have $\mathsf{c}_\lambda \in (\mathcal{J}_{<\lambda^+}(\mathsf{a}) \cap [\mathsf{a}]^{\leq\mu}) \setminus (\mathcal{J}_{<\lambda}(\mathsf{a}) \cap [\mathsf{a}]^{\leq\mu})$. Therefore, by Corollary 3.4.8, the function $(\mathsf{c}_\lambda : \lambda \in \mathrm{pcf}_\mu(\mathsf{a}))$ is an injection from $\mathrm{pcf}_\mu(\mathsf{a})$ into $[\mathsf{a}]^{\leq\mu}$.

From $\mathrm{pcf}(\mathsf{a}) = \mathrm{pcf}_{|\mathsf{a}|}(\mathsf{a})$ we immediately get:

Corollary 3.4.14 *If* a *is progressive, then* $|\mathrm{pcf}(\mathsf{a})| \leq 2^{|\mathsf{a}|}$.

Lemma 3.4.15 *If* a *is progressive and* λ *is a limit cardinal, then*

$$\mathcal{J}_{<\lambda}(\mathsf{a}) = \bigcup\{\mathcal{J}_{<\mu}(\mathsf{a}) : \mu \in \mathrm{CN} \wedge \mu < \lambda\}.$$

Proof By Corollary 3.4.8, we have

$$I := \bigcup\{\mathcal{J}_{<\mu}(\mathsf{a}) : \mu \in \mathrm{CN} \wedge \mu < \lambda\} \subseteq \mathcal{J}_{<\lambda}(\mathsf{a}).$$

Without loss of generality let $I \neq \mathcal{P}(\mathsf{a})$. I is an ideal since it is the union of a chain of ideals. Fix an I-positive set b. There is an ultrafilter D on a extending I which has b as its member. Corollary 3.4.11 says that $\mathrm{cf}(\prod \mathsf{a}/D) \geq \lambda$, hence $\mathsf{b} \notin \mathcal{J}_{<\lambda}(\mathsf{a})$ by the same corollary. So we get $\mathcal{J}_{<\lambda}(\mathsf{a}) \subseteq I$.

The strength of Corollary 3.4.3 is demonstrated by the following applications.

Lemma 3.4.16 *If* a *is a progressive set of regular cardinals,* $(\prod \mathsf{a}, <_I)$ *is* λ*-directed, and* D *is an ultrafilter on* a *extending* I *with* $\mathrm{cf}(\prod \mathsf{a}/D) = \lambda$*, then there is a set* $\mathsf{b} \in D$ *such that* $\mathrm{tcf}(\prod \mathsf{a}/I \restriction \mathsf{b}) = \lambda$.

Proof Since a is progressive, we have $|\mathsf{a}|^+ \leq \min(\mathsf{a}) \leq \lambda$. First let us consider the case that $\lambda = \mathrm{cf}(\prod \mathsf{a}/D) = \min(\mathsf{a})$. Then Lemma 3.3.4 says that D is principal and generated by $\{\min(\mathsf{a})\}$. So we get $\mathsf{b} := \{\min(\mathsf{a})\} \notin I$ and $\mathrm{tcf}(\prod \mathsf{b}/I \cap \mathcal{P}(\mathsf{b})) = \mathrm{cf}(\min(\mathsf{a})) = \min(\mathsf{a}) = \lambda$.

Now assume that $|\mathsf{a}|^+ < \lambda$. Let $\bar{h} = (h_\xi : \xi < \lambda)$ be a sequence which is cofinal in $\prod \mathsf{a}$ modulo D. For an application of Corollary 3.4.3 we define by recursion a sequence $\bar{f} = (f_\xi : \xi < \lambda)$ of members of $\prod \mathsf{a}$ which is modulo I strictly increasing and unbounded in $\prod \mathsf{a}$. Let $\xi < \lambda$, and let the sequence $(f_\zeta : \zeta < \xi)$ with the desired properties be given. Since I is λ-directed, the set $\{f_\zeta : \zeta < \xi\} \cup \{h_\zeta : \zeta \leq \xi\}$ has an upper bound under $<_I$, and we can choose the function f_ξ as such a bound. This completes the definition of $\bar{f}$.

Since $\bar{h}$ is cofinal in $\prod \mathsf{a}$ modulo D and $D \cap I = \emptyset$, the sequence $\bar{f}$ is cofinal in $\prod \mathsf{a}$ modulo D. Therefore it is not bounded in $\prod \mathsf{a}$ modulo I, and consequently Corollary 3.4.3 guarantees, together with $\lambda > |\mathsf{a}|^+$, the existence of a sequence $(\mathsf{b}_\xi : \xi < \lambda)$ of subsets of a such that $\mathrm{tcf}(\prod \mathsf{a}/I \restriction \mathsf{b}_\xi) = \lambda$ for every $\xi < \lambda$ and $\mathsf{b}_\zeta \subseteq_I \mathsf{b}_\xi$ whenever $\zeta < \xi < \lambda$. If at least one b_ξ is a member of D, then we are done.

So let us assume that $\mathsf{b}_\xi \notin D$ for all ordinals $\xi < \lambda$. Then $I \cup \{\mathsf{b}_\xi : \xi < \lambda\}$ generates an ideal J on a such that $D \cap J = \emptyset$. By Corollary 3.4.3, there is a function $g \in \prod \mathsf{a}$ such that g is modulo J an upper bound of the set $\{f_\xi : \xi < \lambda\}$. Then g is also modulo D an upper bound of $\{f_\xi : \xi < \lambda\}$, contradicting the fact that $\bar{f}$ is cofinal in $\prod \mathsf{a}$ modulo D. Therefore there is an ordinal $\alpha < \lambda$ such that $\mathsf{b}_\alpha \in D$, as desired.

Lemma 3.4.17 *If* a *is a progressive set of regular cardinals,* I *is an ideal on* a *and* λ *is a regular cardinal, then*

$$J := I \cup \{\mathsf{b} \in I^+ : \operatorname{tcf}(\prod \mathsf{a}/I \restriction \mathsf{b}) = \lambda\}$$

is an ideal on a *iff* $\mathsf{a} \notin J$.

Proof If $\mathsf{a} \in J$, then $\operatorname{tcf}(\prod \mathsf{a}/I) = \lambda$. For any subset b of a we have $I = I \restriction \mathsf{a} \subseteq I \restriction \mathsf{b}$ and so can conclude that $\operatorname{tcf}(\prod \mathsf{a}/I \restriction \mathsf{b}) = \lambda$. Consequently we get $J = \mathcal{P}(\mathsf{a})$, hence J is not an ideal on a.

Conversely, assume that $\mathsf{a} \notin J$. We will prove that J is an ideal on a. For this let first $\mathsf{b} \in J$ and $\mathsf{c} \subseteq \mathsf{b}$. We must show that $\mathsf{c} \in J$. Without loss of generality we can assume that $\mathsf{c} \notin I$ and thus $\mathsf{b} \notin I$, hence $\operatorname{tcf}(\prod \mathsf{a}/I \restriction \mathsf{b}) = \lambda$. Now $\mathsf{c} \subseteq \mathsf{b}$ yields $I \restriction \mathsf{b} \subseteq I \restriction \mathsf{c}$, and thus any sequence which is cofinal in $\prod \mathsf{a}$ modulo $I \restriction \mathsf{b}$ is cofinal in $\prod \mathsf{a}$ modulo $I \restriction \mathsf{c}$. So we get $\mathsf{c} \in J$.

To prove the second property of an ideal, fix $\mathsf{b}, \mathsf{c} \in J$. If $\mathsf{b}, \mathsf{c} \in I$, then $\mathsf{b} \cup \mathsf{c} \in I$. If $\mathsf{b} \in I^+$ and $\mathsf{c} \in I$, then $I \restriction \mathsf{b} = \{X \subseteq \mathsf{a} : X \cap \mathsf{b} \in I\} = \{X \subseteq \mathsf{a} : (X \cap \mathsf{b}) \cup (X \cap \mathsf{c}) \in I\} = I \restriction (\mathsf{b} \cup \mathsf{c})$, and $\operatorname{tcf}(\prod \mathsf{a}/I \restriction \mathsf{b}) = \lambda$ gives $\mathsf{b} \cup \mathsf{c} \in J$. So let $\mathsf{b}, \mathsf{c} \in I^+$ and choose the sequences $(f_\xi : \xi < \lambda)$ and $(g_\xi : \xi < \lambda)$ cofinal in $\prod \mathsf{a}$ modulo $I \restriction \mathsf{b}$ and modulo $I \restriction \mathsf{c}$, respectively. For all $\xi < \lambda$ and $\nu \in \mathsf{a}$, we define

$$h_\xi(\nu) = \begin{cases} f_\xi(\nu) & \text{if } \nu \in \mathsf{b}, \\ g_\xi(\nu) & \text{otherwise.} \end{cases}$$

To see that $\bar{h} = (h_\xi : \xi < \lambda)$ is cofinal in $\prod \mathsf{a}$ modulo $I \restriction (\mathsf{b} \cup \mathsf{c})$, let $\alpha < \beta < \lambda$. Then $Z := \{\nu \in \mathsf{b} \cup \mathsf{c} : h_\beta(\nu) \leq h_\alpha(\nu)\} \subseteq X \cup Y$, where $X := \{\nu \in \mathsf{b} : f_\beta(\nu) \leq f_\alpha(\nu)\} \in I$ and $Y := \{\nu \in \mathsf{c} \setminus \mathsf{b} : g_\beta(\nu) \leq g_\alpha(\nu)\} \in I$. So we get $Z \in I$ and thus $h_\alpha <_{I \restriction (\mathsf{b} \cup \mathsf{c})} h_\beta$. Consequently $\bar{h}$ is strictly increasing modulo $I \restriction (\mathsf{b} \cup \mathsf{c})$. In the same way one can prove that $\bar{h}$ is unbounded in $\prod \mathsf{a}$ modulo $I \restriction (\mathsf{b} \cup \mathsf{c})$. Now we can conclude that $\operatorname{tcf}(\prod \mathsf{a}/I \restriction (\mathsf{b} \cup \mathsf{c})) = \lambda$, hence $\mathsf{b} \cup \mathsf{c} \in J$. This completes the proof of the lemma.

Theorem 3.4.18 *If* a *is a progressive set of regular cardinals and* I *is an ideal on* a*, then* $\operatorname{tcf}(\prod \mathsf{a}/I) = \lambda$ *iff* $\operatorname{cf}(\prod \mathsf{a}/D) = \lambda$ *for every ultrafilter* D *on* a *such that* $D \cap I = \emptyset$.

Proof If $\operatorname{tcf}(\prod \mathsf{a}/I) = \lambda$, then it is clear by Lemma 3.1.7 that $\operatorname{cf}(\prod \mathsf{a}/D) = \lambda$ is true for every ultrafilter D on a extending I.

Now assume that $\operatorname{cf}(\prod \mathsf{a}/D) = \lambda$ for every ultrafilter D on a such that $D \cap I = \emptyset$. If $\mathsf{b} \notin I$, then there is an ultrafilter D_b on a such that $\mathsf{b} \in D_\mathsf{b}$ and $D_\mathsf{b} \cap I = \emptyset$, hence $\operatorname{cf}(\prod \mathsf{a}/D_\mathsf{b}) = \lambda$. So we can conclude that I is λ-mighty. Theorem 3.4.4 says that $(\prod \mathsf{a}, <_I)$ is λ-directed.

To get a contradiction let us assume that $\mathsf{a} \notin J$ where J is the ideal on a defined in Lemma 3.4.17. We can choose an ultrafilter D on a extending J.

Since $D \cap I = \emptyset$, our assumption gives $\mathrm{cf}\,(\prod \mathsf{a}/D) = \lambda$. Lemma 3.4.16 says that there is a set $\mathsf{b} \in D$ such that $\mathrm{tcf}\,(\prod \mathsf{a}/I \restriction \mathsf{b}) = \lambda$. This implies that $\mathsf{b} \in J$, contradicting $D \cap J = \emptyset$.

Lemma 3.4.19 *If* a *is progressive,* λ *is a cardinal, and* $\mathsf{b} \in \mathcal{J}_{<\lambda^+}(\mathsf{a}) \setminus \mathcal{J}_{<\lambda}(\mathsf{a})$, *then*

$$\mathrm{tcf}\,(\prod \mathsf{a}/\mathcal{J}_{<\lambda}(\mathsf{a}) \restriction \mathsf{b}) = \lambda.$$

Proof Let λ be a cardinal and $\mathsf{b} \in \mathcal{J}_{<\lambda^+}(\mathsf{a}) \setminus \mathcal{J}_{<\lambda}(\mathsf{a})$. Then $I := \mathcal{J}_{<\lambda}(\mathsf{a}) \restriction \mathsf{b}$ is an ideal on a. For an application of Theorem 3.4.18 let D be an arbitrary ultrafilter on a such that $D \cap I = \emptyset$. Then $\mathrm{cf}\,(\prod \mathsf{a}/D) \geq \lambda$ by Corollary 3.4.11. Since $\mathsf{b} \in D \cap \mathcal{J}_{<\lambda^+}(\mathsf{a})$, we get $\mathrm{cf}\,(\prod \mathsf{a}/D) < \lambda^+$ and consequently $\mathrm{cf}\,(\prod \mathsf{a}/D) = \lambda$. Now Theorem 3.4.18 says that $\mathrm{tcf}\,(\prod \mathsf{a}/I) = \lambda$.

Lemma 3.4.20 *If* a *is nonempty and progressive, then* $\mathrm{pcf}(\mathsf{a})$ *possesses a maximum, and we have*

$$\max \mathrm{pcf}(\mathsf{a}) \leq |\prod \mathsf{a}| \leq (\sup(\mathsf{a}))^{|\mathsf{a}|}.$$

In particular, we get with Corollary 3.4.6:

$$\mathcal{J}_{<\lambda}(\mathsf{a}) = \{\mathsf{b} \subseteq \mathsf{a} : \max \mathrm{pcf}(\mathsf{b}) < \lambda\} \cup \{\emptyset\}.$$

If $\mu = \max \mathrm{pcf}(\mathsf{a})$, *then* $\mathcal{J}_{\mu^+}(\mathsf{a}) = \mathcal{P}(\mathsf{a})$.

Proof To get a contradiction let us assume that $\lambda := \sup \mathrm{pcf}(\mathsf{a})$ is not a member of $\mathrm{pcf}(\mathsf{a})$. Then λ is a limit cardinal, and we have $\mathcal{J}_{<\lambda}(\mathsf{a}) = \mathcal{P}(\mathsf{a})$ by Corollary 3.4.6. Lemma 3.4.15 says that $\mathsf{a} \in \mathcal{J}_{<\kappa}(\mathsf{a})$ for some $\kappa < \lambda$. Using Corollary 3.4.6 we get the contradiction $\mathrm{pcf}(\mathsf{a}) \subseteq \kappa$.

The estimate for $\max \mathrm{pcf}(\mathsf{a})$ is easy to verify since every sequence which is cofinal in $\prod \mathsf{a}$ modulo D, where D is an ultrafilter on a, is injective.

The following theorem establishes a connection between the concepts of the true cofinality modulo I and the cofinality of the partial ordering $(\prod \mathsf{a}, \leq_{\mathsf{a}})$.

Theorem 3.4.21 *If* a *is a nonempty progressive set of regular cardinals, then*

$$\max \mathrm{pcf}(\mathsf{a}) = \mathrm{cf}\,(\prod \mathsf{a}, \leq_{\mathsf{a}}).$$

Proof Let $\lambda := \max \mathrm{pcf}(\mathsf{a})$. By Corollary 3.4.6, we have $\mathsf{a} \in \mathcal{J}_{<\lambda^+}(\mathsf{a}) \setminus \mathcal{J}_{<\lambda}(\mathsf{a})$. Therefore Lemma 3.4.19 says that there is a sequence $\bar{f} = (f_\xi : \xi < \lambda)$ which is cofinal in $\prod \mathsf{a}$ modulo $\mathcal{J}_{<\lambda}(\mathsf{a})$.

First let us prove the relation "$\leq$". If $S \subseteq \prod \mathsf{a}$ and $|S| < \lambda$, then every member of S is modulo $\mathcal{J}_{<\lambda}(\mathsf{a})$ smaller than some member of $\bar{f}$, and since λ

is regular we can find an ordinal $\beta < \lambda$ satisfying $h <_{\mathcal{J}_{<\lambda}(\mathsf{a})} f_\beta$, and thus *not* $f_\beta \leq h$, for all functions $h \in S$. Consequently, S is not a cofinal subset of $\prod \mathsf{a}$ under $\leq$. From this we can conclude that $\lambda \leq \mathrm{cf}(\prod \mathsf{a}, \leq_{\mathsf{a}})$.

Now we will prove the reverse inequality by induction on λ. Since $|\mathsf{a}|^+ \leq \min(\mathsf{a}) \leq \lambda$, we can assume without loss of generality that $|\mathsf{a}|^+ < \lambda$, since otherwise we have $\mathsf{a} = \{\lambda\}$ and our assertion is true. By Lemma 3.4.2, applied to id_{a} instead of f, we can choose a function $g \in \prod \mathsf{a}$ such that $g < f$ and, for all $h \in \prod \mathsf{a}$ with $g \leq h < f$, the set $\{\xi < \lambda : \mathrm{B}(g, f_\xi) = \mathrm{B}(h, f_\xi)\}$ is unbounded in λ. There exists an ordinal $\alpha < \lambda$ such that $g <_{\mathcal{J}_{<\lambda}(\mathsf{a})} f_\alpha$. Fix $\xi \in [\alpha, \lambda)_{\mathrm{ON}}$, and let $\mathsf{c}_\xi := \mathrm{B}[f_\xi, g]$. Then $\mathsf{c}_\xi \in \mathcal{J}_{<\lambda}(\mathsf{a})$ and $\max \mathrm{pcf}(\mathsf{c}_\xi) < \lambda$ if $\mathsf{c}_\xi \neq \emptyset$. Thus the inductive hypothesis yields a cofinal subset $\mathcal{F}_{\mathsf{c}_\xi}$ of $\prod \mathsf{c}_\xi$ such that $|\mathcal{F}_{\mathsf{c}_\xi}| < \lambda$; if $\mathsf{c}_\xi = \emptyset$, let $\mathcal{F}_{\mathsf{c}_\xi} = \emptyset$. Now we can define a subset $\mathcal{F}$ of $\prod \mathsf{a}$ by

$$\mathcal{F} := \{f_\xi \restriction (\mathsf{a} \setminus \mathsf{c}_\xi) \cup h : \xi \in [\alpha, \lambda)_{\mathrm{ON}} \wedge h \in \mathcal{F}_{\mathsf{c}_\xi}\}.$$

Obviously we have $|\mathcal{F}| \leq \lambda \cdot \sup\{|\mathcal{F}_{\mathsf{c}_\xi}| : \xi \in [\alpha, \lambda)_{\mathrm{ON}}\} = \lambda$. We will show that $\mathcal{F}$ is a cofinal subset of $\prod \mathsf{a}$; this will give $\mathrm{cf}(\prod \mathsf{a}, \leq_{\mathsf{a}}) \leq \lambda$.

For this, let $k \in \prod \mathsf{a}$ and, without loss of generality, $g \leq k$; otherwise we replace k by the function $\max\{g, k\}$. By the choice of g, there is an ordinal $\beta \in [\alpha, \lambda)_{\mathrm{ON}}$ satisfying $\mathsf{c} := \mathsf{c}_\beta = \mathrm{B}[f_\beta, g] = \mathrm{B}[f_\beta, k]$. If $\mathsf{c} = \emptyset$, then we have $k \leq f_\beta$. Otherwise we can choose a function $h \in \mathcal{F}_{\mathsf{c}}$ such that $k \restriction \mathsf{c} \leq h$. Then obviously $k \leq f_\beta \restriction (\mathsf{a} \setminus \mathsf{c}) \cup h \in \mathcal{F}$, as desired.

Lemma 3.4.22 *Assume that* a *is a progressive set of regular cardinals,* I *is an ideal on* a, *and*

$$\mu := \sup\{\mathrm{cf}(\textstyle\prod \mathsf{a}/D) : D \textit{ is an ultrafilter on } \mathsf{a} \textit{ and } D \cap I = \emptyset\}.$$

Then there is a set $\mathsf{b} \in I$ *such that*

$$\mu = \max \mathrm{pcf}(\mathsf{a} \setminus \mathsf{b}).$$

In particular μ *is regular.*

Proof Consider a fixed member c of I. If D is a ultrafilter on a with $D \cap I = \emptyset$, then $\mathsf{a} \setminus \mathsf{c} \in D$. Therefore $\mathrm{cf}(\prod \mathsf{a}/D) = \mathrm{cf}(\prod(\mathsf{a} \setminus \mathsf{c})/D \cap P(\mathsf{a} \setminus \mathsf{c})) \leq \max \mathrm{pcf}(\mathsf{a} \setminus \mathsf{c})$. Consequently we get $\mu \leq \max \mathrm{pcf}(\mathsf{a} \setminus \mathsf{c})$. The following claim will guarantee the existence of the desired set b.

Claim There exists $\mathsf{b} \in I$ such that $\mathsf{a} \setminus \mathsf{b} \in \mathcal{J}_{<\mu^+}(\mathsf{a})$.
To get a contradiction let us assume that $\mathsf{a} \setminus \mathsf{b} \notin \mathcal{J}_{<\mu^+}(\mathsf{a})$ for every $\mathsf{b} \in I$. If $\mathsf{b} \in I$ and $\mathsf{c} \in \mathcal{J}_{<\mu^+}(\mathsf{a})$ is arbitrary, then $\mathsf{b} \cup \mathsf{c} \neq \mathsf{a}$ since otherwise $\mathsf{a} \setminus \mathsf{b} \subseteq \mathsf{c}$ and hereby $\mathsf{a} \setminus \mathsf{b} \in J_{<\mu^+}(\mathsf{a})$, contradicting our assumption. Consequently, there is an ideal J on a satisfying $I \cup \mathcal{J}_{<\mu^+}(\mathsf{a}) \subseteq J$. If D is an ultrafilter on a with

$D \cap J = \emptyset$, then Corollary 3.4.11 implies $\mathrm{cf}(\prod \mathsf{a}/D) \geq \mu^+$ contradicting the definition of μ. This proves the claim.

So choose a set $\mathsf{b} \in I$ such that $\mathsf{a} \setminus \mathsf{b} \in \mathcal{J}_{<\mu^+}(\mathsf{a})$, and let D be an ultrafilter on a satisfying $\mathsf{a} \setminus \mathsf{b} \in D$ and $\mathrm{cf}(\prod(\mathsf{a} \setminus \mathsf{b})/D \cap \mathcal{P}(\mathsf{a} \setminus \mathsf{b})) = \max \mathrm{pcf}(\mathsf{a} \setminus \mathsf{b})$. Then $\mathrm{cf}(\prod(\mathsf{a} \setminus \mathsf{b})/D \cap \mathcal{P}(\mathsf{a} \setminus \mathsf{b})) = \mathrm{cf}(\prod \mathsf{a}/D)$ and $\mathrm{cf}(\prod \mathsf{a}/D) \leq \mu$ by Lemma 3.4.5. With the first part of the proof we obtain $\mu = \max \mathrm{pcf}(\mathsf{a} \setminus \mathsf{b})$ for this set $\mathsf{b} \in I$, as desired.

Theorem 3.4.23 *Assume that* a *is a progressive set of regular cardinals and* I *is an ideal on* a. *Then*

$$\mathrm{cf}(\prod \mathsf{a}, \leq_I) = \sup\{\mathrm{cf}(\prod \mathsf{a}/D) : D \text{ is an ultrafilter on } \mathsf{a} \text{ and } D \cap I = \emptyset\}.$$

Furthermore the cardinal $\mathrm{cf}(\prod \mathsf{a}, \leq_I)$ *is regular.*

Proof Let $\mu := \sup\{\mathrm{cf}(\prod \mathsf{a}/D) : D$ is an ultrafilter on a and $D \cap I = \emptyset\}$. From Lemma 3.4.22 we know that there is $\mathsf{b} \in I$ such that $\mu = \max \mathrm{pcf}(\mathsf{a} \setminus \mathsf{b})$. In particular μ is regular. Furthermore we have $\mu = \mathrm{cf}(\prod(\mathsf{a} \setminus \mathsf{b}), \leq)$ by Theorem 3.4.21. It is easy to verify that $\mathrm{cf}(\prod \mathsf{a}, \leq_I) \leq \mathrm{cf}(\prod(\mathsf{a} \setminus \mathsf{b}), \leq)$, and so we get $\mu \geq \mathrm{cf}(\prod \mathsf{a}, \leq_I)$. On the other hand, if D is an arbitrary ultrafilter on a with $D \cap I = \emptyset$, then D extends the ideal I, hence $\mathrm{cf}(\prod \mathsf{a}/D) \leq \mathrm{cf}(\prod \mathsf{a}, \leq_I)$. This yields $\mu \leq \mathrm{cf}(\prod \mathsf{a}, \leq_I)$ and altogether $\mu = \mathrm{cf}(\prod \mathsf{a}, \leq_I)$, as desired.

Exercises

1) Assume that a is a progressive set of regular cardinals and I is an ideal on a. Prove: If $\mathrm{tcf}(\prod \mathsf{a}/I) > \sup(\mathsf{a})$, then $\mathrm{dom}(I) = \mathsf{a}$.
2) Assume that a is a set of regular cardinals and I is an ideal on a. Show that I is $\aleph_0$-mighty.
3) Assume that a is a progressive set of regular cardinals and D is an ultrafilter on a with $\mathrm{cf}(\prod \mathsf{a}/D) = \lambda$. Then there is a set $\mathsf{b} \in D$ such that
$$\mathrm{tcf}(\prod \mathsf{a}/\mathcal{J}_{<\lambda}(\mathsf{a}) \restriction \mathsf{b}) = \lambda.$$
4) (S. Neumann) Let a be an infinite progressive set of regular cardinals and I be an ideal on a. Prove that there is a set $\mathsf{b} \in I^+$ such that $\mathrm{tcf}(\prod \mathsf{a}/I \restriction \mathsf{b})$ exists.
5) Assume that a is an infinite progressive set of regular cardinals, I is an ideal on a, μ is an infinite cardinal and $(\prod \mathsf{a}, <_I)$ is μ-directed. Prove: If there is no $\mathsf{b} \in \mathcal{P}(\mathsf{a}) \setminus I$ such that $\mu = \mathrm{tcf}(\prod \mathsf{a}/I \restriction \mathsf{b})$, then the partial ordering $(\prod \mathsf{a}, <_I)$ is μ^+-directed.
   Hint: Use Lemma 3.4.10.

6) If a is an infinite progressive set of regular cardinals, then for every set $\mathsf{b} \in \mathcal{P}(\mathsf{a}) \setminus \{\emptyset\}$ there is a unique regular cardinal λ such that $\mathsf{b} \in \mathcal{J}_{<\lambda^+}(\mathsf{a}) \setminus \mathcal{J}_{<\lambda}(\mathsf{a})$.
7) Assume that a is an infinite progressive set of regular cardinals, μ and λ are infinite cardinals with $\mu < \lambda$, and $(\mathsf{b}_\alpha : \alpha < \mu)$ is a sequence of members of $\mathcal{J}_{<\lambda^+}(\mathsf{a}) \setminus \mathcal{J}_{<\lambda}(\mathsf{a})$. Prove that there exists a modulo $\mathcal{J}_{<\lambda}(\mathsf{a})$ strictly increasing sequence $\bar{f} := (f_\xi : \xi < \lambda)$ of members of $\prod \mathsf{a}$ such that, for every $\alpha < \mu$, $\bar{f}$ is cofinal in $\prod \mathsf{a}$ modulo $\mathcal{J}_{<\lambda}(\mathsf{a}) \restriction \mathsf{b}_\alpha$.
Hint: Use Lemma 3.4.19 and Lemma 3.4.10.
8) Assume that a is an infinite set of regular cardinals such that $|\mathsf{a}|^+ < \min(\mathsf{a})$, μ and λ are infinite cardinals with $\mu < \lambda$, and $(\mathsf{b}_\alpha : \alpha < \mu)$ is a sequence of members of $\mathcal{J}_{<\lambda^+}(\mathsf{a}) \setminus \mathcal{J}_{<\lambda}(\mathsf{a})$. Prove that there exists a sequence $\bar{B} := (B_\xi : \xi < \lambda)$ of members of $\mathcal{P}(\mathsf{a}) \setminus \mathcal{J}_{<\lambda}(\mathsf{a})$ satisfying $B_\zeta \subseteq_{\mathcal{J}_{<\lambda}(\mathsf{a})} B_\xi$ for all ζ and ξ with $\zeta < \xi < \lambda$, such that for every $\gamma < \mu$ there is $\beta < \lambda$ with $\mathsf{b}_\gamma \subseteq_{\mathcal{J}_{<\lambda}(\mathsf{a})} B_\beta$.
Hint: If $\bar{f} = (f_\xi : \xi < \lambda)$ is the sequence from Exercise 7, then $\bar{f}$ is unbounded in $\prod \mathsf{a}$ modulo $\mathcal{J}_{<\lambda}(\mathsf{a})$. Corollary 3.4.3 yields a function $g \in \prod \mathsf{a}$ and a family $(B_\xi : \xi < \lambda)$ of $\mathcal{J}_{<\lambda}(\mathsf{a})$-positive subsets of a with the following properties:
   (1) If $\zeta < \xi < \lambda$, then $B_\zeta \subseteq_{\mathcal{J}_{<\lambda}(\mathsf{a})} B_\xi$.
   (2) For every $\zeta < \lambda$, $\bar{f}$ is cofinal in $\prod \mathsf{a}$ modulo $\mathcal{J}_{<\lambda}(\mathsf{a}) \restriction B_\zeta$.
   (3) If J is an ideal on a extending $\mathcal{J}_{<\lambda}(\mathsf{a})$ such that $B_\xi \in J$ for all $\xi < \lambda$, then g is an upper bound of $\bar{f}$ modulo J.

   Suppose that there is an ordinal $\alpha < \lambda$ such that, for all $\beta < \lambda$, $\mathsf{b}_\alpha \setminus B_\beta \notin \mathcal{J}_{<\lambda}(\mathsf{a})$. For $\beta < \lambda$ let $\mathsf{d}_\beta := \mathsf{b}_\alpha \setminus B_\beta$. By Exercise 2.10 you can choose an ultrafilter D on a such that $D \cap \mathcal{J}_{<\lambda}(\mathsf{a}) = \emptyset$ and $\mathsf{d}_\beta \in D$ for all $\beta < \lambda$. Show that $D \cap \mathcal{J}_{<\lambda}(\mathsf{a}) \restriction \mathsf{b}_\alpha = \emptyset = D \cap J$ and apply properties (2) and (3).
9) Assume that a is an infinite set of regular cardinals with $|\mathsf{a}|^+ < \min(\mathsf{a})$, μ and λ are infinite cardinals with $\mu < \lambda$, and $(\mathsf{b}_\alpha : \alpha < \mu)$ is a sequence of members of $\mathcal{J}_{<\lambda^+}(\mathsf{a}) \setminus \mathcal{J}_{<\lambda}(\mathsf{a})$. Prove that there exists a set $\mathsf{b} \in \mathcal{J}_{<\lambda^+}(\mathsf{a})$ such that $\mathsf{b}_\alpha \in \mathcal{J}_{<\lambda}(\mathsf{a})[\mathsf{b}]$ for all $\alpha < \mu$.
Hint: Let $(B_\xi : \xi < \lambda)$ be the sequence from Exercise 8. Choose, for every $\alpha < \mu$, an ordinal $\beta(\alpha) < \lambda$ such that $\mathsf{b}_\alpha \subseteq_{\mathcal{J}_{<\lambda}(\mathsf{a})} B_{\beta(\alpha)}$. Show for $\delta := \sup\{\beta(\alpha) : \alpha < \mu\}$ that $B_\delta \in \mathcal{J}_{<\lambda^+}(\mathsf{a})$ and $\mathsf{b}_\alpha \subseteq_{\mathcal{J}_{<\lambda}(\mathsf{a})} B_\delta$ for all $\alpha < \mu$.
10) (Shelah) Assume that a is an infinite progressive set of regular cardinals with $2^{|\mathsf{a}|} < \min(\mathsf{a})$ and λ is a cardinal. Prove that there exists a set $\mathsf{b} \in \mathcal{P}(\mathsf{a})$ such that $\mathcal{J}_{<\lambda^+}(\mathsf{a}) = \mathcal{J}_{<\lambda}(\mathsf{a})[\mathsf{b}]$.
Hint: Use the previous question.

11) Assume that a is an infinite progressive set of regular cardinals and $|\mathrm{pcf}(\mathsf{a})| < \min(\mathsf{a})$. Then for all subsets d and e of $\mathrm{pcf}(\mathsf{a})$ the following holds.

    (i) $\mathrm{pcf}(\emptyset) = \emptyset$.
    (ii) $\mathsf{d} \subseteq \mathrm{pcf}(\mathsf{d})$.
    (iii) If $\mathsf{d} \subseteq \mathsf{e}$, then $\mathrm{pcf}(\mathsf{d}) \subseteq \mathrm{pcf}(\mathsf{e})$.
    (iv) $\mathrm{pcf}(\mathsf{d} \cup \mathsf{e}) = \mathrm{pcf}(\mathsf{d}) \cup \mathrm{pcf}(\mathsf{e})$.
    (v) $\mathrm{pcf}(\mathrm{pcf}(\mathsf{d})) = \mathrm{pcf}(\mathsf{d})$.

    In other words: If $|\mathrm{pcf}(\mathsf{a})| < \min(\mathsf{a})$, then pcf is the hull operator of a topology on $\mathrm{pcf}(\mathsf{a})$. This topology is even compact.

12) (Shelah) Assume that a is a nonempty progressive set of regular cardinals and I is an ideal on a. Prove that there is a set $\mathsf{b} \in I$ such that

$$\mathrm{cf}\left(\prod \mathsf{a}, \leq_I\right) = \mathrm{cf}\left(\prod(\mathsf{a} \setminus \mathsf{b}), \leq_{(\mathsf{a}\setminus\mathsf{b})}\right).$$

13) (S. Neumann) Assume that $(P, \leq)$ is a partial ordering and $(P, \preceq)$ is a linear extension of $(P, \leq)$, that means that $(P, \preceq)$ is a linear ordering and $\leq$ is a subset of $\preceq$. Show that $\mathrm{cf}\,(P, \preceq) \leq \mathrm{cf}\,(P, \leq)$.

14) (S. Neumann) Prove: If a is a nonempty progressive interval of regular cardinals, then the set

$$\{\mathrm{cf}\,(\prod \mathsf{a}, \prec) \colon (\prod \mathsf{a}, \prec) \text{ is a linear extension of } (\prod \mathsf{a}, <_{\mathsf{a}})\}$$

    is a closed interval of regular cardinals.

15) Assume that a is a nonempty progressive set of regular cardinals and λ is an infinite cardinal. Show that

$$\mathcal{J}_{<\lambda}(\mathsf{a}) = \{\mathsf{b} \subseteq \mathsf{a} : \mathrm{cf}\left(\prod \mathsf{b}, \leq_{\mathsf{b}}\right) < \lambda\}.$$

16) (Shelah) Let a be an infinite progressive set of regular cardinals and D be an ultrafilter on a. Show that

$$\mathrm{cf}\left(\prod \mathsf{a}/D\right) = \min\{\lambda \in \mathrm{CN} : D \cap \mathcal{J}_{<\lambda^+}(\mathsf{a}) \neq \emptyset\}.$$

Chapter 4

Approximation Sequences

The methods applied in the previous chapters are classical and, in a certain sense, elementary. In this chapter, we will introduce modern model theoretic methods. Notions such as "model of ZFC" and "absoluteness of a formula" are introduced. For any infinite cardinal number Θ we define the set $\mathrm{H}(\Theta)$ of those sets which are hereditarily of cardinality less than Θ. We will show that for all regular uncountable cardinals Θ, $\mathrm{H}(\Theta)$ is a model of all axioms of ZFC except the power set axiom.

An approximation sequence $(N_i : i < \kappa)$ will be any continuous sequence of elementary substructures of $\mathrm{H}(\Theta)$ satisfying $(N_j : j \leq i) \in N_{i+1}$ for all $i < \kappa$. It will be shown in the following chapters that these sequences are a technical tool of great importance.

4.1 The Sets $\mathrm{H}(\Theta)$

Definition *For any infinite cardinal number κ let $\mathrm{H}(\kappa) := \{x : |\mathrm{tc}(x)| < \kappa\}$ be the class of all sets x which are* **hereditarily of cardinality less than κ.**

Remark The class $\mathrm{H}(\kappa)$ is a set since Lemma 4.1.2 shows that $\mathrm{H}(\kappa) \subseteq \mathrm{V}_\kappa$, where V_κ is the κ-th von Neumann level. For any regular cardinal κ, we have $\mathrm{H}(\kappa) = \mathrm{V}_\kappa$ iff $\kappa = \omega$ or κ is a strongly inaccessible cardinal (see Exercise 8). Lemma 4.1.3 represents an alternative definition of $\mathrm{H}(\kappa)$ for regular cardinals κ and is of importance for later applications, since we will consider the sets $\mathrm{H}(\kappa)$ only for such cardinals.

Let $\kappa, \mu \in \mathrm{ICN}$. Clearly we have $\mathrm{H}(\kappa) \subseteq \mathrm{H}(\mu)$ if $\kappa \leq \mu$. The proof of $\mathrm{H}(\kappa) = \bigcup\{\mathrm{H}(\mu^+) : \mu \in \mathrm{ICN} \cap \kappa\}$, for uncountable κ, is left as an exercise. Furthermore $\mathrm{V} = \bigcup\{\mathrm{H}(\kappa) : \kappa \in \mathrm{ICN}\}$ is true, since, for any set x, the definition of $\mathrm{H}(\kappa)$ yields $x \in \mathrm{H}(\lambda)$, where $\lambda := \max\{\aleph_0, |\mathrm{tc}(x)|^+\}$. Together with the fact that $\mathrm{H}(\lambda)$ is transitive (see Lemma 4.1.4), we can conclude in analogy to an assertion about the cumulative hierarchy that, for any set x, there exists $\lambda \in \mathrm{ICN}$ such that $x \subseteq \mathrm{H}(\lambda)$.

Lemma 4.1.1 *For any set x, we have* $\mathrm{tc}(x) = x \cup \bigcup\{\mathrm{tc}(y) : y \in x\}$.

Proof Let $a := x \cup \bigcup\{\mathrm{tc}(y) : y \in x\}$. We will show that a is the smallest transitive superset of x. If $z \in y \in a$, then $y \in x$, hence $z \in y \subseteq \mathrm{tc}(y) \subseteq a$, or $y \in \mathrm{tc}(u)$ for some $u \in x$. Therefore we get $y \subseteq a$, since $\mathrm{tc}(u)$ is transitive. Thus a is a transitive superset of x. If b is any transitive superset of x, and if $y \in x$, then $y \subseteq b$, hence $\mathrm{tc}(y) \subseteq b$. This gives $a \subseteq b$, as desired.

Lemma 4.1.2 *For all $\kappa \in \mathrm{ICN}$, $\mathrm{H}(\kappa)$ is transitive. If V_κ is the corresponding von Neumann level, then $\mathrm{H}(\kappa) \subseteq \mathrm{V}_\kappa$; in particular $\mathrm{H}(\kappa)$ is a set.*

Proof Assume that $x \in \mathrm{H}(\kappa)$. If $y \in x$, then $\mathrm{tc}(y) \subseteq \mathrm{tc}(x)$ and thus $|\mathrm{tc}(y)| \leq |\mathrm{tc}(x)| < \kappa$. So we get $y \in \mathrm{H}(\kappa)$ which implies that $\mathrm{H}(\kappa)$ is transitive.

As it is already known – see Example 1.3.7 – we have $\mathrm{V}_0 = \emptyset$, $\mathrm{V}_{\alpha+1} = \mathcal{P}(\mathrm{V}_\alpha)$, and $\mathrm{V}_\tau = \bigcup\{\mathrm{V}_\beta : \beta < \tau\}$ for limit ordinals τ. Furthermore, $\mathrm{V} = \bigcup\{\mathrm{V}_\alpha : \alpha \in \mathrm{ON}\} = \{x : x = x\}$. For any set x, the rank α of x, written as $\mathrm{rk}(x)$, is the smallest ordinal β such that $x \in \mathrm{V}_{\beta+1}$.

We use transfinite induction on the rank of x to prove that $x \in \mathrm{H}(\kappa)$ implies $x \in \mathrm{V}_\kappa$. If κ is a limit cardinal, then it follows from $|\mathrm{tc}(x)| < \kappa$ that $|\mathrm{tc}(x)| < \lambda^+$ for some cardinal $\lambda < \kappa$; thus $x \in \mathrm{H}(\lambda^+) \subseteq \mathrm{H}(\kappa)$. So we can assume without loss of generality that κ is regular. If $y \in x$, then $\mathrm{rk}(y) < \mathrm{rk}(x)$ and $y \in \mathrm{H}(\kappa)$, which yields $y \in \mathrm{V}_\kappa$ by the inductive hypothesis. So we get $x \subseteq \mathrm{V}_\kappa$. Since κ is regular and $|x| < \kappa$, there is $\alpha < \kappa$ such that $x \subseteq \mathrm{V}_\alpha$, and thus $x \in \mathrm{V}_{\alpha+1} \subseteq \mathrm{V}_\kappa$.

Lemma 4.1.3 *If Θ is a regular cardinal, then* $\mathrm{H}(\Theta) = \{x \subseteq \mathrm{H}(\Theta) : |x| < \Theta\}$.

Proof Fix $x \in \mathrm{H}(\Theta)$. Then $|\mathrm{tc}(x)| < \Theta$. Furthermore Lemma 4.1.2 says that $x \subseteq \mathrm{H}(\Theta)$. Now $x \subseteq \mathrm{tc}(x)$ yields one inclusion. To verify the reverse inclusion, let $x \subseteq \mathrm{H}(\Theta)$ and $|x| < \Theta$. For every $y \in x$, we have $y \in \mathrm{H}(\Theta)$ and thus $|\mathrm{tc}(y)| < \Theta$, hence $|\mathrm{tc}(x)| = |x \cup \bigcup\{\mathrm{tc}(y) : y \in x\}| < \Theta$, since Θ is regular.

Lemma 4.1.4 *For any infinite cardinal number κ, we have*

a) $\mathrm{H}(\kappa)$ *is transitive.*
b) $\kappa \leq \mu \Longrightarrow \mathrm{H}(\kappa) \subseteq \mathrm{H}(\mu)$.
c) $\kappa > \aleph_0 \Longrightarrow \mathrm{H}(\kappa) = \bigcup\{\mathrm{H}(\mu^+) : \mu \in \mathrm{ICN} \cap \kappa\}$.
d) If κ is a limit cardinal, then $\mathrm{H}(\kappa) = \bigcup\{\mathrm{H}(\mu) : \mu \in \mathrm{ICN} \cap \kappa\}$.
e) $y \subseteq x \wedge x \in \mathrm{H}(\kappa) \Longrightarrow y \in \mathrm{H}(\kappa)$.
f) $\mathrm{H}(\kappa) \cap \mathrm{ON} = \kappa$.

g) $x \in \mathrm{H}(\kappa) \implies \bigcup x \in \mathrm{H}(\kappa)$.

h) $x, y \in \mathrm{H}(\kappa) \implies \{x, y\} \in \mathrm{H}(\kappa) \wedge (x, y) \in \mathrm{H}(\kappa) \wedge x \cup y \in \mathrm{H}(\kappa)$.

i) $x, y \in \mathrm{H}(\kappa) \implies x \times y \in \mathrm{H}(\kappa) \wedge (f : x \longrightarrow y \implies f \in \mathrm{H}(\kappa))$.

Proof Part a) is proved in Lemma 4.1.2; the proof of b), c) and d) is left as an exercise. Let $x \in \mathrm{H}(\kappa)$. If $y \subseteq x$, then $\mathrm{tc}(y) \subseteq \mathrm{tc}(x)$, which implies e). To show f), notice that all ordinals α are transitive, hence satisfy $\mathrm{tc}(\alpha) = \alpha$. So we get $|\mathrm{tc}(\alpha)| < \kappa$ iff $\alpha < \kappa$, for all ordinals α, which yields f). For the proof of g) and h), let $x, y \in \mathrm{H}(\kappa)$, i.e., $\max\{|\mathrm{tc}(x)|, |\mathrm{tc}(y)|\} < \kappa$. It is easy to see that $\bigcup x \subseteq \bigcup\{\mathrm{tc}(y) : \ y \in x\} \subseteq \mathrm{tc}(x)$, hence $\mathrm{tc}(\bigcup x) \subseteq \mathrm{tc}(x)$; furthermore we have $\mathrm{tc}(x \cup y) = \mathrm{tc}(x) \cup \mathrm{tc}(y)$ and $\mathrm{tc}(\{x, y\}) = \{x, y\} \cup \mathrm{tc}(x) \cup \mathrm{tc}(y)$, which yields g) and h). The last assertion can be inferred, for regular cardinals κ, from $x \times y \subseteq \mathrm{H}(\kappa)$ and $|x \times y| < \kappa$, using Lemma 4.1.3. If κ is singular and $x, y \in \mathrm{H}(\kappa)$, then we can find a regular cardinal $\mu < \kappa$ such that $\max\{|\mathrm{tc}(x)|, |\mathrm{tc}(y)|\} < \mu$, hence $x, y \in \mathrm{H}(\mu)$ and thus $x \times y \in \mathrm{H}(\mu)$, as we have just shown. With $\mathrm{H}(\mu) \subseteq \mathrm{H}(\kappa)$ we get the first assertion in i). If $f : x \longrightarrow y$ is a function, then we have $f \subseteq x \times y$ and thus $f \in \mathrm{H}(\kappa)$ by e).

Exercises

1) Assume that κ is a regular cardinal. Prove that $x \in \mathrm{H}(\kappa)$ iff, for any $n \in \omega$ and any sequence $f : n \longrightarrow \mathrm{V}$ satisfying $f(n-1) = x$ and $f(j-1) \in f(j)$ for all j with $0 < j < n-1$, we have $|f(0)| < \kappa$.

   Show that one direction of the equivalence is false for any singular cardinal κ.

   Hint: Use Lemma 4.1.1 or the remark in the definition of the transitive closure.

2) Prove, for uncountable cardinals κ, that $\mathrm{H}(\kappa) = \bigcup\{\mathrm{H}(\mu^+) : \mu \in \mathrm{ICN} \cap \kappa\}$. If κ is a limit cardinal, then $\mathrm{H}(\kappa) = \bigcup\{\mathrm{H}(\mu) : \mu \in \mathrm{ICN} \cap \kappa\}$. Why is the last assertion false for successor cardinals?

3) Assume that κ is an infinite cardinal and $f, x, y \in \mathrm{H}(\kappa)$. Prove:

   a) If f is a function, then $\mathrm{dom}(f)$, $\mathrm{ran}(f)$ and, for any $x \in \mathrm{dom}(f)$, $f(x)$ are members of $\mathrm{H}(\kappa)$.

   b) ${}^{x}y \subseteq \mathrm{H}(\kappa)$, where ${}^{x}y$ is the set of functions from x into y.

4) Prove that $\mathrm{V}_\omega = \mathrm{H}(\omega)$.

5) Prove that $\mathrm{H}(\kappa^+) \subsetneqq \mathrm{V}_{\kappa^+}$ for all $\kappa \in \mathrm{ICN}$.

   Hint: $\mathcal{P}(\kappa) \in \mathrm{V}_{\kappa+2}$.

6) Show that the equation $\mathrm{H}(\kappa) = \{x \subseteq \mathrm{H}(\kappa) : |x| < \kappa\}$ is false for $\kappa = \aleph_\omega$.

7) Assume that κ is a regular cardinal. Prove

$$x \in \mathrm{H}(\kappa) \Longleftrightarrow |x| < \kappa \wedge \forall y \in \mathrm{tc}(x)\ |y| < \kappa.$$

Hint: One direction of the equivalence is easy to verify. For the other one, use $\in$-induction and Lemma 4.1.1.

8) Assume that κ is a regular cardinal. Prove

$$\mathrm{V}_\kappa = \mathrm{H}(\kappa) \Longleftrightarrow \kappa = \omega \vee \kappa \text{ is strongly inaccessible }.$$

Hint: If $\kappa > \omega$ is not strongly inaccessible, then there is a cardinal $\lambda < \kappa$ such that $2^\lambda \geq \kappa$. Now use the hint in Exercise 5. Conversely, if κ is strongly inaccessible, use Exercise 1.7.6.

9) Assume that κ is an uncountable regular cardinal. Prove that $\mathrm{H}(\kappa) = \mathrm{V}_\kappa$ iff κ is a fixed point of the beth function.

Hint: See the exercise in Section 1.7 in which the beth function is defined.

10) a) Let κ be an infinite cardinal and $\alpha \leq \kappa^+$. We define $H_\alpha := \mathrm{H}(\kappa^+) \cap \mathrm{V}_\alpha$. Prove for $\alpha < \kappa^+$:

a1) $H_{\alpha+1} = \{X \subseteq H_\alpha : |X| \leq \kappa\}$.

a2) $\alpha \in \mathrm{Lim} \Longrightarrow H_\alpha = \bigcup\{H_\beta : \beta < \alpha\}$.

a3) $H_{\kappa^+} = \mathrm{H}(\kappa^+)$.

a4) $|H_\alpha| \leq 2^\kappa$.

b) Assume that λ is an infinite cardinal. Prove that

$$|\mathrm{H}(\lambda)| = 2^{<\lambda}.$$

Hint: If $\lambda = \kappa^+$, use part a). If λ is a limit cardinal, use Exercise 2.

4.2 Models and Absoluteness

In this section we will represent some fundamental facts from mathematical logic. The required prereqisites are small, and once the basic Lemmas 4.2.12 and 4.2.13 are available, even a reader who is not familiar with the first order predicate calculus will be able to apply them correctly. The formal language $L := \mathcal{L}_{\mathrm{ZF}}$ of ZF has already been introduced in Section 1.1.

A **subformula** of a formula φ of L is a sequence of consecutive symbols in φ which is a formula. For example, the formula

$$(*) \qquad (\exists v_0\ v_0 \in v_1 \wedge \exists v_1\ v_2 \in v_1)$$

has five subformulae, namely $v_0 \in v_1$, $\exists v_0\ v_0 \in v_1$, $v_2 \in v_1$, $\exists v_1\ v_2 \in v_1$ and the formula itself. If there is an occurence of a quantifier $\exists v_j$ or $\forall v_j$ in a formula, then the **scope** of this quantifier is the unique subformula beginning with this quantifier. An occurence of a variable v_j in a formula is called **free** if it does not lie in the scope of a quantifier $\exists v_j$ or $\forall v_j$ acting on that variable; otherwise it is called **bound**. For example, in $(*)$ the first occurence of v_1 is free, but the second and third are bound, whereas v_0 is bound at all of its occurrences and v_2 only occurs free. Any variable which occurs free in φ is called a **free variable** of φ, and a variable which occurs bound in φ is called a **bound variable** of φ.

Intuitively, a formula expresses a property of its free variables, whereas a bound variable, used to make existential or universal statements, can be replaced in the scope of the quantifier binding it by any other variable which does not not occur in the formula, without changing the "meaning" of the formula.

Under the symbols of L are variables v_n for every natural number n. To make the text more readable and to avoid double indices, we will also use syntactical variables a_n, x_n, y_n etc. for variables. Often we will present a formula in the form $\varphi(x_0, \ldots, x_{n-1})$, when its free variables are amongst $x_0, \ldots, x_{n-1}$, to emphasize its dependence on $x_0, \ldots, x_{n-1}$. Often we abbreviate the sequence $x_0, \ldots, x_{n-1}$ by $\vec{x}$ and write $\varphi(\vec{x})$ instead of $\varphi(x_0, \ldots, x_{n-1})$. If $a_0, \ldots, a_{n-1}$ are further variables, then $\varphi(a_0, \ldots, a_{n-1})$ is that formula which results from **substituting** an a_i for every free occurrence of x_i. Such a substitution is called **legitimate** if no free occurence of an x_i is in the scope of a quantifier $\exists a_i$ or $\forall a_i$. Intuitively, the formula $\varphi(a_0, \ldots, a_{n-1})$ says about $a_0, \ldots, a_{n-1}$ the same that $\varphi(x_0, \ldots, x_{n-1})$ says about $x_0, \ldots, x_{n-1}$. So let for example $\varphi(v_1, v_2)$ be the formula in $(*)$. Then the substitution $\varphi(v_2, v_3)$ is legitimate, but the substitution $\varphi(v_0, v_2)$ is not legitimate. Interpreting the symbol $\in$ in this formula, namely in $(\exists v_0\ v_0 \in v_0 \wedge \exists v_1\ v_2 \in v_1)$, as the usual naive membership, it says that some set is an element of itself, whereas the formula in $(*)$ says that v_1 has an element. We will always assume without mentioning that all our substitutions are legitimate. Instead of $\exists v_i(\varphi(v_i) \wedge \forall v_j(\varphi(v_j) \implies v_i = v_j))$ we also write $\exists! v_i \varphi(v_i)$.

A **closed formula** or **sentence** of $\mathcal{L}_{\mathrm{ZF}}$ is a formula in which no variable occurs free. Intuitively, it states an assertion which is either true or false. If φ is a formula, then let a **universal closure** of φ be any sentence $\forall \vec{x}\, \varphi$ obtained by universally quantifying all free variables of φ. Since such a sentence can be obtained in a unique way and all universal closures of fixed formulae are logically equivalent (see below), we will speak of *the* universal closure of φ.

Every formula of $\mathcal{L}_{\mathrm{ZF}}$ is logically equivalent to a formula whose connectives and quantifiers are amongst $\neg$, $\vee$, and $\exists$. Therefore we regard, in definitions and proofs which use induction on the complexity of a formula φ, the formulae $\varphi \wedge \psi$, $\varphi \implies \psi$, $\varphi \iff \psi$, and $\forall x \varphi$ as abbreviations for the formulae $\neg(\neg\varphi \vee \neg\psi)$, $\neg\varphi \vee \psi$, $(\varphi \implies \psi) \wedge (\psi \implies \varphi)$, and $\neg\exists x(\neg\varphi)$ respectively.

If T is a set of sentences of $\mathcal{L}_{\mathrm{ZF}}$, then we also call T a **theory** and its elements **axioms** of T. $T \vdash \varphi$ means that φ is provable with the axioms of T. $T \subseteq \mathrm{ZF}$ means that any formula which is provable with the axioms of T is provable in ZF.

A **structure for $\mathcal{L}_{\mathrm{ZF}}$** is a pair $\mathcal{A} = (A, E)$, where A is a nonempty set, the **universe of $\mathcal{A}$**, and $E = \{(x, y) \in A \times A : x \in y\}$. With L^* or $\mathcal{L}^*_{\mathrm{ZF}}$ we denote that language which results from $\mathcal{L}_{\mathrm{ZF}}$ by adding the binary relation symbol $<$. The notion of a formula is extended correspondingly, adding to the definition of atomic formulae (see Section 1.1) the rule "$v_i < v_j$ is a formula". A **structure for $\mathcal{L}^*_{\mathrm{ZF}}$**, shortly, an L^*-structure, is a triple $(A, E, <^*)$, where again A is a nonempty set, $E = \{(x, y) \in A \times A : x \in y\}$ and $<^* \subseteq A \times A$ is a well-ordering of A. Instead of (A, E) and $(A, E, <^*)$ we use the slightly incorrect but more suggestive notation $(A, \in)$ and $(A, \in, <^*)$, respectively.

Assume that $\mathcal{A}$ is an L^*-structure, $(a_0, \ldots, a_{n-1})$ is a finite sequence of members of the universe A of $\mathcal{A}$, and φ is an L^*-formula with at most the free variables $v_0, \ldots, v_{n-1}$. By induction on the complexity of φ we define the L-formula $\mathcal{A} \models \varphi[a_0, \ldots, a_{n-1}]$. Intuitively, the formula $\mathcal{A} \models \varphi[a_0, \ldots, a_{n-1}]$ is provable iff the property of $a_0, \ldots, a_{n-1} \in A$ formally expressed by φ is true, where "for all x" and "there exists x" are to be read as "for all $x \in A$" and "there exists $x \in A$", and x_i is to be read as a_i.[1] We define all notions for L^*, and it should be obvious that hereby they are also defined for L. In the same way, most properties which we prove for L^*-structures also hold for L-structures.

To be more precise, let $a_0, \ldots, a_{n-1} \in A$. For atomic formulae we define

$$\mathcal{A} \models v_i \in v_j[a_0, \ldots, a_{n-1}] \iff a_i \in a_j,$$

$$\mathcal{A} \models v_i = v_j[a_0, \ldots, a_{n-1}] \iff a_i = a_j,$$

$$\mathcal{A} \models v_i < v_j[a_0, \ldots, a_{n-1}] \iff a_i <^* a_j.$$

Further we agree that

$$\mathcal{A} \models (\neg\psi)[a_0, \ldots, a_{n-1}] \iff \neg\mathcal{A} \models \psi[a_0, \ldots, a_{n-1}],$$

$$\mathcal{A} \models (\psi \vee \chi)[a_0, \ldots, a_{n-1}] \iff \mathcal{A} \models \psi[a_0, \ldots, a_{n-1}] \vee \mathcal{A} \models \chi[a_0, \ldots, a_{n-1}],$$

[1] To be precise, this requires, especially in view of the Löwenheim-Skolem theorem 4.2.3, a so-called arithmetization of our formal language. We may identify the symbols of $\mathcal{L}^*_{\mathrm{ZF}}$ with certain hereditarily finite sets, i.e., with elements of V_ω. Then the formulas of $\mathcal{L}^*_{\mathrm{ZF}}$ are also hereditarily finite sets. $\mathcal{A} \models \varphi[a_0, \ldots, a_{n-1}]$ is defined by induction on the complexity of formulae as a relation between structures $\mathcal{A}$, (arithmetized) formulae φ, and formal n-tuples $(a_0, \ldots, a_{n-1}) \in A^n$, that means (if we identify as usual formal n-tuples with functions which have the domain $n \in \omega$), elements of ${}^nA$. The interested reader is referred to the books of F. R. Drake [Dr] (Chapter 3, §5) and K. Kunen [Ku] (Chapter V, §1), in which he can find a detailed formalization of some first order languages.

$$\mathcal{A} \models \exists v_n \psi[a_0, \ldots, a_{n-1}] \iff \exists a \in A\ \mathcal{A} \models \psi[a_0, \ldots, a_{n-1}, a].$$

With the usual abbreviations we obtain the assertions $\mathcal{A} \models (\psi \wedge \chi)[a_0, \ldots, a_{n-1}]$, $\mathcal{A} \models (\psi \implies \chi)[a_0, \ldots, a_{n-1}]$, $\mathcal{A} \models (\psi \iff \chi)[a_0, \ldots, a_{n-1}]$, $\mathcal{A} \models \forall v_n \psi[a_0, \ldots, a_{n-1}]$, where, for example, the last one is the assertion $\forall a \in A\ \mathcal{A} \models \psi[a_0, \ldots, a_{n-1}, a]$.

If the formula $\mathcal{A} \models \varphi[a_0 \ldots a_{n-1}]$ is *provable* in ZFC, then we say that φ **holds in $\mathcal{A}$ under the valuation $(a_0, \ldots, a_{n-1})$**. If the universal closure $\forall \vec{x} \varphi$ of φ holds in $\mathcal{A}$, then we say that **φ holds in $\mathcal{A}$** or that $\mathcal{A}$ is a **model of φ**, and write $\mathcal{A} \models \varphi$. If T is a set of sentences of L^*, then we call $\mathcal{A}$ a **model of T**, written as $\mathcal{A} \models T$, if $\mathcal{A}$ is a model of φ for every sentence φ in T. If the context is clear, we identify the structure $\mathcal{A}$ with its universe A and write $A \models \varphi[a_0, \ldots, a_{n-1}]$ instead of $\mathcal{A} \models \varphi[a_0, \ldots, a_{n-1}]$. Often we write $\mathcal{A} \models \varphi(a_0, \ldots, a_{n-1})$ instead of $\mathcal{A} \models \varphi[a_0, \ldots, a_{n-1}]$.

Definition *If $\mathcal{A} = (A, \in, <^*_1)$ and $\mathcal{B} = (B, \in, <^*)$ are structures for L^*, then $\mathcal{A}$ is a* **substructure** *of $\mathcal{B}$, written as $\mathcal{A} \subseteq \mathcal{B}$ iff $A \subseteq B$ and $<^*_1 = <^* \cap A \times A$. Then we also write $(A, \in, <^*)$ instead of $(A, \in, <^*_1)$.*

If $\mathcal{A}$ is a substructure of $\mathcal{B}$, then $\mathcal{A}$ is an **elementary substructure** *of $\mathcal{B}$ and $\mathcal{B}$ is an* **elementary extension** *of $\mathcal{A}$, written as $\mathcal{A} \preceq \mathcal{B}$, iff, for every formula φ of L^* and every valuation $a_0, \ldots, a_{n-1} \in A$ of the free variables of φ,*

$$\mathcal{A} \models \varphi[a_0, \ldots, a_{n-1}] \iff \mathcal{B} \models \varphi[a_0, \ldots, a_{n-1}].$$

Remark The reader who is not familiar with these notions may visualize $\mathcal{A} \preceq \mathcal{B}$ as: $\mathcal{A}$ is a substructure of $\mathcal{B}$, and any assertion about objects in A holds in $\mathcal{A}$ iff it holds in $\mathcal{B}$.

Lemma 4.2.1 *If $\mathcal{A} \subseteq \mathcal{B}$, $\mathcal{A} \preceq \mathcal{C}$ and $\mathcal{B} \preceq \mathcal{C}$, then $\mathcal{A} \preceq \mathcal{B}$.*

The applications of the notion of an elementary substructure are simplified by the following lemma.

Lemma 4.2.2 (Tarski, Vaught) *Assume that $\mathcal{A} \subseteq \mathcal{B}$. Then $\mathcal{A} \preceq \mathcal{B}$ iff, for any formula ψ of L^* and any valuation $a_0, \ldots, a_{n-1} \in A$,*

$$\mathcal{B} \models \exists x_n \psi[a_0, \ldots, a_{n-1}] \implies \exists a \in A(\mathcal{B} \models \psi[a_0, \ldots, a_{n-1}, a]).$$

Proof If $\mathcal{A} \preceq \mathcal{B}$ and $\mathcal{B} \models \exists x_n \psi[\vec{a}]$, then, by definition, $\mathcal{A} \models \exists x_n \psi[\vec{a}]$, where $\vec{a}$ abbreviates the sequence $a_0, \ldots, a_{n-1}$. The definition of this formula yields $a \in A$ such that $\mathcal{A} \models \psi[\vec{a}, a]$, and thus, again using $\mathcal{A} \preceq \mathcal{B}$, $a \in A$ such that $\mathcal{B} \models \psi[\vec{a}, a]$.

For the proof of the reverse implication we will verify the equivalence $(\mathcal{A} \models \varphi[\vec{a}] \iff \mathcal{B} \models \varphi[\vec{a}])$ by induction on the complexity of the formula φ. If

φ is atomic, then this follows from $\mathcal{A} \subseteq \mathcal{B}$. The cases $\varphi \equiv \neg\psi$ and $\varphi \equiv \psi \vee \chi$ follow immediately from the inductive hypothesis and the definition of $\models$.

Now assume that φ has the form $\exists x_n \psi$. From $\mathcal{B} \models \exists x_n \psi[\vec{a}]$ we can infer by assumption, that there is $a \in A$ such that $\mathcal{B} \models \psi[\vec{a}, a]$. The inductive hypothesis gives $\mathcal{A} \models \psi[\vec{a}, a]$, hence $\mathcal{A} \models \exists x_n \psi[\vec{a}]$. Conversely, if the last assertion is true, then there is $a \in A$ such that $\mathcal{A} \models \psi[\vec{a}, a]$. Since $\mathcal{A} \subseteq \mathcal{B}$, and since $\mathcal{B} \models \psi[\vec{a}, a]$ holds by inductive hypothesis, we get $\mathcal{B} \models \exists x_n \psi[\vec{a}]$.

Theorem 4.2.3 (Löwenheim-Skolem Theorem) *Assume that $\mathcal{A}$ and $\mathcal{C}$ are L^*-structures, $\mathcal{A} \subseteq \mathcal{C}$, and $Y \subseteq C$. Then there is an elementary substructure $\mathcal{B}$ of $\mathcal{C}$ such that $\mathcal{A} \subseteq \mathcal{B}$, $Y \subseteq B$, and $|B| \leq \max\{\aleph_0, |Y|, |A|\}$.*

Proof Put $\lambda := \max\{\aleph_0, |Y|, |A|\}$ and $X_0 := Y \cup A$. For each formula $\varphi(x_0, \ldots, x_n)$ of L^* and each valuation $(a_0, \ldots, a_{n-1}) \in X_0^n$ satisfying $\mathcal{C} \models \exists x_n \varphi\ [\vec{a}]$, we can choose, using (AC), some $a \in C$ such that $\mathcal{C} \models \varphi\ [\vec{a}, a]$. Take X_0^* as the set of all these "satisfiers". It is easy to verify that $X_0 \subseteq X_0^*$. Since $|X_0^n| \leq \lambda$ and thus $|\bigcup\{X_0^n : n \in \omega\}| \leq \aleph_0 \cdot \lambda = \lambda$, and since L^* and hereby the set of formulae of L^* is countable, we can conclude that $|X_0^*| \leq \aleph_0 \cdot \lambda = \lambda$. So put $X_1 := X_0^*$ and define, if X_n is already defined, $X_{n+1} := X_n^*$ in complete analogy. If we assume that $|X_n| \leq \lambda$, then $|X_{n+1}| \leq \lambda$ follows as above.

Now put $B := \bigcup\{X_n : n \in \omega\}$, and let $\mathcal{B}$ be that structure with universe B whose relations are the restrictions of the corresponding relations of $\mathcal{C}$ to the subset B of C. Clearly we have $\mathcal{B} \subseteq \mathcal{C}$ and $|B| \leq \lambda \cdot \aleph_0 = \lambda$.

For the proof of $\mathcal{B} \preceq \mathcal{C}$ we can apply Lemma 4.2.2. If $\mathcal{C} \models \exists x_n \psi\ [b_0, \ldots, b_{n-1}]$ for certain $b_0, \ldots, b_{n-1} \in B$, then there is $m \in \omega$ such that $b_0, \ldots, b_{n-1} \in X_m$. Our definition of X_{m+1} shows that there exists a satisfier $b \in B$ such that $\mathcal{C} \models \psi[b_0, \ldots, b_{n-1}, b]$, as desired.

We will now introduce the notion of absoluteness of a formula. Since the absoluteness of a formula for a set M depends on its complexity, and since we will build up, starting with simple formulae, a list of absolute formulae, it is suitable to introduce the notion of relativization of a formula φ to a class M.

Definition *If M is a transitive class and φ is a formula of $\mathcal{L}_{\mathrm{ZF}}$, then we define the* **relativization of φ to M**, *denoted by φ^M, by induction on the complexity of φ. $(x \in y)^M$ is the formula $x \in y$, $(x = y)^M$ the formula $x = y$, $(\neg\psi)^M$ the formula $\neg(\psi^M)$, $(\psi \vee \chi)^M$ the formula $\psi^M \vee \chi^M$, and $(\exists x \psi)^M$ the formula $\exists x(x \in M \wedge \psi^M)$.*

If we take again $\psi \wedge \chi$, $\psi \Longrightarrow \chi$, $\psi \Longleftrightarrow \chi$ and $\forall x \psi$ as abbreviations for the formulae $\neg(\neg\psi \vee \neg\chi)$, $\neg\psi \vee \chi$ etc., then $(\psi \wedge \chi)^M$ is the formula $\varphi^M \wedge \psi^M$, $(\varphi \Longrightarrow \psi)^M$ the formula $\varphi^M \Longrightarrow \psi^M$, $(\varphi \Longleftrightarrow \psi)^M$ the formula $\varphi^M \Longleftrightarrow \psi^M$, and $(\forall x \varphi)^M$ the formula $\forall x(x \in M \Longrightarrow \varphi^M)$.

Definition *Let M be a transitive class. A formula $\varphi(\vec{x})$ of $\mathcal{L}_{\mathrm{ZF}}$ is called* **absolute for M**, *written as* φ abs M, *iff the following formula is a theorem of* ZF*:*

$$\forall x_0 \in M \ldots \forall x_{n-1} \in M \left(\varphi \iff \varphi^M\right).$$

The reader is asked to convince himself that the definitions of validity and relativization immediately yield the following lemma that we will always use without mentioning. In the same way we will use the following theorem. It is intuitively clear, but we omit its proof since it requires an exact listing of the logical axioms and logical rules of first-order logic which are tacitly applied in every-day mathematics.

Lemma 4.2.4 *If M is a transitive set and $\varphi(x_0, \ldots, x_{n-1})$ is a formula of $\mathcal{L}_{\mathrm{ZF}}$, then*

a) $\forall a_0 \in M \ldots \forall a_{n-1} \in M(M \models \varphi[a_0, \ldots a_{n-1}] \iff \varphi^M(a_0, \ldots, a_{n-1}))$.

b) φ abs M *iff*
$\forall a_0 \in M \ldots \forall a_{n-1} \in M(\varphi(a_0, \ldots, a_{n-1}) \iff M \models \varphi[a_0, \ldots, a_{n-1}])$.

Theorem 4.2.5 *Assume that $T \subseteq$ ZF is a theory. If M is a model of T and φ is a formula of L which is provable with the axioms of T, then the formula $\forall \vec{a} \in M\ \varphi^M$ is provable in* ZF.

Definition *A* **Δ_0-formula** *is a formula that contains only bounded quantifiers. That means that the Δ_0-formulae are defined by induction in the same way as formulae, with the exception that quantifiers are introduced by the rule "If φ is a Δ_0-formula and if v_i and v_j are variables, then $\exists v_i \in v_j\ \varphi$ and $\forall v_i \in v_j\ \varphi$ are Δ_0-formulae".*

If T is a set of sentences of $\mathcal{L}_{\mathrm{ZF}}$, then φ is called a **Δ_0^T-formula** *iff there is a Δ_0-formula ψ such that the formula $\varphi \iff \psi$ is provable with the axioms of T. For $T = \emptyset$ we will denote, in view of Lemma 4.2.6, also Δ_0^T-formulae as Δ_0-formulae.*

Remark

a) If one fixes exact rules for the introduction of defined and undefined logical symbols, then most formulae which we are handling are not Δ_0-formulae. For example, $x \subseteq y$ is the formula $\forall z \in x(z \in y)$ that contains only bounded quantifiers. But if we take $\exists$, $\vee$ and $\neg$ as the logical symbols of our language, then this string of symbols is an abbreviation for the formula $\neg\exists z \neg(\neg z \in x \vee z \in y)$, which is by no means a Δ_0-formula. On the other hand it is logically equivalent to the Δ_0-formula $\neg\exists z \in x(\neg z \in y)$. This is a typical example for our later way of proceeding. Δ_0-formulae will be absolute, but also formulae which are logically equivalent to Δ_0-formulae. Therefore we do not distinguish between the denotations of these two types of formulae.

b) Formulae which are in "weak" theories T with $M \models T$ equivalent to Δ_0-formulae, will be absolute for M, too. **For later purposes we will need only absoluteness results for the structures $\mathrm{H}(\Theta)$, where $\Theta > \omega$ is regular, and these sets $\mathrm{H}(\Theta)$ with the binary relation $\in$ are models of all axioms of ZFC except the power set axiom.** Therefore we need not take care of such subtle considerations.

Lemma 4.2.6 *Let M be a transitive class and φ and ψ be formulae of L.*

a) Any formula of the form $x = y$ or $x \in y$ is absolute for M.

b) If the formulae φ and ψ are absolute for M, then so are $\neg\varphi$ and $\varphi \diamond \psi$, where $\diamond$ is one of the connectives $\vee$, $\wedge$, $\Longrightarrow$, and $\Longleftrightarrow$.

c) If the formula φ is absolute for M, then so are $\exists x \in y\, \varphi$ and $\forall x \in y\, \varphi$.

d) Every Δ_0-formula is absolute for M.

e) If $T \subseteq \mathrm{ZF}$ is a theory such that $M \models T$ and $T \vdash \forall \vec{x}(\varphi(\vec{x}) \Longleftrightarrow \psi(\vec{x}))$, then φ is absolute for M iff ψ is absolute for M. In particular all Δ_0^T-formulae are absolute for M, if $M \models T$.

f) Assume that $T \subseteq \mathrm{ZF}$ is a theory and M is a model of T. Further let $T \vdash \varphi \Longleftrightarrow \forall x\psi$ and $T \vdash \varphi \Longleftrightarrow \exists x\chi$,[2] where χ and ψ are absolute for M. Then φ is absolute for M.

Proof Part a) follows immediately from the definition of the relativization of atomic formulae, b) follows with arguments from sentential logic.

For c), let $y \in M$. The relativization of the formula $\exists x \in y\, \varphi$ to M is the formula $\exists x \in M(x \in y \wedge \varphi^M)$. Since $y \in M$ and M is transitive, $x \in y$ implies $x \in M$. Furthermore we can infer from $x \in M$ with the absoluteness of φ that $\varphi \Longleftrightarrow \varphi^M$. Therefore the discussed relativization is equivalent to $\exists x \in y\, \varphi$. The proof for the second formula runs similar.

Part d) follows at once from a), b) and c).

Now we prove e). From Theorem 4.2.5 we get $\mathrm{ZF} \vdash \forall \vec{x} \in M(\varphi^M \Longleftrightarrow \psi^M)$. Further, by assumption, we have $\mathrm{ZF} \vdash \forall \vec{x}(\varphi(\vec{x}) \Longleftrightarrow \psi(\vec{x}))$. Thus $\forall \vec{x} \in M(\varphi \Longleftrightarrow \varphi^M)$ is provable in ZF iff $\forall \vec{x} \in M(\psi \Longleftrightarrow \psi^M)$ is provable in ZF. Therefore φ is absolute for M iff ψ is absolute for M.

Finally, we turn to the proof of f). Let $\vec{a}$ be a sequence containing the free variables of the considered formulae, and assume that $\vec{a} \in M$. From φ we get successively $\forall x\psi$, $\forall x \in M\, \psi$ and $\forall x \in M\, \psi^M$, i.e., $(\forall x\psi)^M$, since ψ is absolute for M. Now Theorem 4.2.5 says that $(\forall x\psi)^M \Longrightarrow \varphi^M$, and we have shown

[2] If in addition ψ and χ are Δ_0^T-formulae, then such formulae φ are called **Δ_1^T-formulae.**

that $\varphi \Longrightarrow \varphi^M$ is provable. In complete analogy one proves, using the second equivalence in the assumption and $\exists x \in M\ \chi \Longrightarrow \exists x \chi$, that the implication $\varphi^M \Longrightarrow \varphi$ is provable.

Lemma 4.2.7 *Every nonempty transitive set is a model of the axiom of extensionality, of the empty set axiom, and of the axiom of foundation.*

Proof We must check that $M \models \forall a \forall b(\forall x(x \in a \Longleftrightarrow x \in b) \Longrightarrow a = b)$. So we have to show that

$$(*) \qquad \forall a \in M \forall b \in M(\forall x \in M(x \in a \Longleftrightarrow x \in b) \Longrightarrow a = b).$$

But the fact that M is transitive gives, for $a \in M$, $x \in M \wedge x \in a \Longleftrightarrow x \in a$. Therefore $(*)$ follows from the axiom of extensionality. The corresponding proofs for the other axioms is left to the reader.

Lemma 4.2.8 *Assume that M is a transitive nonempty class. Then the following formulae are absolute for M:*

(1) $x \subseteq y$	(2) $x = \emptyset$	(3) $x = \{a, b\}$
(4) $x = (a, b)$	(5) $\exists u \exists v(x = \{u, v\})$	(6) $\exists u \exists v(x = (u, v))$
(7) $\mathrm{Rel}(x)$	(8) $x \in \mathrm{dom}(u)$	(9) $x = \mathrm{dom}(u)$
(10) $x \in \mathrm{ran}(u)$	(11) $x = \mathrm{ran}(u)$	(12) $x = u \times v$
(13) $x = u \cap v$	(14) $x = u \cup v$	(15) $x = u \setminus v$
(16) $x = \bigcup u$	(17) $\mathrm{Func}(x)$	(18) x *is transitive*
(19) $y = f(x)$	(20) $y = f[x]$	(21) $g = f \restriction x$
(22) $f : x \longrightarrow y$	(23) f *is injective*	(24) $x \in \mathrm{ON}$
(25) x *is a successor ordinal*	(26) *Lim*(x)	(27) $x \in \omega$
(28) $x = \omega$.		

Proof To apply Lemma 4.2.6 we show that the formulae (1) to (23) are Δ_0^T-formulae, where, as required, $T = \emptyset$ or T consists of the extensionality axiom as its unique member.

(1) $x \subseteq y$ is the Δ_0-formula $\forall z \in x(z \in y)$.

(2) $x = \emptyset$ is the Δ_0-formula $\forall z \in x(z \neq z)$.

(3) $x = \{a, b\}$ is the Δ_0-formula $\forall z \in x(z = a \vee z = b) \wedge a \in x \wedge b \in x$.

(4) $x = (a, b)$ is the formula $x = \{\{a\}, \{a, b\}\}$. If we rewrite this formula as in the proof of (3), then the subformulae $\{a\} \in x$ and $\{a, b\} \in x$ still bother us. But the last formula is $\exists y \in x(y = \{a, b\})$, a Δ_0-formula, and so is the first one.

(5) The formula $\exists u \exists v(x = \{u, v\})$ is equivalent to $\exists u \in x \exists v \in x(x = \{u, v\})$. By (3), it is absolute for M.

(6) The formula $\exists u \exists v(x = (u, v))$ is equivalent in T to the formula

$$\exists w \in x \exists z \in x \exists u \in w \exists v \in z(w = \{u\} \wedge z = \{u, v\} \wedge \forall a \in x(a = w \vee a = z)).$$

By (3), it is absolute for M.

(7) $\mathrm{Rel}(x)$ is the formula $\forall z \in x \exists u \exists v(z = (u, v))$. (6) says that it is absolute for M.

(8) The formula $x \in \mathrm{dom}(u)$ is equivalent in T to the formula $\exists v \in u\ \exists w \in v\ \exists y \in w(v = (x, y))$.

(9) $x = \mathrm{dom}(u)$ is equivalent in T to $\forall z \in x(z \in \mathrm{dom}(u)) \wedge \forall z \in \mathrm{dom}(u)(z \in x)$. By (8), we only have to take care of the second formula in the previous conjunction. Now every formula of the form $\forall z \in \mathrm{dom}(u)\ \psi$ is equivalent in T to the formula $\forall v \in u\ \forall w \in v\ \forall y \in w \forall z \in w(v = (z, y) \Longrightarrow \psi)$. Thus, if ψ is equivalent in T to a Δ_0-formula, then this also is true for the formula under consideration.

(10) and (11) The proof is analogous to that of (8) and (9).

(12) The considered formula is equivalent in T to $\forall z \in x\ \exists y \in u\ \exists w \in v(z = (y, w)) \wedge \forall y \in u\ \forall w \in v\ \exists z \in x(z = (y, w))$.

(16) The formula is equivalent in T to $\forall z \in x\ \exists w \in u(z \in w) \wedge \forall w \in u\ \forall z \in w(z \in x)$.

(17) $\mathrm{Func}(x)$ is an abbreviation of the formula $\mathrm{Rel}(x) \wedge \forall z \in \mathrm{dom}(x)\ \forall u, v \in \mathrm{ran}(x)((z, u) \in x \wedge (z, v) \in x \Longrightarrow u = v)$; now see the proof of (9).

(18) "x is transitive" is equivalent to the Δ_0-formula $\forall y \in x \forall z \in y(z \in x)$.

(19) $y = f(x) \iff \mathrm{Func}(f) \wedge \exists z \in f(z = (x, y))$.

(20) $y = f[x] \iff \mathrm{Func}(f) \wedge \forall z \in y\ \exists u \in x(z = f(u)) \wedge \forall u \in x\ (u \in \mathrm{dom}(f) \Longrightarrow \exists z \in y(z = f(u)))$.

(21) $g = f \restriction x \iff \mathrm{Func}(f) \wedge g \subseteq f \wedge \forall z \in x(z \in \mathrm{dom}(f) \Longrightarrow z \in \mathrm{dom}(g)) \wedge \forall z \in \mathrm{dom}(g)(z \in x)$. Now see the proof of (9).

(24) The formula $x \in \mathrm{ON}$ is an abbreviation of "x is transitive and every member of x is transitive". By (18), it is absolute for M.

(25) "x is a successor ordinal" is the formula $x \in \mathrm{ON} \wedge \exists v \in x(x = v \cup \{v\})$. Now apply (3) and (14).

(26) x is a limit ordinal iff x is an ordinal different from $\emptyset$ and x is not a successor ordinal; see (24), (2) and (25).

(27) x is a natural number iff $\neg\mathrm{Lim}(x) \wedge \forall z \in x\ \neg\mathrm{Lim}(z)$. The desired result follows from (26).

(28) $x = \omega$ is the formula

$$\forall z(z \in x \iff z \in \mathrm{ON} \wedge \neg\mathrm{Lim}(z) \wedge \forall u \in z\ \neg\mathrm{Lim}(u)),$$

which is equivalent to the conjunction of the Δ_0^T-formula $\forall z \in x(z \in \mathrm{ON} \wedge \neg\mathrm{Lim}(z) \wedge \forall u \in z\, \neg\mathrm{Lim}(u))$ and the formula $\varphi :\equiv \forall z(z \in \mathrm{ON} \wedge \neg\mathrm{Lim}(z) \wedge \forall u \in z\, \neg\mathrm{Lim}(u) \Longrightarrow z \in x)$. With the previous results we get $\varphi \Longrightarrow \varphi^M$ and thus $x = \omega \Longrightarrow (x = \omega)^M$. Now assume that $x \in M$ and $(x = \omega)^M$, hence φ^M holds. By (24) and (26), we have $\forall z \in x(z \in \mathrm{ON} \wedge \neg\mathrm{Lim}(z) \wedge \forall u \in z\, \neg\mathrm{Lim}(u))$. This gives $x \subseteq \omega$. Next we show that x is transitive. Then x is an ordinal with $x \leq \omega$, and if x would be a natural number, φ^M and $x \in M$ would imply $x \in x$, a contradiction. So assume that $v \in z$ and $z \in x$. Then $v \in M$ and $z \in M$. We have to show that $v \in x$. We have $\neg\mathrm{Lim}(z) \wedge \forall u \in z\, \neg\mathrm{Lim}(u)$. The transitivity of z gives $\neg\mathrm{Lim}(v) \wedge \forall u \in v\, \neg\mathrm{Lim}(u)$. From $v \in M$ and φ^M we can conclude that $v \in x$.

Definition *If $\varphi(x, \vec{y})$ is a formula, then we define the relativization of the class $A = A(\vec{y}) := \{x : \varphi(x, \vec{y})\}$ by $A(\vec{y})^M := \{x \in M : \varphi^M(x, \vec{y})\}$. A is called* **absolute for M** *iff $\forall \vec{y} \in M(A(\vec{y}) = A(\vec{y})^M)$ is provable in* ZF.

Remark If $A(\vec{y}) := \{x : \varphi(x, \vec{y})\}$, φ is absolute for M and $\vec{y} \in M$, then $A(\vec{y})^M = \{x \in M : \varphi^M(x, \vec{y})\} = A(\vec{y}) \cap M$. An important example is $A(y) := \mathcal{P}(y)$. From Lemma 4.2.8 we get $(A(y))^M = \{x \in M : (x \subseteq y)^M\} = \{x \in M : x \subseteq y\} = A(y) \cap M$. Thus the power set is in general **not** absolute for M.

If, for all $\vec{y} \in M$, $A(\vec{y}) \subseteq M$, and if φ is absolute for M, then A is absolute for M. Conversely, if for every $\vec{y} \in M$ we have $\{x : \varphi(x, \vec{y})\}^M = \{x : \varphi(x, \vec{y})\} \cap M$, then φ is absolute for M.

Lemma 4.2.9 *If M is transitive and closed under unordered pairing (i.e., $\{x, y\} \in M$ for all $x, y \in M$), then the following classes are absolute for M.*

$$\{x,y\}, \quad \bigcup x, \quad x \cup y,) \quad \bigcap x \ (\mathit{if}\ x \neq \emptyset), \quad x \cap y, \quad x \setminus y, \quad (x,y),$$

$$\mathrm{dom}(x), \quad \mathrm{ran}(x), \quad \emptyset, \quad \mathrm{fld}(x), \quad a \times b, \quad \in \cap\, a \times b, \quad \{(u,v) : (v,u) \in x\}.$$

If additionally $\omega \subseteq M$, then ω is absolute for M.

Proof The proof is easy if one uses Lemma 4.2.8 and the remark after the definition of an absolute class. For example, $(x, y)^M = \{z \in M : z = \{x\} \vee z = \{x, y\}\} \subseteq (x, y)$. By assumption we have $\{x\}, \{x, y\} \in M$ for all $x, y \in M$. Therefore we get $(x, y)^M = \{z : z = \{x\} \vee z = \{x, y\}\} = (x, y)$. Or, to give a further example, $\mathrm{dom}(x)^M = \{z \in M : \exists u \in M\ (z, u) \in x\} \subseteq \mathrm{dom}(x)$. If $(z, u) \in x$ and $x \in M$, then we infer from the fact that M is transitive that $u \in M$ and $z \in M$. This gives $z \in \mathrm{dom}(x)^M$. The other absoluteness assertions are verified analogously.

Remark 4.2.10

a) The following formulae and classes are in general not *absolute:*

$$y = \mathcal{P}(x),\ \mathcal{P}(x),\ |y| = |x|,\ |x|,\ x \textit{ is a cardinal number},\ \mathrm{cf}\,(x),\ x \textit{ is regular}.$$

However, by Lemma 4.2.11, the property of being a cardinal number is preserved in M. Intuitively, this is clear, since with fewer sets there are more cardinals, because there is less chance of a one-one correspondence.

b) For uncountable cardinals Θ, our models $\mathrm{H}(\Theta)$ satisfy very strong absoluteness properties, see Lemma 4.2.12.

Lemma 4.2.11 *If M is transitive, then*

$$\forall x \in M (x \in \mathrm{CN} \Longrightarrow M \models x \in \mathrm{CN}).$$

Proof Let $\varphi(x, f)$ be the formula

$$\mathrm{Func}(f) \wedge f \text{ is injective } \wedge \mathrm{dom}(f) = x \wedge \mathrm{ran}(f) \in x.$$

φ is absolute for M by Lemma 4.2.8. Further we have by the definition of a cardinal: $x \in \mathrm{CN} \iff x \in \mathrm{ON} \wedge \neg\exists f\ \varphi(x, f)$. Thus from $x \in \mathrm{CN}$ we get $x \in \mathrm{ON} \wedge \neg\exists f \in M\ \varphi(x, f)$, and from Lemma 4.2.8 we can infer $(x \in \mathrm{ON})^M \wedge \neg\exists f \in M\ \varphi^M(x, f)$. But this is the formula $(x \in \mathrm{CN})^M$.

Lemma 4.2.12 *If Θ is an uncountable regular cardinal, then $(\mathrm{H}(\Theta), \in)$ is a model of the theory ZFC^-, that is ZFC without the power set axiom. Furthermore, for arbitrary uncountable cardinals κ, the following formulae are absolute for $\mathrm{H}(\kappa)$:*

a) $x \in \mathrm{CN}$.
b) x is a regular cardinal.
c) $x \in \mathrm{CN} \wedge y \in \mathrm{CN} \wedge y = x^+$.
d) x is a limit ordinal and $y = \mathrm{cf}\,(x)$.

Proof We will prove that $\mathrm{H}(\Theta)$ is a model of all axioms of ZFC^-. For this we will use Lemma 4.1.3, Lemma 4.1.4 and the previous results on absoluteness. The validity of (A1), (A2), (A4) and (A5) is clear. As an exercise, we demonstrate this for (A5). The formula $y = \bigcup x$ is absolute for $\mathrm{H}(\Theta)$. Therefore we get

$$\begin{aligned} &\mathrm{H}(\Theta) \models \mathrm{A5} \iff \mathrm{H}(\Theta) \models \forall x \exists y \forall z (z \in y \iff \exists u \in x (z \in u)) \\ &\iff \mathrm{H}(\Theta) \models \forall x \exists y (y = \textstyle\bigcup x) \iff \forall x \in \mathrm{H}(\Theta) \exists y \in \mathrm{H}(\Theta) (\mathrm{H}(\Theta) \models y = \textstyle\bigcup x) \iff \\ &\forall x \in \mathrm{H}(\Theta) \exists y \in \mathrm{H}(\Theta) (y = \textstyle\bigcup x) \iff \forall x \in \mathrm{H}(\Theta) (\textstyle\bigcup x \in \mathrm{H}(\Theta)), \end{aligned}$$

where the last formula holds by Lemma 4.1.4 g).

(A7) holds in $\mathrm{H}(\Theta)$, since $\Theta > \aleph_0$ and thus, by Lemma 4.1.4, ω is a member of $\mathrm{H}(\Theta)$.

For the comprehension axioms (A3), let $\varphi(z,\vec{p})$ be a formula of $\mathcal{L}_{\mathrm{ZF}}$, and let $\vec{p} \in \mathrm{H}(\Theta)$. We want to show that $\mathrm{H}(\Theta) \models \forall x \exists y(y = \{z \in x : \varphi(z,\vec{p})\})$, that means

$$\forall x \in \mathrm{H}(\Theta)\ \exists y \in \mathrm{H}(\Theta)(\mathrm{H}(\Theta) \models y = \{z \in x : \varphi(z,\vec{p})\}).$$

Since $\mathrm{H}(\Theta)$ is transitive, the assertion $\mathrm{H}(\Theta) \models y = \{z \in x : \varphi(z,\vec{p})\}$ is true for $x, y \in \mathrm{H}(\Theta)$ iff $y = \{z \in x : \mathrm{H}(\Theta) \models \varphi(x,\vec{p})\}$. By Lemma 4.1.4 e), the set $\{z \in x : \mathrm{H}(\Theta) \models \varphi(x,\vec{p})\}$ is a member of $\mathrm{H}(\Theta)$ since it is a subset of $x \in \mathrm{H}(\Theta)$.

For the replacement axioms (A8), let $\varphi(x,y)$ be a formula of $\mathcal{L}_{\mathrm{ZF}}$. We have to show that $\mathrm{H}(\Theta) \models \forall x \forall y \forall z(\varphi(x,y) \wedge \varphi(x,z) \implies y = z)$ implies $\mathrm{H}(\Theta) \models \forall x \exists y \forall z(z \in y \iff \exists u \in x\ \varphi(u,z))$. The last assertion is equivalent to $\forall x \in \mathrm{H}(\Theta) \exists y \in \mathrm{H}(\Theta)(y = \{z \in \mathrm{H}(\Theta) : \exists u \in x\ \varphi^{\mathrm{H}(\Theta)}(u,z)\})$. Fix $x \in \mathrm{H}(\Theta)$. The premiss says that the formula $\varphi^{\mathrm{H}(\Theta)}$ is functional on $\mathrm{H}(\Theta)$. So we can apply a replacement axiom and obtain a set y such that $y = \{z \in \mathrm{H}(\Theta) : \exists u \in x\ \varphi^{\mathrm{H}(\Theta)}(u,z)\}$. Then $y \subseteq \mathrm{H}(\Theta)$ and $|y| < \Theta$, since $|y| \leq |x| < \Theta$. Since Θ is regular, Lemma 4.1.3 yields $y \in \mathrm{H}(\Theta)$.

For the axiom of choice, we observe that the formula "f is a choice function for x" is absolute for $\mathrm{H}(\Theta)$. Let $x \in \mathrm{H}(\Theta)$ be a set of nonempty sets. Using (AC), we obtain a choice function f for x. Since $f \subseteq x \times \bigcup x$, Lemma 4.1.4 yields $f \subseteq \mathrm{H}(\Theta)$. Now we can infer from $|f| \leq |x|$ together with Lemma 4.1.3 that $f \in \mathrm{H}(\Theta)$.

Now we turn to the proof of the desired absoluteness results. For the proof of part a), fix $x \in \mathrm{H}(\kappa)$. Note that $(x \in \mathrm{CN} \implies \mathrm{H}(\kappa) \models x \in \mathrm{CN})$ by Lemma 4.2.11. For the converse, assume that $\mathrm{H}(\kappa) \models x \in \mathrm{CN}$. Then we have $x \in \mathrm{ON}$. Assume, to get a contradiction, that $x \notin \mathrm{CN}$. If we take the formula $\varphi_1(x,f)$ as "f is a one-one function $\wedge$ $\mathrm{dom}(f) = x \wedge \mathrm{ran}(f) \in x$", then we can infer $\exists f\ \varphi_1(x,f)$. Since $\mathrm{ran}(f) \in x$, we have $\mathrm{ran}(f) \in \mathrm{H}(\kappa)$, which gives $f \in \mathrm{H}(\kappa)$ by Lemma 4.1.4. Now we come to the contradiction $\mathrm{H}(\kappa) \models x \notin \mathrm{CN}$, since φ_1 is absolute for $\mathrm{H}(\kappa)$.

For part b), observe that the definition of regularity says that x is a cardinal and not regular iff

$$x \in \mathrm{ON} \wedge \neg\exists f\ \varphi_1(x,f) \wedge \exists y \in x \exists g(g : y \longrightarrow x \wedge \bigcup \mathrm{ran}(g) = x).$$

Now we can proceed as in the proof of a).

For c), let $x, y \in \mathrm{H}(\kappa)$ and observe that $x \in \mathrm{CN} \wedge y \in \mathrm{CN} \wedge y = x^+$ is the formula

$$x \in \mathrm{CN} \wedge y \in \mathrm{CN} \wedge x \in y \wedge \forall z\ (z \in \mathrm{CN} \wedge x \in z \implies y \leq z),$$

which we abbreviate as $\varphi_2(x,y)$. From Lemma 4.2.11 we know that $(\varphi_2(x,y) \implies \mathrm{H}(\kappa) \models \varphi_2(x,y))$. Assume that $\mathrm{H}(\kappa) \models \varphi_2(x,y)$. Then, by a), x and y are cardinal numbers. If $z \in \mathrm{CN}$ and $x \in z$, then $y \leq z$, since $z \in y$ would give $z \in \mathrm{H}(\kappa)$ and therefore $y \leq z$. This proves c).

For the proof of d), let $x, y \in \mathrm{H}(\kappa)$ and

$$\varphi_3(y,x) :\equiv x, y \in \mathrm{ON} \wedge y \leq x \wedge \exists g\, (g : y \longrightarrow x \wedge \bigcup \mathrm{ran}(g) = x).$$

Then "x is a limit ordinal and $y = \mathrm{cf}(x)$" is formalized by $\varphi_4(x, y)$ where

$$\varphi_4(y,x) :\equiv \mathrm{Lim}(x) \wedge \varphi_3(y,x) \wedge \forall z\, (\varphi_3(z,x) \Longrightarrow y \leq z).$$

Assume that $\mathrm{H}(\kappa) \models \varphi_4(y,x)$. Our absoluteness results yield $\varphi_3(y,x)$ and $\mathrm{Lim}(x)$. If $z \in \mathrm{ON}$ such that $\varphi_3(z,x)$, then $z \leq x$ gives $z \in \mathrm{H}(\kappa)$ and thus $y \leq z$. So we have shown that $(\mathrm{H}(\kappa) \models \varphi_4(y,x)) \Longrightarrow \varphi_4(y,x)$. Conversely assume $\varphi_4(y,x)$. Then there is a function $g : y \longrightarrow x$ such that $\bigcup \mathrm{ran}(g) = x$. Lemma 4.1.4 says that $g \in \mathrm{H}(\kappa)$, our absoluteness results give $\mathrm{H}(\kappa) \models \varphi_4(y,x)$, and d) is proved.

Lemma 4.2.13 *Assume that Θ is a regular cardinal, N is a set and $a_0, \ldots, a_{n-1}$ are members of N. Then the following holds.*

a) If the $\mathcal{L}_{\mathrm{ZF}}$-formula $\varphi(x_0, \ldots, x_n)$ is absolute for $\mathrm{H}(\Theta)$, N is an elementary substructure of $\mathrm{H}(\Theta)$ and $\varphi(\vec{a}, a)$ holds for some $a \in \mathrm{H}(\Theta)$, then there is $b \in N$ such that $\varphi(\vec{a}, b)$. In particular we have: If there is a unique $a \in \mathrm{H}(\Theta)$ such that $\varphi(\vec{a}, a)$, then $a \in N$.

If $\varphi(x_0, \ldots, x_n)$ is an arbitrary formula, and if there is a unique member u of $\mathrm{H}(\Theta)$ such that $\mathrm{H}(\Theta) \models \varphi(a_0, \ldots, a_{n-1}, u)$, then $u \in N$.

b) If the $\mathcal{L}_{\mathrm{ZF}}$-formula $\varphi(x_0, \ldots, x_n)$ is absolute for $\mathrm{H}(\Theta)$, $<^*$ is a well-ordering of $\mathrm{H}(\Theta)$, $(N, \in, <^*)$ is an elementary substructure of $(\mathrm{H}(\Theta), \in, <^*)$, and b is the $<^*$-least element of $\mathrm{H}(\Theta)$ such that $\varphi(\vec{a}, b)$, then $b \in N$.

c) If $\varphi(x_0, \ldots, x_n)$ is a formula of $\mathcal{L}^*_{\mathrm{ZF}}$, $<^*$ is a well-ordering of $\mathrm{H}(\Theta)$, $(N, \in, <^*)$ is an elementary substructure of $(\mathrm{H}(\Theta), \in, <^*)$, and b is the $<^*$-least element of $\mathrm{H}(\Theta)$ such that $\mathrm{H}(\Theta) \models \varphi(\vec{a}, b)$, then $b \in N$.

Proof For the proof of a), we observe first that $\mathrm{H}(\Theta) \models \exists x_n \varphi[\vec{a}]$ since $\varphi(\vec{x}, x_n)$ is absolute for $\mathrm{H}(\Theta)$. From $N \preceq \mathrm{H}(\Theta)$, Lemma 4.2.2 and the absoluteness of φ again we can conclude that the first assertion in a) is true. The second assertion follows from Lemma 4.2.2.

Since b) follows immediately from c), let us turn to the proof of part c). Let ψ be the $\mathcal{L}^*_{\mathrm{ZF}}$-formula $\exists x_n(\varphi(\vec{x}, x_n) \wedge \forall y(y < x_n \Longrightarrow \neg\varphi(\vec{x}, y)))$. Then we have by assumption

$$(*) \qquad \mathrm{H}(\Theta) \models \psi[\vec{a}].$$

Furthermore, b is the unique member of $\mathrm{H}(\Theta)$ with the property

$$\mathrm{H}(\Theta) \models (\varphi(\vec{x}, x_n) \wedge \forall y(y < x_n \Longrightarrow \neg\varphi(\vec{x}, y)))\ [\vec{a}, b].$$

Since $(N, \in, <^*)$ is an elementary substructure of $(\mathrm{H}(\Theta), \in, <^*)$, we can infer from $(*)$ and Lemma 4.2.2 that there is $c \in N$ such that

$$\mathrm{H}(\Theta) \models (\varphi(\vec{x}, x_n) \wedge \forall y(y < x_n \Longrightarrow \neg\varphi(\vec{x}, y)))\ [\vec{a}, c].$$

Obviously this gives $b = c \in N$.

Definition *If $\mathcal{N} = (N, \in)$ is an elementary substructure of $\mathrm{H}(\Theta)$, $\varphi(x, \vec{p})$ is a formula of $\mathcal{L}_{\mathrm{ZF}}$, $p_0, \ldots, p_{n-1}$ are members of N, and the set b is defined by*

$$b := \{x \in \mathrm{H}(\Theta) : \mathrm{H}(\Theta) \models \varphi[x, \vec{p}]\},$$

then we say that b is **definable** *in $\mathrm{H}(\Theta)$ by a formula* **with parameters from $\mathcal{N}$**.

Lemma 4.2.14 *If $\mathcal{N} = (N, \in)$ is an elementary substructure of $\mathrm{H}(\Theta)$, and $b \in \mathrm{H}(\Theta)$ is definable in $\mathrm{H}(\Theta)$ by a formula of $\mathcal{L}_{\mathrm{ZF}}$ with parameters $\vec{p}$ from N, then $b \in N$.*

Furthermore, there is a formula ψ such that b is the unique member u of $\mathrm{H}(\Theta)$ which satisfies $\mathrm{H}(\Theta) \models \psi[u, \vec{p}]$.

Proof Let $b := \{a \in \mathrm{H}(\Theta) : \mathrm{H}(\Theta) \models \varphi(x, \vec{y})[a, \vec{p}]\}$ where the parameters of φ are members of N, and let $b \in \mathrm{H}(\Theta)$. Obviously we have $\mathrm{H}(\Theta) \models \exists z \forall x(x \in z \Longleftrightarrow \varphi(x))[\vec{p}]$, since $b \in \mathrm{H}(\Theta)$. Thus there is a member $c \in N$ such that $\mathrm{H}(\Theta) \models \forall x(x \in c \Longleftrightarrow \varphi(x))[\vec{p}]$. Now we get $c = \{a \in \mathrm{H}(\Theta) : \mathrm{H}(\Theta) \models \varphi(x, \vec{y})[a, \vec{p}]\}$, and $b = c$ gives $b \in N$. ψ can be taken as the formula $\forall x(x \in z \Longleftrightarrow \varphi(x, \vec{y}))$.

Example 4.2.15 With the help of the following examples, the reader should become familiar with the previous results. At the same time, they are important components of various proofs in later sections and chapters, and we will often refer to them. Θ denotes a regular cardinal. Often we use the absoluteness results from Lemma 4.2.8 without mentioning.

a) Assume that N is an elementary substructure of $\mathrm{H}(\Theta)$.

First we will show that the formulae $|x| \leq |y|$ and $|x| = |y|$ are absolute for $\mathrm{H}(\Theta)$. In the theory ZFC^-, the formula $|y| \leq |x|$ is equivalent to the formula

$$\psi :\equiv y = \emptyset \vee \exists f(f : x \longrightarrow y \wedge \mathrm{ran}(f) = y).$$

We know that the formula $f : x \longrightarrow y \wedge \mathrm{ran}(f) = y$ is absolute for $\mathrm{H}(\Theta)$. So, assuming $x, y \in \mathrm{H}(\Theta)$, we get $(\mathrm{H}(\Theta) \models \psi) \Longrightarrow \psi$; furthermore we have $f \in \mathrm{H}(\Theta)$ by Lemma 4.1.4 for any function $f : x \longrightarrow y$. Thus we get $\psi \Longrightarrow (\mathrm{H}(\Theta) \models \psi)$, hence the formula ψ is absolute for $\mathrm{H}(\Theta)$. Since $\mathrm{H}(\Theta)$ is a model of ZFC^-, Lemma 4.2.6 e) says that the formula $|y| \leq |x|$

is also absolute for $\mathrm{H}(\Theta)$. An analogous argument works for the formula $|x| = |y|$.

Now fix $b \in N$. We want to show that $|b| \in N$. The previous results show that the formula

$$|y| \le |x| \wedge x \in \mathrm{ON} \wedge \forall z \in x\ \neg(|y| \le |z|)$$

is absolute for $\mathrm{H}(\Theta)$. Since it is equivalent in the theory ZFC^- to $x = |y|$, this formula is also absolute for $\mathrm{H}(\Theta)$. Since $b \in \mathrm{H}(\Theta)$ gives $|b| \in \mathrm{H}(\Theta)$, there is a unique $x \in \mathrm{H}(\Theta)$ satisfying $x = |b|$. Thus Lemma 4.2.13 a) tells us that $|b| \in N$. Consequently we have shown:

Every elementary substructure of $\mathrm{H}(\Theta)$ has as members all cardinalities of its elements.

b) Let N be an elementary substructure of $\mathrm{H}(\Theta)$. Using complete induction it is easy to see that $\omega \subseteq N$ and $\mathrm{V}_\omega \subseteq N$.

From the Lemmas 4.2.12 and 4.2.13 we can conclude that the following sets are members of N *if they are members of* $\mathrm{H}(\Theta)$.

ω;

$\alpha + 1$ for any ordinal $\alpha \in N$; α for any ordinal $\alpha + 1 \in N$.

κ^+ for any cardinal $\kappa \in N$, hence $\aleph_n$ for $n \in \omega$; κ for any cardinal $\kappa^+ \in N$.

$\mathrm{cf}(\alpha)$ for any ordinal $\alpha \in N$.

$\max(b)$, $\min(b)$, $\sup(b)$ for any $b \in N$ such that $b \subseteq \mathrm{ON}$.

$\mathrm{dom}(f)$, $\mathrm{ran}(f)$, $f \restriction b$ for any $b \in N$ and any relation $f \in N$.

$f(b)$ for any function $f \in N$ and any $b \in N \cap \mathrm{dom} f$.

c) Assume that the L^*-structure $(M, \in, <^*)$ is an elementary substructure of $(\mathrm{H}(\Theta), \in, <^*)$, and let $b \in M$. By a), we have $|b| \in M$. Any function from b onto $|b|$ is an element of $\mathrm{H}(\Theta)$ by Lemma 4.1.4, and thus M has a member which is a bijection from b onto $|b|$. Now Lemma 4.2.8 says that the formula $\varphi(x, y, z) :\equiv$ (x is a bijection from y onto z) is absolute for $\mathrm{H}(\Theta)$. So we can infer, using Lemma 4.2.13 b):

$$\boldsymbol{b \in M \wedge h \in \mathrm{H}(\Theta)} \textbf{ is the } \boldsymbol{<^*}\textbf{-least bijection from } \boldsymbol{b} \textbf{ onto } \boldsymbol{|b| \Longrightarrow h \in M}.$$

d) Assume that N is an L-structure and an elementary substructure of $\mathrm{H}(\Theta)$. First we get, using b) and c):

If $\alpha \in N$ is a limit ordinal such that $\mathrm{cf}(\alpha) \subseteq N$, then $\sup(N \cap \alpha) = \alpha$.

To see this, observe first that $\mathrm{cf}(\alpha) \in N$. Since N is an elementary substructure of $\mathrm{H}(\Theta)$, there exists a function $f : \mathrm{cf}(\alpha) \longrightarrow \alpha$ such that $f \in N$, $\sup(\mathrm{ran}(f)) = \alpha$ and $\mathrm{ran}(f) \subseteq N$.

Now assume that $h, b, c \in N$ have the property that h is a bijection from b onto c. In addition to c) we want to show some further results concerning N.

Since $\mathrm{H}(\Theta) \models \varphi[h, b, c]$, we get $N \models \varphi[h, b, c]$, since the "parameters" are members of N and N is an elementary substructure of $\mathrm{H}(\Theta)$. We "translate" this according to the definition of $\models$. First, the assertion $N \models \forall u \in b \exists! v \in c(h(u) = v)$ is equivalent to $\forall u \in N \cap b \exists! v \in N \cap c(h(u) = v)$, and similarly $N \models \forall v \in c \exists! u \in b(h(u) = v)$ is equivalent to $\forall v \in N \cap c \exists! u \in N \cap b(h(u) = v)$. Thus, for any $h, b, c \in N$, the following holds.

$$\boldsymbol{h} \textbf{ is a bijection from } \boldsymbol{b} \textbf{ onto } \boldsymbol{c} \iff$$

$$\boldsymbol{h \restriction (N \cap b)} \textbf{ is a bijection from } \boldsymbol{b \cap N} \textbf{ onto } \boldsymbol{c \cap N}.$$

e) In part c), let $(N, \in, <^*)$ be a further elementary substructure of $(\mathrm{H}(\Theta), \in, <^*)$ and take $b = \gamma \in \mathrm{ON} \cap M \cap N$ and $\gamma \geq \aleph_0$. Then $\kappa := |\gamma| \in M \cap N$. If we assume that $M \cap \kappa = N \cap \kappa$, then we can infer from d) for the $<^*$-least bijection $h : \kappa \longrightarrow \gamma$ that $h \in M \cap N$ and $M \cap \gamma = h[M \cap \kappa] = h[N \cap \kappa] = N \cap \gamma$. Thus we have shown under the given assumptions: **If $\boldsymbol{\kappa}$ is a cardinal such that $\boldsymbol{\kappa \in M \cap N}$ and $\boldsymbol{M \cap \kappa = N \cap \kappa}$, then $\boldsymbol{M \cap \gamma = N \cap \gamma}$ for all $\boldsymbol{\gamma \in M \cap N \cap \kappa^+}$.**

f) The proof of the following assertion is a prototype of our procedure in later chapters. For the set $\mathrm{pcf}(\mathsf{a})$ it makes precise the following intuition: *Given finitely many sets,* for example $\mathcal{J}_{<\lambda}(\mathsf{a})$, $\max \mathrm{pcf}(\mathsf{b})$, $\mathrm{pcf}(\mathsf{c})$ etc., *we can choose a regular cardinal* Θ *"large enough" such that these sets are members of any elementary substructure N of* $\mathrm{H}(\Theta)$ *which contains the parameters* λ, a, b, c *as members.* The properties of $\mathrm{H}(\Theta)$ will guarantee that the considered sets are members of $\mathrm{H}(\Theta)$, and if we can characterize them uniquely by a formula which is absolute in a restricted sense (see below) and thus allows the proof that the sets have the form $\{x \in \mathrm{H}(\Theta) : \mathrm{H}(\Theta) \models \varphi[x, \vec{p}]\}$ with $\vec{p} \in N$, then they will be definable in $\mathrm{H}(\Theta)$ with parameters from N. So Lemma 4.2.14 will say that they are members of N.

Claim If N is an elementary substructure of $\mathrm{H}(\Theta)$, $\mathsf{a} \in N$ is a set of regular cardinals, and $2^{\sup(\mathsf{a})} < \Theta$, then $\mathrm{pcf}(\mathsf{a}) \in N$. If in addition $\lambda \in \mathrm{pcf}(\mathsf{a}) \cap N$, then $\mathcal{J}_{<\lambda}(\mathsf{a}) \in N$.

We want to show that Θ is large enough such that $\mathcal{P}(\mathsf{a})$, $\prod \mathsf{a}$ and $\mathrm{pcf}(\mathsf{a})$ are members of $\mathrm{H}(\Theta)$. Then all objects occurring in the definition of $\mathrm{pcf}(\mathsf{a})$ will be members of $\mathrm{H}(\Theta)$. We repeatedly apply Lemma 4.1.3 and Lemma 4.1.4. Since $\mathrm{H}(\Theta) \cap \mathrm{ON} = \Theta$, we get $x \subseteq \mathrm{H}(\Theta)$ for any $x \subseteq \mathsf{a}$, $\mathcal{P}(\mathsf{a}) \subseteq \mathrm{H}(\Theta)$ and thus $\mathcal{P}(\mathsf{a}) \in \mathrm{H}(\Theta)$ since $|\mathcal{P}(\mathsf{a})| \leq 2^{\sup(\mathsf{a})} < \Theta$, and also $I \in \mathrm{H}(\Theta)$ for any ideal I on a. The fact that $\mathsf{a}, \sup(\mathsf{a}) \in \mathrm{H}(\Theta)$, hence

$\mathsf{a} \times \sup(\mathsf{a}) \in \mathrm{H}(\Theta)$, yields in the same way that $\mathcal{P}(\mathsf{a} \times \sup(\mathsf{a})) \in \mathrm{H}(\Theta)$. Since $\prod \mathsf{a}$ is a subset of this set, we get $\prod \mathsf{a} \in \mathrm{H}(\Theta)$. Next we observe that $\mathrm{pcf}(\mathsf{a})$ is a member of $\mathrm{H}(\Theta)$, since for every $\lambda \in \mathrm{pcf}(\mathsf{a})$ we have $\lambda < |\prod \mathsf{a}| + 1 \in \mathrm{H}(\Theta)$; furthermore this gives $f \in \mathrm{H}(\Theta)$ for any $f : \lambda \longrightarrow \prod \mathsf{a}$ by Lemma 4.1.4.

Now let $\varphi(x, a)$ be the formula

$$x \in \mathrm{Reg} \wedge \exists I \exists z \exists f (z = \prod a \wedge I \text{ is an ideal on } a \wedge \mathrm{cof}(f, x, z, I)),$$

where the formula $\mathrm{cof}(f, x, z, I)$ is given by

$$f : x \longrightarrow z \wedge \forall \alpha < x \forall \beta < x (\alpha < \beta \Longrightarrow f_\alpha <_I f_\beta) \wedge \forall g \in z \exists \alpha < x (g <_I f_\alpha).$$

The reader may convince himself that any subformula of $\varphi(x, a)$ in which no quantifier occurs is absolute for $\mathrm{H}(\Theta)$. But we have shown above that all objects under quantification are members of $\mathrm{H}(\Theta)$ if we consider $a = \mathsf{a}$. So we have proved:

$$\begin{aligned} & a \subseteq \mathrm{Reg} \wedge 2^{\sup(a)} < \Theta \\ & \quad \Rightarrow \mathrm{pcf}(a) \in \mathrm{H}(\Theta) \wedge \forall x \in \mathrm{H}(\Theta)(\varphi(x, a) \Leftrightarrow \mathrm{H}(\Theta) \models \varphi(x, a)). \end{aligned}$$

Since $\mathrm{pcf}(\mathsf{a}) = \{x \in \mathrm{H}(\Theta) : \varphi(x, \mathsf{a})\}$, we can infer using our assumption on Θ that

$$\mathrm{pcf}(\mathsf{a}) = \{x \in \mathrm{H}(\Theta) : \mathrm{H}(\Theta) \models \varphi(x, \mathsf{a})\}.$$

Now Lemma 4.2.14 says that $\mathrm{pcf}(\mathsf{a}) \in N$ since $\mathsf{a} \in N$.

To prove the second assertion of the claim, let $\lambda \in \mathrm{pcf}(\mathsf{a}) \cap N$. Under the assumptions on a and Θ we have shown that $\mathsf{b} \subseteq \mathsf{a} \Longrightarrow (\mathrm{pcf}(\mathsf{b}) \subseteq \lambda \Longleftrightarrow \mathrm{H}(\Theta) \models \mathrm{pcf}(\mathsf{b}) \subseteq \lambda)$. Consequently, $\mathcal{J}_{<\lambda}(\mathsf{a}) = \{\mathsf{b} \subseteq \mathsf{a} : \mathrm{pcf}(\mathsf{b}) \subseteq \lambda\} = \{\mathsf{b} \in \mathrm{H}(\Theta) : \mathrm{H}(\Theta) \models (\mathsf{b} \subseteq \mathsf{a} \wedge \mathrm{pcf}(\mathsf{b}) \subseteq \lambda)\}$ is definable in $\mathrm{H}(\Theta)$ with parameters from N and thus a member of N.

Lemma 4.2.16 (Collection Principle) *Let $\varphi(x, \vec{y})$ be a formula of $\mathcal{L}_{\mathrm{ZF}}$. Then*

$$\mathrm{ZF} \vdash \forall S_0 \exists S \supseteq S_0 \, \forall \vec{y} \in S (\exists x \varphi(x, \vec{y}) \Longrightarrow \exists x \in S \; \varphi(x, \vec{y})).$$

With the axiom of choice, we have in addition $|S| \leq \aleph_0 \cdot |S_0|$.

Proof If A is a class, then let $A^* := \{x \in A : \forall y \in A(\mathrm{rk}(x) \leq \mathrm{rk}(y))\}$ be the set of members of A which have minimal rank in the von Neumann hierarchy. A^* is set, since there is an ordinal α such that $A^* \subseteq \mathrm{V}_\alpha$. In particular, for our given formula φ and for every $\vec{y}$, the class $A^*_{\vec{y}}$ is also a set where $A_{\vec{y}} := \{x : \varphi(x, \vec{y})\}$. For any class X we put $\phi(X) := X \cup \bigcup\{A^*_{\vec{y}} : \vec{y} \in X\}$. If X is a set, then $\phi(X)$ is also a set. Now fix a set S_0 and let, for each $n \in \omega$,

$S_{n+1} := \phi(S_n)$ and $S := \bigcup\{S_n : n \in \omega\}$. Clearly we have $S_0 \subseteq S$. Consider $\vec{y} \in S$, that means $\vec{y} \in S_n$ for some $n \in \omega$, such that $\exists x \varphi(x, \vec{y})$. Then $A^*_{\vec{y}} \neq \emptyset$, und furthermore every $z \in A^*_{\vec{y}}$ is a member of $\phi(S_n) = S_{n+1} \subseteq S$. Thus there exists $z \in S$ such that $\varphi(z, \vec{y})$, as desired.

By (AC), we can take a choice function f for $\mathcal{P}(S) \setminus \{\emptyset\}$ and define $\phi'(X) := X \cup \{f(A^*_{\vec{y}}) : \vec{y} \in X\}$, $S'_0 := S_0$, $S'_{n+1} := \phi'(S'_n)$ and $S' := \bigcup\{S'_n : n \in \omega\}$. It is easy to see by induction that $|S'_n| \leq \aleph_0 \cdot |S_0|$ for all $n \in \omega$. So the set S' instead of S has the desired properties.

Remark Let α be an ordinal. We can choose the set S from the collection principle as a von Neumann level V_β such that β is a limit ordinal with $\alpha < \beta$. To see this, take in the construction of S an ordinal β_0 such that $\alpha < \beta_0$ and $S_0 \subseteq V_{\beta_0}$. If β_n is defined, choose $\beta_{n+1} > \beta_n$ such that $\phi(V_{\beta_n}) \subseteq V_{\beta_{n+1}}$ and put $\beta = \sup\{\beta_n : n \in \omega\}$.

Corollary 4.2.17 *If φ and ψ are formulae whose free variables occur in the sequence $\vec{y}$, then the following sentence is a theorem of* ZF*:*

$$\begin{aligned}&\forall S_0 \exists S \supseteq S_0 \ \forall \vec{y} \in S((\exists x \varphi(x, \vec{y}) \\ &\Rightarrow \exists x \in S \ \varphi(x, \vec{y})) \wedge (\exists x \psi(x, \vec{y}) \Rightarrow \exists x \in S \ \psi(x, \vec{y}))).\end{aligned}$$

An analogous assertion is provable for finitely many formulae $\varphi_1, \ldots, \varphi_n$.

Proof We make the same construction as above with

$$\phi(X) := X \cup \bigcup\{A^*_{\vec{y}} : \vec{y} \in X\} \cup \bigcup\{B^*_{\vec{y}} : \vec{y} \in X\},$$

where $B^*_{\vec{y}} := \{x : \psi(x, \vec{y})\}$.

Theorem 4.2.18 (Reflection Principle) *If φ is a formula of $\mathcal{L}_{\mathrm{ZF}}$, then*

$$\mathrm{ZF} \vdash \forall M_0 \exists M (M_0 \subseteq M \wedge \forall \vec{y} \in M(\varphi(\vec{y}) \Longleftrightarrow M \models \varphi(\vec{y}))).$$

Then we say that $\boldsymbol{M}$ **reflects** $\boldsymbol{\varphi}$.

Proof Let $\varphi_1, \ldots, \varphi_n$ be the subformulae of φ such that φ_n is the formula φ. Furthermore choose M according to Corollary 4.2.17 such that

$$\vec{y} \in M \Longrightarrow (\exists z \varphi_i(z, \vec{y}) \Longleftrightarrow \exists z \in M \ \varphi_i(z, \vec{y}))$$

for $i = 1, \ldots, n$. By induction on the definition of a formula we prove that M reflects every formula φ_i. Firstly, every atomic formula is reflected by every set since it is identical with its relativization. If φ_i and φ_j are reflected by M, then we can conclude at once that M reflects $\neg\varphi_i$ and $(\varphi_i \vee \varphi_j)$. Now assume that φ_i has the form $\exists z \varphi_j$. By the choice of M we have for any $\vec{y} \in M$:

$$\exists z \varphi_j(z, \vec{y}) \Longleftrightarrow \exists z \in M \ \varphi_j(z, \vec{y}).$$

The inductive hypothesis gives

$$\exists z \in M\ \varphi_j(z,\vec{y}) \Longleftrightarrow \exists z \in M(M \models \varphi_j(z,\vec{y})) \Longleftrightarrow M \models \exists z \varphi_j(z,\vec{y}) \Longleftrightarrow M \models \varphi_i.$$

Thus we get $\varphi_i \Longleftrightarrow (M \models \varphi_i)$ for all $\vec{y} \in M$, as desired. Now the assertion of the theorem follows immediately if we take φ_i as the subformula φ of φ.

Corollary 4.2.19 *In the preceding theorem, we can choose the set M as a suitable set V_α for some $\alpha \in \mathrm{ON}$.*

Exercises

1) Prove Lemma 4.2.1.
2) Give an example for isomorphic structures $\mathcal{A}$ and $\mathcal{B}$ such that $\mathcal{A} \subseteq \mathcal{B}$ and $\mathcal{A} \not\preceq \mathcal{B}$.
   Hint: Consider $(\omega \setminus \{0\}, \in)$ and $(\omega, \in)$ and take the formula $\varphi(v_0) \equiv \neg\exists v_1(v_1 \in v_0)$.
3) If $\mathcal{A}$ and $\mathcal{B}$ are structures for L, then we say that $\mathcal{A}$ is **elementary equivalent** to $\mathcal{B}$ (or that $\mathcal{A}$ and $\mathcal{B}$ are elementary equivalent) iff, for every sentence φ of L, $\mathcal{A} \models \varphi \Longleftrightarrow \mathcal{B} \models \varphi$. If $\mathcal{A}$ is an elementary substructure of $\mathcal{B}$, then clearly $\mathcal{A}$ and $\mathcal{B}$ are elementary equivalent.
   a) Show that, for any ordinals α and β such that α is a successor and β is a limit ordinal, $(\alpha, \in)$ is not elementary equivalent to $(\beta, \in)$.
   b) Assume that β is a limit ordinal with $\beta > \omega$. Show that $(\omega, \in)$ is not elementary equivalent to $(\beta, \in)$.
   c) Show that $(\omega \cdot \omega, \in)$ is not elementary equivalent to $(\aleph_1, \in)$.
   d) Show that $(\omega^\omega, \in)$ is not elementary equivalent to $(\aleph_1, \in)$.
4) Prove that, for regular cardinals Θ, the formula $z = \prod a$ is absolute for $\mathrm{H}(\Theta)$.
5) Prove: If $2^{\aleph_0} = \aleph_{\omega+1}$, then the formula $y = \mathrm{pcf}(x)$ is not absolute for $\mathrm{H}(\aleph_{\omega+1})$.
6) Prove: If $\mathcal{N} = (N, \in)$ is an elementary substructure of $\mathrm{H}(\Theta)$, $b \in N$, and Θ is a regular cardinal such that $2^{|b|} < \Theta$, then $\mathcal{P}(b) \in N$.
7) Show that $\mathrm{H}(\aleph_\omega)$ is not a model of all replacement axioms.
   Hint: Let $\aleph = \{(u,v) : \varphi(u,v)\}$ be the aleph function. Show first, using transfinite induction, that $\varphi(u,v)$ is absolute for $\mathrm{H}(\kappa)$, where κ is an uncountable cardinal. Then consider the function defined by $\psi(u,v) :\Longleftrightarrow u \in \omega \wedge \varphi(u,v)$, i.e., $\psi(u,v) \Longleftrightarrow u \in \omega \wedge v = \aleph_u$. If $\mathrm{H}(\aleph_\omega)$ was a model for the replacement axiom for ψ, then we would have $\{\aleph_n : n \in \omega\} \in \mathrm{H}(\aleph_\omega)$ which yields $\aleph_\omega \in \mathrm{H}(\aleph_\omega)$, contradicting Lemma 4.1.4.
8) The beth function $\beth : \mathrm{ON} \longrightarrow \mathrm{ICN}$ is known from Section 1.7. Prove: If $\kappa > \omega$ is a cardinal, then $\mathrm{H}(\kappa)$ is a model of the power set axiom iff

$\kappa = \beth(\gamma)$ for some limit ordinal γ. Show further that $H(\beth(\omega))$ is not a model of all replacement axioms.

Hint: For the second assertion, follow the hint in the previous question and take the beth function instead of the aleph function. From the first assertion one can conclude that the formula defining the beth function is absolute for $H(\aleph_\omega)$.

9) Assume that κ is an inaccessible cardinal number. Prove in ZF (in ZFC) that V_κ is a model of all axioms of ZF (of ZFC).

Hint: See the exercises in Section 1.3 and Exercise 1.7.6 and prove first that, for all sets x, we have $x \in V_\kappa \iff x \subseteq V_\kappa \wedge |x| < \kappa$. Then follow the proof of Lemma 4.2.12.

10) Assume that $\kappa > \omega$ is a regular cardinal. Prove in ZFC that κ is inaccessible iff $H(\kappa)$ is a model of ZFC.

Hint: Use, for example, a previous question and Exercise 1.7.7.

11) Let λ be a singular and κ be an infinite cardinal. Prove that $H(\lambda)$ is not elementary equivalent to $H(\kappa^+)$.

Hint: Consider the formula $\exists y(y \in \mathrm{CN} \wedge \forall z(z \in \mathrm{CN} \implies z \neq y^+))$ and use the absoluteness results from Lemma 4.2.12.

12) We call a formula ψ of ZF a **Σ_1-formula** iff it has the form $\exists x_0 \ldots \exists x_{n-1}\, \varphi$ for some Δ_0-formula φ. The case $n = 0$ shows that every Δ_0-formula is a Σ_1-formula. Prove that for each Σ_1-formula ψ there is a Δ_0-formula φ such that the formula $\exists u\varphi \iff \psi$ is provable in ZF without using the power set axiom.

13) Assume that $\varphi(u, x_1, \ldots, x_n)$ is a Δ_0-formula. Prove: If κ is an uncountable cardinal, then the formula $\exists u\varphi$ is absolute for $H(\kappa)$.

Hint: Observe that it suffices to prove

$$\forall x_1 \ldots \forall x_n \in H(\lambda^+)(\exists u\varphi \implies \exists u \in H(\lambda^+)\, \varphi).$$

For this, use the reflection principle (which yields, for given sets u and $a_1, \ldots, a_n \in H(\lambda^+)$, some V_α such that $V_\alpha \models \varphi(u, a_1, \ldots, a_n)$), the Löwenheim-Skolem theorem (to obtain a sufficiently small elementary substructure A of V_α) and the Mostowski lemma (to get an $\in$-isomorphism from A onto a transitive set $B \in H(\lambda^+)$).

14) Assume that κ is an uncountable cardinal. Prove that each Σ_1-formula is absolute for $H(\kappa)$.

15) **(Levy)** Assume that $\psi(x, x_1, \ldots, x_n)$ is a Σ_1-formula. Prove: If κ is an uncountable cardinal, then

$$\forall x_1 \in H(\kappa) \ldots \forall x_n \in H(\kappa)(\exists x\psi \implies \exists x \in H(\kappa)\, \psi).$$

Hint: Use the previous questions.

4.3 Approximation Sequences

Definition *A sequence* $(\mathcal{A}_i : i < \gamma)$ *of L-structures or* L^**-structures is called a* **chain** *iff* $\mathcal{A}_i \subseteq \mathcal{A}_j$ *for all i and j with* $i < j < \gamma$. *It is called an* **elementary chain** *iff* $\mathcal{A}_i \preceq \mathcal{A}_j$ *for all i and j with* $i < j < \gamma$. *If* $(\mathcal{A}_i : i < \gamma)$ *is a chain and* $\mathcal{A}_i = (A_i, \in, <_i^*)$ *for all* $i < \gamma$, *we denote the* **union** *of the chain by* $\bigcup\{\mathcal{A}_i : i < \gamma\}$. *It is defined as that structure* $\mathcal{A} = (A, \in, <^*)$ *for which* $A = \bigcup\{A_i : i < \gamma\}$ *and* $<^* = \bigcup\{<_i^* : i < \gamma\}$. *We say that* $(\mathcal{A}_i : i < \gamma)$ *is* **continuous** *iff* $\mathcal{A}_j = \bigcup\{\mathcal{A}_i : i < j\}$ *for all limit ordinals* $j < \gamma$.

Analogously, for any family $(\mathcal{A}_i : i \in I)$ *of substructures of* $(\mathrm{H}(\Theta), \in, <^*)$, *we call* $\bigcap\{\mathcal{A}_i : i \in I\}$ *the* **intersection** *of this family. It is defined as that structure* $\mathcal{A} = (A, \in, <_A^*)$ *for which* $A = \bigcap\{A_i : i \in I\}$ *and* $<_A^* = <^* \cap (A \times A)$.

Definition *For any set M, we define the* **characteristic function of** M, *written as* χ_M, *by*

$$\chi_M(\alpha) := \sup(M \cap \alpha) \quad \textit{for every } \alpha \in \mathrm{ON}.$$

A sequence $(\mathcal{N}_i : i < \gamma)$ *of elementary substructures of* $\mathrm{H}(\Theta)$ *is called an* **approximation sequence in H(Θ)** *iff*

(1) If $i + 1 < \gamma$, *then* $\mathcal{N}_i$ *is an elementary substructure of* $\mathcal{N}_{i+1}$.

(2) If $i < \gamma$ *is a limit ordinal, then* $\mathcal{N}_i = \bigcup\{\mathcal{N}_j : j < i\}$.

(3) If $i + 1 < \gamma$, *then* $(N_j : j \leq i) \in N_{i+1}$.

The first two properties just say that every approximation sequence is a continuous elementary chain.

Often, we will also denote the structure $\mathcal{N}$ *by* N. *The context will make it clear which notion will be meant.*

Lemma 4.3.1 *Assume that* Θ *is a regular cardinal,* μ *is an infinite cardinal with* $\mu < \Theta$, *and* σ *is an ordinal satisfying* $\sigma \leq \mu^+$. *Then there exists, for each* $Y \in \mathrm{H}(\Theta)$ *with* $|Y| \leq \mu$ *and each sequence* $(X_i : i < \sigma) \in \mathrm{H}(\Theta)$ *satisfying* $|X_i| \leq \mu$ *for all* $i < \sigma$, *an approximation sequence* $(N_i : i \leq \sigma)$ *such that* $Y \subseteq N_0$, $X_i \subseteq N_{i+1}$ *and* $|N_i| \leq \mu$ *for all* $i < \sigma$.

Proof The assertion follows easily by recursion from the Löwenheim-Skolem theorem 4.2.3. This yields first an elementary substructure N_0 of $\mathrm{H}(\Theta)$ such that $Y \subseteq N_0$ and $|N_0| \leq \max\{|Y|, \aleph_0\}$. Assume that the sequence $(N_j : j < i)$ is already defined and has the desired properties. If i is a limit ordinal, then, by Lemma 4.2.2, the union N_i of all structures N_j, $j < i$, is an elementary substructure of $\mathrm{H}(\Theta)$ and satisfies our claims. If $i = j + 1$, then $i \in \mathrm{H}(\Theta)$, since $i < \mu^+ \leq \Theta$, and thus we can infer from Lemmas 4.1.3 and 4.1.4 that also the sequence $(N_k : k < i)$ is a member of $\mathrm{H}(\Theta)$. Again Theorem 4.2.3 tells us that there is an elementary substructure N_{j+1} of $\mathrm{H}(\Theta)$ such that $(N_k : k \leq j) \in$

N_{j+1}, $X_j \subseteq N_{j+1}$, $|N_{j+1}| \leq \max\{|N_j|, |X_j|, \aleph_0\} \leq \mu$, and $N_j \subseteq N_{j+1}$. Since N_j is an elementary substructure of $\mathrm{H}(\Theta)$, Lemma 4.2.1 says that it is also an elementary substructure of N_{j+1}. This completes the definition of the desired sequence.

Remark 4.3.2 Assume that Θ is a regular cardinal, κ is an uncountable regular cardinal, and σ is a limit ordinal such that $\sigma < \kappa$ and $\kappa \in \mathrm{H}(\Theta)$. Further let $Y \in \mathrm{H}(\Theta)$, $|Y| < \kappa$, and $(X_i : i < \sigma) \in \mathrm{H}(\Theta)$ be a sequence satisfying $|X_i| < \kappa$ for all $i < \sigma$. Then we can define, similar to the previous construction, an approximation sequence $(N_i : i \leq \sigma)$ such that $Y \subseteq N_0$, $X_j \subseteq N_{j+1}$, $|N_i| < \kappa$ and $N_i \cap \kappa \in \kappa$ for all $i \leq \sigma$.

To prove this, we will construct N_{j+1} from N_j. N_0 is defined analogously, and the limit step is the same as above. N_{j+1} will be the union of a countable chain $(M_n : n \in \omega)$ of elementary substructures of $\mathrm{H}(\Theta)$ obtained by a standard construction, a so-called ω-iteration, which we already used in the proof of the Löwenheim-Skolem theorem. Choose $M_0 \preceq \mathrm{H}(\Theta)$ according to Theorem 4.2.3 as a superstructure of N_j such that $X_j \subseteq M_0$ and $|M_0| < \kappa$, and let M_n with $|M_n| < \kappa$ be already defined. Since κ is regular, we have $\alpha_n := \sup(M_n \cap \kappa) < \kappa$, and again Theorem 4.2.3 yields a superstructure M_{n+1} of M_n such that $\alpha_n + 1 \subseteq M_{n+1}$, $M_{n+1} \preceq \mathrm{H}(\Theta)$ and $|M_{n+1}| < \kappa$. Then the structure $N_{j+1} := \bigcup\{M_n : n \in \omega\}$ is an elementary substructure of $\mathrm{H}(\Theta)$ as the union of the elementary chain $(M_n : n \in \omega)$. Furthermore we get $|N_{j+1}| \leq \aleph_0 \cdot \sup\{|M_n| : n \in \omega\} < \kappa$. If $\alpha \in \beta \in N_{j+1} \cap \kappa$, then there is some $n \in \omega$ such that $\beta \in M_n \cap \kappa$. By the definition of M_{n+1}, we have $\alpha \in M_{n+1} \subseteq N_{j+1}$. Therefore $N_{j+1} \cap \kappa$ is an ordinal, since it is a transitive set of ordinals.

The following lemma presents some simple properties of approximation sequences. Part d) will be a practical tool for the construction of characterizing functions.

Lemma 4.3.3 *Assume that M is a set, Θ is a regular cardinal, σ is a limit ordinal with $\sigma < \Theta$, $(N_i : i \leq \sigma)$ is an approximation sequence in $\mathrm{H}(\Theta)$, and $N := N_\sigma$. Then*

a) If $|M| < \mathrm{cf}(\alpha)$, then $\chi_M(\alpha) < \alpha$.

b) For all $i < \sigma$, we have $N_i \in N_{i+1}$, $\sigma \subseteq N$, and $\{N_i : i < \sigma\} \subseteq N$.

c) If $i < \sigma$ and $\alpha \in N_i$, then $\chi_{N_i}(\alpha) \in N_{i+1}$; if in addition $\chi_{N_i}(\alpha) < \alpha$, then $\chi_{N_i}(\alpha) < \chi_{N_{i+1}}(\alpha)$.

d) If $\tau \in N_0$ is a limit ordinal such that $|N_i| < \mathrm{cf}(\tau)$ for all $i < \sigma$, then $(\chi_{N_i}(\tau) : i < \sigma)$ is a normal function satisfying $\sup\{\chi_{N_i}(\tau) : i < \sigma\} = \chi_N(\tau)$, all of whose values are members of $N \cap \tau$. In particular we have

$$\mathrm{cf}(\chi_N(\tau)) = \mathrm{cf}(\sigma).$$

Proof Part a) is trivial. For b), fix $i < \sigma$. Then, by definition, we have $f := (N_j : j \leq i) \in N_{i+1}$, and Example 4.2.15 gives $i + 1 = \mathrm{dom}(f) \in N_{i+1}$, $i \in N_{i+1}$, and $N_i = f(i) \in N_{i+1}$, as desired.

For the proof of c), let $i < \sigma$ and $\alpha \in N_i$. Then we have $\alpha \in N_{i+1}$, and $N_i \in N_{i+1}$ by b). Since $N_{i+1} \preceq N$ and there is a unique member $\delta \in \mathrm{H}(\Theta)$ such that $\delta = \sup(N_i \cap \alpha)$, we get $\chi_{N_i}(\alpha) = \sup(N_i \cap \alpha) \in N_{i+1}$ by Lemma 4.2.13. Furthermore, $\delta \in N_{i+1}$ implies $\delta + 1 \in N_{i+1}$, hence $\chi_{N_i}(\alpha) < \chi_{N_{i+1}}(\alpha)$ if $\chi_{N_i}(\alpha) < \alpha$.

For d), we have just shown that $\chi_{N_i}(\tau)$ is smaller than $\chi_{N_{i+1}}(\tau)$. If $i \leq \sigma$ is a limit ordinal, we can infer with the continuity of approximation sequences that

$$\begin{aligned} \sup\{\chi_{N_j}(\tau) : j < i\} &= \bigcup\{\bigcup(N_j \cap \tau) : j < i\} \\ &= \bigcup(\bigcup\{N_j : j < i\} \cap \tau) = \bigcup(N_i \cap \tau) = \chi_{N_i}(\tau). \end{aligned}$$

Thus $(\chi_{N_i}(\tau) : i < \sigma)$ is a normal function which is cofinal in $\chi_N(\tau)$ and has all its values in $N \cap \tau$.

A typical application of this lemma is the following important lemma which we need for later purposes.

Lemma 4.3.4 *Assume that* a *is an infinite set of regular cardinals satisfying* $|\mathsf{a}|^+ < \min(\mathsf{a})$, Θ *is a regular cardinal,* $\mu \in (|\mathsf{a}|, \min(\mathsf{a}))_{\mathrm{reg}}$ *and* $(N_i : i \leq \mu)$ *is an approximation sequence in* $\mathrm{H}(\Theta)$ *such that* $\mathsf{a} \cup \{\mathsf{a}\} \subseteq N_0$ *and* $|N_i| \leq \mu$ *for all* $i < \mu$. *Further let* τ *be a limit ordinal such that* $\mu < \mathrm{cf}(\tau)$ *and* $\tau \in N_0$, $N := N_\mu$, *and* $(f_\xi : \xi < \tau) \in N_0$ *be a sequence of members of* $\prod \mathsf{a}$ *such that*

$$f_{\chi_N(\tau)}(\nu) = \min\{\sup\{f_\beta(\nu) : \beta \in C\} : C \textit{ is a club in } \chi_N(\tau)\}$$

for every $\nu \in \mathsf{a}$. *Then the following holds.*

a) There is a strictly increasing continuous sequence $(i(\beta) : \beta < \mu)$ *of limit ordinals in* μ *such that* $f_{\chi_N(\tau)}(\nu) = \sup\{f_{\chi_{N_{i(\beta)}}(\tau)}(\nu) : \beta < \mu\}$ *for all* $\nu \in \mathsf{a}$, *and we have*

$$f_{\chi_N(\tau)} \leq_{\mathsf{a}} \chi_N \restriction \mathsf{a}.$$

b) If $2^{\sup(\mathsf{a})} < \Theta$ *and* $(f_\xi : \xi < \tau)$ *is cofinal in* $\prod \mathsf{a}$ *modulo some ultrafilter* D *on* a *with* $D \in N_0$, *then for any* $i < \mu$ *there is an ordinal* $\xi_0 \in N_{i+1} \cap \tau$ *such that*

$$\chi_{N_i} \restriction \mathsf{a} <_D f_{\xi_0} <_D f_{\chi_{N_{i+1}}(\tau)}.$$

c) If $2^{\sup(\mathsf{a})} < \Theta$, $\mathsf{b} \in \mathcal{P}(\mathsf{a}) \cap N_0$, *and* $\{f_\xi \restriction \mathsf{b} : \xi < \tau\}$ *is cofinal in* $(\prod \mathsf{b}, \leq_{\mathsf{b}})$, *then for any* $i < \mu$ *there is an ordinal* $\xi_0 \in N_{i+1} \cap \tau$ *such that*

$$\chi_{N_i} \restriction \mathsf{b} <_{\mathsf{b}} f_{\xi_0} \restriction \mathsf{b}.$$

Proof For a), observe that $\mathrm{cf}(\chi_N(\tau)) = \mathrm{cf}(\sup\{\chi_{N_i}(\tau) : i < \mu\}) = \mu$ by the preceding lemma, and thus $\chi_N(\tau) < \tau$. Let, for each $\nu \in \mathsf{a}$, C_ν be a club in $\chi_N(\tau)$ such that $f_{\chi_N(\tau)}(\nu) = \sup\{f_\beta(\nu) : \beta \in C_\nu\}$. Then $C := \bigcap\{C_\nu : \nu \in \mathsf{a}\}$ is also a club in $\chi_N(\tau)$. If E is any subclub of C in $\chi_N(\tau)$, then the assumption on $f_{\chi_N(\tau)}(\nu)$ yields that $f_{\chi_N(\tau)}(\nu) = \sup\{f_\beta(\nu) : \beta \in E\}$ for all $\nu \in \mathsf{a}$. So we can assume without loss of generality that $\mathrm{otp}(C) = \mu$. Lemma 4.3.3 tells us that $\{\chi_{N_i}(\tau) : i < \mu\}$ is a club in $\chi_N(\tau)$. Taking its intersection with C, we obtain a strictly increasing continuous sequence $(i(\beta) : \beta < \mu)$ of limit ordinals in μ such that for all $\nu \in \mathsf{a}$

$$(*) \qquad f_{\chi_N(\tau)}(\nu) = \sup\{f_{\chi_{N_{i(\beta)}}(\tau)}(\nu) : \beta < \mu\}.$$

Now we have $(f_\xi : \xi < \tau) \in N_0$, hence $\mathrm{dom}((f_\xi : \xi < \tau)) = \tau \in N$. Since $\chi_{N_{i(\beta)}}(\tau) \in N$, we conclude that $f_{\chi_{N_{i(\beta)}}(\tau)} \in N$, hence $f_{\chi_{N_{i(\beta)}}(\tau)}(\nu) \in N \cap \nu$ for all $\nu \in \mathsf{a}$, since $\mathsf{a} \subseteq N$. This gives $f_{\chi_{N_{i(\beta)}}(\tau)}(\nu) \leq \sup(N \cap \nu) = \chi_N(\nu)$, as desired.

For the proof of b), fix $i < \mu$. Then $N_i \in N_{i+1}$, and from $\mathsf{a} \in N_{i+1}$ and $|N_i| \leq \mu < \min(\mathsf{a})$ it follows, using Lemma 4.2.14, that $\chi_{N_i} \restriction \mathsf{a} \in N_{i+1} \cap \prod \mathsf{a}$. Furthermore we have $D \in N_{i+1}$. Since $\Theta > 2^{\sup(\mathsf{a})}$, Example 4.2.15 says that $\prod \mathsf{a} \in \mathrm{H}(\Theta)$ and that the property of being cofinal in $\prod \mathsf{a}$ modulo D is absolute for $\mathrm{H}(\Theta)$. So we can infer from the assumption, Lemma 4.2.13 and $N_{i+1} \preceq \mathrm{H}(\Theta)$ that there exists an ordinal $\xi_0 \in N_{i+1} \cap \tau$ such that $\chi_{N_i} \restriction \mathsf{a} <_D f_{\xi_0}$. Clearly $\xi_0 < \chi_{N_{i+1}}(\tau)$ and thus $\chi_{N_i} \restriction \mathsf{a} <_D f_{\xi_0} <_D f_{\chi_{N_{i+1}}(\tau)}$.

c) can be proved in analogy to the proof of b), considering the function $\chi_{N_i} \restriction \mathsf{b} + 1$.

4.4 The Skolem Hull in $\mathrm{H}(\Theta)$

In this section, Θ denotes a regular cardinal. We associate with each L^*-formula $\varphi(x_0, \ldots, x_n)$ an L^*-formula φ^* as follows.

$$\begin{aligned} &\varphi^*(x_0, x_1, \ldots, x_n) \\ &:\equiv \varphi(x_0, \ldots, x_n) \wedge \forall x_{n+1}(x_{n+1} < x_0 \implies \neg\varphi(x_{n+1}, x_1, \ldots, x_n)). \end{aligned}$$

Consequently, if $<^*$ is a well-ordering of $\mathrm{H}(\Theta)$, $a_0, \ldots, a_n \in \mathrm{H}(\Theta)$, and if $\mathrm{H}(\Theta) \models \varphi^*(a_0, \ldots, a_n)$, then a_0 is the $<^*$-least element $a \in \mathrm{H}(\Theta)$ so that $\mathrm{H}(\Theta) \models \varphi(a, a_1, \ldots, a_n)$.

Further let $\bigwedge_{i=1}^n \varphi_i$ be the conjunction $\varphi_1 \wedge \ldots \wedge \varphi_n$ and $\bigvee_{i=1}^n \varphi_i$ be the disjunction $\varphi_1 \vee \ldots \vee \varphi_n$ of the formulae $\varphi_1, \ldots, \varphi_n$. The property "elementary substructure" refers in this section always to the language L^*.

Lemma 4.4.1 *If $\mathcal{E}$ is a nonempty set of elementary substructures of* $\mathrm{H}(\Theta)$*, then* $\bigcap\mathcal{E}$ *is an elementary substructure of* $\mathrm{H}(\Theta)$.

Proof Certainly we have $\bigcap\mathcal{E} \neq \emptyset$, since for example the empty set is a member of every elementary substructure of $\mathrm{H}(\Theta)$. Furthermore, $\bigcap\mathcal{E}$ is a substructure of $\mathrm{H}(\Theta)$.

Consider $a_1, \ldots, a_n \in \bigcap\mathcal{E}$ such that

$$\exists b \in \mathrm{H}(\Theta)(\mathrm{H}(\Theta) \models \varphi(b, a_1, \ldots, a_n)),$$

for a fixed L^*-formula φ. To show that $\bigcap\mathcal{E}$ is an elementary substructure of $\mathrm{H}(\Theta)$ it suffices to show, using Lemma 4.2.2, that there is some "satisfier" b in $\bigcap\mathcal{E}$.

Now there is a unique $a \in \mathrm{H}(\Theta)$ such that $\mathrm{H}(\Theta) \models \varphi^*(a, a_1, \ldots, a_n)$. If $N \in \mathcal{E}$, then $a_1, \ldots, a_n \in N$, and thus Lemma 4.2.13 yields $a \in N$, since $N \preceq \mathrm{H}(\Theta)$. This gives $a \in \bigcap\mathcal{E}$.

This lemma justifies the following definition.

Definition *If* $X \subseteq \mathrm{H}(\Theta)$*, then let*

$$\mathrm{Skol}_\Theta(X) := \bigcap\{N : N \preceq \mathrm{H}(\Theta) \land X \subseteq N\}$$

be the **Skolem hull** *of* X. *Obviously* Skol_Θ *is a hull operator on* $\mathrm{H}(\Theta)$*, and* $\mathrm{Skol}_\Theta(X)$ *is the smallest elementary substructure of* $\mathrm{H}(\Theta)$ *whose universe is a superset of* X.

For each L^**-formula* $\varphi(x_0, \ldots, x_n)$ *in which* x_n *occurs free, let* f_φ *be that function from* $\mathrm{H}(\Theta)^n$ *into* $\mathrm{H}(\Theta)$ *which assigns to each tupel* $(a_1, \ldots, a_n)$ *the* $<^*$*-least element* $a \in \mathrm{H}(\Theta)$ *with* $\mathrm{H}(\Theta) \models \varphi(a, a_1, \ldots, a_n)$*, that is the unique* $a \in \mathrm{H}(\Theta)$ *such that* $\mathrm{H}(\Theta) \models \varphi^*(a, a_1, \ldots, a_n)$*, if there is such an* a*; otherwise, we take* 0 *as the value of* f_φ *at* $(a_1, \ldots, a_n)$*. We say that* f_φ *is a* **Skolem function for** φ.

If F *is an* n*-ary function (i.e.,* F *is a function such that* $\mathrm{dom}(F) \subseteq \mathrm{V}^n$*) and* A *is a set, then we put*

$$F''A := F[\,A^n \cap \mathrm{dom}(F)\,].$$

Remark This definition is a bit ambiguous since every function is certainly 1-ary on an appropriate set. However, if there are $n, m \in \omega \setminus \{0\}$ with $m \neq n$, such that F is m-ary *and* n-ary, then $m = 1$ or $n = 1$ (if $F \neq \emptyset$). In this case the previous definition refers to the greater one of the numbers n and m.

Next we will obtain a simple method to determine Skolem hulls.

Lemma 4.4.2 *If $X \subseteq \mathrm{H}(\Theta)$, then*

$$\mathrm{Skol}_\Theta(X) = \bigcup\{f''_\varphi X : \varphi \text{ is an } L^*\text{-formula}\}.$$

Proof Put $X_0 := X$, $X_{n+1} := \bigcup\{f''_\varphi X_n : \varphi \text{ is an } L^*\text{-formula}\}$ for all $n < \omega$, and $X^+ := \bigcup\{X_n : n < \omega\}$. Then we know that $(X_n : n < \omega)$ is an ascending chain of sets and X^+ is an elementary substructure of $\mathrm{H}(\Theta)$; this is just the usual proof of the Löwenheim-Skolem theorem if we take as a satisfier always the $<^*$-smallest satisfier. By the definition of the Skolem functions and Lemma 4.2.13, we even have $X^+ = \mathrm{Skol}_\Theta(X)$. Therefore it suffices to show that $X_2 = X_1$, because this yields $X_n = X_1$ for all $n \geq 2$ and thus $X^+ = X_1$, as desired.

Consider $a_1, \ldots, a_n \in X_1$, and let $a = f_\varphi(a_1, \ldots, a_n)$ for some L^*-formula φ. Then we can find, for $i = 1, \ldots, n$, members $a^i_1, \ldots, a^i_{m_i}$ of X_0 and an L^*-formula φ_i such that $a_i = f_{\varphi_i}(a^i_1, \ldots, a^i_{m_i})$. Now put $s_0 := 0$ and $s_i := \sum_{k=1}^i m_k$ for $i = 1, \ldots, n$, and let ψ be the formula

$$\exists x_{s_n+1} \ldots \exists x_{s_n+n} \, (\varphi^*(x_0, x_{s_n+1}, \ldots, x_{s_n+n}) \wedge \bigwedge_{i=1}^n \varphi_i^*(x_{s_n+i}, x_{s_{i-1}+1}, \ldots, x_{s_i})),$$

Then we have $a = f_\psi(a^1_1, \ldots, a^1_{m_1}, a^2_1, \ldots, a^2_{m_2}, \ldots, a^n_1, \ldots, a^n_{m_n}) \in X_1$.

Lemma 4.4.3 *Assume that N is an elementary substructure of $\mathrm{H}(\Theta)$ and that $\gamma, A \in N$ satisfy $|A| < \mathrm{cf}(\gamma)$. If N' is the Skolem hull of $N \cup A$, then*

$$\chi_{N'}(\gamma) = \chi_N(\gamma).$$

Proof We can assume without loss of generality that γ is a limit ordinal, since otherwise $A = \emptyset$, hence $N = N'$. Clearly we have $\chi_{N'}(\gamma) \geq \chi_N(\gamma)$.

To prove the reverse inequality, let $\beta \in N' \cap \gamma$. By the above lemma, there are $a_1, \ldots, a_n \in N \cup A$ and an L^*-formula $\varphi(x_0, \ldots, x_n)$ such that $\beta = f_\varphi(a_1, \ldots, a_n)$. Let $\{b_1, \ldots, b_m\} := \{a_1, \ldots, a_n\} \cap N$; then $\beta \in A^+ := f''_\varphi(A \cup \{b_1, \ldots, b_m\})$. Note that $A^+ \subseteq \mathrm{H}(\Theta)$ and $|A| < \Theta$, hence $A^+ \in \mathrm{H}(\Theta)$. Now let ψ be the formula

$$\exists y_1 \ldots \exists y_n \, (\varphi^*(x_0, y_1, \ldots, y_n) \wedge \bigwedge_{i=1}^n (y_i \in x_1 \vee \bigvee_{j=1}^m y_i = x_{j+1})).$$

Then $a \in A^+$ iff $\mathrm{H}(\Theta) \models \psi[a, A, b_1, \ldots, b_m]$, i.e., A^+ is definable in $\mathrm{H}(\Theta)$ with parameters from N, hence $A^+ \in N$ by Lemma 4.2.14.

From $|A| < \mathrm{cf}(\gamma)$ we can infer that $A^+ \cap \gamma$ is a bounded subset of γ. Therefore there is a $<^*$-minimal ordinal $\zeta \in \gamma$ such that $A^+ \cap \gamma \subseteq \zeta$. Lemma 4.2.13

shows that $\zeta \in N$, hence $\beta < \zeta < \chi_N(\gamma)$. So we get $\sup(N' \cap \gamma) \leq \chi_N(\gamma)$, as desired.

With the previous lemma we now can conclude that we can enlarge elementary substructures of $\mathrm{H}(\Theta)$ without changing the values of their characteristic functions at sufficiently large arguments.

Corollary 4.4.4 *Assume that N is an elementary substructure of $\mathrm{H}(\Theta)$, $\kappa \in N$ is a regular cardinal, and $\alpha < \chi_N(\kappa)$. If N' is the Skolem hull of $N \cup \{\alpha\}$, then*

$$\chi_{N'}(\kappa) = \chi_N(\kappa).$$

Proof Choose some $\zeta \in N \cap \kappa$ with $\alpha < \zeta$. Applying the above lemma with $A = \zeta$ and $\gamma = \kappa$, we get $\chi_{\mathrm{Skol}_\Theta(N \cup \zeta)}(\kappa) = \chi_N(\kappa)$. Since N' is a substructure of $\mathrm{Skol}_\Theta(N \cup \zeta)$, the desired assertion is proved.

4.5 ◇-Club Sequences

For regular cardinals μ and κ with $\mu < \kappa$, we got acquainted in Lemma 1.8.4 with the stationary subset $\Sigma_{\kappa,\mu}$ of κ:

$$\Sigma_{\kappa,\mu} := \{\alpha \in \kappa : \mathrm{cf}(\alpha) = \mu\}.$$

Definition *If $A \subseteq \mathrm{ON}$, then $\overline{A} := \{\bigcup b : b \subseteq A \wedge b \neq \emptyset\}$ is called the* **hull** *of A with respect to the order topology on* ON.

Remark If $A \subseteq \mathrm{ON}$ is a set whose order type is a limit ordinal, then A has the same order type as the set $A^* := \overline{A} \setminus \{\sup(A)\}$. To see this, define the function $f : A^* \longrightarrow A$ by $f(\alpha) := \min\{\beta \in A : \beta > \alpha\}$. First we notice that every element α of A^* is smaller than $\sup(A)$, hence $f(\alpha) \in A$ is defined. If $\alpha_1, \alpha_2 \in A^*$ and $\alpha_1 < \alpha_2$, then, by the definition of $\overline{A}$ (we have $\alpha_2 = \sup(b)$ for some $b \subseteq A$), there is $\beta \in A$ such that $\alpha_1 < \beta \leq \alpha_2$. Therefore $f(\alpha_1) \leq \beta \leq \alpha_2 < f(\alpha_2)$, and f turns out to be strictly monotone. Together with $A \subseteq \overline{A}$ this gives $\mathrm{otp}(A^*) \leq \mathrm{otp}(A) \leq \mathrm{otp}(\overline{A}) = \mathrm{otp}(A^*)+1$. So we can infer by assumption that $\mathrm{otp}(A^*) = \mathrm{otp}(A)$.

Definition *Assume that μ and κ are uncountable regular cardinals with $\mu < \kappa$ and S is a subset of $\Sigma_{\kappa^+,\mu}$ which is stationary in κ^+. We call a sequence $(C_\alpha : \alpha \in S)$ an* **ecl-sequence**[3] *for S iff, for all $\alpha, \beta \in S$,*

(i) C_α is a club in α of order type μ.

(ii) If γ is a common limit point of C_α and C_β, then $C_\alpha \cap \gamma = C_\beta \cap \gamma$.

[3] Equal at common limit points.

Remark To prove the existence of ecl-sequences, we use for the first time the method of approximation sequences to get a powerful combinatorial result.

Lemma 4.5.1 *Let κ and Θ be uncountable regular cardinals such that $\kappa^+ < \Theta$ and $\tau < \kappa$ a limit ordinal. Further assume that $(N_i : i \leq \tau)$ is an approximation sequence in $\mathrm{H}(\Theta)$ such that $\kappa \in N_0$, $|N_i| < \kappa$ for all $i < \tau$, and $N_i \cap \kappa \in \mathrm{ON}$ for all $i \leq \tau$ (such a sequence exists by Remark 4.3.2). Then all limit points of N_τ in κ^+ whose cofinality differs from $\mathrm{cf}(\tau)$ are members of N_τ.*

Proof Let $N := N_\tau$, and let γ be such a limit point, that means that $\gamma < \kappa^+$, $\mathrm{cf}(\tau) \neq \mathrm{cf}(\gamma)$ and $\sup(N \cap \gamma) = \gamma$. By the definition of an approximation sequence, we have $\gamma = \sup\{\sup(N_i \cap \gamma) : i < \tau\}$, hence $\mathrm{cf}(\gamma) \leq \mathrm{cf}(\tau)$ and thus $\mathrm{cf}(\gamma) < \mathrm{cf}(\tau)$. Consequently, having chosen a sequence of length $\mathrm{cf}(\gamma)$ cofinal in $N \cap \gamma$, we observe that all its members are already members of a set N_α for some $\alpha < \tau$, and furthermore we get $\sup(N_\alpha \cap \gamma) = \gamma$. Since α, N_α and κ^+ are, by Lemma 4.3.3 and Example 4.2.15, members of N, we can conclude from Lemma 4.2.13 that also $A := \overline{N_\alpha \cap \kappa^+} \in N$, since the hull of x is absolute for $\mathrm{H}(\Theta)$. Again by Example 4.2.15, the cardinal $\lambda := |A|$ and a bijection h from λ onto A turn out to be members of N. Since $|N_\alpha| < \kappa$, we can infer with the remark following the definition of the hull of a set that $\lambda < \kappa$. Thus, if $\lambda \in N_\beta$ for some β, our assumption $N_\beta \cap \kappa \in \mathrm{ON}$ gives $\lambda \subseteq N_\beta \subseteq N$. From 4.2.15 we get $A = h[\lambda] \subseteq N$, and $\sup(N_\alpha \cap \gamma) = \gamma$ yields $\gamma \in A$, hence $\gamma \in N$.

Lemma 4.5.2 *Assume that μ and κ are regular cardinals with $\max\{\aleph_1, \mu\} < \kappa$. Then there is a set $S \subseteq \sum_{\kappa^+,\mu}$ stationary in κ^+ and an ecl-sequence for S.*

Proof First let us explain why Example 4.2.15 e) motivates us to investigate sets $N_i \cap i$ for suitable elementary substructures N_i of $\mathrm{H}(\Theta)$, where $i \in \Sigma_{\kappa^+,\mu}$, $\kappa \in N_i$ and Θ is large enough. If $|N_i| < \kappa$ and $N_i \cap \kappa \in \mathrm{ON}$ for all $i \in \Sigma_{\kappa^+,\mu}$, then $N_i \cap \kappa \in \kappa$, and $\sum_{\kappa^+,\mu}$ is the union of the sets $X_\sigma := \{i \in \sum_{\kappa^+,\mu} : N_i \cap \kappa = \sigma\}$, $\sigma < \kappa$. Since $\Sigma_{\kappa^+,\mu}$ is stationary in κ^+, Lemma 1.8.4 says that one of these sets is stationary in κ^+, too; call it $S_1 := X_{\sigma_0}$. If $\alpha, \beta \in S_1$, then the mentioned example shows that $N_\alpha \cap \gamma = N_\beta \cap \gamma$ for all $\gamma \in N_\alpha \cap N_\beta \cap \kappa^+$. With Example 4.2.15 d) we can infer for $\alpha \in S_1$ that $N_\alpha \cap \alpha$ is unbounded in α, if we can only realize $\alpha \in N_\alpha$ and $\mu = \mathrm{cf}(\alpha) \subseteq N_\alpha$. However, $N_\alpha \cap \alpha$ will not be a club in α. On the other hand, if we pass to the hull of this set to obtain a club, we will leave N_α, and our starting idea will be no more applicable. But Lemma 4.5.1 takes care that an appropriate N_α has enough limit points amongst its members. Namely, we can choose N_α as the union of an approximation sequence of length $\aleph_1 < \kappa$ satisfying the assumptions in 4.5.1, such that in addition $\{\alpha, \kappa\} \subseteq N_\alpha$ and $\mu = \mathrm{cf}(\alpha) \subseteq N_\alpha$ (which is possible by Lemma 4.3.1 since $\mu < \kappa$). Then N_α contains as members all those limit points of N_α in κ^+ which have countable cofinality. Furthermore it is not difficult to verify that, for the proof of property (ii) in the definition of a ecl-sequence, it suffices to show this property for limit points with countable cofinality. Namely, if γ is a common limit point of the clubs C_α and C_β with $\mathrm{cf}(\gamma) > \aleph_0$, then γ is the supremum of a sequence of common limit points of countable cofinality (see the "hand-over-hand" argument in the proof of Lemma 1.8.2 a)), and the validity of (ii) for these points yields at once $C_\alpha \cap \gamma = C_\beta \cap \gamma$.

Let us choose N_α as above and take $D_\alpha := \overline{N_\alpha \cap \alpha} \setminus \{\alpha\}$ for $\alpha \in S_1$. Certainly D_α is a club in α, but unfortunately D_α has not yet order type μ. However, with the remark following the definition of the hull, we get $\mathrm{otp}(D_\alpha) = \mathrm{otp}(N_\alpha \cap \alpha) < \kappa$, since $|N_\alpha| < \kappa$. Again the set S_1 is stationary in κ^+ and is the union of the sets $Y_\sigma := \{\alpha \in S_1 : \mathrm{otp}(N_\alpha \cap \alpha) = \sigma\}$, $\sigma < \kappa$. Therefore one of these sets, say $S := Y_\delta$, is stationary in κ^+. Take $\alpha \in S$. Then $\mathrm{otp}(N_\alpha \cap \alpha) = \delta$, $\alpha \in S_1 \subseteq \Sigma_{\kappa^+,\mu}$, hence $\mathrm{cf}(\alpha) = \mu$, and $N_\alpha \cap \alpha$ is unbounded in α. This gives $\mathrm{cf}(\delta) = \mu$. Consequently there is a club D in δ of order type μ. If we denote by f_α, for $\alpha \in S$, the unique order isomorphism from δ onto D_α, then the set $C_\alpha := f_\alpha[D]$ is also a club in α of order type μ. Consequently, the club sequence $(C_\alpha : \alpha \in S)$ has property (i).

If $\alpha, \beta \in S$, and if $\gamma \leq \min\{\alpha, \beta\}$ is a common limit point of C_α and C_β with countable cofinality (see above), then γ is also a common limit point of D_α and D_β, and a limit point of N_α and of N_β. Now Lemma 4.5.1 says that γ is a member of $N_\alpha \cap N_\beta$ since N_α and N_β are unions of approximation sequences of length $\aleph_1$. From Example 4.2.15 e) we can conclude that $N_\alpha \cap \gamma = N_\beta \cap \gamma$, which gives $D_\alpha \cap \gamma = D_\beta \cap \gamma$. Since f_α and f_β are order isomorphisms, we have $\tau := f_\alpha^{-1}[D_\alpha \cap \gamma] = f_\beta^{-1}[D_\beta \cap \gamma]$. Now we can infer that $C_\alpha \cap \gamma = f_\alpha[D \cap \tau] = f_\beta[D \cap \tau] = C_\beta \cap \gamma$, as desired. Thus $(C_\alpha : \alpha \in S)$ has property (ii).

Remember that a sequence $(C_\alpha : \alpha \in A)$ where A is a set of limit ordinals is called a club sequence iff, for every $\alpha \in A$, C_α is a club in α of order type $\mathrm{cf}(\alpha)$. If μ and κ are regular cardinals such that $\mu < \kappa$, and if in particular $A \subseteq \Sigma_{\kappa,\mu}$, then C_α is a club in α of order type μ for all $\alpha \in A$. Note that then any subset C'_α of C_α which is unbounded in α also has the order type μ: We have $\mathrm{otp}(C'_\alpha) \leq \mu$ since $C'_\alpha \subseteq C_\alpha$, and $\mathrm{otp}(C'_\alpha) < \mu$ is impossible since C'_α is unbounded in α. The following definition yields an important tool for the next chapter.

Definition *Assume that μ and κ are regular cardinals, $\mu < \kappa$, and $S \subseteq \Sigma_{\kappa,\mu}$ is a stationary subset of κ. Any sequence $\bar{T} = (T_\alpha : \alpha \in S)$ is called a* $\diamondsuit_{\mathbf{club}}$**-sequence** *for S iff $\bar{T}$ is a club sequence and*

(cc) *For every club C in κ, $\{\alpha \in S : T_\alpha \subseteq C\}$ is stationary in κ.*

In general the existence of $\diamondsuit_{\mathrm{club}}$-sequences for $\Sigma_{\kappa,\mu}$ is not provable in ZFC; for example, this holds for $\Sigma_{\omega_2,\omega_1}$.

The following result, however, is provable in ZFC.

Lemma 4.5.3 *Let κ and μ be uncountable regular cardinals with $\mu^+ < \kappa$. Further assume that $S \subseteq \Sigma_{\kappa,\mu}$ is a stationary subset of κ and $(F_\alpha : \alpha \in S)$ is a fixed club sequence. If $\bar{T} = (T_\alpha : \alpha \in S)$ is a club sequence, then there is a club C in κ such that the sequence $\bar{T}^*(C) := (T^*_\alpha : \alpha \in S)$, given by*

$$T^*_\alpha = \begin{cases} T_\alpha \cap C & \textit{if } T_\alpha \cap C \textit{ is a club in } \alpha, \\ F_\alpha & \textit{otherwise,} \end{cases}$$

is a $\diamondsuit_{\text{club}}$-sequence for S.

Proof[4] Assume that S is a stationary subset of $\Sigma_{\kappa,\mu}$ and $\bar{T} = (T_\alpha : \alpha \in S)$ is a club sequence. To get a contradiction let us assume that, for any club C in κ, the sequence $\bar{T}^*(C) = (T^*_\alpha : \alpha \in S)$ defined in the assertion of the lemma is not a $\diamondsuit_{\text{club}}$-sequence for S.

First we note that any intersection of less than κ clubs in κ is a club in κ. Consequently, given a sequence $(D_\xi : \xi < \gamma)$ of clubs in κ of limit length γ less than κ, we obtain a sequence $(E_\xi : \xi < \gamma)$ by $E_0 := \kappa$ and $E_\xi := \bigcap_{\zeta<\xi} D_\zeta$ for $0 < \xi < \gamma$ which is decreasing under $\subseteq$ and has the same intersection as $(D_\xi : \xi < \gamma)$. Let E be a club in κ such that $E \subseteq \bigcap\{E_\xi : \xi < \gamma\}$. Assume that we can arrange that $T^\xi_\alpha := T_\alpha \cap E_\xi$ is a club in α for every member α of the stationary set $E \cap S$ and every $\xi \in \gamma$, and denote this property by $(*)$. Since $(E_\xi : \xi < \gamma)$ is decreasing, the sequence $\bar{T}_\alpha = (T^\xi_\alpha : \xi < \gamma)$ is decreasing, and $|T_\alpha| = \mu$ for all $\alpha \in S$. Thus, if $\gamma := \mu^+$, $\bar{T}_\alpha$ must be constant from some ordinal $\xi(\alpha) < \gamma$ on, and the function $(\xi(\alpha) : \alpha \in E \cap S \cap (\kappa \setminus \gamma))$ is regressive. Since $E \cap S \cap (\kappa \setminus \gamma)$ is stationary in κ, we obtain by Fodor's theorem an ordinal $\beta < \gamma$ such that $Q = \{\alpha < \kappa : \alpha \in E \cap S \wedge \xi(\alpha) = \beta\}$ is stationary in κ. If $\alpha \in Q$, then $T^\beta_\alpha = T^{\beta+1}_\alpha = T_\alpha \cap E_{\beta+1} = T_\alpha \cap \bigcap\{D_\zeta : \zeta < \beta+1\} = T_\alpha \cap E_\beta \cap D_\beta = T^\beta_\alpha \cap D_\beta$, hence $T^\beta_\alpha \subseteq D_\beta$. So the set $\{\alpha \in S : T^\beta_\alpha \subseteq D_\beta\}$ is stationary in κ. The preceding considerations have lead us to the key of the proof: If we can realize $(*)$ and can arrange for $\alpha \in S$ that $\bar{T}_\alpha := (T_\alpha \cap E_\xi : \xi < \mu^+)$ is a club sequence such that for every $\xi < \mu^+$ the set $\{\alpha \in S : T^\xi_\alpha \subseteq D_\xi\}$ is not stationary in κ, we will have obtained the desired contradiction.

Clearly D_ξ will be given by our assumption on $\bar{T}$, if we will have $\bar{T}_\alpha = \bar{T}^*(E_\xi)$. This fixes our definition of the sequence $\bar{T}_\alpha$:

$$T^\xi_\alpha := \begin{cases} T_\alpha \cap E_\xi & \text{if } T_\alpha \cap E_\xi \text{ is a club in } \alpha, \\ F_\alpha & \text{otherwise.} \end{cases}$$

There remains the definition of the clubs E and D_ξ for any $\xi < \mu^+$. By recursion we will realize the above properties and require

[4] We use ideas from [BM].

$(\mathrm{ns})_\xi$ The set $\{\alpha \in S : T^\xi_\alpha \subseteq D_\xi\}$ is not stationary in κ.
$(\mathrm{ic})_\xi$ The set $\{\alpha < \kappa : \alpha \notin S$ or $T_\alpha \cap E_\xi$ is a club in $\alpha\}$ includes a club in κ.

Then $(\mathrm{ic})_\xi$ will guarantee that $(*)$ holds. For $\alpha \in S$ we have $T^0_\alpha = T_\alpha$ since $E_0 = \kappa$, and clearly $(\mathrm{ic})_0$ is true. By our assumption, $\bar{T}^*(\kappa)$ is not a $\diamondsuit_{\mathrm{club}}$-sequence, and there is a club D_0 in κ such that $(\mathrm{ns})_0$ is satisfied.

Now fix $\xi < \mu^+$ and let, for $\zeta < \xi$, clubs D_ζ in κ be already defined such that $E_\zeta = \bigcap\{D_\delta : \delta < \zeta\}$ (for $\zeta \neq 0$) and $(T^\zeta_\alpha : \alpha \in S)$ satisfy $(\mathrm{ns})_\zeta$ and $(\mathrm{ic})_\zeta$. Then $E_\xi := \bigcap\{D_\zeta : \zeta < \xi\}$ is a club in κ. Let T^ξ_α be defined as above. Then T^ξ_α is a club in α of order type μ since $T_\alpha \cap E_\xi$ is unbounded in α, as stated above. Our supposition says that property (cc) does not hold for the club sequence $\bar{T}^*(E_\xi) = (T^\xi_\alpha : \alpha \in S)$; thus there is a club D_ξ in κ, such that $(\mathrm{ns})_\xi$ is fulfilled. The set E'_ξ of limit points of E_ξ is also a club in κ. To verify $(\mathrm{ic})_\xi$ it suffices to show that $E'_\xi \cap S \subseteq \{\alpha \in \kappa : T_\alpha \cap E_\xi$ is a club in $\alpha\}$. Fix $\alpha \in E'_\xi \cap S$. Then $E_\xi \cap \alpha$ is a club in α, and therefore $T_\alpha \cap (E_\xi \cap \alpha) = T_\alpha \cap E_\xi$ is also a club in α, since $\mu = \mathrm{cf}(\alpha) > \aleph_0$, as desired.

Now put

$$E := \bigcap\{D_\xi : \xi < \mu^+\} \cap \bigcap\{E'_\xi : \xi < \mu^+\}.$$

Since $\mu^+ < \kappa$, E is a club in κ, $E \subseteq \bigcap\{E_\xi : \xi < \mu^+\}$, and for every $\alpha \in E \cap S$ we have $\alpha \in E'_\xi$, hence $T_\alpha \cap E_\xi$ is a club in α. Now all requirements from above are fulfilled, and we obtain the desired contradiction, which proves the lemma.

Lemma 4.5.4 *Assume that μ and κ are regular cardinals such that $\aleph_0 < \mu < \kappa$. Then there is a set $S \subseteq \sum_{\kappa^+,\mu}$ stationary in κ^+ and an ecl-sequence for S, which is a $\diamondsuit_{\mathrm{club}}$-sequence for S as well.*

Proof By Lemma 4.5.2, there is a stationary set $T \subseteq \sum_{\kappa^+,\mu}$ and a sequence $(C_\alpha : \alpha \in T)$ such that

(i) $\forall \alpha \in T\ (C_\alpha$ is a club in $\alpha \wedge \mathrm{otp}(C_\alpha) = \mu)$.
(ii) $\forall \alpha, \beta \in T\ \forall \gamma \leq \min\{\alpha, \beta\}$ (γ is a limit point of C_α and of $C_\beta \Longrightarrow C_\alpha \cap \gamma = C_\beta \cap \gamma)$.

Lemma 4.5.3, applied to κ^+, μ, T, and $(C_\alpha : \alpha \in T)$, yields a club C_0 (note that $0 \notin T$) in κ^+, such that the sequence $(S^*_\alpha : \alpha \in T)$ defined by

$$S^*_\alpha := \begin{cases} C_\alpha \cap C_0 & \text{if } C_\alpha \cap C_0 \text{ is a club in } \alpha, \\ C_\alpha & \text{otherwise} \end{cases}$$

has the following properties:

(a) $\forall \alpha \in T\ (S^*_\alpha$ is a club in α of order type $\mu)$.
(b) $\forall C \subseteq \kappa^+\ (C$ is a club in $\kappa^+ \Longrightarrow \{\alpha \in T : S^*_\alpha \subseteq C\}$ is stationary in $\kappa^+)$.

The set C_0' of limit points of the club C_0 is also a club in κ^+. If $\alpha \in T \cap C_0'$, then $C_0 \cap \alpha$ and C_α are clubs in α, and therefore $C_\alpha \cap (C_0 \cap \alpha) = C_\alpha \cap C_0$ is a club in α, since $\text{cf}(\alpha) = \mu > \aleph_0$. This gives $T \cap C_0' \subseteq \{\alpha \in T : C_\alpha \cap C_0 \text{ is a club in } \alpha\}$, and consequently the set

$$S := \{\alpha \in T : C_\alpha \cap C_0 \ \text{ is a club in } \alpha\}$$

is stationary in κ^+ as a superset of a stationary set. For each $\alpha \in S$,

$$E_\alpha := C_\alpha \cap C_0 = S_\alpha^*$$

is a club in α such that $\text{otp}(E_\alpha) = \mu$. If $\alpha, \beta \in S$ and $\gamma \leq \min\{\alpha, \beta\}$, and if γ is a limit point of E_α and of E_β, then $E_\alpha \cap \gamma = E_\beta \cap \gamma$ by (ii), hence $(E_\alpha : \alpha \in S)$ is a ecl-sequence for S.

Finally, if C is a club in κ^+, then $C_0' \cap \{\alpha \in T : S_\alpha^* \subseteq C\}$ is stationary in κ^+ by (b), and so the set $\{\alpha \in S : E_\alpha \subseteq C\}$ is also stationary in κ^+, as desired.

Chapter 5

Generators of $\mathcal{J}_{<\lambda^+}(\mathsf{a})$

In this chapter, a denotes an infinite set of regular cardinals. We will prove the following assertion, due to S. Shelah: If $|\mathsf{a}|^+ < \min(\mathsf{a})$ and $\lambda \in \mathrm{pcf}(\mathsf{a})$, then there is a set $\mathsf{b}_\lambda \in \mathcal{J}_{<\lambda^+}(\mathsf{a})$ which generates the ideal $\mathcal{J}_{<\lambda^+}(\mathsf{a})$ over $\mathcal{J}_{<\lambda}(\mathsf{a})$, which means that for each set $\mathsf{c} \in \mathcal{J}_{<\lambda^+}(\mathsf{a})$ there is a set $\mathsf{d} \in \mathcal{J}_{<\lambda}(\mathsf{a})$ such that $\mathsf{c} \subseteq \mathsf{b}_\lambda \cup \mathsf{d}$. Such a set b_λ is called a **generator** of $\mathcal{J}_{<\lambda^+}(\mathsf{a})$ over $\mathcal{J}_{<\lambda}(\mathsf{a})$. In Section 5.1 we show under the previous assumptions that there exists a universal sequence $(f_\xi : \xi < \lambda)$. It has the property that, for every ultrafilter D on a satisfying $\mathrm{cf}(\prod \mathsf{a}/D) = \lambda$, it is cofinal in $\prod \mathsf{a}$ modulo D. With the results of Section 4.5, we can prove in the second section the existence of the desired sequence $(\mathsf{b}_\lambda : \lambda \in \mathrm{pcf}(\mathsf{a}))$ of generators.[1]

5.1 Universal Sequences

Definition *If λ and μ are regular cardinals with $\mu < \lambda$, then any sequence $(f_\xi : \xi < \lambda)$ of members of $\prod \mathsf{a}$ is called* ***μ-continuous*** *iff*

$$f_\xi(\nu) = \min\{\sup\{f_\zeta(\nu) : \zeta \in C\} : C \textit{ is a club in } \xi\}$$

for all $\xi < \lambda$ with $\mathrm{cf}(\xi) = \mu$ and all $\nu \in \mathsf{a}$.

Lemma 5.1.1 *Assume that λ and μ are regular cardinals with $|\mathsf{a}| < \mu < \lambda$, and let the sequence $(f_\xi : \xi < \lambda)$ of elements of $\prod \mathsf{a}$ be μ-continuous.*

a) There is a club C in ξ such that $f_\xi = \sup\{f_\zeta : \zeta \in C\}$ for every $\xi < \lambda$ with $\mathrm{cf}(\xi) = \mu$.

b) If I is an ideal on a such that $f_\eta <_I f_{\eta+1}$ and $f_\eta <_I f_\xi$ for all $\eta < \lambda$ and all limit ordinals ξ with $\eta < \xi < \lambda$ and $\mathrm{cf}(\xi) \neq \mu$, then $(f_\xi : \xi < \lambda)$ is strictly increasing modulo I.

[1] In this chapter, we owe much to the representation in [BM].

Proof For part a), consider a fixed $\xi < \lambda$ with $\mathrm{cf}(\xi) = \mu$. For each $\nu \in \mathsf{a}$, we choose a club C_ν in ξ such that $f_\xi(\nu) = \sup\{f_\zeta(\nu) : \zeta \in C_\nu\}$. Then we can conclude from Lemma 1.8.2, together with $|\mathsf{a}| < \mathrm{cf}(\xi) = \mu$, that $C := \bigcap\{C_\nu : \nu \in \mathsf{a}\}$ is also a club in ξ. By the definition of μ-continuity, we get $f_\xi = \sup\{f_\zeta : \zeta \in C\}$.

For the proof of part b), let $\xi < \lambda$ and assume, as an inductive hypothesis, that the sequence $(f_\zeta : \zeta < \xi)$ is strictly increasing modulo I. We will show that $f_\zeta <_I f_\xi$ for all ζ with $\zeta < \xi$. If $\xi = \eta + 1$, then this follows from $f_\eta <_I f_{\eta+1}$ and from the inductive hypothesis, and if ξ is a limit ordinal with $\mathrm{cf}(\xi) \neq \mu$, then it holds by assumption. So we can assume that $\mathrm{cf}(\xi) = \mu$. We choose a club C in ξ as in part a). Then there is some $\eta \in C$ with $\zeta < \eta < \xi$. Clearly we have $f_\eta \leq_{\mathsf{a}} f_\xi$ by the choice of C, and from $f_\zeta <_I f_\eta$ we get $f_\zeta <_I f_\xi$, as desired.

Corollary 5.1.2 *Assume that λ and μ are regular cardinals with $|\mathsf{a}| < \mu < \lambda$, I is an ideal on a, and the partial order $(\prod \mathsf{a}, <_I)$ is λ-directed. Then there is a μ-continuous sequence $(f_\xi : \xi < \lambda)$ of elements of $\prod \mathsf{a}$ which is strictly increasing modulo I.*

Proof We define the requested sequence by recursion and choose $f_0 \in \prod \mathsf{a}$ arbitrarily. If ξ is a successor ordinal or a limit ordinal with cofinality different from μ, then we obtain the function f_ξ with the help of the assumption that $(\prod \mathsf{a}, <_I)$ is λ-directed. If $\mathrm{cf}(\xi) = \mu$, then we define f_ξ according to the definition of μ-continuity. Now the assertion follows from Lemma 5.1.1 b).

Lemma 5.1.3 *If $|\mathsf{a}|^+ < \min(\mathsf{a})$ and $\lambda \in \mathrm{pcf}(\mathsf{a})$, then there exists a sequence $(f_\xi : \xi < \lambda)$ of elements of $\prod \mathsf{a}$ which is strictly increasing modulo $\mathcal{J}_{<\lambda}(\mathsf{a})$ and is cofinal in $\prod \mathsf{a}$ modulo D for every ultrafilter D on a with $\mathrm{cf}(\prod \mathsf{a}/D) = \lambda$.*

Proof To get a contradiction, let us assume that the assertion of the lemma is false. Then there exists a cardinal $\lambda \in \mathrm{pcf}(\mathsf{a})$ such that, for any modulo $\mathcal{J}_{<\lambda}(\mathsf{a})$ strictly increasing sequence $(f_\xi : \xi < \lambda)$ of members of $\prod \mathsf{a}$, there is some ultrafilter D on a with $\mathrm{cf}(\prod \mathsf{a}/D) = \lambda$ such that the sequence $(f_\xi : \xi < \lambda)$ is not cofinal in $\prod \mathsf{a}$ modulo D.

The following proof will again demonstrate the strength of the method of approximation sequences. Assume that Θ is a regular cardinal with $2^{\sup(\mathsf{a})} < \Theta$ and $(N_i : i \leq |\mathsf{a}|^+)$ is an approximation sequence in $\mathrm{H}(\Theta)$ such that $\mathsf{a} \cup \{\mathsf{a}\} \subseteq N_0$ and $|N_i| = |\mathsf{a}|^+$ for all $i < |\mathsf{a}|^+$. Further let $N := N_{|\mathsf{a}|^+}$. Since $|N_i| < \mathrm{cf}(\lambda) = \lambda$ for all $i < |\mathsf{a}|^+$, Lemma 4.3.3 says that $\chi_N(\lambda) = \sup\{\chi_{N_i}(\lambda) : i < |\mathsf{a}|^+\}$ and $\mathrm{cf}(\chi_N(\lambda)) = |\mathsf{a}|^+ < \lambda$. To apply our assumption, assume that the sequence $\bar{f} = (f_\xi : \xi < \lambda)$ of members of $\prod \mathsf{a}$ is strictly increasing modulo $\mathcal{J}_{<\lambda}(\mathsf{a})$ and is a member of N_0. Then there is an ultrafilter D on a such that $\mathrm{cf}(\prod \mathsf{a}/D) = \lambda$ and $\bar{f}$ is not cofinal in $\prod \mathsf{a}$ modulo D. Corollary 3.4.11 says that $D \cap \mathcal{J}_{<\lambda}(\mathsf{a}) =$

$\emptyset$, and thus we conclude from Lemma 2.1.3 that $\bar{f}$ is also strictly increasing modulo D. Consequently, the set $\{f_\xi : \xi < \lambda\}$ has an upper bound g_0' in $\prod \mathsf{a}$ under the relation $<_D$. We can "extend" g_0' to a sequence $(g_\xi' : \xi < \lambda)$ which is cofinal in $\prod \mathsf{a}$ modulo D since $(\prod \mathsf{a}, <_D)$ is λ-directed. Since we want to apply the assumption of our indirect proof also to the sequence $\bar{g}$ we are going to construct, we must take care that it is strictly increasing modulo $\mathcal{J}_{<\lambda}(\mathsf{a})$. Therefore we define g_ξ by induction on ξ and assume, as usual, that we know g_ζ for any $\zeta < \xi$. If $\mathrm{cf}(\xi) \neq |\mathsf{a}|^+$, then we can choose g_ξ as an upper bound of the set $\{g_\zeta : \zeta < \xi\} \cup \{f_\xi, g_\xi'\}$ under the relation $<_{\mathcal{J}_{<\lambda}(\mathsf{a})}$ with the additional property that $\max\{f_\xi, g_\xi'\} \leq_{\mathsf{a}} g_\xi$. If $\mathrm{cf}(\xi) = |\mathsf{a}|^+$, we ensure that the new sequence is $|\mathsf{a}|^+$-continuous and define $g_\xi(\nu) := \min\{\sup\{g_\zeta(\nu) : \zeta \in C\} : C \text{ is a club in } \xi\}$. Let $\bar{g} = (g_\xi : \xi < \lambda)$ and assume that $\bar{g} \in N_0$. Using Lemma 4.3.4 and the fact that $\bar{g}$ is strictly increasing under $<_D$, we obtain

$$(*) \qquad g_{\chi_N(\lambda)} \leq_{\mathsf{a}} \chi_N \restriction \mathsf{a}.$$

$$(**) \qquad \forall i, j < |\mathsf{a}|^+ (i < j \Longrightarrow \chi_{N_i} \restriction \mathsf{a} <_D g_{\chi_{N_j}(\lambda)}).$$

Now let $\mathsf{b} := \{\nu \in \mathsf{a} : f_{\chi_N(\lambda)}(\nu) = \chi_N(\nu)\}$ and $\mathsf{c} := \{\nu \in \mathsf{a} : g_{\chi_N(\lambda)}(\nu) = \chi_N(\nu)\}$.

Since $f_{\chi_N(\lambda)} \leq_{\mathsf{a}} g_{\chi_N(\lambda)}$, we have $\mathsf{b} \subseteq \mathsf{c}$ by $(*)$. We want to show that $\mathsf{b} \subsetneqq \mathsf{c}$, since our indirect proof will allow the definition of $|\mathsf{a}|^+$ sequences like $\bar{f}$ and $\bar{g}$, and the existence of a strictly increasing sequence of subsets of a of length $|\mathsf{a}|^+$ will yield the desired contradiction.

Since $\bar{g}$ is $|\mathsf{a}|^+$-continuous, we obtain from Lemma 4.3.4 a strictly increasing continuous sequence $(i(\beta) : \beta < |\mathsf{a}|^+)$ of limit ordinals in $|\mathsf{a}|^+$ such that $g_{\chi_N(\lambda)}(\nu) = \sup\{g_{\chi_{N_{i(\beta)}}(\lambda)}(\nu) : \beta < |\mathsf{a}|^+\}$ for all $\nu \in \mathsf{a}$.

Fix $\beta < |\mathsf{a}|^+$. Using $(*)$ and $f_{\chi_N(\lambda)} <_D g_\xi$ for all $\xi < \lambda$, we get $f_{\chi_N(\lambda)} <_D \chi_N \restriction \mathsf{a}$, and $\chi_{N_{i(\beta)}} \restriction \mathsf{a} <_D g_{\chi_{N_{i(\beta+1)}}(\lambda)}$ follows from $(**)$. Therefore there exists a cardinal $\nu_\beta \in \mathsf{a}$ such that

(i) $f_{\chi_N(\lambda)}(\nu_\beta) < \chi_N(\nu_\beta)$ and
(ii) $\chi_{N_{i(\beta)}}(\nu_\beta) < g_{\chi_{N_{i(\beta+1)}}(\lambda)}(\nu_\beta)$.

Since $|\mathsf{a}|^+$ is regular, there is an ordinal $\nu \in \mathsf{a}$ such that the set $A = \{\beta < |\mathsf{a}|^+ : \nu_\beta = \nu\}$ has cardinality $|\mathsf{a}|^+$. (i) tells us that $\nu \notin \mathsf{b}$. We will show that $\nu \in \mathsf{c}$, i.e. we prove $g_{\chi_N(\lambda)}(\nu) = \chi_N(\nu)$.

$(*)$ says that it suffices to prove the relation "$\geq$". From (ii) we conclude that

$$\chi_N(\nu) = \sup\{\chi_{N_{i(\beta)}}(\nu) : \beta \in A\} \leq \sup\{g_{\chi_{N_{i(\beta+1)}}(\lambda)}(\nu) : \beta \in A\} \leq$$

$$\sup\{g_{\chi_{N_{i(\beta+1)}}(\lambda)}(\nu) : \beta < |\mathsf{a}|^+\} \leq \sup\{g_{\chi_{N_{i(\beta)}}(\lambda)}(\nu) : \beta < |\mathsf{a}|^+\} = g_{\chi_N(\lambda)}(\nu).$$

This proves $\mathsf{b} \subsetneqq \mathsf{c}$, as desired.

The rest of the proof consists of an appropriate recursive definition. We will define, for each $\alpha < |\mathsf{a}|^+$, sequences $\bar{f}_\alpha = (f^\alpha_\xi : \xi < \lambda)$ of members of $\prod \mathsf{a}$ and ultrafilters D_α on a with $\operatorname{cf}(\prod \mathsf{a}/D_\alpha) = \lambda$ such that, for all $\alpha, \gamma < |\mathsf{a}|^+$,

$\bar{f}_\alpha$ is strictly increasing modulo $\mathcal{J}_{<\lambda}(\mathsf{a})$.
If $\gamma < \alpha$, then $f^\gamma_\xi \leq_{\mathsf{a}} f^\alpha_\xi$ for all $\xi < \lambda$.
$f^\alpha_\xi <_{D_\alpha} f^{\alpha+1}_0$ for all $\xi < \lambda$.
$\bar{f}_{\alpha+1}$ is cofinal in $\prod \mathsf{a}$ modulo D_α.
$\bar{f}_\alpha$ is $|\mathsf{a}|^+$-continuous.

Then, for each $\xi < |\mathsf{a}|^+$, the functions $\bar{f}_\alpha$ and $\bar{f}_{\alpha+1}$ can play the above role of $\bar{f}$ and $\bar{g}$. If we choose N_0 in such a way that

$$\mathsf{a} \cup \{\mathsf{a}\} \cup \{\bar{f}_\alpha : \alpha < |\mathsf{a}|^+\} \cup \{D_\alpha : \alpha < |\mathsf{a}|^+\} \subseteq N_0,$$

and put

$$\mathsf{c}_\alpha := \{\nu \in \mathsf{a} : f^\alpha_{\chi_N(\lambda)}(\nu) = \chi_N(\nu)\},$$

then we have just shown that $\mathsf{c}_\alpha \subsetneq \mathsf{c}_{\alpha+1}$ for every $\alpha < |\mathsf{a}|^+$. Since $f^\gamma_\xi \leq_{\mathsf{a}} f^\alpha_\xi$ for all $\xi < \lambda$ and all $\gamma < \alpha$, we get $\mathsf{c}_\gamma \subseteq \mathsf{c}_\alpha$ whenever $\gamma < \alpha < |\mathsf{a}|^+$, and obtain a strictly increasing sequence of subsets of a of length $|\mathsf{a}|^+$, yielding the desired contradiction.

If $\alpha = 0$, we can choose by Corollary 5.1.2 a $|\mathsf{a}|^+$-continuous sequence $(f^0_\xi : \xi < \lambda)$ which is strictly increasing modulo $\mathcal{J}_{<\lambda}(\mathsf{a})$, using the fact that, by Lemma 3.4.10, $(\prod \mathsf{a}, <_{\mathcal{J}_{<\lambda}(\mathsf{a})})$ is λ-directed. The ultrafilter D_0 will be defined in the successor step.

Assume that $\alpha < |\mathsf{a}|^+$ is a limit ordinal, and that, for each $\gamma < \alpha$, the sequence $(f^\gamma_\xi : \xi < \lambda)$ is already defined and has the above properties. We define the sequence $(f^\alpha_\xi : \xi < \lambda)$ by induction on $\xi < \lambda$. Let f^α_ζ be defined for every $\zeta < \xi$. If $\operatorname{cf}(\xi) \neq |\mathsf{a}|^+$, then let h^α_ξ be given by $h^\alpha_\xi := \sup\{f^\gamma_\xi : \gamma < \alpha\}$. Since $|\mathsf{a}|^+ < \min(\mathsf{a})$, we have $h^\alpha_\xi \in \prod \mathsf{a}$. Now take $f^\alpha_\xi \in \prod \mathsf{a}$ as an upper bound of the set $\{f^\alpha_\zeta : \zeta < \xi\} \cup \{h^\alpha_\xi\}$ under the partial ordering $<_{\mathcal{J}_{<\lambda}(\mathsf{a})}$, satisfying in addition $h^\alpha_\xi \leq_{\mathsf{a}} f^\alpha_\xi$. If $\operatorname{cf}(\xi) = |\mathsf{a}|^+$, then we have no other choice than putting, for all $\nu \in \mathsf{a}$, $f^\alpha_\xi(\nu) := \min\{\sup\{f^\alpha_\zeta(\nu) : \zeta \in C\} : C \text{ is a club in } \xi\}$. Again, the ultrafilter D_α will be defined in the successor step.

But we have already carried through this step: If the sequence $\bar{f} = (f^\alpha_\xi : \xi < \lambda)$ with the above properties is given, we take $(f^{\alpha+1}_\xi : \xi < \lambda) := \bar{g}$ and $D_\alpha := D$ as above. This completes our definition.

We now verify the above properties. Using Lemma 5.1.1 b), it is easy to see that $\bar{f}_\alpha$ is strictly increasing modulo $\mathcal{J}_{<\lambda}(\mathsf{a})$. The last three properties are satisfied by definition. To show the second property, consider a fixed limit ordinal α and $\gamma < \alpha$. By induction on $\xi < \lambda$ we show that $f^\gamma_\xi \leq_{\mathsf{a}} f^\alpha_\xi$. If cf $(\xi) \neq |\mathsf{a}|^+$, this is true by definition. Thus assume that cf $(\xi) = |\mathsf{a}|^+$ and, according to Lemma 5.1.1 a), $f_\xi^\alpha = \sup\{f_\zeta^\alpha : \zeta \in C\}$ for some club C in ξ. The inductive hypothesis gives $f_\zeta^\gamma \leq_{\mathsf{a}} f_\zeta^\alpha$ for all $\zeta \in C$, and thus we have, using the definition of f_ξ^γ,

$$f_\xi^\gamma \leq_{\mathsf{a}} \sup\{f_\zeta^\gamma : \zeta \in C\} \leq_{\mathsf{a}} \sup\{f_\zeta^\alpha : \zeta \in C\} = f_\xi^\alpha .$$

The case that α is a successor ordinal follows from the definition and the inductive hypothesis. This completes the proof of the lemma.

5.2 The Existence of Generators

Lemma 5.2.1 *Assume that μ and λ are regular cardinal numbers such that $\aleph_0 < \mu$ and $\mu^+ < \lambda$. Further let $S \subseteq \Sigma_{\mu^{++},\mu}$ be a stationary subset of μ^{++} and $(C_\alpha : \alpha \in S)$ be an ecl-sequence for S which is a $\diamondsuit_{\text{club}}$-sequence for S as well. Then there is a sequence $(R_\zeta : \zeta < \lambda)$, satisfying $R_\zeta \subseteq [\zeta]^{\leq\mu}$ and $|R_\zeta| \leq \lambda$ for all $\zeta < \lambda$, which has the property that, for every regular cardinal $\Theta > \lambda$ and every approximation sequence $(N_\xi : \xi < \mu^{++})$ in $\mathrm{H}(\Theta)$ such that*

(i) $|N_\xi| < \lambda$ for all $\xi < \mu^{++}$,

(ii) $\{\lambda, (C_\alpha : \alpha \in S)\} \subseteq N_0$,

there exists a normal sequence $(\xi(i) : i < \mu)$ of ordinals less than λ such that

$$\{\chi_{N_{\xi(j)}}(\lambda) : j < i\} \in N_{\xi(i+1)} \cap R_{\chi_{N_{\xi(i)}}(\lambda)}$$

for all limit ordinals $i < \mu$.

Corollary 5.2.2 *If μ and λ are regular cardinals with $\aleph_0 < \mu$ and $\mu^+ < \lambda$, then there is a set $R \subseteq [\lambda]^{\leq\mu}$ with $|R| \leq \lambda$ such that, for every regular cardinal $\Theta > \lambda$, there exists an elementary substructure N of $\mathrm{H}(\Theta)$ and a set $U \subseteq N \cap \lambda$ with the following properties:*

(i) U is cofinal in $N \cap \lambda$ and $\mathrm{otp}(U) = \mu$.

(ii) If $\alpha \in U$ and $\mathrm{otp}(U \cap \alpha)$ is a limit ordinal, then $U \cap \alpha \in N \cap R$.

Proof Assume that $\Theta > \lambda$ is regular. Lemma 4.5.4 guarantees the existence of a set $S \subseteq \Sigma_{\mu^{++},\mu}$ which is stationary in μ^{++}, and of a sequence $(C_\alpha : \alpha \in S)$ which is an ecl-sequence and a $\diamondsuit_{\text{club}}$-sequence for S as well. Let $(R_\zeta : \zeta < \lambda)$ be a sequence with the properties from Lemma 5.2.1 and $R := \bigcup\{R_\zeta : \zeta < \lambda\}$. By Lemma 4.3.1 we can choose an approximation sequence $(N_i : i < \mu^{++})$ in $\mathrm{H}(\Theta)$ such that $\lambda, (C_\alpha : \alpha \in S) \in N_0$ and $|N_i| < \mu^{++}$ for all $i < \mu^{++}$. Now we take, according to Lemma 5.2.1, a normal sequence

$(\xi(i) : i < \mu)$ of members of μ^{++} and put $N := \bigcup\{N_{\xi(i)} : i < \mu\}$. Then we know from Lemma 4.3.3 that $(\chi_{N_{\xi(i)}}(\lambda) : i < \mu)$ is a normal sequence of members of $N \cap \lambda$, and furthermore

$$U := \{\chi_{N_{\xi(i)}}(\lambda) : i < \mu\}$$

is a subset of $N \cap \lambda$ of order type μ which is cofinal in $N \cap \lambda$. If $\alpha \in U$ and $\mathrm{otp}(U \cap \alpha) = i$ for some limit ordinal $i < \mu$, then $\alpha = \chi_{N_{\xi(i)}}(\lambda)$, and Lemma 5.2.1 says that

$$U \cap \alpha = \{\chi_{N_{\xi(j)}} : j < i\} \in N_{\xi(i+1)} \cap R_{\chi_{N_{\xi(i)}}(\lambda)} \subseteq N \cap R.$$

Proof of Lemma 5.2.1 Let $S \subseteq \Sigma_{\mu^{++},\mu}$ be a stationary subset of μ^{++} and $(C_\alpha : \alpha \in S)$ be an ecl-sequence for S which is a $\diamondsuit_{\mathrm{club}}$-sequence for S as well. Then we have

(1) $\forall \alpha \in S\ (C_\alpha \text{ is a club in } \alpha \wedge \mathrm{otp}(C_\alpha) = \mu)$.

(2) $\forall \alpha, \beta \in S\ \forall \gamma \leq \min\{\alpha, \beta\}\ (\gamma \text{ is a limit point of } C_\alpha \text{ and of } C_\beta \Longrightarrow C_\alpha \cap \gamma = C_\beta \cap \gamma)$.

(3) $\forall C \subseteq \mu^{++}\ (C \text{ is a club in } \mu^{++} \Longrightarrow \{\alpha \in S : C_\alpha \subseteq C\} \text{ is stationary in } \mu^{++})$.

For each $\beta \leq \lambda$ with $\mathrm{cf}(\beta) = \mu^{++}$, we choose a club E_β in β so that $\mathrm{otp}(E_\beta) = \mu^{++}$. If k_β is the $\in$-isomorphism from μ^{++} onto E_β, we put $E_{\beta,\alpha} := k_\beta[C_\alpha]$ for $\alpha \in S$. Then $(E_{\beta,\alpha} : \alpha \in S)$ is a copy of $(C_\alpha : \alpha \in S)$. If $\lambda = \mu^{++}$, we can take $E_\lambda = \lambda$, hence $k_\lambda = \mathrm{id}$. For each $\zeta < \lambda$ let

$$R_\zeta := \{E_{\beta,\alpha} \cap \zeta : \beta \leq \lambda,\ \mathrm{cf}(\beta) = \mu^{++},\ \alpha \in S\}.$$

From $\mathrm{otp}(C_\alpha) = \mu$ we can infer that $|E_{\beta,\alpha}| = \mu$, hence $|E_{\beta,\alpha} \cap \zeta| \leq \mu$. Since $\lambda \cdot |S| = \lambda \cdot \mu^{++} = \lambda$, we get $|R_\zeta| \leq \lambda$.

Now let $\Theta > \lambda$ be a regular cardinal and $(N_\xi : \xi < \mu^{++})$ be an approximation sequence in $\mathrm{H}(\Theta)$, satisfying

$$|N_\xi| < \lambda \ \text{ for all } \xi < \mu^{++} \text{ and}$$

$$\{\lambda, (C_\alpha : \alpha \in S)\} \subseteq N_0.$$

Let $\delta_\xi := \chi_{N_\xi}(\lambda)$ for all $\xi < \mu^{++}$. The sequence $(\delta_\xi : \xi < \mu^{++})$ is a normal sequence of members of λ, and thus $F := \{\delta_\xi : \xi < \mu^{++}\}$ is a club in $\beta_0 := \sup(F)$. Note that $\beta_0 \leq \lambda$ and $\mathrm{cf}(\beta_0) = \mu^{++}$. Clearly, also $F \cap E_{\beta_0}$ is a club in β_0. Furthermore, by definition, $k_{\beta_0} : \mu^{++} \longrightarrow E_{\beta_0}$ is an isomorphism. We put

$$F^* := \{\sigma \in \beta_0 : \exists \xi \in \mu^{++}\ (\delta_\xi = \sigma = k_{\beta_0}(\xi))\}.$$

Then it is easy to verify that F^* is a club in β_0 such that $F^* \subseteq F \cap E_{\beta_0}$. Consequently, $F^{**} := k_{\beta_0}^{-1}[F^*] = \{\xi \in \mu^{++} : \delta_\xi = k_{\beta_0}(\xi)\}$ is a club in μ^{++}. By (3), the set $\{\alpha \in S : C_\alpha \subseteq F^{**}\}$ is stationary in μ^{++}; in particular, it is nonempty. So we can choose $\alpha_0 \in S$ with $C_{\alpha_0} \subseteq F^{**}$. Further we take some order preserving enumeration $(\tau_i : i < \mu)$ of C_{α_0}, i.e.

$$C_{\alpha_0} = \{\tau_i : i < \mu\}.$$

Since $C_{\alpha_0} \subseteq F^{**}$, we have $k_{\beta_0}[C_{\alpha_0}] = E_{\beta_0,\alpha_0} \subseteq k_{\beta_0}[F^{**}] = F^* \subseteq F$. Thus there is a normal sequence $(\xi(i) : i < \mu)$ of members of λ satisfying $E_{\beta_0,\alpha_0} = \{\delta_{\xi(i)} : i < \mu\}$.

For any $i < \mu$, we have

$$\{\delta_{\xi(j)} : j < i\} = E_{\beta_0,\alpha_0} \cap \delta_{\xi(i)} \in R_{\delta_{\xi(i)}}.$$

There remains to be shown that $\{\delta_{\xi(j)} : j < i\} \in N_{\xi(i+1)}$ for all limit ordinals $i < \mu$. So let $i < \mu$ be a limit ordinal and $j \le i$. Since $\tau_j \in C_{\alpha_0} \subseteq F^{**}$, we get

$$k_{\beta_0}(\tau_j) = \delta_{\tau_j}\,.$$

On the other hand, $\delta_{\xi(j)}$ is the j-th element of E_{β_0,α_0}, and so we can conclude, using $E_{\beta_0,\alpha_0} = k_{\beta_0}[C_{\alpha_0}]$, that

$$k_{\beta_0}(\tau_j) = \delta_{\xi(j)}.$$

This gives $\delta_{\xi(j)} = \delta_{\tau_j}$, hence

$$\xi(j) = \tau_j \quad \text{for all } j \le i,$$

in particular $\xi(i) = \tau_i < \alpha_0$. Now we have $\tau_i \in N_{\xi(i+1)}$, and τ_i is a limit point of C_{α_0}. Furthermore we have $S \in N_{\xi(i+1)}$, and thus we can infer from

$$\mathrm{H}(\Theta) \models \exists\gamma\ (\gamma \in S \wedge \gamma > \tau_i \wedge \tau_i \text{ is a limit point of } C_\gamma)$$

and $N_{\xi(i+1)} \preceq \mathrm{H}(\Theta)$ that there is an ordinal $\gamma \in N_{\xi(i+1)} \cap S$ with $\gamma > \tau_i$ such that τ_i is a limit point of C_γ. From property (2) of the sequence $(C_\alpha : \alpha \in S)$ we get $\{\tau_j : j < i\} = C_{\alpha_0} \cap \tau_i = C_\gamma \cap \tau_i \in N_{\xi(i+1)}$. From $\lambda \in N_{\xi(i+1)}$ and $g := (N_{\xi'} : \xi' \le \xi(i)) \in N_{\xi(i+1)}$ we can infer for $f := (\sup(N_{\xi'} \cap \lambda) : \xi' \le \xi(i))$, that $f \in N_{\xi(i+1)}$, since $f = \{(x,y) \in \mathrm{H}(\Theta) : \mathrm{H}(\Theta) \models x \in \mathrm{dom}(g) \wedge y = \sup(g(x) \cap \lambda)\}$ is definable in $\mathrm{H}(\Theta)$ with parameters from $N_{\xi(i+1)}$. Thus $\{\tau_j : j < i\} = \{\xi(j) : j < i\} \in N_{\xi(i+1)}$ implies that $\mathrm{ran}(f \restriction \{\xi(j) : j < i\}) = \{\delta_{\xi(j)} : j < i\} \in N_{\xi(i+1)}$. So we get $\{\delta_{\xi(j)} : j < i\} \in N_{\xi(i+1)}$, which completes the proof of the lemma.

Theorem 5.2.3 (Main Theorem, S. Shelah) *If* a *is an infinite set of regular cardinals,* $|\mathsf{a}|^+ < \min(\mathsf{a})$*, and* λ *is a cardinal, then there is a subset* c *of* a *which generates the ideal* $\mathcal{J}_{<\lambda^+}(\mathsf{a})$ *over* $\mathcal{J}_{<\lambda}(\mathsf{a})$.

Proof The case that $\lambda \notin \mathrm{pcf}(\mathsf{a})$ is trivial since then $\mathcal{J}_{<\lambda^+}(\mathsf{a}) = \mathcal{J}_{<\lambda}(\mathsf{a})$ and we can choose $\mathsf{c} = \emptyset$.

So let us assume that $\lambda \in \mathrm{pcf}(\mathsf{a})$. If $\lambda = \max \mathrm{pcf}(\mathsf{a})$, then $\mathcal{J}_{<\lambda^+}(\mathsf{a}) = \mathcal{P}(\mathsf{a})$, and the assertion of the theorem is true since we can take $\mathsf{c} := \mathsf{a}$. If $\lambda = \min \mathrm{pcf}(\mathsf{a})$, then $\mathcal{J}_{<\lambda^+}(\mathsf{a}) = \{\emptyset, \{\min(\mathsf{a})\}\}$, and $\mathcal{J}_{<\lambda^+}(\mathsf{a})$ is generated by $\{\min(\mathsf{a})\}$ over $\{\emptyset\}$ which proves the theorem for this trivial case. So we can assume that $\min \mathrm{pcf}(\mathsf{a}) < \lambda < \max \mathrm{pcf}(\mathsf{a})$. Note that $\mathcal{J}_{<\lambda^+}(\mathsf{a}) \neq \mathcal{P}(\mathsf{a})$.

Choose a universal sequence $\bar{f} = (f_\xi : \xi < \lambda)$ according to Lemma 5.1.3. Then $\bar{f}$ is strictly increasing modulo $\mathcal{J}_{<\lambda}(\mathsf{a})$ and, for every ultrafilter D on a such that $\mathrm{cf}(\prod \mathsf{a}/D) = \lambda$, $\bar{f}$ is cofinal in $\prod \mathsf{a}$ modulo D. Let D be an ultrafilter on a such that $\mathrm{cf}(\prod \mathsf{a}/D) = \lambda$. From Corollary 3.4.11 we get $D \cap \mathcal{J}_{<\lambda}(\mathsf{a}) = \emptyset$, and thus $\bar{f}$ is not bounded in $\prod \mathsf{a}$ modulo $\mathcal{J}_{<\lambda}(\mathsf{a})$. Now we can apply Corollary 3.4.3 and obtain a sequence $(\mathsf{b}_\xi : \xi < \lambda)$ of $\mathcal{J}_{<\lambda}(\mathsf{a})$-positive subsets of a with the following properties:

(a) If $\zeta < \xi < \lambda$, then $\mathsf{b}_\zeta \subseteq_{\mathcal{J}_{<\lambda}(\mathsf{a})} \mathsf{b}_\xi$.
(b) For all $\xi < \lambda$, $\bar{f}$ is cofinal in $\prod \mathsf{a}$ modulo $\mathcal{J}_{<\lambda}(\mathsf{a}) \restriction \mathsf{b}_\xi$, and thus $\mathrm{tcf}(\prod \mathsf{a}/\mathcal{J}_{<\lambda}(\mathsf{a}) \restriction \mathsf{b}_\xi) = \lambda$.
(c) There is a function $g \in \prod \mathsf{a}$ such that g is an upper bound of $\bar{f}$ modulo I if $\mathcal{J}_{<\lambda}(\mathsf{a}) \cup \{\mathsf{b}_\xi : \xi < \lambda\}$ generates the ideal I on a.

We will show that $\mathsf{b}_\xi \in \mathcal{J}_{<\lambda^+}(\mathsf{a})$ for all $\xi < \lambda$. For this let us assume that $\mathsf{b}_\xi \notin \mathcal{J}_{<\lambda^+}(\mathsf{a})$ for some $\xi < \lambda$. Then Lemma 3.4.5 says that there is an ultrafilter D on a such that $\mathsf{b}_\xi \in D$ and $\mathrm{cf}(\prod \mathsf{a}/D) \geq \lambda^+$. Corollary 3.4.11 tells us that $D \cap \mathcal{J}_{<\lambda^+}(\mathsf{a}) = \emptyset$, and thus all the more $D \cap \mathcal{J}_{<\lambda}(\mathsf{a}) = \emptyset$, hence $D \cap \mathcal{J}_{<\lambda}(\mathsf{a}) \restriction \mathsf{b}_\xi = \emptyset$. From (b) we get $\mathrm{tcf}(\prod \mathsf{a}/\mathcal{J}_{<\lambda}(\mathsf{a}) \restriction \mathsf{b}_\xi) = \lambda$. This gives $\mathrm{cf}(\prod \mathsf{a}/D) = \lambda$, since D extends $\mathcal{J}_{<\lambda}(\mathsf{a}) \restriction \mathsf{b}_\xi$, contradicting $\mathrm{cf}(\prod \mathsf{a}/D) \geq \lambda^+$.

So we can conclude that $I \subseteq \mathcal{J}_{<\lambda^+}(\mathsf{a})$. To show that $I = \mathcal{J}_{<\lambda^+}(\mathsf{a})$ assume, to get a contradiction, that $\mathsf{b} \in \mathcal{J}_{<\lambda^+}(\mathsf{a}) \setminus I$ for some b. Then there exists an ultrafilter D on a such that $\mathsf{b} \in D$ and $D \cap I = \emptyset$. From $\mathcal{J}_{<\lambda}(\mathsf{a}) \subseteq I$ we get $D \cap \mathcal{J}_{<\lambda}(\mathsf{a}) = \emptyset$, hence $\mathrm{cf}(\prod \mathsf{a}/D) \geq \lambda$. Furthermore we have $\mathrm{cf}(\prod \mathsf{a}/D) < \lambda^+$, since $\mathsf{b} \in \mathcal{J}_{<\lambda^+}(\mathsf{a})$, and altogether we can conclude that $\mathrm{cf}(\prod \mathsf{a}/D) = \lambda$. Now $\bar{f}$ is cofinal in $\prod \mathsf{a}$ modulo D, but on the other hand we can infer from (c) and the fact that D extends I that the function g is an upper bound of $\bar{f}$ modulo D. This contradiction and our assumption that $\mathcal{J}_{<\lambda^+}(\mathsf{a}) \neq \mathcal{P}(\mathsf{a})$ show that $\mathcal{J}_{<\lambda^+}(\mathsf{a})$ is the ideal generated by $\mathcal{J}_{<\lambda}(\mathsf{a}) \cup \{\mathsf{b}_\xi : \xi < \lambda\}$. Together with (a) we get

$$(*) \qquad I = \{X \subseteq \mathsf{a} : \exists Y \in \mathcal{J}_{<\lambda}(\mathsf{a}) \exists \xi < \lambda (X \subseteq Y \cup \mathsf{b}_\xi)\}.$$

The inclusion "$\supseteq$" is obvious. Conversely, if $X \in I$, then there are ordinals $\xi_0 < \xi_1 < \cdots < \xi_{n-1}$ and a set $Y_1 \in \mathcal{J}_{<\lambda}(\mathsf{a})$ such that $X \subseteq Y_1 \cup \mathsf{b}_{\xi_0} \cup \cdots \cup \mathsf{b}_{\xi_{n-1}}$. Now we have $X \subseteq Y \cup \mathsf{b}_{\xi_{n-1}}$ where $Y := Y_1 \cup (\mathsf{b}_{\xi_0} \setminus \mathsf{b}_{\xi_1}) \cup \cdots \cup (\mathsf{b}_{\xi_{n-2}} \setminus \mathsf{b}_{\xi_{n-1}})$ is a member of $\mathcal{J}_{<\lambda}(\mathsf{a})$ by (a).

Claim There is an ordinal $\xi < \lambda$ such that $\mathcal{J}_{<\lambda^+}(\mathsf{a})$ is generated by $\mathcal{J}_{<\lambda}(\mathsf{a}) \cup \{\mathsf{b}_\xi\}$.

Let us assume that the claim is false. If $\xi < \lambda$, then the assertion $\mathsf{b}_\delta \subseteq_{\mathcal{J}_{<\lambda}(\mathsf{a})} \mathsf{b}_\xi$ is not true for all δ with $\xi < \delta < \lambda$, since otherwise it would follow from (a) and $(*)$ that $\mathcal{J}_{<\lambda}(\mathsf{a}) \cup \{\mathsf{b}_\xi\}$ generates the ideal $\mathcal{J}_{<\lambda^+}(\mathsf{a})$. On the other hand, if $\mathsf{b}_\delta \subseteq_{\mathcal{J}_{<\lambda}(\mathsf{a})} \mathsf{b}_\xi$ holds for δ with $\xi < \delta < \lambda$, then $(*)$ shows that b_δ is superfluous as a generating element of the ideal $\mathcal{J}_{<\lambda^+}(\mathsf{a})$ and thus can be omitted in the sequence $(\mathsf{b}_\xi : \xi < \lambda)$. Therefore we can assume without loss of generality that

$$(**) \qquad \forall \delta \forall \xi (\xi < \delta < \lambda \Longrightarrow \neg(\mathsf{b}_\delta \subseteq_{\mathcal{J}_{<\lambda}(\mathsf{a})} \mathsf{b}_\xi)).$$

Put $\mu := |\mathsf{a}|^+$. Since $\lambda > \min(\mathsf{a})$, and, by assumption, $\min(\mathsf{a}) > |\mathsf{a}|^+$, we have $\aleph_0 < \mu < \mu^+ < \lambda$ and can apply Corollary 5.2.2. So let us choose a set $R \subseteq [\lambda]^{\leq\mu}$ with the properties from this corollary.

For each $X \in [\lambda]^{\leq\mu}$ let $f_X \in \prod \mathsf{a}$ be defined by

$$f_X := \sup\{f_\xi : \xi \in X\}.$$

Using $|X| \leq \mu = |\mathsf{a}|^+ < \min(\mathsf{a})$ we can infer that $f_X \in \prod \mathsf{a}$. Since $(\prod \mathsf{a}, <_{\mathcal{J}_{<\lambda^+}(\mathsf{a})})$ is λ^+-directed and $|R| \leq \lambda$, the set $\{f_X : X \in R\}$ has an upper bound $h \in \prod \mathsf{a}$ modulo $\mathcal{J}_{<\lambda^+}(\mathsf{a})$. We will show that this leads to a contradiction, using an appropriate elementary substructure N of $\mathrm{H}(\Theta)$, where Θ is a large enough regular cardinal. The crucial point in the proof is the existence of $N \preceq \mathrm{H}(\Theta)$ and of a set U which is cofinal in $N \cap \lambda$ and has order type μ, such that every initial segment Z of U of limit order type is a member of $N \cap R$. We can choose N and U with these properties by Corollary 5.2.2. By starting the construction of the structure N in this corollary with $N_0 \supseteq \{\mathsf{a}, h, (\mathsf{b}_\xi : \xi < \lambda), \bar{f}\}$, and taking $\Theta > 2^{\sup(\mathsf{a})}$, we can arrange, as demonstrated in Example 4.2.15, that λ, $\prod \mathsf{a}$, $\mathcal{J}_{<\lambda}(\mathsf{a})$, $\bar{f}$ and $(\mathsf{b}_\xi : \xi < \lambda)$ are members of $\mathrm{H}(\Theta)$ and thus of N. Let $(\delta_\alpha : \alpha < \mu)$ be the $\in$-isomorphism from $\mu = \mathrm{otp}(U)$ onto U, and put

$$U_\tau := \{\delta_\alpha : \alpha < \tau\}$$

for every limit ordinal $\tau < \mu$ and

$$A := \{\nu \in \mathsf{a} : f_U(\nu) > h(\nu)\}.$$

Since $|\mathsf{a}| < \mu$, there is a limit ordinal $\tau < \mu$ such that $A = \{\nu \in \mathsf{a} : f_Z(\nu) > h(\nu)\}$, where $Z := U_\tau \in N \cap R$, and thus $f_Z \leq_{\mathcal{J}_{<\lambda^+}(\mathsf{a})} h$ implies $A \in \mathcal{J}_{<\lambda^+}(\mathsf{a})$. By $(*)$, there exists $Y \in \mathcal{J}_{<\lambda}(\mathsf{a})$ and an ordinal $\gamma < \lambda$ such that $A \subseteq Y \cup \mathsf{b}_\gamma$.

We will show below, as a routine proof, that $A \in N$ and that there exists such an ordinal γ which is a member of N. Consequently, we have $\gamma + 1 \in N$ and thus $(\mathsf{b}_\xi : \xi < \lambda)(\gamma+1) = \mathsf{b}_{\gamma+1} \in N$. Now the fact that in (b), for $\xi = \gamma+1$, all considered objects are members of N and the formula is absolute for $\mathrm{H}(\Theta)$, hence holds in $\mathrm{H}(\Theta)$, yields an ordinal $\beta \in N \cap \lambda$ such that $h <_{\mathcal{J}_{<\lambda}(\mathsf{a}) \restriction \mathsf{b}_{\gamma+1}} f_\beta$. Since U is cofinal in $N \cap \lambda$ and $\bar{f}$ is also strictly increasing modulo $\mathcal{J}_{<\lambda}(\mathsf{a}) \restriction \mathsf{b}_{\gamma+1}$, we can assume that $\beta \in U$. By the definition of f_U, this gives

$$h <_{\mathcal{J}_{<\lambda}(\mathsf{a}) \restriction \mathsf{b}_{\gamma+1}} f_\beta \leq_{\mathsf{a}} f_U.$$

So the set $X := \{\nu \in \mathsf{b}_{\gamma+1} : h(\nu) \geq f_U(\nu)\}$ is a member of $\mathcal{J}_{<\lambda}(\mathsf{a})$. But now we get

$$\mathsf{b}_{\gamma+1} \setminus \mathsf{b}_\gamma \subseteq X \cup \{\nu \in \mathsf{b}_{\gamma+1} \setminus \mathsf{b}_\gamma : h(\nu) < f_U(\nu)\} \subseteq X \cup (A \setminus \mathsf{b}_\gamma)$$

and can infer that $\mathsf{b}_{\gamma+1} \setminus \mathsf{b}_\gamma \in \mathcal{J}_{<\lambda}(\mathsf{a})$, hence $\mathsf{b}_{\gamma+1} \subseteq_{\mathcal{J}_{<\lambda}(\mathsf{a})} \mathsf{b}_\gamma$, which contradicts (**).

This will prove the claim and complete the proof of the theorem if we show that $A \in N$ and there is an ordinal $\gamma \in N$ such that A is a member of the ideal J generated by $\mathcal{J}_{<\lambda}(\mathsf{a}) \cup \{\mathsf{b}_\gamma\}$. First let us note that

$$f_Z = \{(\nu, y) \in \mathrm{H}(\Theta) : \mathrm{H}(\Theta) \models \nu \in \mathsf{a} \wedge y = \sup\{\bar{f}(\delta)(\nu) : \delta \in Z\}\},$$

hence f_Z is definable in $\mathrm{H}(\Theta)$ with parameters from N and thus a member of N. From $f_Z, h \in N$ we can infer that $A = \{\nu \in \mathsf{a} : f_Z(\nu) > h(\nu)\} \in N$. Finally, since

$$\mathrm{H}(\Theta) \models \exists\delta\exists w \in \mathcal{J}_{<\lambda}(\mathsf{a})\ (A \subseteq w \cup (\mathsf{b}_\xi : \xi < \lambda)(\delta))$$

and $N \preceq \mathrm{H}(\Theta)$, there is some $\gamma \in N \cap \lambda$ such that A is a member of the ideal J generated by $\mathcal{J}_{<\lambda}(\mathsf{a}) \cup \{\mathsf{b}_\gamma\}$, as desired.

5.3 Properties of Generators

From Corollary 3.4.12, Lemma 3.4.19 and Corollary 3.4.6 we obtain immediately the following lemma.

Lemma 5.3.1 *Assume that* a *is progressive,* $\lambda \in \mathrm{pcf}(\mathsf{a})$, *and* b_λ *is a subset of* a *generating the ideal* $\mathcal{J}_{<\lambda^+}(\mathsf{a})$ *over* $\mathcal{J}_{<\lambda}(\mathsf{a})$. *Then* $\max \mathrm{pcf}(\mathsf{b}_\lambda) = \lambda$.

Generators yield a further result that illustrates the importance of the ideal $\mathcal{J}_{<\lambda}(\mathsf{a})$.

Theorem 5.3.2 *Assume that* a *is progressive,* $\lambda \in \mathrm{pcf}(\mathsf{a})$ *and* b_λ *is a subset of* a *generating* $\mathcal{J}_{<\lambda^+}(\mathsf{a})$ *over* $\mathcal{J}_{<\lambda}(\mathsf{a})$. *Then* $\mathcal{J}_{<\lambda}(\mathsf{a}) \restriction \mathsf{b}_\lambda$ *is the smallest ideal* I *on* a *such that* $\mathrm{tcf}(\prod \mathsf{a}/I) = \lambda$.

Proof First Lemma 3.4.19 tells us that

$$\operatorname{tcf}(\prod \mathsf{a}/\mathcal{J}_{<\lambda}(\mathsf{a}) \restriction \mathsf{b}_\lambda) = \lambda.$$

Consider a fixed ideal I on a satisfying $\operatorname{tcf}(\prod \mathsf{a}/I) = \lambda$. For each I-positive set b, the ideal $I \restriction \mathsf{b}$ can be extended to an ultrafilter D on a, clearly satisfying $\operatorname{cf}(\prod \mathsf{a}/D) = \lambda$. From Lemma 3.4.5 we infer that $\mathsf{b} \notin \mathcal{J}_{<\lambda}(\mathsf{a})$. Thus we have shown that $\mathcal{J}_{<\lambda}(\mathsf{a}) \subseteq I$.

It remains to be shown that $\mathsf{a} \setminus \mathsf{b}_\lambda \in I$ since then $\mathcal{J}_{<\lambda}(\mathsf{a}) \restriction \mathsf{b}_\lambda = \mathcal{J}_{<\lambda}(\mathsf{a})[\mathsf{a} \setminus \mathsf{b}_\lambda] \subseteq I$. We assume that this assertion is false, and choose an ultrafilter D on a such that $D \cap I = \emptyset$ and $\mathsf{a} \setminus \mathsf{b}_\lambda \in D$. Then $\operatorname{cf}(\prod \mathsf{a}/D) = \lambda$, and Corollary 3.4.11 says that there is $\mathsf{b} \in \mathcal{J}_{<\lambda^+}(\mathsf{a}) \cap D$. By the choice of b_λ, we have $\mathsf{b} \subseteq_{\mathcal{J}_{<\lambda}(\mathsf{a})} \mathsf{b}_\lambda$. The first part of the proof shows that D, extending I, is also an extension of $\mathcal{J}_{<\lambda}(\mathsf{a})$, and hence $\mathsf{b} \subseteq_D \mathsf{b}_\lambda$. It follows that $\mathsf{b}_\lambda \in D$ which contradicts $\mathsf{a} \setminus \mathsf{b}_\lambda \in D$.

Corollary 5.3.3 *Assume that* a *is progressive,* $\lambda \in \operatorname{pcf}(\mathsf{a})$, *and* b_λ *is a subset of* a *generating* $\mathcal{J}_{<\lambda^+}(\mathsf{a})$ *over* $\mathcal{J}_{<\lambda}(\mathsf{a})$. *Then* $\lambda \notin \operatorname{pcf}(\mathsf{a} \setminus \mathsf{b}_\lambda)$.

Proof We assume, to get a contradiction, that $\lambda \in \operatorname{pcf}(\mathsf{a} \setminus \mathsf{b}_\lambda)$. Then there is an ultrafilter D on a satisfying $\mathsf{a} \setminus \mathsf{b}_\lambda \in D$ and $\operatorname{cf}(\prod \mathsf{a}/D) = \lambda$. The previous theorem says that $D \cap \mathcal{J}_{<\lambda}(\mathsf{a}) \restriction \mathsf{b}_\lambda = \emptyset$ on account of the minimality of $\mathcal{J}_{<\lambda}(\mathsf{a}) \restriction \mathsf{b}_\lambda$, and $\mathsf{a} \setminus \mathsf{b}_\lambda \in \mathcal{J}_{<\lambda}(\mathsf{a}) \restriction \mathsf{b}_\lambda$ and thus $\mathsf{a} \setminus \mathsf{b}_\lambda \notin D$ leads to a contradiction.

Definition *If* c *is a set of cardinal numbers, then we call any sequence* $(\mathsf{b}_\lambda : \lambda \in \mathsf{c})$ *a* **sequence of generators for a** *iff, for every* $\lambda \in \mathsf{c}$, *the ideal* $\mathcal{J}_{<\lambda^+}(\mathsf{a})$ *is generated by* $\{\mathsf{b}_\lambda\} \cup \mathcal{J}_{<\lambda}(\mathsf{a})$.

Corollary 5.3.4 *If* a *is progressive and* $(\mathsf{b}_\lambda : \lambda \in \operatorname{pcf}(\mathsf{a}))$ *is a sequence of generators for* a, *then, for each subset* c *of* a, *there is* $n \in \omega$ *and* $\lambda_1, \ldots, \lambda_n \in \operatorname{pcf}(\mathsf{c})$ *such that*

a) $\max\{\lambda_1, \ldots, \lambda_n\} = \max \operatorname{pcf}(\mathsf{c})$ *and*

b) $\mathsf{c} \subseteq \mathsf{b}_{\lambda_1} \cup \ldots \cup \mathsf{b}_{\lambda_n}$.

Proof Put $\mathsf{c}_0 := \mathsf{c}$ and let, for $n < \omega$,

$$\mathsf{c}_{n+1} := \begin{cases} \mathsf{c}_n \setminus \mathsf{b}_{\max \operatorname{pcf}(\mathsf{c}_n)} & \text{if } \mathsf{c}_n \neq \emptyset, \\ \emptyset & \text{if } \mathsf{c}_n = \emptyset. \end{cases}$$

Corollary 5.3.3 tells us that $\max \operatorname{pcf}(\mathsf{c}_{n+1}) < \max \operatorname{pcf}(\mathsf{c}_n)$ if $\mathsf{c}_{n+1} \neq \emptyset$. Consequently, there must be a smallest $n < \omega$ such that $\mathsf{c}_n = \emptyset$. Putting $\lambda_{i+1} := \max \operatorname{pcf}(\mathsf{c}_i)$ for all $i < n$, we get the assertion of the corollary.

Corollary 5.3.5 *If* a *is progressive,* $(\mathsf{b}_\lambda : \lambda \in \mathrm{pcf}(\mathsf{a}))$ *is a sequence of generators for* a*, and* D *is an ultrafilter on* a*, then*

$$\mathrm{cf}\,(\prod \mathsf{a}/D) = \min\{\lambda \in \mathrm{pcf}(\mathsf{a}) : \mathsf{b}_\lambda \in D\}.$$

Proof For $\mu := \mathrm{cf}\,(\prod \mathsf{a}/D)$ we can infer from Corollary 3.4.11 that $D \cap J_{<\mu}(\mathsf{a}) = \emptyset$ and $D \cap J_{<\mu^+}(\mathsf{a}) \neq \emptyset$. Taking a member b of $D \cap J_{<\mu^+}(\mathsf{a})$, we get $\mathsf{b} \setminus \mathsf{b}_\mu \in J_{<\mu}(\mathsf{a})$ and thus $\mathsf{b}_\mu \in D$. If $\mathsf{b}_\lambda \in D$, then $\mu = \mathrm{cf}\,(\prod \mathsf{a}/D) \leq \lambda$ since b_λ is a member of $\mathcal{J}_{<\lambda^+}(\mathsf{a})$. This proves the assertion.

Chapter 6

The Supremum of $\mathrm{pcf}_\mu(\mathsf{a})$

In this chapter we introduce first the essential notion of a control function. Its importance will come out when, in Chapter 8, we determine the cofinalities of partial orderings of the form $([\lambda]^{\leq\kappa}, \subseteq)$. There we will prove the following theorem of Shelah:
If κ is an infinite cardinal number and α is an ordinal such that $1 \leq \alpha < \kappa$, then

$$\mathrm{cf}\,([\kappa^{+\alpha}]^{\leq\kappa}, \subseteq) = \max\ \mathrm{pcf}(\{\kappa^{+(\beta+1)} : \beta < \alpha\}).$$

Furthermore the notion of a control function will be applied in the proof of a central theorem, due to Shelah:
There is *no* sequence $(\kappa_i : i < |\mathsf{a}|^+)$ of elements of $\mathrm{pcf}(\mathsf{a})$ such that

$$\max\ \mathrm{pcf}(\{\kappa_j : j < i\}) < \kappa_i \text{ for all } i < |\mathsf{a}|^+.$$

As the main result of this chapter we will get:
If a is a nonempty progressive interval of regular cardinals without a maximum, μ is a cardinal, and $\min(\mathsf{a})$ is μ-strong, then

$$\sup\ \mathrm{pcf}_\mu(\mathsf{a}) = (\sup(\mathsf{a}))^{|\mathsf{a}|}.$$

In the introduction of Chapter 3 we have already described the crucial role this theorem plays in cardinal arithmetic.

6.1 Control Functions

In the whole chapter, a *denotes an infinite progressive set of regular cardinals.*

Definition *For any set* b *of regular cardinals with* $\mathsf{b} \subseteq \mathsf{a}$, *we put*

$$\lambda_{\mathsf{b}} := \begin{cases} \max\ \mathrm{pcf}(\mathsf{b}) & \textit{if } \mathsf{b} \neq \emptyset, \\ 0 & \textit{if } \mathsf{b} = \emptyset. \end{cases}$$

If $\mu \in (|\mathsf{a}|, \min(\mathsf{a}))_{\text{reg}}$ *is a cardinal number, and, for each subset* b *of* a, $\bar{f}^{\mathsf{b}} = (f^{\mathsf{b}}_\xi : \xi < \lambda_{\mathsf{b}})$ *is a sequence of members of* $\prod \mathsf{a}$ *such that*

(a) $\bar{f}^{\mathsf{b}}$ *is strictly increasing modulo* $\mathcal{J}_{<\lambda_{\mathsf{b}}}(\mathsf{a})$,

(b) $\bar{f}^{\mathsf{b}}$ *is* μ*-continuous, and*

(c) $\{f^{\mathsf{b}}_\xi \restriction \mathsf{b} : \xi < \lambda_{\mathsf{b}}\}$ *is cofinal in* $(\prod \mathsf{b}, \leq_{\mathsf{b}})$,

then we call $\bar{f} := (\bar{f}^{\mathsf{b}} : \mathsf{b} \subseteq \mathsf{a})$ *a* **special sequence for a and** μ.

Lemma 6.1.1 *There exists a sequence* $\bar{f} = (\bar{f}^{\mathsf{b}} : \mathsf{b} \subseteq \mathsf{a})$, *which is special for* a *and* μ *for all cardinals* $\mu \in (|\mathsf{a}|, \min(\mathsf{a}))_{\text{reg}}$.

Proof Fix a subset b of a. Using Theorem 3.4.21, we can find a subset G of $\prod \mathsf{b}$ with $|G| = \lambda_{\mathsf{b}}$, which is cofinal in $\prod \mathsf{b}$ under $\leq_{\mathsf{b}}$. Let $(g_\xi : \xi < \lambda_{\mathsf{b}})$ be an enumeration of G. We will define the desired sequence $\bar{f}^{\mathsf{b}} := (f^{\mathsf{b}}_\xi : \xi < \lambda_{\mathsf{b}})$ by recursion on ξ. Choose $f^{\mathsf{b}}_0 \in \prod \mathsf{a}$ and define in the successor step, for every $\xi < \lambda_{\mathsf{b}}$ and $\nu \in \mathsf{a}$,

$$f^{\mathsf{b}}_{\xi+1}(\nu) := \begin{cases} \max\{f^{\mathsf{b}}_\xi(\nu) + 1, g_\xi(\nu)\} & \text{if } \nu \in \mathsf{b}, \\ f^{\mathsf{b}}_\xi(\nu) + 1 & \text{if } \nu \notin \mathsf{b}. \end{cases}$$

Now let $\xi < \lambda_{\mathsf{b}}$ be a limit ordinal and $(f^{\mathsf{b}}_\zeta : \zeta < \xi)$ with the desired properties be already defined. If $\mathrm{cf}(\xi) \in (|\mathsf{a}|, \min(\mathsf{a}))_{\text{reg}}$, then we put for every $\nu \in \mathsf{a}$

$$f^{\mathsf{b}}_\xi(\nu) := \min\{\sup\{f^{\mathsf{b}}_\zeta(\nu) : \zeta \in C\} : C \text{ is a club in } \xi\}.$$

If $\mathrm{cf}(\xi) \notin (|\mathsf{a}|, \min(\mathsf{a}))_{\text{reg}}$, then we can choose, using the fact that $(\prod \mathsf{a}, <_{\mathcal{J}_{<\lambda_{\mathsf{b}}}(\mathsf{a})})$ is λ_{b}-directed, the function $f^{\mathsf{b}}_\xi \in \prod \mathsf{a}$ such that $f^{\mathsf{b}}_\zeta <_{\mathcal{J}_{<\lambda_{\mathsf{b}}}(\mathsf{a})} f^{\mathsf{b}}_\xi$ for all $\zeta < \xi$. This completes the recursive definition.

The construction shows that $\bar{f}^{\mathsf{b}}$ is μ-continuous for all $\mu \in (|\mathsf{a}|, \min(\mathsf{a}))_{\text{reg}}$. Furthermore $\{f^{\mathsf{b}}_\xi \restriction \mathsf{b} : \xi < \lambda_{\mathsf{b}}\}$ is cofinal in $\prod \mathsf{b}$ because, for each $h \in \prod \mathsf{b}$, there is $\alpha < \lambda_{\mathsf{b}}$ such that $h \leq g_\alpha$, hence $h \leq f^{\mathsf{b}}_{\alpha+1} \restriction \mathsf{b}$. Finally Lemma 5.1.1 tells us that $\bar{f}^{\mathsf{b}}$ is strictly increasing modulo $\mathcal{J}_{<\lambda_{\mathsf{b}}}(\mathsf{a})$.

Lemma 6.1.2 *Assume that* $\bar{f} = (\bar{f}^{\mathsf{b}} : \mathsf{b} \subseteq \mathsf{a})$ *is a special sequence for* a *and* μ *and* Θ *is a regular cardinal such that* $2^{2^{\sup(\mathsf{a})}} < \Theta$. *Then* $\bar{f}$ *is a member of* $\mathrm{H}(\Theta)$.

Proof Let $\lambda := \max \mathrm{pcf}(\mathsf{a})$. Obviously we have

$$(*) \qquad \bar{f} \subseteq \mathcal{P}(\mathsf{a}) \times \bigcup\{{}^{\lambda_{\mathsf{b}}}\prod \mathsf{a} : \mathsf{b} \subseteq \mathsf{a}\}.$$

In Example 4.2.15 f) we have shown that $\mathcal{P}(\mathsf{a}) \in \mathrm{H}(\Theta)$ and ${}^{\lambda_{\mathsf{b}}}\prod \mathsf{a} \subseteq \mathrm{H}(\Theta)$ for every $\mathsf{b} \subseteq \mathsf{a}$. From the assumption on Θ we can infer that $|{}^{\lambda_{\mathsf{b}}}\prod \mathsf{a}| \leq |\prod \mathsf{a}|^{|\prod \mathsf{a}|} \leq$

$2^{|\prod \mathsf{a}|} \leq 2^{\sup(\mathsf{a})^{|\mathsf{a}|}} \leq 2^{2^{\sup(\mathsf{a})}} < \Theta$. Since $|\bigcup\{{}^{\lambda_\mathsf{b}}\prod \mathsf{a} : \mathsf{b} \subseteq \mathsf{a}\}| \leq |\mathcal{P}(\mathsf{a})| \cdot |{}^{\lambda}\prod \mathsf{a}| < \Theta$, we know from Lemma 4.1.4 that $\mathcal{P}(\mathsf{a}) \times \bigcup\{{}^{\lambda_\mathsf{b}}\prod \mathsf{a} : \mathsf{b} \subseteq \mathsf{a}\}$ is a member of $\mathrm{H}(\Theta)$. Using $(*)$, we get $\bar{f} \in \mathrm{H}(\Theta)$.

Lemma 6.1.3 *Let Θ be a regular cardinal with $\Theta > 2^{2^{\sup(\mathsf{a})}}$, $\mu \in (|\mathsf{a}|, \min(\mathsf{a}))_{\mathrm{reg}}$, $(N_i : i \leq \mu)$ be an approximation sequence in $\mathrm{H}(\Theta)$, and $N := N_\mu$. If $\bar{f} := (\bar{f^\mathsf{b}} : \mathsf{b} \subseteq \mathsf{a})$ is a special sequence for a and μ, $|N_i| < \min(\mathsf{a})$ for all $i < \mu$, and $\mathsf{a} \cup \{\mathsf{a}, \bar{f}\} \subseteq N_0$, then, for all $\mathsf{b} \in N \cap \mathcal{P}(\mathsf{a})$:*

a) If $\mathsf{b} \neq \emptyset$, then $f^\mathsf{b}_{\chi_N(\lambda_\mathsf{b})} \leq_\mathsf{a} \chi_N \restriction \mathsf{a}$.

b) (i) There is a set $\mathsf{c} \in N \cap J_{<\lambda_\mathsf{b}}(\mathsf{b})$ such that

$$f^\mathsf{b}_{\chi_N(\lambda_\mathsf{b})} \restriction (\mathsf{b} \setminus \mathsf{c}) = \chi_N \restriction (\mathsf{b} \setminus \mathsf{c}).$$

Moreover, c has the form $\mathsf{c} = \mathrm{B}[f^\mathsf{b}_\beta, f^\mathsf{b}_\alpha] \cap \mathsf{b}$ for some $\alpha, \beta \in N \cap \lambda_\mathsf{b}$ with $\alpha < \beta$.

(ii) There is a club C in μ, such that, for all $i, j \in C$ with $i < j$, the assertion

$$f^\mathsf{b}_{\chi_N(\lambda_\mathsf{b})} \restriction (\mathsf{b} \setminus \mathsf{c}_{i,j}) = \chi_N \restriction (\mathsf{b} \setminus \mathsf{c}_{i,j})$$

holds for $\mathsf{c}_{i,j} := \mathrm{B}[f^\mathsf{b}_{\chi_{N_j}(\lambda_\mathsf{b})}, \chi_{N_i}] \cap \mathsf{b}$. The sets $\mathsf{c}_{i,j}$ are definable in $\mathrm{H}(\Theta)$ by a formula with the parameters $\bar{f}$, b, N_i, and N_j from N. Furthermore,

$$\mathsf{c}_{i,j} \in N \cap \mathcal{J}_{<\lambda_\mathsf{b}}(\mathsf{b}).$$

Proof Consider $\mathsf{b} \in N \cap \mathcal{P}(\mathsf{a})$. We want to apply Lemma 4.3.4. Without loss of generality we can assume that b is a member of N_0 since otherwise, if $\mathsf{b} \in N_{j_0}$, we can replace the given approximation sequence by its final segment $(N_i : j_0 \leq i \leq \mu)$. Since

$$\lambda_\mathsf{b} = \max \mathrm{pcf}(\mathsf{b}) \leq \max \mathrm{pcf}(\mathsf{a}) \leq \sup(\mathsf{a})^{|\mathsf{a}|} \leq 2^{\sup(\mathsf{a})} < 2^{2^{\sup(\mathsf{a})}} < \Theta,$$

Lemma 4.1.4 tells us that $\lambda_\mathsf{b} \in \mathrm{H}(\Theta)$. $\mathsf{b} \in N_0$ and $N_0 \preceq \mathrm{H}(\Theta)$ yield $\lambda_\mathsf{b} \in N_0$, see Example 4.2.15 f), and $\{\mathsf{b}, \bar{f}\} \subseteq N_0$ gives $\bar{f^\mathsf{b}} \in N_0$. Now, together with property (b) from the definition of a special sequence, all assumptions of Lemma 4.3.4 are fulfilled, and we get $f^\mathsf{b}_{\chi_N(\lambda_\mathsf{b})} \leq_\mathsf{a} \chi_N \restriction \mathsf{a}$, which proves a). Futhermore, we obtain a normal sequence $(i(\xi) : \xi < \mu)$ of limit ordinals in μ such that, for every $\nu \in \mathsf{a}$,

$$(*) \qquad f^\mathsf{b}_{\chi_N(\lambda_\mathsf{b})}(\nu) = \sup\{f^\mathsf{b}_{\chi_{N_{i(\xi)}}(\lambda_\mathsf{b})}(\nu) : \xi < \mu\}.$$

Clearly the set $C_1 := \{i(\xi) : \xi < \nu\}$ is a club in μ. Together with property (c) from the definition of a special sequence, Lemma 4.3.4 implies that, for each $i \in C_1$, there is an ordinal $\beta_i \in N_{i+1} \cap \lambda_\mathsf{b}$ such that

$$(**) \qquad \chi_{N_i} \restriction \mathsf{b} <_\mathsf{b} f^\mathsf{b}_{\beta_i} \restriction \mathsf{b}.$$

For any $i \leq \mu$ let $\delta_i := \chi_{N_i}(\lambda_{\mathsf{b}})$. If we put

$$\mathsf{c}_i := \{\nu \in \mathsf{b} : \chi_{N_i}(\nu) \geq f^{\mathsf{b}}_{\delta_\mu}(\nu)\},$$

for $i < \mu$, then the sequence $(\mathsf{c}_i : i < \mu)$ of subsets of a is increasing. Since $\mu > |\mathsf{a}|$ is regular, there must be an ordinal $i_1 < \mu$ such that $\mathsf{c}_{i_1} = \mathsf{c}_i$ for all $i \in [i_1, \mu)_{\mathrm{ON}}$.

We put $C := C_1 \setminus i_1$. Let $i \in C$ and $\nu \in \mathsf{b} \setminus \mathsf{c}_i$. For every $j \in [i, \mu)_{\mathrm{ON}}$ we have $\nu \in \mathsf{b} \setminus \mathsf{c}_j$, hence $\chi_{N_j}(\nu) < f^{\mathsf{b}}_{\delta_\mu}(\nu)$; this gives $\chi_N(\nu) \leq f^{\mathsf{b}}_{\delta_\mu}(\nu)$. Thus we can conclude from a) that

$$f^{\mathsf{b}}_{\delta_\mu} \upharpoonright (\mathsf{b} \setminus \mathsf{c}_i) = \chi_N \upharpoonright (\mathsf{b} \setminus \mathsf{c}_i).$$

Fix $i, j \in C$ such that $i < j$. Since $\beta_i \in N_{i+1}$, we have $\beta_i \in N_j$, hence $\beta_i < \delta_j$, and property (a) in the definition of a special sequence yields, together with $\mathcal{J}_{<\lambda}(\mathsf{b}) = \mathcal{J}_{<\lambda}(\mathsf{a}) \cap \mathcal{P}(\mathsf{b})$:

$$\mathsf{c} := \{\nu \in \mathsf{b} : f^{\mathsf{b}}_{\beta_i}(\nu) \geq f^{\mathsf{b}}_{\delta_j}(\nu)\} = \mathrm{B}[f^{\mathsf{b}}_{\delta_j}, f^{\mathsf{b}}_{\beta_i}] \cap \mathsf{b} \in J_{<\lambda_{\mathsf{b}}}(\mathsf{b}).$$

We put

$$\mathsf{c}_{i,j} := \{\nu \in \mathsf{b} : \chi_{N_i}(\nu) \geq f^{\mathsf{b}}_{\delta_j}(\nu)\} = \mathrm{B}[f^{\mathsf{b}}_{\delta_j}, \chi_{N_i}] \cap \mathsf{b}.$$

Then $\mathsf{c}_{i,j} \subseteq \mathsf{c}$, since $\chi_{N_i} \upharpoonright \mathsf{b} <_{\mathsf{b}} f^{\mathsf{b}}_{\beta_i} \upharpoonright \mathsf{b}$ by (**), and thus $\mathsf{c}_{i,j} \in J_{<\lambda_{\mathsf{b}}}(\mathsf{b})$. Now (*) implies $\mathsf{c}_i \subseteq \mathsf{c}_{i,j}$, since $j \in C$, and therefore we can infer that

$$f^{\mathsf{b}}_{\delta_\mu} \upharpoonright (\mathsf{b} \setminus \mathsf{c}) = \chi_N \upharpoonright (\mathsf{b} \setminus \mathsf{c}) \quad \text{and} \quad f^{\mathsf{b}}_{\delta_\mu} \upharpoonright (\mathsf{b} \setminus \mathsf{c}_{i,j}) = \chi_N \upharpoonright (\mathsf{b} \setminus \mathsf{c}_{i,j}).$$

The sets $\mathsf{c}_{i,j}$ and c are members of $\mathrm{H}(\Theta)$, since they are subsets of $\mathrm{H}(\Theta)$ and have a cardinality less than Θ. Furthermore the above definition of the sets $\mathsf{c}_{i,j}$ shows that they are definable in $\mathrm{H}(\Theta)$ by a formula φ with the parameters $\bar{f}$, b, N_i and N_j from N, and thus they are also members of N. Analogously one can see, together with $\beta_i \in N$, that $\mathsf{c} \in N$, which completes the proof of the lemma.

Definition *If $\bar{f}$ is a special sequence for a and μ for all $\mu \in (|\mathsf{a}|, \min(\mathsf{a}))_{\mathrm{reg}}$ (the existence of such a sequence has been established in Lemma 6.1.1), we define by recursion on λ_{b} the sets*

$$\begin{aligned} \mathcal{F}_{\mathsf{b}} \;:=\; & \{f^{\mathsf{b}}_{\xi} \upharpoonright (\mathsf{b} \setminus \mathsf{c}) \cup g \upharpoonright \mathsf{c} : \xi < \lambda_{\mathsf{b}} \wedge g \in \mathcal{F}_{\mathsf{c}} \wedge \\ & \exists \alpha, \beta < \lambda_{\mathsf{b}}\, (\alpha < \beta \wedge \mathsf{c} = \{\nu \in \mathsf{b} : f^{\mathsf{b}}_{\alpha}(\nu) \geq f^{\mathsf{b}}_{\beta}(\nu)\})\} \end{aligned} \tag{6.1}$$

for each $\mathsf{b} \subseteq \mathsf{a}$. Since $\bar{f}^{\mathsf{b}}$ is strictly increasing modulo $\mathcal{J}_{<\lambda_{\mathsf{b}}}(\mathsf{a})$, we have $\mathsf{c} \in \mathcal{J}_{<\lambda_{\mathsf{b}}}(\mathsf{a})$ for the set c in Equation (6.1). Lemma 3.4.20 tell us that $\lambda_{\mathsf{c}} < \lambda_{\mathsf{b}}$, hence $\mathcal{F}_{\mathsf{c}}$ is already defined.

Any function $\mathcal{F}$ given by

$$\mathcal{F} := \begin{cases} \mathcal{P}(\mathsf{a}) & \longrightarrow & \bigcup\{\mathcal{P}(\prod \mathsf{b}) : \mathsf{b} \subseteq \mathsf{a}\} \\ \mathsf{b} & \longmapsto & \mathcal{F}_\mathsf{b} \end{cases}$$

is called a **control function for a generated by $\bar{f}$**; $\mathcal{F}_\mathsf{b}$ *is called a* **control set** *for* b.

Lemma 6.1.4 *If $\mathcal{F}$ is a control function for* a, *and* b *is a nonempty subset of* a, *then*

$$|\mathcal{F}_\mathsf{b}| \leq \max \mathrm{pcf}(\mathsf{b}).$$

Proof The assertion can easily be verified by induction on λ_b using the recursive definition in Equation (6.1): There are λ_b choices for ξ, α and β, and, by the inductive hypothesis, for each such choice $|\mathcal{F}_\mathsf{c}| \leq \lambda_\mathsf{c} < \lambda_\mathsf{b}$ many choices of g.

The following theorem allows the construction of an elementary substructure of $\mathrm{H}(\Theta)$ whose characteristic functions on subsets of a we know in advance since they are members of a given control set. The cardinality of a control set can be estimated with the previous lemma. Thus a very strong assertion is obtained.

Theorem 6.1.5 *Assume that* a *is a progressive set of regular cardinals,* $\mathsf{b} \subseteq \mathsf{a}$, *and $\mathcal{F}$ is a control function for* a *generated by $\bar{f}$. Further let $\Theta > 2^{2^{\sup(\mathsf{a})}}$ be regular, $\mu \in (|\mathsf{a}|, \min(\mathsf{a}))_{\mathrm{reg}}$, and $(N_i : i \leq \mu)$ be an approximation sequence in $\mathrm{H}(\Theta)$ such that $\{\bar{f}, \mathsf{a}, \mathsf{b}\} \cup \mathsf{a} \subseteq N_0$ and $|N_i| < \min(\mathsf{a})$ for all $i < \mu$. Then*

$$\chi_{N_\mu} \restriction \mathsf{b} \in \mathcal{F}_\mathsf{b}.$$

Proof We use transfinite induction on λ_b. Choose the set c as in part b) (i) of Lemma 6.1.3. Without loss of generality we can assume that $\mathsf{c} \in N_0$; otherwise we consider a corresponding final segment of the approximation sequence. So we can apply the inductive hypothesis to c and conclude that $\chi_{N_\mu} \restriction \mathsf{c} \in \mathcal{F}_\mathsf{c}$, hence

$$\chi_{N_\mu} \restriction \mathsf{b} = f^\mathsf{b}_{\chi_{N_\mu}(\lambda_\mathsf{b})} \restriction (\mathsf{b} \setminus \mathsf{c}) \cup \chi_{N_\mu} \restriction \mathsf{c} \in \mathcal{F}_\mathsf{b}.$$

6.2 The Supremum of $\mathrm{pcf}_\mu(\mathsf{a})$

Theorem 6.2.1 *There is no sequence $(\kappa_i : i < |\mathsf{a}|^+)$ of members of* pcf(a) *such that*

$$\max \mathrm{pcf}(\{\kappa_j : j < i\}) < \kappa_i$$

for all $i < |\mathsf{a}|^+$.

Remark Observe that in the previous theorem, for any $i < |\mathsf{a}|^+$, the maximum of

$$\mathrm{pcf}(\{\kappa_j : j < i\})$$

exists since the set $\{\kappa_j : j < i\}$ is progressive.

Proof[1] Let us assume, to get a contradiction, that there exists a sequence $(\kappa_i : i < |\mathsf{a}|^+)$ of members of pcf(a) such that

$$\max\ \mathrm{pcf}(\{\kappa_j : j < i\}) < \kappa_i$$

for all $i < |\mathsf{a}|^+$. Without loss of generality we can assume that $|\mathsf{a}|^{++} < \min(\mathsf{a})$. We put $\mathsf{a}^* := \mathsf{a} \cup \{\kappa_i : i < |\mathsf{a}|^+\}$ and $\mu := |\mathsf{a}^*|^+$. Then clearly a^* is progressive. Let $\bar{f} = (\bar{f}^{\mathsf{b}} : \mathsf{b} \subseteq \mathsf{a}^*)$ be a special sequence for a^* and μ. Further let

$$\bar{\mathsf{b}} := (\mathsf{b}_\nu : \nu \in \mathrm{pcf}(\mathsf{a}))$$

be a sequence of generators for a.

Choose $\Theta > 2^{2^{\sup(\mathsf{a}^*)}}$ regular, and let $(N_i : i \leq \mu)$ be an approximation sequence in $\mathrm{H}(\Theta)$, $N := N_\mu$, such that $|N_i| < \min(\mathsf{a})$ for all $i < \mu$ and

$$\mathsf{a}^* \cup |\mathsf{a}|^+ \cup \{\mathsf{a}, (\kappa_i : i < |\mathsf{a}|^+), \bar{\mathsf{b}}, \bar{f}\} \subseteq N_0.$$

First we claim that for each $i < |\mathsf{a}|^+$ there is a sequence $(\mathsf{c}^i_k : k \leq n_i) \in N$ such that $\emptyset = \mathsf{c}^i_{n_i} \subseteq \mathsf{c}^i_{n_i-1} \subseteq \ldots \subseteq \mathsf{c}^i_0 = \{\kappa_j : j < i\}$,

$$f^{\mathsf{c}^i_k}_{\chi_N(\lambda_{\mathsf{c}^i_k})} \leq_{\mathsf{c}^i_k} \chi_N \restriction \mathsf{a}^*$$

for any $k < n_i$, and

$$f^{\mathsf{c}^i_k}_{\chi_N(\lambda_{\mathsf{c}^i_k})} \restriction (\mathsf{c}^i_k \setminus \mathsf{c}^i_{k+1}) = \chi_N \restriction (\mathsf{c}^i_k \setminus \mathsf{c}^i_{k+1})$$

for any $k < n_i$. To show its existence, we need only apply Lemma 6.1.3 several times. Since $\mathsf{c}^i_{k+1} \in \mathcal{J}_{<\lambda_{\mathsf{c}^i_k}}(\mathsf{c}^i_k)$, we have $\max\ \mathrm{pcf}(\mathsf{c}^i_{k+1}) = \lambda_{\mathsf{c}^i_{k+1}} < \lambda_{\mathsf{c}^i_k} = \max\ \mathrm{pcf}(\mathsf{c}^i_k)$ by Lemma 3.4.20, and thus the descending chain of sets becomes constant after finitely many steps.

Put $f_{\nu,\xi} := f^{\mathsf{b}_\nu}_\xi \restriction \mathsf{a}$ for all $\nu \in \mathrm{pcf}(\mathsf{a})$ and $\xi < \lambda_{\mathsf{b}_\nu}$. We define

$$g_i := \sup\{f_{\kappa_j, \chi_N(\kappa_j)} : j < i\}$$

for $i < |\mathsf{a}|^+$. Since $\chi_N(\kappa_j) < \kappa_j$ by Lemma 4.3.3 and $\kappa_j = \max\ \mathrm{pcf}(\mathsf{b}_{\kappa_j}) = \lambda_{\mathsf{b}_{\kappa_j}}$ by Lemma 5.3.1, g_i is well-defined, and $|i| \leq |\mathsf{a}| < \min(\mathsf{a})$ gives $g_i \in \prod \mathsf{a}$. It is

[1] We follow the proof in [Sh5], VIII, 3.4.

easy to see that we get a special sequence for a and μ by restricting $\bar{f}$ to $\mathcal{P}(\mathsf{a})$. Consequently, Lemma 6.1.3 yields $g_i \leq \chi_N \restriction \mathsf{a}$. Put for $i < |\mathsf{a}|^+$

$$B_i := \{\nu \in \mathsf{a} : g_i(\nu) = \chi_N(\nu)\}.$$

Clearly we have $g_j \leq_{\mathsf{a}} g_i$ for $j < i < |\mathsf{a}|^+$, hence $B_j \subseteq B_i$. So there must be $i_0 < |\mathsf{a}|^+$ such that $B_i = B_{i_0}$ for all $i \in [i_0, |\mathsf{a}|^+)_{\mathrm{ON}}$. If $j < i < |\mathsf{a}|^+$, then Lemma 6.1.3 says that

$$\{\nu \in \mathsf{b}_{\kappa_j} : f_{\kappa_j, \chi_N(\kappa_j)}(\nu) < \chi_N(\nu)\} \in \mathcal{J}_{<\kappa_j}(\mathsf{a}) \tag{6.2}$$

and thus

$$\{\nu \in \mathsf{b}_{\kappa_j} : g_i(\nu) < \chi_N(\nu)\} \in \mathcal{J}_{<\kappa_j}(\mathsf{a}). \tag{6.3}$$

Putting $\lambda_{i,k} := \lambda_{\mathsf{c}_k^i}$ for $i < |\mathsf{a}|^+$ and $k < n_i$, we get

$$\begin{aligned}(\chi_N(\kappa_j) : j < i_0) &= \chi_N \restriction \{\kappa_j : j < i_0\} = \chi_N \restriction \mathsf{c}_0^{i_0} \\ &= f^{\mathsf{c}_0^{i_0}}_{\chi_N(\lambda_{i_0,0})} \restriction (\mathsf{c}_0^{i_0} \setminus \mathsf{c}_1^{i_0}) \cup \ldots \cup f^{\mathsf{c}^{i_0}_{n_{i_0}-1}}_{\chi_N(\lambda_{i_0,n_{i_0}-1})} \restriction \mathsf{c}^{i_0}_{n_{i_0}-1}.\end{aligned}$$

Now we have $\bar{f} \in N$, $i_0 \in |\mathsf{a}|^+ \subseteq N$, $(\mathsf{c}_k^{i_0} : k \leq n_{i_0}) \in N$, and $(\kappa_i : i < |\mathsf{a}|^+) \in N$, hence $\kappa_{i_0} \in N$. Therefore the above representation of the sequence $(\chi_N(\kappa_j) : j < i_0)$ shows that this sequence and thus the function $g_{i_0} = \sup\{f_{\kappa_j, \chi_N(\kappa_j)} : j < i_0\}$ are members of

$$N^* := \mathrm{Skol}_\Theta(N \cup \{\chi_N(\lambda_{i_0,k}) : k < n_{i_0}\}).$$

The assumption on the sequence $(\kappa_i : i < |\mathsf{a}|^+)$ yields, for $k < n_{i_0}$, $\lambda_{i_0,k} = \max \mathrm{pcf}(\mathsf{c}_k^{i_0}) \leq \max \mathrm{pcf}(\mathsf{c}_0^{i_0}) < \kappa_{i_0}$. Example 4.2.15 tells us that $\lambda_{i_0,k} \in N$, hence $\chi_N(\lambda_{i_0,k}) < \chi_N(\kappa_{i_0})$. Since $\kappa_{i_0} \in N$, Corollary 4.4.4 says that $\chi_N(\kappa_{i_0}) = \chi_{N^*}(\kappa_{i_0})$.

Now, by the choice of $\bar{f}$, the ordinal

$$\delta := \min\{\xi < \kappa_{i_0} : g_{i_0} \leq_{\mathcal{J}_{<\kappa_{i_0}}(\mathsf{a}) \restriction \mathsf{b}_{\kappa_{i_0}}} f_{\kappa_{i_0}, \xi}\}$$

exists and is a member of N^*, since every parameter in the definition of δ is a member of N^*, and thus $\delta < \chi_N(\kappa_{i_0})$. We conclude that

$$\mathsf{e}_1 := \{\nu \in \mathsf{b}_{\kappa_{i_0}} : f_{\kappa_{i_0}, \delta}(\nu) \geq f_{\kappa_{i_0}, \chi_N(\kappa_{i_0})}(\nu)\} \in \mathcal{J}_{<\kappa_{i_0}}(\mathsf{a}).$$

From 6.2 and 6.3, with $i_0, i_0 + 1$ instead of j, i, together with $f_{\kappa_{i_0}, \chi_N(\kappa_{i_0})} \leq_{\mathsf{a}} g_{i_0+1} \leq_{\mathsf{a}} \chi_N(\nu) \restriction \mathsf{a}$, we can infer that

$$\mathsf{e}_2 := \{\nu \in \mathsf{b}_{\kappa_{i_0}} : \neg(f_{\kappa_{i_0}, \chi_N(\kappa_{i_0})}(\nu) = \chi_N(\nu) = g_{i_0+1}(\nu))\} \in \mathcal{J}_{<\kappa_{i_0}}(\mathsf{a}).$$

If ν is a member of the $\mathcal{J}_{<\kappa_{i_0}}(\mathsf{a})$-positive set $\mathsf{b}_{\kappa_{i_0}} \setminus (\mathsf{e}_1 \cup \mathsf{e}_2)$, then

$$f_{\kappa_{i_0},\delta}(\nu) < f_{\kappa_{i_0},\chi_N(\kappa_{i_0})}(\nu) = \chi_N(\nu) = g_{i_0+1}(\nu) = g_{i_0}(\nu).$$

To justify the last equation we observe first that $\nu \in B_{i_0+1}$. By the choice of i_0, we have $B_{i_0+1} = B_{i_0}$, and thus $\nu \in B_{i_0}$, hence $g_{i_0+1}(\nu) = g_{i_0}(\nu)$. Altogether we get

$$\neg(g_{i_0} \leq_{J_{<\kappa_{i_0}}(\mathsf{a}) \restriction \mathsf{b}_{\kappa_{i_0}}} f_{\kappa_{i_0},\delta}),$$

contradicting the choice of δ. This completes the proof of the theorem.

Definition *Assume that M is an arbitrary set and b is a set of ordinals. We say that a function* **D characterizes M over b** *iff*

(1) $M \cap b \subseteq \mathrm{dom}(D)$ and
(2) For any $\delta \in M \cap b$, $D(\delta)$ is a cofinal subset of $M \cap [\min(b), \delta)_{\mathrm{ON}}$.

The following lemma shows the importance of this definition.

Lemma 6.2.2 *Assume that Θ is an uncountable regular cardinal, $a \subseteq \mathrm{H}(\Theta) \cap$ ICN is nonempty such that $a \setminus \{\min(a)\}$ is an interval of regular cardinals, and M and N are elementary substructures of $\mathrm{H}(\Theta)$ with the following properties:*

(1) $M \cap \min(a) \subseteq N$.
(2) $M \cap a \subseteq N$.
(3) There is a family of subsets of N characterizing M over a.

Then $M \cap \sup(a) \subseteq N$.

Proof Let $(D_\delta : \delta \in a \cap M)$ be a characterizing function for M over a such that $D_\delta \subseteq N$ for all $\delta \in a \cap M$. By induction on λ we will show that $M \cap \lambda \subseteq N$ for all cardinals $\lambda \leq \sup(a)$.

For $\lambda = \min(a)$, this assertion is true by (1), and for limit cardinals λ it follows from the inductive hypothesis. So assume that $\lambda \in [\min(a), \sup(a))_{\mathrm{CN}}$ and $M \cap \lambda \subseteq N$. We will show that $M \cap \lambda^+ \subseteq N$.

Let $\alpha \in M \cap \lambda^+$ and, without loss of generality, $\alpha \geq \lambda$. From Example 4.2.15 we can infer that $\lambda = |\alpha| \in M$ and $\lambda^+ \in M$. Clearly λ^+ is an element of a. By (2) we have $\lambda^+ \in N$, hence $\lambda \in N$. From assumption (3) we obtain an ordinal $\beta \in D_{\lambda^+}$ such that $\alpha < \beta$ and $\beta \in M \cap N \cap \lambda^+$. Since $|\beta| = \lambda$, there is a $<^*$-least bijection h from λ onto β, which is a member of both M and N by Example 4.2.15. Again 4.2.15 tells us that $h \restriction (M \cap \lambda)$ is a bijection from $\lambda \cap M$ onto $\beta \cap M$. Analogously this holds for N. Using the inductive hypothesis, we can conclude that

$$\alpha \in M \cap \beta = h[M \cap \lambda] \subseteq h[N \cap \lambda] = N \cap \beta,$$

and thus $\alpha \in N$, as desired.

The set $M \cap \sup(a)$ here is sufficiently characterized by $M \cap \min(a)$, $M \cap a$ and a corresponding characterizing sequence. This permits an estimate for the number of sets of the form $M \cap \sup(a)$.

Lemma 6.2.3 *Assume that* $\mathsf{a} \subseteq \mathrm{H}(\Theta)$ *is a nonempty interval of regular cardinals,* μ *is an infinite cardinal, and* $\mathcal{D}$ *is a set. We say that an elementary substructure* M *of* $\mathrm{H}(\Theta)$ *is* **$\mathcal{D}$-nice** *or* **nice** *iff* $|M| \leq \mu$ *and there is a function which characterizes* M *over* a *and is an element of* $\mathcal{D}$*. Then*

$$|\{M \cap \sup(\mathsf{a}) : M \text{ is nice}\}| \leq |\min(\mathsf{a})|^\mu \cdot |\mathsf{a}|^\mu \cdot |\mathcal{D}|.$$

Proof For every nice M we can choose, by the assumption on a, a member D_M of $\mathcal{D}$ characterizing M over a. For any nice elementary substructures M_1, M_2 of $\mathrm{H}(\Theta)$ let $M_1 \sim M_2$ iff $M_1 \cap \sup(\mathsf{a}) = M_2 \cap \sup(\mathsf{a})$, and let $\mathcal{R}$ be a set of representatives for the equivalence classes of $\sim$. Define, for $M \in \mathcal{R}$,

$$(*) \qquad g(M \cap \sup(\mathsf{a})) := (M \cap \min(\mathsf{a}), M \cap \mathsf{a}, D_M).$$

If $M, N \in \mathcal{R}$ and $(M \cap \min(\mathsf{a}), M \cap \mathsf{a}, D_M) = (N \cap \min(\mathsf{a}), N \cap \mathsf{a}, D_N)$, then Lemma 6.2.2 says that $M \cap \sup(\mathsf{a}) = N \cap \sup(\mathsf{a})$. Consequently, equation $(*)$ defines an injective function g from $\{M \cap \sup(\mathsf{a}) : M \text{ is nice}\}$ into the set $[\min(\mathsf{a})]^{\leq\mu} \times [\mathsf{a}]^{\leq\mu} \times \mathcal{D}$.

Remark If μ is a cardinal, then $\sup \mathrm{pcf}_\mu(\mathsf{a}) \leq (\sup(\mathsf{a}))^\mu$ by Lemma 3.3.4.

Theorem 6.2.4 *Assume that* a *is a nonempty progressive interval of regular cardinals without a maximum,* μ *is an infinite cardinal number, and* $\min(\mathsf{a})$ *is* μ*-strong. Then*

$$\sup \mathrm{pcf}_\mu(\mathsf{a}) = (\sup(\mathsf{a}))^\mu.$$

Proof In view of the preceding remark we only have to prove the inequality "$\geq$". Put $\lambda := \min(\mathsf{a})$. Since λ is μ-strong, we have $\mu < \lambda$ and get $\lambda^\mu = \lambda$. If λ is a successor cardinal, this follows from Hausdorff's formula, and if λ is a limit cardinal, hence weakly inaccessible, this follows from Lemma 1.6.15. Therefore λ^+ is also μ-strong. Since the assertion of the theorem is the same if we consider $\mathsf{a} \setminus \{\min(\mathsf{a})\}$ instead of a, we can assume without loss of generality that $\lambda = \kappa^+$, $\kappa > |\mathsf{a}|$ is regular, and $\mu < \kappa$. Now a has no members which are limit cardinals, since any limit cardinal $\nu \in [\min(\mathsf{a}), \sup(\mathsf{a}))_{\mathrm{CN}}$ obviously is the supremum of $\mathsf{a} \cap \nu$, thus we have $\mathrm{cf}(\nu) \leq |\mathsf{a}| < \min(\mathsf{a}) \leq \nu$, and ν is singular. Since $|\mathsf{a}| < \lambda$ and λ is μ-strong, we further conclude that

$$|[\mathsf{a}]^{\leq\mu}| = |\mathsf{a}|^\mu < \lambda. \tag{6.4}$$

Fix a regular cardinal $\Theta > 2^{2^{\sup(\mathsf{a})}}$, and let $\mathcal{D}$ be a set of functions. If M is an elementary substructure of $\mathrm{H}(\Theta)$ and M is nice according to the definition in Lemma 6.2.3, then $|\mathcal{P}(M)| \leq 2^\mu$. Assume that we can choose $\mathcal{D}$ in such a way that we can find, for each $x \subseteq \sup(\mathsf{a})$ with $|x| = \mu$, a nice M such that $x \subseteq M$. Then Lemma 6.2.3 says that

$$(\sup(\mathsf{a}))^\mu = |[\sup(\mathsf{a})]^\mu| \leq 2^\mu \cdot |\{M \cap \sup(\mathsf{a}) : M \text{ is nice}\}| \\ \leq |\min(\mathsf{a})|^\mu \cdot |\mathsf{a}|^\mu \cdot |\mathcal{D}| \leq \lambda \cdot |\mathcal{D}|.$$

This will complete the proof of the theorem if we can define in addition the set $\mathcal{D}$ in such a way that $|\mathcal{D}| \leq \sup \mathrm{pcf}_\mu(\mathsf{a})$.

Let $\mathcal{F} = (\mathcal{F}_\mathsf{b} : \mathsf{b} \subseteq \mathsf{a})$ be a control function for a generated by a function $\bar{h}$ which is special for a and all $\nu \in (|\mathsf{a}|, \min(\mathsf{a}))_{\mathrm{reg}}$. For each subset b of a we put

$$\mathcal{F}'_\mathsf{b} := \{f \in \mathcal{F}_\mathsf{b} : \forall \nu \in \mathsf{b}\ \mathrm{cf}(f(\nu)) = \kappa\}$$

and

$$\mathcal{G} := \bigcup\{\mathcal{F}'_A : A \in [\mathsf{a}]^{\leq\mu}\}.$$

From Lemma 6.1.4 and (6.4) we can infer that $|\mathcal{G}| \leq \sup \mathrm{pcf}_\mu(\mathsf{a})$. We will see later on that $\mathcal{G}$ is not empty. Now we choose for each limit ordinal $\alpha < \sup(\mathsf{a})$ with $\mathrm{cf}(\alpha) = \kappa$ a club C_α in α of order type κ, and define for each $f \in \mathcal{G}$ a set $\mathcal{D}_f$ by

$$\mathcal{D}_f := \{(D_\nu : \nu \in \mathrm{dom}(f)) : \forall \nu \in \mathrm{dom}(f)\ D_\nu \in [C_{f(\nu)}]^{\leq\aleph_0}\}.$$

Finally we put

$$\mathcal{D} := \bigcup\{\mathcal{D}_f : f \in \mathcal{G}\}.$$

If $f \in \mathcal{G}$, then $A := \mathrm{dom}(f) \in [\mathsf{a}]^{\leq\mu}$, and, for all $\nu \in A$, $\mathrm{cf}(f(\nu)) = \kappa$. This means that the club $C_{f(\nu)}$ is defined. Furthermore we have

$$|\mathcal{D}_f| \leq |\{\phi : \phi : \mathrm{dom}(f) \longrightarrow [\bigcup\{C_{f(\nu)} : \nu \in \mathrm{dom}(f)\}]^{\leq\aleph_0}\}| \leq |^\mu([\kappa]^{\leq\aleph_0})| < \lambda.$$

Consequently, we get $|\mathcal{D}| \leq \sup \mathrm{pcf}_\mu(\mathsf{a})$, and there remains to be shown that $\mathcal{D}$ has the desired properties.

So fix $x \subseteq \sup(\mathsf{a})$ with $|x| = \mu$. We will construct a nice structure M such that $x \subseteq M$. For this, we choose first an approximation sequence $(N_i : i \leq \kappa)$ in $\mathrm{H}(\Theta)$ such that

$$x \cup \mathsf{a} \cup [\mathsf{a}]^{\leq\mu} \cup \{\bar{h}, \mathsf{a}\} \subseteq N_0$$

and $|N_i| = \kappa$ for all $i \leq \kappa$. By Theorem 6.1.5, $\chi_{N_\kappa} \restriction A$ is a member of $\mathcal{F}_A$ and thus, using Lemma 4.3.3, also a member of $\mathcal{F}'_A$ for any $A \in [\mathsf{a}]^{\leq\mu}$. The same lemma yields, for each $\nu \in \mathsf{a}$, a club $C^*_\nu \subseteq N_\kappa \cap \nu$ in $\chi_{N_\kappa}(\nu)$ of order type κ. Without loss of generality we can assume that $C^*_\nu \subseteq C_{\chi_{N_\kappa}(\nu)}$, since otherwise we can take the intersection of the clubs which is possible since $\kappa > \omega$.

Next we define by recursion an ascending sequence $(M_n : n < \omega)$ of elementary substructures of N_κ having cardinality μ. First, applying the Löwenheim-Skolem theorem, let M_0 be an arbitrary elementary substructure of N_κ such that $x \subseteq M_0$ and $|M_0| = \mu$. Now assume that $n < \omega$, M_n is already defined, and $\nu \in M_n \cap [\min(\mathsf{a}), \sup(\mathsf{a}))_{\mathrm{CN}}$. Then $\chi_{M_n}(\nu^+) < \chi_{N_\kappa}(\nu^+)$, since $|M_n| = \mu < \kappa = \mathrm{cf}(\chi_{N_\kappa}(\nu^+))$. Therefore there is an ordinal, say $\sigma^n_{\nu^+}$, in $C^*_{\nu^+}$ such that $\chi_{M_n}(\nu^+) < \sigma^n_{\nu^+}$.

Take M_{n+1} as an elementary substructure of N_κ satisfying $M_n \subseteq M_{n+1}$, $|M_{n+1}| = \mu$ and

$$\{\sigma^n_{\nu^+} : \nu \in M_n \cap [\min(\mathsf{a}), \sup(\mathsf{a}))_{\mathrm{CN}}\} \subseteq M_{n+1}.$$

This completes the definition of the structures M_n, and we put

$$M := \bigcup\{M_n : n < \omega\}$$

and $f := \chi_{N_\kappa} \upharpoonright (M \cap \mathsf{a})$. The previous remark shows that $f \in \mathcal{F}'_{M\cap\mathsf{a}} \subseteq \mathcal{G}$.

Assume that $\rho \in M \cap \mathsf{a}$. If $\rho = \min(\mathsf{a})$, we put $\Sigma_\rho := \emptyset$. Otherwise we have shown that $\rho = \nu^+$ for some $\nu \in [\min(\mathsf{a}), \sup(\mathsf{a}))_{\mathrm{CN}}$, hence $\nu \in M$. There is $n_\nu < \omega$ with $\nu \in M_{n_\nu}$. Let

$$\Sigma_\rho := \{\sigma^n_\rho : n_\nu \le n < \omega\}.$$

Our construction shows that Σ_ρ is a countable subset of C^*_ρ and thus of $C_{f(\rho)}$. Therefore the sequence $\Sigma := (\Sigma_\rho : \rho \in M \cap \mathsf{a})$ is a member of $\mathcal{D}_f$ and thus of $\mathcal{D}$. It remains to be shown that Σ characterizes the set M over a, that means that Σ_ρ is unbounded in $M \cap [\min(\mathsf{a}), \rho)_{\mathrm{ON}}$ for all $\rho \in M \cap \mathsf{a}$.

To see this, let $\rho = \nu^+ \in M \cap \mathsf{a}$ and $\gamma \in M \cap \rho$. Without loss of generality we can assume that $\nu \ge \min(\mathsf{a})$, and, since $\nu \in M \cap [\min(\mathsf{a}), \rho)_{\mathrm{ON}}$ and $\nu < \chi_{M_{n_\nu}}(\nu^+) < \sigma^{n_\nu}_{\nu^+}$, that $\gamma \ge \nu$. Then there is $n < \omega$ such that $n \ge n_\nu$ and $\gamma \in M_n \cap [\nu, \nu^+)_{\mathrm{ON}}$. By construction, σ^n_ρ is greater than γ and a member of Σ_ρ, as desired. This completes the proof of the theorem.

Corollary 6.2.5 *If* a *is a nonempty interval of regular cardinals which has no maximum and satisfies* $(\min(\mathsf{a}))^{|\mathsf{a}|} < \sup(\mathsf{a})$, *then*

$$\max \mathrm{pcf}(\mathsf{a}) = |\prod \mathsf{a}|.$$

Proof Put $\kappa := ((\min(\mathsf{a}))^{|\mathsf{a}|})^+$. If $\mu < \kappa$ is a cardinal, then

$$\mu^{|\mathsf{a}|} \le ((\min(\mathsf{a}))^{|\mathsf{a}|})^{|\mathsf{a}|} = (\min(\mathsf{a}))^{|\mathsf{a}|} < \kappa,$$

i.e., κ is $|\mathsf{a}|$-strong. We put $\mathsf{b} := [\kappa, \sup(\mathsf{a}))_{\mathrm{reg}}$. Then $\prod_{\nu \in \mathsf{a}\setminus\mathsf{b}} \nu \le \kappa^{|\mathsf{a}|} = \kappa$. Since $\sup(\mathsf{a}) \le |\prod \mathsf{a}|$ and $\sup(\mathsf{a})$ is a limit cardinal, the assumption yields $\kappa <$

$\prod_{\nu\in\mathsf{a}}\nu$. From $\prod_{\nu\in\mathsf{a}}\nu = \prod_{\nu\in\mathsf{a}\setminus\mathsf{b}}\nu \cdot \prod_{\nu\in\mathsf{b}}\nu$ we conclude that $\prod_{\nu\in\mathsf{a}}\nu = \prod_{\nu\in\mathsf{b}}\nu$. Theorem 6.2.4, applied to the set b, implies

$$\max \mathrm{pcf}(\mathsf{a}) \geq \max \mathrm{pcf}(\mathsf{b}) = \sup \mathrm{pcf}_{|\mathsf{b}|}(\mathsf{b}) = (\sup(\mathsf{b}))^{|\mathsf{b}|} \geq |\prod \mathsf{b}| = |\prod \mathsf{a}|.$$

Now $\max \mathrm{pcf}(\mathsf{a}) \leq |\prod \mathsf{a}|$ yields the desired assertion.

Remark Under the assumptions of the previous corollary, we can infer that

$$\mathrm{cf}\left(\prod \mathsf{a}, \leq_{\mathsf{a}}\right) = |\prod \mathsf{a}|,$$

using Theorem 3.4.21.

Chapter 7

Local Properties

In his book [Sh5], Shelah introduces the operator pcf_* which satisfies $\mathrm{pcf}_*(\mathrm{pcf}_*(\mathsf{a})) = \mathrm{pcf}_*(\mathsf{a})$. If b is a set of regular cardinals, which is not necessarily progressive, and if every limit point of b is a singular cardinal, then $\mathrm{pcf}_*(\mathsf{b}) = \mathrm{pcf}(\mathsf{b})$. Progressive sets a of regular cardinals have only singular limit points, so they satisfy $\mathrm{pcf}_*(\mathsf{a}) = \mathrm{pcf}(\mathsf{a})$. In this case, we even get $\mathrm{pcf}(\mathsf{a}) = \mathrm{pcf}_*(\mathrm{pcf}(\mathsf{a}))$. If in addition every limit point of $\mathrm{pcf}(\mathsf{a})$ is singular, which is for example guaranteed by the condition $|\mathrm{pcf}(\mathsf{a})| < \min(\mathsf{a})$, then $\mathrm{pcf}(\mathrm{pcf}(\mathsf{a})) = \mathrm{pcf}(\mathsf{a})$. A central tool for the proof of the hull property of pcf_* is the so-called localisation theorem: If $\mathsf{c} \subseteq \mathrm{pcf}(\mathsf{a})$ is progressive and $\lambda \in \mathrm{pcf}(\mathsf{c})$, then there exists a set d such that $\mathsf{d} \subseteq \mathsf{c}$, $|\mathsf{d}| \leq |\mathsf{a}|$, and $\lambda \in \mathrm{pcf}(\mathsf{d})$. This theorem will also be applied in the proof of a main result of pcf-theory: If a is a progressive interval of regular cardinals, then $|\mathrm{pcf}(\mathsf{a})| < |\mathsf{a}|^{+4}$. The importance of this result will be demonstrated in Section 8.1.

7.1 The Ideals $\mathcal{J}_*(\mathsf{b})$ and $\mathcal{J}^p_{<\lambda}(\mathsf{a})$

In [Sh5], VIII §3, S. Shelah introduces several new concepts which we will investigate and characterize in this chapter[1]. Instead of progressive sets a, we will consider the sets $\mathrm{pcf}(\mathsf{a})$. The ideal $\mathcal{J}_{<\lambda}(\mathrm{pcf}(\mathsf{a}))$ exists, but many results of the preceding chapters cannot be applied to $\mathrm{pcf}(\mathsf{a})$, since it is not known whether $\mathrm{pcf}(\mathsf{a})$ is progressive.

One can extend the ideal $\mathcal{J}_{<\lambda}(\mathsf{a})$ on a to $\mathrm{pcf}(\mathsf{a})$ to obtain on this set the ideal $\mathcal{J}^p_{<\lambda}(\mathsf{a})$. Considering not all subsets of $\mathrm{pcf}(\mathsf{a})$ but only those being members of the set $\mathcal{J}_*(\mathrm{pcf}(\mathsf{a}))$, we will obtain a series of analogies to well-

[1] In addition, we use ideas from Shelah's note "An answer to complaints of E. Weitz on [Sh:371]", communicated to us in July 1995.

known results on progressive sets. In the same way this is true if we replace the pcf-operator by the operator pcf_*.

For the rest of this chapter, a denotes a progressive set of regular cardinals.

The first lemma contains a definition. Note that this definition makes sense since, by Theorem 5.2.3, a sequence of generators $(\mathsf{b}_\mu : \mu \in \mathrm{pcf}(\mathsf{a}))$ exists.

Lemma 7.1.1 *Let* $\lambda \leq \max \mathrm{pcf}(\mathsf{a})$ *be a cardinal. If* $(\mathsf{b}_\mu : \mu \in \mathrm{pcf}(\mathsf{a}))$ *and* $(\mathsf{b}^*_\mu : \mu \in \mathrm{pcf}(\mathsf{a}))$ *are sequences of generators for* a*, then the sets* $\{\mathrm{pcf}(\mathsf{b}_\mu) : \mu \in \lambda \cap \mathrm{pcf}(\mathsf{a})\}$ *and* $\{\mathrm{pcf}(\mathsf{b}^*_\mu) : \mu \in \lambda \cap \mathrm{pcf}(\mathsf{a})\}$ *generate the same ideal on* $\mathrm{pcf}(\mathsf{a})$. *We denote it by* $\mathcal{J}^p_{<\lambda}(\mathsf{a})$.

Proof Consider $\mu_1, \ldots, \mu_n \in \lambda \cap \mathrm{pcf}(\mathsf{a})$ with $\mu_1 < \ldots < \mu_n$. Then Lemma 5.3.1 says that

$$\max(\mathrm{pcf}(\mathsf{b}_{\mu_1}) \cup \ldots \cup \mathrm{pcf}(\mathsf{b}_{\mu_n})) = \max \mathrm{pcf}(\mathsf{b}_{\mu_n}) = \mu_n < \lambda,$$

hence $\mathrm{pcf}(\mathsf{b}_{\mu_1}) \cup \ldots \cup \mathrm{pcf}(\mathsf{b}_{\mu_n}) \neq \mathrm{pcf}(\mathsf{a})$. Therefore $\{\mathrm{pcf}(\mathsf{b}_\mu) : \mu \in \lambda \cap \mathrm{pcf}(\mathsf{a})\}$ generates an ideal on $\mathrm{pcf}(\mathsf{a})$.

Now let $\mu \in \lambda \cap \mathrm{pcf}(\mathsf{a})$. By Corollary 5.3.4, there are cardinals $\nu_1, \ldots, \nu_n \in \mathrm{pcf}(\mathsf{b}_\mu)$ such that $\mathsf{b}_\mu \subseteq \mathsf{b}^*_{\nu_1} \cup \ldots \cup \mathsf{b}^*_{\nu_n}$, hence $\mathrm{pcf}(\mathsf{b}_\mu) \subseteq \mathrm{pcf}(\mathsf{b}^*_{\nu_1}) \cup \ldots \cup \mathrm{pcf}(\mathsf{b}^*_{\nu_n})$. This shows that the ideal generated by $\{\mathrm{pcf}(\mathsf{b}_\mu) : \mu \in \lambda \cap \mathrm{pcf}(\mathsf{a})\}$ is a subset of the ideal generated by $\{\mathrm{pcf}(\mathsf{b}^*_\mu) : \mu \in \lambda \cap \mathrm{pcf}(\mathsf{a})\}$. So the assertion is proved by symmetry arguments.

If λ is a cardinal satisfying $\lambda > \max \mathrm{pcf}(\mathsf{a})$, then in addition to the definition of the previous lemma we define

$$\mathcal{J}^p_{<\lambda}(\mathsf{a}) := \mathcal{P}(\mathrm{pcf}(\mathsf{a})).$$

For the rest of this chapter, $(\mathsf{b}_\lambda : \lambda \in \mathrm{pcf}(\mathsf{a}))$ is a sequence of generators for a such that $\mathsf{b}_{\max \mathrm{pcf}(\mathsf{a})} := \mathsf{a}$.

First we collect some results which easily follow from the definition of $\mathcal{J}^p_{<\lambda}(\mathsf{a})$ (see the exercises of this chapter). The first four lemmata are the analogies to 3.4.8, 3.4.12, 3.4.15 and 5.2.3, respectively.

Lemma 7.1.2 *If* κ *and* λ *are cardinals with* $\kappa < \lambda$, *then* $\mathcal{J}^p_{<\kappa}(\mathsf{a}) \subseteq \mathcal{J}^p_{<\lambda}(\mathsf{a})$.

Lemma 7.1.3 *For every cardinal* λ,

$$\lambda \in \mathrm{pcf}(\mathsf{a}) \iff \mathcal{J}^p_{<\lambda}(\mathsf{a}) \subsetneq \mathcal{J}^p_{<\lambda^+}(\mathsf{a}).$$

Lemma 7.1.4 *If* λ *is a limit cardinal, then* $\mathcal{J}^p_{<\lambda}(\mathsf{a}) = \bigcup\{\mathcal{J}^p_{<\mu}(\mathsf{a}) : \mu \in \mathrm{CN} \wedge \mu < \lambda\}$.

Lemma 7.1.5 *If $\lambda \in \mathrm{pcf}(\mathsf{a})$, then the set $\mathrm{pcf}(\mathsf{b}_\lambda)$ generates the ideal $\mathcal{J}^p_{<\lambda^+}(\mathsf{a})$ over $\mathcal{J}^p_{<\lambda}(\mathsf{a})$.*

Lemma 7.1.6 *For any cardinal λ, we have $\mathcal{J}_{<\lambda}(\mathsf{a}) \subseteq \mathcal{J}^p_{<\lambda}(\mathsf{a})$.*

Lemma 7.1.7 *If $\mathsf{b} \subseteq \mathsf{a}$ and λ is a cardinal, then $\mathcal{J}^p_{<\lambda}(\mathsf{b}) \subseteq \mathcal{J}^p_{<\lambda}(\mathsf{a})$.*

Proof The sequence $(\mathsf{b}_\mu \cap \mathsf{b} : \mu \in \mathrm{pcf}(\mathsf{b}))$ is a sequence of generators for b.

Definition *We specify a function assigning members of $\prod \mathrm{pcf}(\mathsf{a})$ to the members of $\prod \mathsf{a}$. Fix, for each $\lambda \in \mathrm{pcf}(\mathsf{a})$, a sequence $(f^\lambda_\xi : \xi < \lambda)$ of members of $\prod \mathsf{a}$ such that*

(a) $(f^\lambda_\xi : \xi < \lambda)$ is strictly increasing modulo $\mathcal{J}_{<\lambda}(\mathsf{a})$ and

(b) $\{f^\lambda_\xi \restriction \mathsf{b}_\lambda : \xi < \lambda\}$ is cofinal in $(\prod \mathsf{b}_\lambda, \leq_{\mathsf{b}_\lambda})$.

Such a sequence can be obtained if we take a special sequence $\bar{f}$ as in Section 6.1[2] *and put $f^\lambda_\xi := f^{\mathsf{b}_\lambda}_\xi$ for each $\lambda \in \mathrm{pcf}(\mathsf{a})$ and $\xi < \lambda$, since $\lambda_{\mathsf{b}_\lambda} =$* $\max \mathrm{pcf}(\mathsf{b}_\lambda) = \lambda$.

To each ordinal function $f \in \prod \mathsf{a}$ we assign an ordinal function $f^* \in \prod \mathrm{pcf}(\mathsf{a})$ extending f, putting for $\nu \in \mathrm{pcf}(\mathsf{a})$

$$f^*(\nu) := \begin{cases} f(\nu) & \text{if } \nu \in \mathsf{a}, \\ \min\{\gamma < \nu : f \leq_{\mathcal{J}_{<\nu}(\mathsf{a}) \restriction \mathsf{b}_\nu} f^\nu_\gamma\} & \text{if } \nu \notin \mathsf{a}. \end{cases}$$

Remark Obviously, we have $(f^\lambda_\xi)^*(\lambda) = \xi$ for any $\xi < \lambda$ and $\lambda \in \mathrm{pcf}(\mathsf{a}) \setminus \mathsf{a}$.

Now we can state that relations between ordinal functions on a are transferred to the associated functions on $\mathrm{pcf}(\mathsf{a})$. The next lemma is immediate from the definition.

Lemma 7.1.8 *If f and g are members of $\prod \mathsf{a}$ such that $f \leq g$, then $f^* \leq g^*$.*

Lemma 7.1.9 *If $f, g \in \prod \mathsf{a}$ and λ is a cardinal such that $f \leq_{\mathcal{J}_{<\lambda}(\mathsf{a})} g$, then*

$$f^* \leq_{\mathcal{J}^p_{<\lambda}(\mathsf{a})} g^*.$$

Proof By assumption, we have $\mathsf{c} := \mathrm{B}(g, f) = \{\nu \in \mathsf{a} : g(\nu) < f(\nu)\} \in \mathcal{J}_{<\lambda}(\mathsf{a})$. Therefore Corollary 5.3.4 tells us that there are elements $\mu_1, \ldots, \mu_n$ of $\lambda \cap \mathrm{pcf}(\mathsf{a})$ such that c is a subset of $\mathsf{b} := \bigcup\{\mathsf{b}_{\mu_i} : 1 \leq i \leq n\}$. Now, if $\nu \in \mathrm{pcf}(\mathsf{a}) \setminus \mathsf{a}$ and $\mathsf{b}_\nu \cap \mathsf{b} \in \mathcal{J}_{<\nu}(\mathsf{a})$, then $\mathsf{b}_\nu \cap \mathsf{c} \in \mathcal{J}_{<\nu}(\mathsf{a})$, which gives $f \leq_{\mathcal{J}_{<\nu}(a) \restriction \mathsf{b}_\nu} g$ and thus $f^*(\nu) \leq g^*(\nu)$.

[2]We need not refer to the cardinal μ since we do not need the fact that the functions f^{b} are μ-continuous.

So we have shown that

$$\mathrm{B}(g^*, f^*) \setminus \mathsf{a} \subseteq \mathsf{d} := \{\nu \in \mathrm{pcf}(\mathsf{a}) : \mathsf{b}_\nu \cap \mathsf{b} \notin \mathcal{J}_{<\nu}(\mathsf{a})\}.$$

In view of Lemma 7.1.6 and since $\mathrm{B}(g, f) \in \mathcal{J}_{<\lambda}(\mathsf{a})$, it suffices to prove that $\mathsf{d} \in \mathcal{J}^p_{<\lambda}(\mathsf{a})$. We have

$$\mathrm{pcf}(\mathsf{b}) = \bigcup\{\mathrm{pcf}(\mathsf{b}_{\mu_i}) : 1 \leq i \leq n\} \in \mathcal{J}^p_{<\lambda}(\mathsf{a}),$$

so we show that $\mathsf{d} \subseteq \mathrm{pcf}(\mathsf{b})$.

If $\nu \in \mathsf{d}$, then there is an ultrafilter D on a satisfying $\mathsf{b}_\nu \cap \mathsf{b} \in D$ and $\mathcal{J}_{<\nu}(\mathsf{a}) \cap D = \emptyset$. Therefore we get $\mathrm{cf}(\prod \mathsf{a}/D) \geq \nu$ by Corollary 3.4.11 and $\mathrm{cf}(\prod \mathsf{a}/D) \leq \nu$ by Lemma 5.3.5, since $\mathsf{b}_\nu \in D$. Furthermore, $\mathsf{b} \in D$ yields $\mathsf{b}_{\mu_j} \in D$ for some $j \in \{1, \ldots, n\}$. This gives

$$\nu = \mathrm{cf}\left(\prod \mathsf{b}_{\mu_j}/D \cap \mathcal{P}(\mathsf{b}_{\mu_j})\right) \in \mathrm{pcf}(\mathsf{b}).$$

Definition *If b is a, not necessarily progressive, set of regular cardinals, we define*

$$\mathcal{J}_*(\mathsf{b}) := \{\mathsf{c} \subseteq \mathsf{b} : \forall \rho(\rho = \sup(\rho \cap \mathsf{c}) \implies \rho \textit{ is not regular})\}.$$

It is easy to see that either $\mathcal{J}_*(\mathsf{b})$ is an ideal on b or $\mathcal{J}_*(\mathsf{b}) = \mathcal{P}(\mathsf{b})$. Especially observe that every progressive subset of b is a member of $\mathcal{J}_*(\mathsf{b})$, and that $\mathcal{J}_*(\mathsf{c}) = \mathcal{J}_*(\mathsf{b}) \cap \mathcal{P}(\mathsf{c})$ for all $\mathsf{c} \subseteq \mathsf{b}$.

Lemma 7.1.10 *If b is a set of regular cardinals, then $\mathcal{J}_*(\mathsf{b})$ is the smallest subset J of $\mathcal{P}(\mathsf{b})$ with the following properties:*

a) $\{\nu\} \in J$ *for all* $\nu \in \mathsf{b}$.

b) If κ is a cardinal and $(\mathsf{c}_i : i < \kappa)$ is a sequence of members of J satisfying $\kappa < \min(\mathsf{c}_i)$ for all $i < \kappa$, then $\bigcup\{\mathsf{c}_i : i < \kappa\} \in J$.

Proof Certainly $\mathcal{J}_*(\mathsf{b})$ has property a). To verify that it has property b), let $(\mathsf{c}_i : i < \kappa)$ be a sequence of members of $\mathcal{J}_*(\mathsf{b})$ such that $\kappa < \min(\mathsf{c}_i)$ for all $i < \kappa$, and $\mathsf{c} := \bigcup\{\mathsf{c}_i : i < \kappa\}$. Suppose, contrary to our hopes, that there is a regular cardinal ρ such that

$$\rho = \sup(\rho \cap \mathsf{c}) = \sup(\bigcup\{\rho \cap \mathsf{c}_i : i < \kappa\}) = \sup\{\sup(\rho \cap \mathsf{c}_i) : i < \kappa\}.$$

By assumption we have $\sup(\rho \cap \mathsf{c}_i) < \rho$ for all $i < \kappa$, hence $\rho = \mathrm{cf}(\rho) \leq \kappa$. This contradicts $\rho > \min(\mathsf{c}) > \kappa$.

Now consider a fixed set $J \subseteq \mathcal{P}(\mathsf{b})$ with the properties a) and b). We show that $\mathcal{J}_*(\mathsf{b}) \subseteq J$. Note first that from b) it follows that J is closed under finite unions. Let $\mathsf{c} \in \mathcal{J}_*(\mathsf{b})$. We prove $\mathsf{c} \in J$ using transfinite induction on the cardinal

$$\nu := \sup\{\mu^+ : \mu \in \mathsf{c}\}.$$

If $\nu = 0$, then $\mathsf{c} = \emptyset$, and we have $\mathsf{c} \in J$, using b) for $\kappa = 0$. If $\nu = \mu^+$ for some μ, then $\mu = \max(\mathsf{c})$. With the inductive hypothesis we get $\mathsf{c} \setminus \{\mu\} \in J$. Since $\{\mu\} \in J$, this gives $\mathsf{c} \in J$. If ν is a limit cardinal, then $\nu = \sup(\mathsf{c})$. By the definition of $\mathcal{J}_*(\mathsf{b})$, ν is singular. Put $\kappa := \mathrm{cf}(\nu)$, and let the sequence $(\nu_i : i < \kappa)$ be cofinal in ν such that $\kappa < \nu_0$. From the inductive hypothesis we can conclude that the members $\mathsf{c} \cap \nu_0$ and $\mathsf{c} \cap [\nu_i, \nu_{i+1})_{\mathrm{ON}}$ of $\mathcal{J}_*(\mathsf{b})$, for $i < \kappa$, are members of J. b) yields

$$\mathsf{c} \cap [\nu_0, \nu)_{\mathrm{ON}} = \bigcup\{\mathsf{c} \cap [\nu_i, \nu_{i+1})_{\mathrm{ON}} : i < \kappa\} \in J,$$

and thus $\mathsf{c} \in J$, as desired.

This characterization of $\mathcal{J}_*(\mathsf{b})$ enables us to prove the following important result.

Lemma 7.1.11 *If* $g \in \prod \mathrm{pcf}(\mathsf{a})$ *and* $\mathsf{b} \in \mathcal{J}_*(\mathrm{pcf}(\mathsf{a}))$, *then there is a function* $f \in \prod \mathsf{a}$ *such that*

$$g \restriction \mathsf{b} \leq f^* \restriction \mathsf{b}.$$

Proof Fix $g \in \prod \mathrm{pcf}(\mathsf{a})$, and let

$$J := \{\mathsf{b} \subseteq \mathrm{pcf}(\mathsf{a}) : \exists f \in \prod \mathsf{a} \quad g \restriction \mathsf{b} \leq f^* \restriction \mathsf{b}\}$$

We will show, using Lemma 7.1.10, that $\mathcal{J}_*(\mathrm{pcf}(\mathsf{a})) \subseteq J$.

First consider $\nu \in \mathrm{pcf}(\mathsf{a})$ and $\beta := g(\nu)$. To prove property a) of the lemma, we must find a function $f \in \prod \mathsf{a}$ such that $\beta \leq f^*(\nu)$. If $\nu \in \mathsf{a}$, then choose $f \in \prod \mathsf{a}$ with $f(\nu) = \beta$. If $\nu \in \mathrm{pcf}(\mathsf{a}) \setminus \mathsf{a}$, put $f := f^\nu_\beta$.

Now we prove property b). Assume that κ is a cardinal and $(\mathsf{c}_i : i < \kappa)$ is a sequence of members of J satisfying $\kappa < \min(\mathsf{c}_i)$ for all $i < \kappa$. For each $i < \kappa$, choose a function $f_i \in \prod \mathsf{a}$ according to the definition of J, such that $g \restriction \mathsf{c}_i \leq f_i^* \restriction \mathsf{c}_i$. Since $(\prod \mathsf{a}, <_{\mathcal{J}_{<\kappa^+}(\mathsf{a})})$ is κ^+-directed, there is a function $f \in \prod \mathsf{a}$ such that $f_i <_{\mathcal{J}_{<\kappa^+}(\mathsf{a})} f$ for all $i < \kappa$. Without loss of generality we can assume that $g \restriction \mathsf{a} \leq f$, since otherwise we can replace f by $\max\{f, g \restriction \mathsf{a}\}$. We verify $g \restriction \mathsf{c} \leq f^* \restriction \mathsf{c}$ for the set $\mathsf{c} := \bigcup\{\mathsf{c}_i : i < \kappa\}$.

Let $\nu \in \mathsf{c}$. If $\nu \in \mathsf{a}$, then $g(\nu) \leq f(\nu)$ by the choice of f. If $\nu \notin \mathsf{a}$ and $i < \kappa$ is fixed with $\nu \in \mathsf{c}_i$, then $\kappa < \nu$ and $g(\nu) \leq f_i^*(\nu)$. $\kappa < \nu$ gives $\mathcal{J}_{<\kappa^+}(\mathsf{a}) \subseteq \mathcal{J}_{<\nu}(\mathsf{a})$, hence $f_i <_{\mathcal{J}_{<\nu}(\mathsf{a})} f$, which implies $f_i \leq_{\mathcal{J}_{<\nu}(\mathsf{a}) \restriction \mathsf{b}_\nu} f$. So we get $g(\nu) \leq f_i^*(\nu) \leq f^*(\nu)$.

Now we will prove a number of results concerning the ideal $\mathcal{J}^p_{<\lambda}(\mathsf{a})$, and indicate in brackets which results from the preceding chapters are generalized.

Theorem 7.1.12 (3.4.4) *If λ is a cardinal, $\lambda \leq \max \mathrm{pcf}(\mathsf{a})$, and $\mathsf{b} \in \mathcal{J}_*(\mathrm{pcf}(\mathsf{a})) \setminus \mathcal{J}^p_{<\lambda}(\mathsf{a})$, then $(\prod \mathrm{pcf}(\mathsf{a}), <_{\mathcal{J}^p_{<\lambda}(\mathsf{a}) \restriction \mathsf{b}})$ is λ-directed.*

Proof Let S be a subset of $\prod \mathrm{pcf}(\mathsf{a})$ such that $|S| < \lambda$. Using Lemma 7.1.11, we can choose for each $g \in S$ a function $f_g \in \prod \mathsf{a}$ such that $g \restriction \mathsf{b} \leq f^*_g \restriction \mathsf{b}$. Since $(\prod \mathsf{a}, <_{\mathcal{J}_{<\lambda}(\mathsf{a})})$ is λ-directed, there is a function $f \in \prod \mathsf{a}$ satisfying $f_g \leq_{\mathcal{J}_{<\lambda}(\mathsf{a})} f$ for all $g \in S$. Now Lemma 7.1.9 says that $f^*_g \leq_{\mathcal{J}^p_{<\lambda}(\mathsf{a})} f^*$ for all $g \in S$, and this together with $g \restriction \mathsf{b} \leq f^*_g \restriction \mathsf{b}$ yields $\{\nu \in \mathsf{b} : f^*(\nu) < g(\nu)\} \in \mathcal{J}^p_{<\lambda}(\mathsf{a})$. So we get $g \leq_{\mathcal{J}^p_{<\lambda}(\mathsf{a}) \restriction \mathsf{b}} f^*$ for all $g \in S$.

Theorem 7.1.13 *If $\lambda = \max \mathrm{pcf}(\mathsf{a})$ and $\mathsf{b} \in \mathcal{J}_*(\mathrm{pcf}(\mathsf{a})) \setminus \mathcal{J}^p_{<\lambda}(\mathsf{a})$, then*

$$\mathrm{tcf}\left(\prod \mathrm{pcf}(\mathsf{a}) / \mathcal{J}^p_{<\lambda}(\mathsf{a}) \restriction \mathsf{b}\right) = \lambda.$$

Proof For each $\xi < \lambda$, we put $h_\xi := (f^\lambda_\xi)^*$. (Remember the fixed sequence from page 235. In particular here we have $\mathsf{b}_\lambda = \mathsf{a}$.) By Lemma 7.1.9, the sequence $\bar{h} := (h_\xi : \xi < \lambda)$ is increasing modulo $\mathcal{J}^p_{<\lambda}(\mathsf{a})$, hence also modulo $\mathcal{J}^p_{<\lambda}(\mathsf{a}) \restriction \mathsf{b}$.

Now consider a fixed function $g \in \prod \mathrm{pcf}(\mathsf{a})$. Lemma 7.1.11 says that there is a function $f \in \prod \mathsf{a}$ such that

$$(g+1) \restriction \mathsf{b} \leq f^* \restriction \mathsf{b}.$$

There is an ordinal $\alpha < \lambda$ such that $f \leq f^\lambda_\alpha$, since $(f^\lambda_\xi : \xi < \lambda)$ is pointwise cofinal in $\prod \mathsf{b}_\lambda = \prod \mathsf{a}$. Lemma 7.1.8 tells us that $f^* \leq h_\alpha$, hence $g \restriction \mathsf{b} < h_\alpha \restriction \mathsf{b}$. Together with $h_\alpha \leq_{\mathcal{J}^p_{<\lambda}(\mathsf{a}) \restriction \mathsf{b}} h_{\alpha+1}$ we can conclude that $g <_{\mathcal{J}^p_{<\lambda}(\mathsf{a}) \restriction \mathsf{b}} h_{\alpha+1}$, which shows that a suitable subsequence of $\bar{h}$ is cofinal in $\prod \mathrm{pcf}(\mathsf{a})$ modulo $\mathcal{J}^p_{<\lambda}(\mathsf{a}) \restriction \mathsf{b}$.

Corollary 7.1.14 (3.4.19) *Assume that $\lambda \in \mathrm{pcf}(\mathsf{a})$ and $\mathsf{b} \in \mathcal{J}_*(\mathrm{pcf}(\mathsf{a})) \cap \mathcal{J}^p_{<\lambda^+}(\mathsf{a}) \setminus \mathcal{J}^p_{<\lambda}(\mathsf{a})$. Then*

$$\mathrm{tcf}\left(\prod \mathrm{pcf}(\mathsf{a}) / \mathcal{J}^p_{<\lambda}(\mathsf{a}) \restriction \mathsf{b}\right) = \lambda.$$

Proof By assumption and Lemma 7.1.5, we have $\mathsf{b} \setminus \mathrm{pcf}(\mathsf{b}_\lambda) \in \mathcal{J}^p_{<\lambda}(\mathsf{a})$, and thus $\mathcal{J}^p_{<\lambda}(\mathsf{a}) \restriction \mathsf{b} = \mathcal{J}^p_{<\lambda}(\mathsf{a}) \restriction (\mathrm{pcf}(\mathsf{b}_\lambda) \cap \mathsf{b})$. So we can assume without loss of generality that $\mathsf{b} \subseteq \mathrm{pcf}(\mathsf{b}_\lambda)$. Applying Theorem 7.1.13 to b_λ instead of a, we can conclude that

$$\mathrm{tcf}\left(\prod \mathrm{pcf}(\mathsf{b}_\lambda) / \mathcal{J}^p_{<\lambda}(\mathsf{b}_\lambda) \restriction \mathsf{b}\right) = \lambda.$$

Now we choose a sequence of length λ which is cofinal in $\prod \mathrm{pcf}(\mathsf{b}_\lambda)$ modulo $\mathcal{J}^p_{<\lambda}(\mathsf{b}_\lambda) \restriction \mathsf{b}$ and extend its members arbirarily to $\mathrm{pcf}(\mathsf{a})$. By Lemma 7.1.7, the resulting sequence is cofinal in $\prod \mathrm{pcf}(\mathsf{a})$ modulo $\mathcal{J}^p_{<\lambda}(\mathsf{a}) \restriction \mathsf{b}$.

Corollary 7.1.15 (5.3.5) *If D is an ultrafilter on* $\mathrm{pcf}(\mathsf{a})$ *such that* $D \cap \mathcal{J}_*(\mathrm{pcf}(\mathsf{a})) \neq \emptyset$, *then*

$$\begin{aligned}\mathrm{cf}\,(\prod \mathrm{pcf}(\mathsf{a})/D) &= \min\{\lambda \in \mathrm{pcf}(\mathsf{a}) : D \cap \mathcal{J}^p_{<\lambda^+}(\mathsf{a}) \neq \emptyset\} \\ &= \min\{\lambda \in \mathrm{pcf}(\mathsf{a}) : \mathrm{pcf}(\mathsf{b}_\lambda) \in D\}.\end{aligned}$$

Proof Let D be an ultrafilter on $\mathrm{pcf}(\mathsf{a})$ such that $D \cap \mathcal{J}_*(\mathrm{pcf}(\mathsf{a})) \neq \emptyset$. Clearly there is a least member, say λ, of $\mathrm{pcf}(\mathsf{a})$ satisfying $D \cap \mathcal{J}^p_{<\lambda^+}(\mathsf{a}) \neq \emptyset$, since $\mathcal{J}^p_{<\kappa}(\mathsf{a}) = \mathcal{P}(\mathrm{pcf}(\mathsf{a}))$ for all $\kappa > \max \mathrm{pcf}(\mathsf{a})$. Then for all $\mu \in \lambda \cap \mathrm{pcf}(\mathsf{a})$ we have $\mathrm{pcf}(\mathsf{b}_\mu) \notin D$

Consider a fixed set $\mathsf{d} \in D \cap \mathcal{J}^p_{<\lambda^+}(\mathsf{a})$. Without loss of generality we can assume that $\mathsf{d} \in \mathcal{J}_*(\mathrm{pcf}(\mathsf{a}))$, since otherwise we could take the intersection of d and some member of $D \cap \mathcal{J}_*(\mathrm{pcf}(\mathsf{a}))$. By the definition of $\mathcal{J}^p_{<\lambda^+}(\mathsf{a})$, there are cardinals $\mu_1, \ldots, \mu_n \in \lambda + 1$, which are members of $\mathrm{pcf}(\mathsf{a})$ and satisfy $\mathsf{d} \subseteq \bigcup\{\mathrm{pcf}(\mathsf{b}_{\mu_i}) : 1 \leq i \leq n\}$. Now there is $k \in \{1, \ldots, n\}$ such that $\mathrm{pcf}(\mathsf{b}_{\mu_k}) \in D$. The minimality of λ shows that $\mu_k = \lambda$. This yields the second equation of the corollary.

Furthermore Corollary 7.1.14 says, together with the choice of λ, that

$$\mathrm{cf}\,(\prod \mathrm{pcf}(\mathsf{a})/D) = \mathrm{tcf}\,(\prod \mathrm{pcf}(\mathsf{a})/\mathcal{J}^p_{<\lambda}(\mathsf{a}) \restriction \mathsf{d}) = \lambda,$$

which proves the first equation of the corollary.

Corollary 7.1.16 (3.4.11) *Assume that D is an ultrafilter on* $\mathrm{pcf}(\mathsf{a})$ *such that* $D \cap \mathcal{J}_*(\mathrm{pcf}(\mathsf{a})) \neq \emptyset$, *and λ is a cardinal. Then*

$$\mathrm{cf}\,(\prod \mathrm{pcf}(\mathsf{a})/D) < \lambda \iff D \cap \mathcal{J}^p_{<\lambda}(\mathsf{a}) \neq \emptyset.$$

Proof First we observe that, by the definition of $\mathcal{J}^p_{<\lambda}(\mathsf{a})$, $D \cap \mathcal{J}^p_{<\lambda}(\mathsf{a}) \neq \emptyset$ holds iff $D \cap \mathcal{J}^p_{<\kappa^+}(\mathsf{a}) \neq \emptyset$ for some $\kappa < \lambda$. By Corollary 7.1.15, the last assertion is equivalent to $\mathrm{cf}\,(\prod \mathrm{pcf}(\mathsf{a})/D) \leq \kappa$.

Corollary 7.1.17 *For any cardinal λ, we have*

$$\mathcal{J}^p_{<\lambda}(\mathsf{a}) \cap \mathcal{J}_*(\mathrm{pcf}(\mathsf{a})) = \{\mathsf{b} \in \mathcal{J}_*(\mathrm{pcf}(\mathsf{a})) : \mathrm{pcf}(\mathsf{b}) \subseteq \lambda\}.$$

Proof Consider $\mathsf{b} \in \mathcal{J}_*(\mathrm{pcf}(\mathsf{a}))$. First let us assume that $\mathsf{b} \in \mathcal{J}^p_{<\lambda}(\mathsf{a})$ and D is an ultrafilter on $\mathrm{pcf}(\mathsf{a})$ such that $\mathsf{b} \in D$. Since $\mathsf{b} \in D \cap \mathcal{J}_*(\mathrm{pcf}(\mathsf{a}))$, we have $\mathrm{cf}\,(\prod \mathrm{pcf}(\mathsf{a})/D) = \mathrm{cf}\,(\prod \mathsf{b}/D \cap \mathcal{P}(\mathsf{b})) < \lambda$ by Corollary 7.1.16. This gives $\mathrm{pcf}(\mathsf{b}) \subseteq \lambda$. Conversely, if $\mathsf{b} \notin \mathcal{J}^p_{<\lambda}(\mathsf{a})$, then we can choose an ultrafilter D on $\mathrm{pcf}(\mathsf{a})$ such that $\mathsf{b} \in D$ and $D \cap \mathcal{J}^p_{<\lambda}(\mathsf{a}) = \emptyset$. Corollary 7.1.16 implies that $\mathrm{cf}\,(\prod \mathsf{b}/D \cap \mathcal{P}(\mathsf{b})) \geq \lambda$, and thus $\mathrm{pcf}(\mathsf{b})$ is not a subset of λ.

Corollary 7.1.18 *If* $\mathsf{b} \in \mathcal{J}_*(\mathrm{pcf}(\mathsf{a}))$ *and* λ *is a cardinal, then*

$$\mathcal{J}_{<\lambda}(\mathsf{b}) = \mathcal{J}^{\mathcal{P}}_{<\lambda}(\mathsf{a}) \cap \mathcal{P}(\mathsf{b}).$$

In particular, $\mathcal{J}_{<\lambda}(\mathsf{a}) = \mathcal{J}^{\mathcal{P}}_{<\lambda}(\mathsf{a}) \cap \mathcal{P}(\mathsf{a})$.

Proof Use the fact that $\mathcal{J}_{<\lambda}(\mathsf{b}) = \{\mathsf{c} \subseteq \mathsf{b} : \mathrm{pcf}(\mathsf{c}) \subseteq \lambda\}$.

Corollary 7.1.19 (3.4.20) *If* $\mathsf{b} \in \mathcal{J}_*(\mathrm{pcf}(\mathsf{a}))$, *then* $\mathrm{pcf}(\mathsf{b})$ *is a subset of* $\mathrm{pcf}(\mathsf{a})$ *and has a maximum.*

Proof The first assertion follows immediately from Corollary 7.1.15. To get a contradiction, let us suppose that $\mathsf{b} \in \mathcal{J}_*(\mathrm{pcf}(\mathsf{a}))$ and $\lambda := \sup \mathrm{pcf}(\mathsf{b})$ is not a member of $\mathrm{pcf}(\mathsf{b})$. Then λ is a limit cardinal, and Corollary 7.1.17 says that $\mathsf{b} \in \mathcal{J}^{\mathcal{P}}_{<\lambda}(\mathsf{a})$, hence $\mathsf{b} \in \mathcal{J}^{\mathcal{P}}_{<\mu}(\mathsf{a})$ for some $\mu < \lambda$. But this gives $\mathrm{pcf}(\mathsf{b}) \subseteq \mu$, contradicting the choice of λ.

Lemma 7.1.20 *Assume that* $\mathsf{c} \in \mathcal{J}_*(\mathrm{pcf}(\mathsf{a}))$ *and* $\mathsf{b} \subseteq \mathrm{pcf}(\mathsf{c})$ *is progressive. If* D *is an ultrafilter on* b *and, for each* $\nu \in \mathsf{b}$, D_ν *is an ultrafilter on* c *such that* $\nu = \mathrm{cf}\left(\prod \mathsf{c}/D_\nu\right)$, *then*

$$D^* := \{X \subseteq \mathsf{c} : \{\nu \in \mathsf{b} : X \in D_\nu\} \in D\}$$

is an ultrafilter on c *satisfying* $\mathrm{cf}\left(\prod \mathsf{c}/D^*\right) = \mathrm{cf}\left(\prod \mathsf{b}/D\right)$.

Proof One easily verifies that D^* is an ultrafilter on c. Take, for each $\nu \in \mathsf{b}$, a sequence $(g^\nu_\alpha : \alpha < \nu)$ which is cofinal in $\prod \mathsf{c}$ modulo D_ν. To apply Lemma 3.1.10, we define a function $G : \prod \mathsf{c} \longrightarrow \prod \mathsf{b}$ by

$$G(f)(\nu) := \min\{\alpha < \nu : f \leq_{D_\nu} g^\nu_\alpha\}$$

for any $f \in \prod \mathsf{c}$ and any $\nu \in \mathsf{b}$.

Now we show that G satisfies the assumptions of Lemma 3.1.10. Fix $h \in \prod \mathsf{b}$. Since b is progressive, hence $|\mathsf{b}| < \min(\mathsf{b})$, we have $\mathsf{c} \notin \mathcal{J}^{\mathcal{P}}_{<|\mathsf{b}|^+}(\mathsf{a})$, since otherwise Corollary 7.1.17 would yield the contradiction $\mathsf{b} \subseteq \mathrm{pcf}(\mathsf{c}) \subseteq |\mathsf{b}|^+$. From Theorem 7.1.12, applied to $|\mathsf{b}|^+$ and c, we can conclude together with Lemma 2.1.4 that there is a function $p \in \prod \mathsf{c}$ such that

$$g^\nu_{h(\nu)} <_{\mathcal{J}^{\mathcal{P}}_{<|\mathsf{b}|^+}(\mathsf{a}) \cap \mathcal{P}(\mathsf{c})} p$$

for all $\nu \in \mathsf{b}$. If $\nu \in \mathsf{b}$, then $|\mathsf{b}|^+ \leq \nu = \mathrm{cf}\left(\prod \mathsf{c}/D_\nu\right)$, hence $D_\nu \cap \mathcal{J}^{\mathcal{P}}_{<|\mathsf{b}|^+}(\mathsf{a}) = \emptyset$ by Corollary 7.1.16; obviously, to apply this corollary, it is no problem to extend D_ν to $\mathrm{pcf}(\mathsf{a})$. So we get $g^\nu_{h(\nu)} <_{D_\nu} p$, which yields $h <_{\mathsf{b}} G(p)$ by the definition of G.

Assume that $q \in \prod \mathsf{c}$ and $p \leq_{D^*} q$. It remains to be shown that $h <_D G(q)$. First we get $\mathrm{B}[p,q] \in D^*$, that means $\mathrm{B}[p,q] \in D_\nu$ or, equivalently, $p \leq_{D_\nu} q$ for almost all $\nu \in \mathsf{b}$ modulo D. The definition of G shows that $G(p) \leq_D G(q)$. So we get $h <_D G(q)$ from the previous part of the proof. Now we can infer from Lemma 3.1.10 that $\mathrm{cf}(\prod \mathsf{c}/D^*) = \mathrm{cf}(\prod \mathsf{b}/D)$, as desired.

Corollary 7.1.21 *If* $\mathsf{c} \in \mathcal{J}_*(\mathrm{pcf}(\mathsf{a}))$ *and* $\mathsf{b} \subseteq \mathrm{pcf}(\mathsf{c})$ *is progressive, then* $\mathrm{pcf}(\mathsf{b})$ *is a subset of* $\mathrm{pcf}(\mathsf{c})$.

Lemma 7.1.22 (5.3.3) *If* $\mathsf{c} \in \mathcal{J}_*(\mathrm{pcf}(\mathsf{a}))$ *and* $\lambda \in \mathrm{pcf}(\mathsf{a})$, *then*

$$\lambda \notin \mathrm{pcf}(\mathsf{c} \setminus \mathrm{pcf}(\mathsf{b}_\lambda)).$$

Proof If $\mathsf{c} \in \mathcal{J}^p_{<\lambda^+}(\mathsf{a})$, then $\mathsf{c}\setminus\mathrm{pcf}(\mathsf{b}_\lambda) \in \mathcal{J}^p_{<\lambda}(\mathsf{a})$, and the assertion follows from Corollary 7.1.17. So assume that $\mathsf{c} \notin \mathcal{J}^p_{<\lambda^+}(\mathsf{a})$. Then the set $\mathsf{c}' := \mathsf{c}\setminus\mathrm{pcf}(\mathsf{b}_\lambda)$ is not a member of $\mathcal{J}^p_{<\lambda}(\mathsf{a})$.

Let us assume that the lemma is false. Then there is an ultrafilter D on $\mathrm{pcf}(\mathsf{a})$ such that $\mathsf{c}' \in D$ and $\mathrm{cf}(\prod \mathrm{pcf}(\mathsf{a})/D) = \lambda$, hence $D \cap \mathcal{J}^p_{<\lambda}(\mathsf{a}) = \emptyset$ by Corollary 7.1.16. We choose a sequence $(f_\xi : \xi < \lambda)$ which is cofinal in $\prod \mathrm{pcf}(\mathsf{a})$ modulo D. Since $(\prod \mathrm{pcf}(\mathsf{a}), <_{\mathcal{J}^p_{<\lambda^+}(\mathsf{a}) \restriction \mathsf{c}})$ is λ^+-directed by Theorem 7.1.12, there is a function $f \in \prod \mathrm{pcf}(\mathsf{a})$ such that

$$f_\xi \leq_{\mathcal{J}^p_{<\lambda^+}(\mathsf{a}) \restriction \mathsf{c}} f$$

for all $\xi < \lambda$. But then

$$\mathrm{B}(f, f_\xi) \cap \mathsf{c}' = (\mathrm{B}(f, f_\xi) \cap \mathsf{c}) \setminus \mathrm{pcf}(\mathsf{b}_\lambda) \in \mathcal{J}^p_{<\lambda}(\mathsf{a})$$

for all $\xi < \lambda$, and thus f is an upper bound of $(f_\xi : \xi < \lambda)$ modulo $\mathcal{J}^p_{<\lambda}(\mathsf{a}) \restriction \mathsf{c}'$ and, consequently, also modulo D. This contradicts the fact that the sequence $(f_\xi : \xi < \lambda)$ is cofinal in $\prod \mathrm{pcf}(\mathsf{a})$ modulo D.

Definition *If* b *is a set of regular cardinals, then we put*

$$\mathrm{pcf}_*(\mathsf{b}) := \{\mathrm{cf}(\prod \mathsf{b}/D) : D \textit{ is an ultrafilter on } \mathsf{b} \wedge D \cap \mathcal{J}_*(\mathsf{b}) \neq \emptyset\}.$$

With the help of Lemma 3.3.4, the following assertions are easy consequences of the definition of the operator pcf_*.

Lemma 7.1.23 (3.3.4) *If* b *and* c *are sets of regular cardinals, then*

a) $\mathrm{pcf}_*(\mathsf{b}) = \bigcup\{\mathrm{pcf}(\mathsf{d}) : \mathsf{d} \in \mathcal{J}_*(\mathsf{b})\}$.

b) *If* $\mathsf{b} \in \mathcal{J}_*(\mathsf{b})$, *then* $\mathrm{pcf}_*(\mathsf{b}) = \mathrm{pcf}(\mathsf{b})$. *In particular, we have* $\mathrm{pcf}_*(\mathsf{b}) = \mathrm{pcf}(\mathsf{b})$ *for any progressive set* b.

c) $\mathrm{pcf}_*(\mathsf{b})$ *is a set of regular cardinals.*
d) $\mathsf{b} \subseteq \mathrm{pcf}_*(\mathsf{b})$.
e) $\min \mathrm{pcf}_*(\mathsf{b}) = \min(\mathsf{b})$.
f) $\mathsf{b} \subseteq \mathsf{c}$ *implies* $\mathrm{pcf}_*(\mathsf{b}) \subseteq \mathrm{pcf}_*(\mathsf{c})$.
g) $\mathrm{pcf}_*(\mathsf{b} \cup \mathsf{c}) = \mathrm{pcf}_*(\mathsf{b}) \cup \mathrm{pcf}_*(\mathsf{c})$.

Theorem 7.1.24 (Localisation Theorem) *If* a *is a progressive set of regular cardinals,* c *is a subset of* $\mathrm{pcf}(\mathsf{a})$*, and* $\lambda \in \mathrm{pcf}_*(\mathsf{c})$*, then there exists a subset* d *of* c *such that* $|\mathsf{d}| \leq |\mathsf{a}|$ *and* $\lambda \in \mathrm{pcf}(\mathsf{d})$.

Proof We suppose, contrary to our hopes, that the assertion of the theorem is false. By Lemma 7.1.23 a), there is a set $\mathsf{c} \in \mathcal{J}_*(\mathrm{pcf}(\mathsf{a}))$ and a cardinal $\lambda \in \mathrm{pcf}(\mathsf{c})$, such that $\lambda \notin \mathrm{pcf}(\mathsf{d})$ for all $\mathsf{d} \in [\mathsf{c}]^{\leq|\mathsf{a}|}$. We choose λ minimal with this property, and for this λ a set $\mathsf{c} \in \mathcal{J}_*(\mathrm{pcf}(\mathsf{a}))$ satisfying

$$\lambda \in \mathrm{pcf}(\mathsf{c}) \setminus \bigcup\{\mathrm{pcf}(\mathsf{d}) : \mathsf{d} \subseteq \mathsf{c} \wedge |\mathsf{d}| \leq |\mathsf{a}|\},$$

such that the cardinal

$$\kappa := \sup\{\mu^+ : \mu \in \mathrm{pcf}(\mathsf{c}) \wedge \mu < \max \mathrm{pcf}(\mathsf{c})\}$$

is minimal. Then Corollary 7.1.19 says that $\mathrm{pcf}(\mathsf{c})$ is a subset of $\mathrm{pcf}(\mathsf{a})$, hence $\lambda \in \mathrm{pcf}(\mathsf{a})$. Lemma 7.1.22 yields $\lambda \in \mathrm{pcf}(\mathsf{c} \cap \mathrm{pcf}(\mathsf{b}_\lambda))$. Therefore we can assume without loss of generality that $\mathsf{c} \subseteq \mathrm{pcf}(\mathsf{b}_\lambda)$. Since $\mathsf{c} \in \mathcal{J}^p_{<\lambda^+}(\mathsf{a})$, $\mathrm{pcf}(\mathsf{c})$ is a subset of λ^+ by Corollary 7.1.17. From this and $\lambda \in \mathrm{pcf}(\mathsf{c})$ we get $\lambda = \max \mathrm{pcf}(\mathsf{c})$.

Suppose that the set $\lambda \cap \mathrm{pcf}(\mathsf{c})$ has a maximum, say μ. Then $\kappa = \mu^+$. From Lemma 7.1.22 we infer that $\mu \notin \mathrm{pcf}(\mathsf{c} \setminus \mathsf{b}_\mu)$. Corollary 7.1.17 together with $\mathsf{c} \cap \mathrm{pcf}(\mathsf{b}_\mu) \in \mathcal{J}^p_{<\mu^+}(\mathsf{a}) \cap \mathcal{J}_*(\mathrm{pcf}(\mathsf{a}))$ tells us that $\mathrm{pcf}(\mathsf{c} \cap \mathrm{pcf}(\mathsf{b}_\mu)) \subseteq \mu + 1$, hence $\lambda \in \mathrm{pcf}(\mathsf{c} \setminus \mathrm{pcf}(\mathsf{b}_\mu))$. This property of $\mathsf{c} \setminus \mathrm{pcf}(\mathsf{b}_\mu)$ contradicts the choice of c and the minimality of κ. Therefore $\kappa = \sup(\lambda \cap \mathrm{pcf}(\mathsf{c}))$ is a limit cardinal such that $\kappa \notin \lambda \cap \mathrm{pcf}(\mathsf{c})$.

Since $\mathrm{pcf}(\mathsf{c}) \subseteq \mathrm{pcf}(\mathsf{a})$, $\mathsf{d} \subseteq \mathrm{pcf}(\mathsf{c})$ implies $\mathsf{d} \subseteq \mathrm{pcf}(\mathsf{a})$. If in addition $|\mathsf{d}| \leq |\mathsf{a}|$, then the set d is progressive, and thus $\mathrm{pcf}(\mathsf{d})$ has a maximum. Furthermore $\mathrm{pcf}(\mathsf{d})$ is, by Corollary 7.1.21, a subset of $\mathrm{pcf}(\mathsf{c})$. If we prove the

Claim $\mathrm{pcf}(\mathsf{d}) \subseteq \kappa$ for all $\mathsf{d} \subseteq \lambda \cap \mathrm{pcf}(\mathsf{c})$ with $|\mathsf{d}| \leq |\mathsf{a}|$,

then the proof of the theorem will be completed, since we can construct by recursion a sequence $(\kappa_i : i < |\mathsf{a}|^+)$ of elements of $\lambda \cap \mathrm{pcf}(\mathsf{c})$, hence of $\mathrm{pcf}(\mathsf{a})$, in such a way that its existence contradicts Theorem 6.2.1.

To give an indirect proof of the claim, let us assume that there is a set $\mathsf{d} \subseteq \lambda \cap \mathrm{pcf}(\mathsf{c})$ such that $|\mathsf{d}| \leq |\mathsf{a}|$ and

$$\kappa^* := \max \mathrm{pcf}(\mathsf{d}) \geq \kappa.$$

Now $\mathrm{pcf}(\mathsf{d}) \subseteq \mathrm{pcf}(\mathsf{c})$ by Corollary 7.1.21, hence $\max\mathrm{pcf}(\mathsf{d}) \leq \lambda$. If $\max\mathrm{pcf}(\mathsf{d}) < \lambda$, then $(\kappa^*)^+ < \kappa$, contradicting $\kappa \leq \kappa^*$. So we have $\kappa^* = \lambda$. For each $\nu \in \mathsf{d}$ we obtain, by the choice of λ, a set $\mathsf{e}_\nu \subseteq \mathsf{c}$ satisfying $\nu \in \mathrm{pcf}(\mathsf{e}_\nu)$ and $|\mathsf{e}_\nu| \leq |\mathsf{a}|$. With $\mathsf{e} := \bigcup\{\mathsf{e}_\nu : \nu \in \mathsf{d}\}$ we get $\mathsf{d} \subseteq \mathrm{pcf}(\mathsf{e})$, which gives again $\mathrm{pcf}(\mathsf{d}) \subseteq \mathrm{pcf}(\mathsf{e})$ by Corollary 7.1.21, and in particular $\lambda \in \mathrm{pcf}(\mathsf{e})$. But this together with $\mathsf{e} \subseteq \mathsf{c}$ and $|\mathsf{e}| \leq |\mathsf{a}|$ contradicts the choice of c and λ.

Lemma 7.1.25 $\mathrm{pcf}(\mathsf{a}) = \mathrm{pcf}_*(\mathrm{pcf}(\mathsf{a}))$.

Proof This follows from Lemma 7.1.23 and Corollary 7.1.19.

Theorem 7.1.26 *If* $\mathsf{b} \subseteq \mathrm{pcf}(\mathsf{a})$, *then* $\mathrm{pcf}_*(\mathsf{b}) = \mathrm{pcf}_*(\mathrm{pcf}_*(\mathsf{b}))$.

Proof Fix $\lambda \in \mathrm{pcf}_*(\mathrm{pcf}_*(\mathsf{b}))$. From the assumption, Lemma 7.1.23 and Lemma 7.1.25 we get $\mathrm{pcf}_*(\mathsf{b}) \subseteq \mathrm{pcf}_*(\mathrm{pcf}(\mathsf{a})) = \mathrm{pcf}(\mathsf{a})$, and Theorem 7.1.24 yields $\lambda \in \mathrm{pcf}(\mathsf{d})$ for some $\mathsf{d} \subseteq \mathrm{pcf}_*(\mathsf{b})$ satisfying $|\mathsf{d}| \leq |\mathsf{a}|$.

For each $\nu \in \mathsf{d} \subseteq \mathrm{pcf}_*(\mathsf{b})$ we obtain, again by Theorem 7.1.24, a set $\mathsf{d}_\nu \subseteq \mathsf{b}$ such that $|\mathsf{d}_\nu| \leq |\mathsf{a}|$ and $\nu \in \mathrm{pcf}(\mathsf{d}_\nu)$. The estimate $|\mathsf{d}| \leq |\mathsf{a}| < \min(\mathsf{a}) \leq \min\mathrm{pcf}(\mathsf{a}) \leq \min\mathrm{pcf}_*(\mathsf{b}) \leq \min(\mathsf{d})$ shows that d is progressive. So we can conclude that

$$\lambda \in \mathrm{pcf}(\mathsf{d}) = \mathrm{pcf}_*(\mathsf{d}) \subseteq \mathrm{pcf}_*(\bigcup\{\mathrm{pcf}(\mathsf{d}_\nu) : \nu \in \mathsf{d}\}) \subseteq \mathrm{pcf}_*(\mathrm{pcf}(\bigcup\{\mathsf{d}_\nu : \nu \in \mathsf{d}\}))$$
$$= \mathrm{pcf}(\bigcup\{\mathsf{d}_\nu : \nu \in \mathsf{d}\}) = \mathrm{pcf}_*(\bigcup\{\mathsf{d}_\nu : \nu \in \mathsf{d}\}) \subseteq \mathrm{pcf}_*(\mathsf{b}).$$

The last steps in this chain of inclusions follow from Lemma 7.1.25 and Lemma 7.1.23 respectively, since $|\bigcup\{\mathsf{d}_\nu : \nu \in \mathsf{d}\}| \leq |\mathsf{a}| < \min(\mathsf{a}) \leq \min(\bigcup\{\mathsf{d}_\nu : \nu \in \mathsf{d}\})$ and thus the set $\bigcup\{\mathsf{d}_\nu : \nu \in \mathsf{d}\}$ is progressive.

Now we can state an analogy to Lemma 3.4.20 better than Corollary 7.1.19.

Lemma 7.1.27 *If* $\mathsf{b} \subseteq \mathrm{pcf}(\mathsf{a})$, *then* $\mathrm{pcf}_*(\mathsf{b})$ *is a subset of* $\mathrm{pcf}(\mathsf{a})$ *and has a maximum.*

Proof The first assertion follows from Lemma 7.1.23 and Corollary 7.1.19. To get a contradiction, we assume that the set $\mathrm{pcf}_*(\mathsf{b})$ has no maximum and claim that, for any $\mathsf{c} \subseteq \mathrm{pcf}_*(\mathsf{b})$ with $|\mathsf{c}| \leq |\mathsf{a}|$, there is $\lambda \in \mathrm{pcf}_*(\mathsf{b})$ such that $\max\mathrm{pcf}(\mathsf{c}) < \lambda$. This again gives rise to the construction of a sequence of elements of $\mathrm{pcf}_*(\mathsf{b})$ having length $|\mathsf{a}|^+$ whose existence contradicts Theorem 6.2.1.

To prove the claim, assume that $\mathsf{c} \subseteq \mathrm{pcf}_*(\mathsf{b})$ and $|\mathsf{c}| \leq |\mathsf{a}|$. As in the proof of Theorem 7.1.26 for the set d, we can verify that c is progressive. So we can conclude from Theorem 7.1.26 that

$$\mathrm{pcf}(\mathsf{c}) = \mathrm{pcf}_*(\mathsf{c}) \subseteq \mathrm{pcf}_*(\mathrm{pcf}_*(\mathsf{b})) = \mathrm{pcf}_*(\mathsf{b}).$$

Furthermore $\mathrm{pcf}(\mathsf{c})$ has a maximum by Lemma 3.4.20. But from our supposition on $\mathrm{pcf}_*(\mathsf{b})$ we can infer that there is $\lambda \in \mathrm{pcf}_*(\mathsf{b})$ greater than this maximum, which proves the claim and completes the proof of the lemma.

Corollary 7.1.28 (5.3.1) *For any* $\lambda \in \mathrm{pcf}(\mathsf{a})$, $\max \mathrm{pcf}_*(\mathrm{pcf}(\mathsf{b}_\lambda)) = \lambda$.

Proof Since b_λ is progressive as a subset of the progressive set a, we get

$$\mathrm{pcf}_*(\mathrm{pcf}(\mathsf{b}_\lambda)) = \mathrm{pcf}_*(\mathrm{pcf}_*(\mathsf{b}_\lambda)) = \mathrm{pcf}_*(\mathsf{b}_\lambda) = \mathrm{pcf}(\mathsf{b}_\lambda),$$

and together with $\lambda = \max \mathrm{pcf}(\mathsf{b}_\lambda)$ the corollary is proved.

Exercises

1) Prove Lemma 7.1.3.
Hint: See the proof of Lemma 7.1.1.
2) Prove Lemma 7.1.4.
3) Prove Lemma 7.1.5.
4) Prove Lemma 7.1.6.
5) If $\mathsf{c} \subseteq \mathrm{pcf}(\mathsf{a})$, then there are $n \in \omega$ and $\lambda_1, \ldots, \lambda_n \in \mathrm{pcf}_*(\mathsf{c})$ such that
   (a) $\max\{\lambda_1, \ldots, \lambda_n\} = \max \mathrm{pcf}_*(\mathsf{c})$ and
   (b) $\mathsf{c} \subseteq \mathrm{pcf}(\mathsf{b}_{\lambda_1}) \cup \ldots \cup \mathrm{pcf}(\mathsf{b}_{\lambda_n})$.
6) For any cardinal λ,
$$\mathcal{J}^{p}_{<\lambda}(\mathsf{a}) = \{\mathsf{c} \subseteq \mathrm{pcf}(\mathsf{a}) : \mathrm{pcf}_*(\mathsf{c}) \subseteq \lambda\}.$$
7) Prove Lemma 7.1.23.
8) If λ is a cardinal and $\mathsf{b} \subseteq \mathsf{a}$, then
$$\mathcal{J}^{p}_{<\lambda}(\mathsf{b}) = \mathcal{J}^{p}_{<\lambda}(\mathsf{a}) \cap \mathcal{P}(\mathrm{pcf}(\mathsf{b})).$$

7.2 Intervals in pcf(a)

Definition *In analogy to Lemma 1.8.9 regarding normal ideals, we now define the notion of a weakly normal ideal. Let A be a set of ordinals such that* $\sup(A)$ *is not a member of A. We call an ideal I on A* **weakly normal** *iff, for every I-positive subset S of A and every regressive function f on S, there is an I-positive subset S_0 of S and an ordinal $\gamma <$* $\sup(A)$ *such that $f[S_0] \subseteq \gamma$.*

Let α be a limit ordinal of uncountable cofinality and C be a subset of α such that $\sup(C) = \alpha$. *Then we know that the set*

$$\mathrm{I_d}(C) := \{X \subseteq C : X \textit{ is not stationary in } \alpha\}$$

of thin subsets of C is an ideal on C.

If E is a set of cardinals, then we put $E^+ := \{\nu^+ : \nu \in E\}$.

Lemma 7.2.1 *If* $\mathrm{cf}(\alpha) > \omega$ *and* C *is a club in* α *of order type* $\mathrm{cf}(\alpha)$, *then the ideal* $\mathrm{I_d}(C)$ *on* α *is weakly normal.*

Proof From Section 1.8 we know that $\mathrm{I_d}(C)$ is an ideal on α. Let Γ be the $\in$-isomorphism from $\kappa := \mathrm{cf}(\alpha)$ onto C. Further assume that S is an $\mathrm{I_d}(C)$-positive subset of C and f is regressive on S. Then the set

$$S' := \Gamma^{-1}[S] \cap \{i < \kappa : i \text{ is a limit ordinal}\}$$

is a stationary subset of κ. For each $i \in S'$ let $g(i)$ be the least $j < \kappa$ such that $f(\Gamma(i)) < \Gamma(j)$. Since all members i of S' are limit ordinals and f is regressive, g is also regressive. Fodor's theorem says that there is a stationary subset $S_0' \subseteq S'$ and an ordinal $i_0 < \kappa$ such that $g[S_0'] = \{i_0\}$. Now $S_0 := \Gamma[S_0']$ is an $\mathrm{I_d}(C)$-positive subset of S satisfying $f[S_0] \subseteq \Gamma(i_0) < \alpha$.

Lemma 7.2.2 *If* I *is a weakly normal ideal on a set* A *and* X *is an* I*-positive subset of* A, *then the ideal* $I \restriction X$ *is weakly normal.*

Proof Let $X \subseteq A$ be I-positive and $Y \subseteq A$. Further assume that Y is $(I \restriction X)$-positive and f is a regressive function on Y. Then f is regressive on the I-positive set $Y \cap X$, and thus $f[Z] \subseteq \gamma$ for some I-positive set $Z \subseteq Y \cap X$ and some $\gamma < \sup(A)$. This gives $Z \cap X = Z \notin I$, hence Z is $(I \restriction X)$-positive.

Theorem 7.2.3 *Assume that* λ *is a singular cardinal with uncountable cofinality. Then there is a progressive club* C_0 *in* λ *of cardinal numbers such that* $\mathrm{otp}(C_0) = \mathrm{cf}(\lambda)$ *and*

$$\max \mathrm{pcf}(C_0^+) = \lambda^+.$$

Proof From Lemma 1.6.11 we obtain a club C in λ such that all members of C are singular cardinals and $\mathrm{otp}(C) = \mathrm{cf}(\lambda) < \min(C)$. Let

$$\mathsf{a} := C^+.$$

Obviously a is a progressive set of regular cardinals. Theorem 5.2.3 yields a set $\mathsf{b}_{\lambda^+} \subseteq \mathsf{a}$ generating the ideal $\mathcal{J}_{<\lambda^{++}}(\mathsf{a})$ over $\mathcal{J}_{<\lambda^+}(\mathsf{a})$; if $\lambda^+ \notin \mathrm{pcf}(\mathsf{a})$, then let $\mathsf{b}_{\lambda^+} = \emptyset$. It suffices to prove that there is a club $C_0 \subseteq C$ in λ such that $C_0^+ \subseteq \mathsf{b}_{\lambda^+}$. To see this, note that then we can conclude from Lemma 3.3.4 and Lemma 5.3.1, together with the fact that C_0^+ is progressive, that $\lambda \le \max \mathrm{pcf}(C_0^+) \le \max \mathrm{pcf}(\mathsf{b}_{\lambda^+}) = \lambda^+$. Since λ is singular, this will give $\max \mathrm{pcf}(C_0^+) = \lambda^+$.

To get a contradiction, let us assume that such a club does not exist. Let b_0 be that subset of C with $b_0^+ = \mathsf{b}_{\lambda^+}$. By our supposition, b_0 does not include a club in λ, and thus $C \setminus b_0$ is stationary in λ. Consequently, by the previous lemmas,

$$J := \mathrm{I_d}(C) \restriction (C \setminus b_0) = \acute{\mathrm{I}}_{\mathrm{d}}(C)[b_0]$$

is a weakly normal ideal on C. We transfer its ideal structure to a, putting

$$I := \{X^+ : X \in J\}.$$

Clearly I is an ideal on a. Since $\mathrm{I_d}(C)$ has all bounded subsets of C amongst its members, all bounded subsets of a are members of I. We can infer from Corollary 3.4.12, together with the fact that λ is singular, that $\mathcal{J}_{<\lambda}(\mathsf{a}) = \mathcal{J}_{<\lambda^+}(\mathsf{a})$. So we get $\mathcal{J}_{<\lambda^+}(\mathsf{a}) \subseteq I$ from Lemma 3.4.9 and $\lambda = \sup(\mathsf{a})$. Furthermore we have $\lambda = \lim_I(\mathsf{a})$. $b_0 \in J$ gives $\mathsf{b}_{\lambda^+} \in I$, hence $\mathcal{J}_{<\lambda^{++}}(\mathsf{a}) \subseteq I$ by the choice of b_{λ^+}. Consequently, $(\prod \mathsf{a}, <_I)$ is λ^{++}-directed.

Lemma 3.2.4 says that there is a sequence of members of $\prod \mathsf{a}$ which is strictly increasing modulo I, has length λ^+, and is κ-rapid for all $\kappa \in [\aleph_1, \lambda)_{\mathrm{reg}}$. By Theorem 3.2.1 and Lemma 3.2.3, this sequence has a supremum g. Since $(\prod \mathsf{a}, <_I)$ is λ^{++}-directed, $\prod \mathsf{a}$ contains an $\leq_I$-upper bound of the sequence. So we can assume that $g \in \prod \mathsf{a}$. If $\rho \in C$, then $g(\rho^+) < \rho^+$, hence $h(\rho) := \mathrm{cf}\,(g(\rho^+)) < \rho$ since ρ is singular. Now J is weakly normal and the function h is regressive on the J-positive set C, and so there is a J-positive set $Y \subseteq C$ and an ordinal $\gamma < \lambda = \sup(C)$ such that $\mathrm{cf}\,(g(\rho^+)) < \gamma$ for all $\rho \in Y$. Using Lemma 3.2.2, we get the contradiction

$$\lambda = \lim_I(\mathrm{cf} \circ g) \leq \lim_{(I \restriction Y^+)}(\mathrm{cf} \circ g) \leq \gamma.$$

Before applying this theorem to intervals which are subsets of $\mathrm{pcf}(\mathsf{a})$, we will shortly demonstrate that we have also proved a representation theorem. If A is a nonempty set of ordinals such that $\sup(A) \notin A$, then we know that

$$\mathrm{I_b}(A) := \{X \subseteq A : \sup(X) < \sup(A)\}$$

is an ideal on A.

Theorem 7.2.4 *If λ is a singular cardinal with uncountable cofinality, then there is a club C_0 in λ of cardinal numbers such that* $\mathrm{otp}(C) = \mathrm{cf}\,(\lambda)$ *and*

$$\mathrm{tcf}\,(\prod C_0^+/\mathrm{I_b}(C_0^+)) = \lambda^+.$$

Proof Choose C_0 as in Theorem 7.2.3. From Lemma 3.4.9 and the fact that λ is singular we can infer that the ideal $\mathcal{J}_{<\lambda^+}(C_0^+)$ is a subset of $\mathrm{I_b}(C_0^+)$. Now we apply Lemma 3.4.19 with $\mathsf{a} = \mathsf{b} = C_0^+$ and get $\mathrm{tcf}\,(\prod C_0^+/\mathrm{I_b}(C_0^+)) = \mathrm{tcf}\,(\prod C_0^+/\mathcal{J}_{<\lambda^+}(C_0^+)) = \lambda^+$, as desired.

Lemma 7.2.5 *Assume that $\kappa \geq \omega$ is a cardinal, $\tau > \kappa$ is an ordinal, and F is a function from $\mathcal{P}(\tau)$ into* ON, *such that for all $X, Y \in \mathcal{P}(\tau)$*

(1) $X \subseteq Y \implies F(X) \leq F(Y)$.

(2) $F(X) \geq \sup(X)$.

(3) If $\gamma < \tau$ and $\mathrm{cf}(\gamma) = \kappa^{+3}$, then there is a club E in γ such that $F(E) = \gamma$.

(4) If X has order type κ^+, then there is $\beta < \sup(X)$ such that $F(X \cap \beta) \geq \sup(X)$.

Then we have $\tau < \kappa^{+4}$.

Proof Consider the cardinal $\kappa \geq \omega$. To get a contradiction, let us suppose that there is a function F and an ordinal τ with the properties (1)-(4) such that $\tau \geq \kappa^{+4}$. By Lemma 4.5.3, we can choose a $\diamondsuit_{\mathrm{Club}}(\kappa^{+3}, \kappa^+)$-sequence $\bar{C} := (C_\alpha : \alpha \in \Sigma_{\kappa^{+3},\kappa^+})$ for $\Sigma_{\kappa^{+3},\kappa^+}$. Choose the regular cardinal Θ large enough and an approximation sequence $(N_i : i \leq \kappa^{+3})$ in $\mathrm{H}(\Theta)$ such that $|N_i| = \kappa^{+3}$ for all $i < \kappa^{+3}$ and $\kappa^{+3} \cup \{\bar{C}, F, \kappa^{+4}\} \subseteq N_0$. Let $\delta_i := \chi_{N_i}(\kappa^{+4})$ for all $i \leq \kappa^{+3}$.

Lemma 4.3.3 says that $(\delta_i : i < \kappa^{+3})$ is normal sequence which is cofinal in $\delta_{\kappa^{+3}}$, and that $\delta_{\kappa^{+3}} < \kappa^{+4}$. By property (3), there is a club E in $\delta_{\kappa^{+3}}$ such that $F(E) = \delta_{\kappa^{+3}}$. Then $E^* := E \cap \{\delta_i : i < \kappa^{+3}\}$ is a club in $\delta_{\kappa^{+3}}$ and $C := \{j \in \kappa^{+3} : \delta_j \in E^*\}$ is a club in κ^{+3}. Furthermore we have $F(E^*) = \delta_{\kappa^{+3}}$, since (1) gives $F(E^*) \leq F(E)$, and (2) yields $F(E^*) \geq \sup(E^*) = \delta_{\kappa^{+3}}$. So let, without loss of generality, $E = \{\delta_i : i \in C\}$. By the definition of a $\diamondsuit_{\mathrm{Club}}(\kappa^{+3}, \kappa^+)$-sequence, we can find an ordinal $\alpha \in \Sigma_{\kappa^{+3},\kappa^+}$ such that $C_\alpha \subseteq C$, and furthermore the club C_α has order type κ^+. Therefore this also holds for the set $X := \{\delta_i : i \in C_\alpha\}$, since $(\delta_i : i < \kappa^{+3})$ is normal. We have $\sup(X) = \delta_\alpha$, since C_α is unbounded in α. Using (4) and (1), we obtain a limit ordinal $\sigma < \sup(X) = \delta_\alpha$ such that $F(X \cap \sigma) \geq \delta_\alpha$. We put $\beta := \sup\{i < \kappa^{+3} : \delta_i < \sigma\}$. Then we have $\delta_\beta \leq \sigma$, hence $\beta < \alpha$ and

$$\zeta := F(\{\delta_i : i \in C_\alpha \cap \beta\}) \geq \sup\{\delta_i : i \in C_\alpha\} = \delta_\alpha.$$

From $\kappa^{+3} \subseteq N_0$ and $\bar{C} \in N_0$ we can infer that α and thus C_α and β are members of N_0. Furthermore we have $(N_i : i < \beta) \in N_{\beta+1} \subseteq N_\alpha$ by the definition of an approximation sequence. Altogether we get $\{\delta_i : i \in C_\alpha \cap \beta\} \in N_\alpha$, hence $\zeta \in N_\alpha$ since $F \in N_0$. By (1) we have $\zeta \leq F(E) = \delta_{\kappa^{+3}} < \kappa^{+4}$. This yields the contradiction $\zeta < \chi_{N_\alpha}(\kappa^{+4}) = \delta_\alpha$.

Now we can prove one of the most important results of pcf-theory.

Theorem 7.2.6 (Shelah) *If* a *is a progressive set of regular cardinals and* c *is an interval of regular cardinal numbers such that* $\mathsf{c} \subseteq \mathrm{pcf}(\mathsf{a})$, *then* $|\mathsf{c}| \leq |\mathsf{a}|^{+3}$.

Proof Let us assume that the assertion of the theorem is false. Then there is an interval $\mathsf{c} \subseteq \mathrm{pcf}(\mathsf{a})$ of regular cardinals of order type $|\mathsf{a}|^{+4}$. Clearly a is infinite and c has no maximum. Since $|\mathsf{a}| < \min(\mathsf{a}) \leq \min(\mathsf{c})$ and $|\mathsf{c}| = |\mathsf{a}|^{+4}$, we can assume without loss of generality that c is progressive. Hence, by Lemma 3.3.3, all its members are successor cardinals. To get a contradiction, we will construct a function F which satisfies properties (1) to (4) in Lemma 7.2.5 for $\kappa := |\mathsf{a}|$ and $\tau := \kappa^{+4}$.

Let Γ be the $\in$-isomorphism from τ onto c. We define, for every $X \subseteq \tau$,

$$F(X) := \begin{cases} \Gamma^{-1}(\max \operatorname{pcf}(\Gamma[X])) & \text{if } \operatorname{pcf}(\Gamma[X]) \subseteq \mathsf{c}, \\ \tau & \text{otherwise.} \end{cases}$$

Property (1) is satisfied since we have $\operatorname{pcf}(\Gamma[X]) \subseteq \operatorname{pcf}(\Gamma[Y])$ by Lemma 3.3.4, whenever $X \subseteq Y \subseteq \tau$. Furthermore $\max \operatorname{pcf}(\Gamma[X]) \geq \sup(\Gamma[X])$ gives (2). For the proof of (3) assume that $\gamma < \tau$ and $\operatorname{cf}(\gamma) = \kappa^{+3}$, and let λ be that cardinal with $\lambda^+ = \Gamma(\gamma)$. Then λ is a limit cardinal and $\lambda = \sup(\Gamma[\gamma])$. This gives $\operatorname{cf}(\lambda) = \operatorname{cf}(\gamma)$, and in particular λ is singular. By Theorem 7.2.3, there is a club C_0 in λ all of whose elements are cardinals such that $\operatorname{otp}(C_0) = \operatorname{cf}(\lambda) = \kappa^{+3}$ and $\max \operatorname{pcf}(C_0^+) = \lambda^+$. Then we have $\mu \in \mathsf{c}$ for every member μ of C_0^+ which is greater than $\min(\mathsf{c})$, and $\operatorname{otp}(C_0^+) = \kappa^{+3}$. We put $E := \Gamma^{-1}[C_0^+ \cap \mathsf{c}]$. By Lemma 3.3.4, the maximum of $\operatorname{pcf}(C_0^+ \cap \mathsf{c})$ is not less than λ. It is different from λ, since λ is singular, and at most λ^+, since $C_0^+ \cap \mathsf{c} \subseteq C_0^+$. So we get $F(E) = \gamma$. Furthermore, E is a club in γ. To see this, we observe first that $\Gamma^{-1}[C_0^+ \cap \mathsf{c}]$ is unbounded in γ since $C_0^+ \cap \mathsf{c}$ is unbounded in $\lambda = \sup([\Gamma[\gamma])$. We must further show for an arbitrary limit ordinals $\sigma < \gamma$ in which $E \cap \sigma$ is unbounded that it is a member of E. We have $\operatorname{cf}(\sigma) < \kappa^{+3}$ since $\operatorname{otp}(E) = \kappa^{+3}$, and $\Gamma[E \cap \sigma]$ is unbounded in the cardinal $\mu := \sup(\Gamma[\sigma])$. But μ is also the supremum of the predecessors of members of $\Gamma[\sigma]$ and thus an element of C_0. Furthermore we have $\mu \notin \mathsf{c}$ since no member of c is of cofinality less than κ^{+3}. So we get $\Gamma(\sigma) = \mu^+ \in C_0^+ \cap \mathsf{c}$, and σ is a member of E.

Finally we prove (4). Assume that X is a subset of τ of order type κ^+. As above, we have $\nu := \max \operatorname{pcf}(\Gamma[X]) > \sup(\Gamma[X])$. Since $\Gamma[X]$ is progressive, Theorem 7.1.24 tells us that there is a set $\mathsf{d} \subseteq \Gamma[X]$, such that $|\mathsf{d}| \leq |\mathsf{a}|$ and $\nu \in \operatorname{pcf}(\mathsf{d})$. With the regularity of κ^+ we obtain an ordinal $\beta < \sup(X)$ such that $\Gamma^{-1}[\mathsf{d}] \subseteq \beta$. So we can conclude, using (1) and the definition of F, that $F(X \cap \beta) \geq F(\Gamma^{-1}[\mathsf{d}]) \geq \sup(X)$, as desired.

Now Lemma 7.2.5 implies $\tau < \kappa^{+4}$ which contradicts the definition of τ.

Corollary 7.2.7 *If* a *is a progressive interval of regular cardinals, then*

$$|\operatorname{pcf}(\mathsf{a})| \leq |\mathsf{a}|^{+3}.$$

Proof By Theorem 3.3.6, $\operatorname{pcf}(\mathsf{a})$ is an interval of regular cardinals. Theorem 7.2.6 says that $|\operatorname{pcf}(\mathsf{a})| \leq |\mathsf{a}|^{+3}$.

Chapter 8

Applications of pcf-Theory

As a first application of pcf-theory, we will prove in this section two estimates for cardinal numbers. Initially, if δ is a limit ordinal and μ is a cardinal with $\mu < \aleph_\delta$, then

$$\aleph_\delta^\mu < \aleph_{(|\delta|^\mu)^+}.$$

This result of Shelah generalizes the Galvin-Hajnal theorem to cardinals $\aleph_\delta$ with countable cofinality. Secondly we will show for limit ordinals δ that

$$\aleph_\delta^{|\delta|} < \max\{\aleph_{|\delta|^{+4}}, (2^{|\delta|})^+\}.$$

A further application of pcf-theory guarantees the existence of a Jónsson-Algebra on the cardinal $\aleph_{\omega_1}$.

If λ is a cardinal number, then we write $\mathrm{p}(\lambda)$ for the following assertion: There is a function $f : [\lambda]^2 \longrightarrow \lambda$ such that any subset A of λ of cardinality λ satisfies $|\{f(x) : x \in [A]^2\}| = \lambda$. A theorem of Todorčević will be proved which says: If λ is a singular cardinal and $\mu < \lambda$ is a cardinal satisfying $\mathrm{p}(\kappa)$ for every regular cardinal $\kappa \in [\mu, \lambda)_{\mathrm{reg}}$, then $\mathrm{p}(\lambda^+)$. Finally, we will get the following theorem concerning the cofinalities of partial orderings of the form $([\lambda]^{\leq\kappa}, \subseteq)$: If κ is an infinite cardinal number, α is an ordinal such that $1 \leq \alpha < \kappa$, and $\lambda := \kappa^{+\alpha}$, then

$$\mathrm{cf}\,([\lambda]^{\leq\kappa}, \subseteq) = \max\,\mathrm{pcf}(\{\kappa^{+(\beta+1)} : \beta < \alpha\}).$$

8.1 Cardinal Estimates

Lemma 8.1.1 *Let $\aleph_\delta$ be a limit cardinal and μ be an infinite cardinal. Further assume that α is an ordinal with $\alpha < \delta$ such that $\aleph_\alpha$ is μ-strong and $\mathsf{a} := [\aleph_\alpha, \aleph_\delta)_{\mathrm{reg}}$ is a progressive interval of regular cardinals. Then*

$$\aleph_\delta^\mu < \aleph_{\alpha+|\mathrm{pcf}_\mu(\mathsf{a})|^+}.$$

Proof Theorem 6.2.4 says that sup $\mathrm{pcf}_\mu(\mathsf{a}) = \aleph_\delta^\mu$, and Theorem 3.3.6 tells us that $\mathrm{pcf}_\mu(\mathsf{a})$ is an interval of regular cardinals. Furthermore we have $\min(\mathsf{a}) = \min \mathrm{pcf}_\mu(\mathsf{a})$. It is obvious that any interval b of cardinals with $\min(b) = \aleph_\alpha$ satisfies $\sup(b) < \aleph_{\alpha+|b|^+}$. Altogether we get the desired inequality.

Theorem 8.1.2 *If δ is a limit ordinal and μ is a cardinal with $\mu < \aleph_\delta$, then*

$$\aleph_\delta^\mu < \aleph_{(|\delta|^\mu)^+}.$$

Proof Obviously the inequality is true if μ is finite. If $\delta = \aleph_\delta$, then $|\delta| = \aleph_\delta$, hence $\aleph_\delta^\mu = |\delta|^\mu < \aleph_{(|\delta|^\mu)^+}$. So we can assume that $\delta < \aleph_\delta$ and μ is infinite. If $\aleph_\delta \leq 2^\mu$, then $\aleph_\delta^\mu \leq 2^\mu \leq |\delta|^\mu < \aleph_{(|\delta|^\mu)^+}$.

Thus let us assume in addition that $2^\mu < \aleph_\delta$. We distinguish two cases.

Case 1 $\aleph_\delta$ is μ-strong.

Put $\aleph_\alpha := (|\delta|^\mu)^+$. Then we have $\aleph_\alpha < \aleph_\delta$, since $|\delta| \leq \delta < \aleph_\delta$, δ is a limit ordinal and $\aleph_\delta$ is μ-strong. Now $\mathsf{a} := [\aleph_\alpha, \aleph_\delta)_{\mathrm{reg}}$ is a progressive interval of regular cardinals since $|\mathsf{a}| \leq |\delta| < (|\delta|^\mu)^+$. Furthermore $\aleph_\alpha$ is μ-strong. With $\aleph_\alpha < \aleph_\delta$ we get $\alpha < \delta$, and thus $\aleph_\delta^\mu < \aleph_{\alpha+|\mathrm{pcf}_\mu(\mathsf{a})|^+}$ by Lemma 8.1.1. Theorem 3.4.13 says that $|\mathrm{pcf}_\mu(\mathsf{a})| \leq |\mathsf{a}|^\mu$. So $|\mathsf{a}| \leq |\delta|$ gives $|\mathrm{pcf}_\mu(\mathsf{a})|^+ \leq (|\delta|^\mu)^+$. With $\alpha < \delta$ we conclude that $\alpha < |\delta|^+$, hence $\alpha < (|\delta|^\mu)^+$. On the other hand, Lemma 1.5.9 says that every infinite cardinal is a principal number of addition, hence $\alpha + (|\delta|^\mu)^+ = (|\delta|^\mu)^+$. Now it is easy to check that $\aleph_\delta^\mu < \aleph_{(|\delta|^\mu)^+}$.

Case 2 $\aleph_\delta$ is not μ-strong.

Since $2^\mu < \aleph_\delta^\mu$ and $\mu < \aleph_\delta$, we can infer from Lemma 1.7.3 that there exists an ordinal $\gamma < \delta$ such that $\aleph_\gamma$ is a μ-strong singular cardinal satisfying $\mathrm{cf}(\gamma) \leq \mu < \aleph_\gamma$ and $\aleph_\delta^\mu = \aleph_\gamma^\mu$. If $\gamma = \aleph_\gamma$, then $\aleph_\delta^\mu = \aleph_\gamma^\mu = |\gamma|^\mu < \aleph_{(|\gamma|^\mu)^+} \leq \aleph_{(|\delta|^\mu)^+}$. So we can assume that $\gamma < \aleph_\gamma$. If $\aleph_\gamma \leq 2^\mu$, then $\aleph_\gamma^\mu \leq 2^\mu \leq |\gamma|^\mu < \aleph_{(|\gamma|^\mu)^+}$, and thus $\aleph_\delta^\mu < \aleph_{(|\delta|^\mu)^+}$. So let in addition $2^\mu < \aleph_\gamma$. From Case 1 we get $\aleph_\gamma^\mu < \aleph_{(|\gamma|^\mu)^+}$, and thus $\aleph_\delta^\mu < \aleph_{(|\delta|^\mu)^+}$, as desired.

Corollary 8.1.3 *If $\aleph_\delta$ is a singular strong limit cardinal, then $2^{\aleph_\delta} < \aleph_{(2^{|\delta|})^+}$.*

Proof From Lemma 1.6.15 we get $2^{\aleph_\delta} = \aleph_\delta^{\mathrm{cf}(\aleph_\delta)} = \aleph_\delta^{\mathrm{cf}(\delta)}$. Thus we have $2^{\aleph_\delta} < \aleph_{(|\delta|^{\mathrm{cf}(\delta)})^+} \leq \aleph_{(2^{|\delta|})^+}$.

Example $\aleph_\alpha^{\aleph_0} < \aleph_{(2^{\aleph_0})^+}$ holds for all ordinals $\alpha < (2^{\aleph_0})^+$.
To see this, put $\delta := \alpha + \omega$. Then $\mathrm{cf}(\delta) = \omega$ and $|\delta|^{\aleph_0} \leq (2^{\aleph_0})^{\aleph_0} = 2^{\aleph_0}$. From Theorem 8.1.2 we can conclude that

$$\aleph_\alpha^{\aleph_0} \leq \aleph_\delta^{\aleph_0} < \aleph_{(|\delta|^{\aleph_0})^+} \leq \aleph_{(2^{\aleph_0})^+}.$$

Theorem 8.1.4 *If δ is a limit ordinal, then*

$$\aleph_\delta^{|\delta|} < \max\{\aleph_{|\delta|^{+4}}, (2^{|\delta|})^+\}.$$

In particular, we have

$$2^{\aleph_0} < \aleph_\omega \implies \aleph_\omega^{\aleph_0} < \aleph_{\omega_4}.$$

Proof If $\aleph_\delta \leq 2^{|\delta|}$, then $\aleph_\delta^{|\delta|} \leq 2^{|\delta|}$. So assume that $2^{|\delta|} < \aleph_\delta$. Then $(2^{|\delta|})^+ < \aleph_\delta$, hence $\mathsf{a} := [\aleph_\alpha, \aleph_\delta)_{\text{reg}}$ with $\aleph_\alpha := (2^{|\delta|})^+$ is a progressive interval of regular cardinals. Lemma 8.1.1 says that $\aleph_\delta^{|\delta|} < \aleph_{\alpha+|\text{pcf}_{|\delta|}(\mathsf{a})|^+}$, and from Corollary 7.2.7 we can conclude that $|\text{pcf}(\mathsf{a})| < |\mathsf{a}|^{+4}$. $|\mathsf{a}| \leq |\delta|$ gives $\text{pcf}_{|\delta|}(\mathsf{a}) = \text{pcf}(\mathsf{a})$. Furthermore we have $\alpha < \delta$ and thus $\alpha < |\delta|^+ < |\delta|^{+4}$. Every infinite cardinal is a principal number of addition, and therefore $\alpha + |\delta|^{+4} = |\delta|^{+4}$. Altogether we get $\aleph_\delta^{|\delta|} < \aleph_{|\delta|^{+4}}$.

Remark

a) If $2^{\aleph_0} < \aleph_\omega$, then $\aleph_\omega^{\aleph_0}$ is a regular cardinal.
   To see this, put $\mathsf{a} := [\aleph_1, \aleph_\omega)_{\text{CN}}$. Then $\aleph_\omega^{\aleph_0} = |\prod \mathsf{a}|$ by Lemma 1.6.15. Hausdorff's formula yields $\aleph_1^{\aleph_0} = \aleph_1 \cdot 2^{\aleph_0}$.
   So our assumption gives $\min(\mathsf{a})^{|\mathsf{a}|} < \sup(\mathsf{a})$. Now Corollary 6.2.5 says that $|\prod \mathsf{a}| = \max \text{pcf}(\mathsf{a})$, and therefore $\aleph_\omega^{\aleph_0}$ is regular since it is the maximum of $\text{pcf}(\mathsf{a})$.

b) If $2^{\aleph_0} = \aleph_1$, then $\aleph_\omega^{\aleph_0} < \aleph_{\omega_2}$ by Theorem 8.1.2 and $\aleph_\omega^{\aleph_0} < \aleph_{\omega_4}$ by Theorem 8.1.4. On the other hand, if $2^{\aleph_0} = \aleph_4$, we get $\aleph_\omega^{\aleph_0} < \aleph_{\omega_5}$ from Theorem 8.1.2 and $\aleph_\omega^{\aleph_0} < \aleph_{\omega_4}$ from Theorem 8.1.4.

8.2 Jónsson Algebras

A Jónsson algebra is a structure $(A, (f_k : k < \omega))$ with countably many functions f_k which has no proper substructure of the same cardinality. We will obtain results on Jónsson algebras by using suitable elementary substructures of $\text{H}(\Theta)$ (for suitable cardinals Θ). So it is necessary to generalize the notions *language*, *structure* and *model* from Section 1.1 and Chapter 4.

A **language L of first order logic** is given by its alphabet, its terms and its formulae[1].

The **alphabet** of L consists of the following objects which we call the **symbols** of L:

[1] To be precise, the following objects must actually be defined in ZF as sets, for example as members of V_ω. The interested reader is referred to the book of F. R. Drake [Dr], in which he can find a detailed formalization of some first order languages.

variables v_n for each natural number n;

the logical symbols $\neg$, $\vee$, $\exists$ and $=$ (the further symbols $\Longrightarrow$, $\Longleftrightarrow$, $\wedge$ and $\forall$ are introduced as abbreviations, as in Section 4.2);

bracket symbols (,);

function symbols $f_0^{n_0}, f_1^{n_1}, \ldots$, where $n_0, n_1, \ldots$ are natural numbers, and

relation symbols $R_0^{n_0}, R_1^{n_1}, \ldots$, where $n_0, n_1, \ldots$ are positive natural numbers.

We also use f^n and R^n as syntactical variables for function symbols and relation symbols, respectively.

If $f_j^{n_j}$ and $R_j^{n_j}$ are function symbols and relation symbols, respectively, then n_j is called the **arity** of these symbols. 0-ary function symbols are called **constant symbols.**

An **expression** is a finite sequence of symbols. The **concatenation** of expressions is that sequence which results from writing the given expressions one behind the other.

The **terms** of L are defined by recursion on the number of symbols:

(T1) Every variable is a term, and every constant symbol is a term.

(T2) If $n_j > 0$, $t_1, \ldots, t_{n_j}$ are terms, and $f_j^{n_j}$ is a function symbol, then $f_j^{n_j} t_1 \ldots t_{n_j}$ is a term (resulting from the concatenation of the expressions $f_j^{n_j}, t_1, \ldots, t_{n_j}$).

The **formulae** of L are also defined by recursion:

(F1) If t_1 and t_2 are terms, then $(t_1 = t_2)$ is a formula; if $R_j^{n_j}$ is a relation symbol and $t_1, \ldots, t_n$ are terms, then $(R_j^{n_j} t_1 \ldots t_{n_j})$ is a formula (resulting, analogously to (T2), from concatenation).

(F2) If φ and ψ are formulae, then so are $(\neg\varphi)$, $(\varphi \vee \psi)$ and $(\exists v_j\ \varphi)$ for any natural number j.

A **structure** $\mathcal{A}$ for the language L consists of

a) a nonempty set A, called the **universe of** $\mathcal{A}$;

b) a function assigning to each n-ary function symbol f of L, $n > 0$, a function f_A from A^n into A; if f is a constant symbol, then it assigns to f a member f_A of A;

c) a function assigning to each n-ary relation symbol R^n a relation $R_A^n \subseteq A^n$.

Usually we write f and R instead of f_A and R_A. From the context it will be clear whether f denotes a function symbol or a function. For structures $\mathcal{A}$ we use the notation

$$(A, (f_i : i \in \alpha), (R_i : i \in \beta)),^2$$

where $\alpha, \beta \leq \omega$. The **cardinality of a structure** is the cardinality of its universe.

If L has only finitely many function and relation symbols, then we characterize as usual a structure $\mathcal{A}$ as an n-tupel, listing its functions and relations. For example, the ordered field of real numbers is the structure $(\mathbb{R}, 0, 1, +, \cdot, <)$.

If the language L has $\in$ as nonlogical relation symbol, then, for every structure $\mathcal{A}$ for L, the binary relation assigned to $\in$ is $\{(x, y) \in A \times A : x \in y\}$.

The free variables of a formula have already been defined in Section 4.2. If φ is a formula of the language L whose free variables occur in the sequence $(v_0, \ldots, v_{n-1})$, then we define by induction on the complexity of φ, for finite sequences $(a_0, \ldots, a_{n-1})$ of elements of A, the relation $\mathcal{A} \models \varphi[a_0, \ldots, a_{n-1}]$ [3]. For this, we need the **interpretation of a term under a valuation** $(a_0, \ldots, a_{n-1})$, written as $t[a_0, \ldots, a_{n-1}]$ for terms t. $t[a_0, \ldots, a_{n-1}]$ is defined as a member of A by induction on the complexity of t as follows:

a) If t is a variable v_i with $i < n$, then $t[a_0, \ldots, a_{n-1}] := a_i$; if t is a constant symbol f^0, then $t[a_0, \ldots, a_{n-1}] := f^0_{\mathcal{A}}$.
b) If t is the term $f^k t_1 \ldots t_k$ where f^k is a function symbol and $t_1, \ldots, t_k$ are terms, and if all variables occurring in $t_1, \ldots, t_k$ are members of the sequence $(v_0, \ldots, v_{n-1})$, then

$$t[a_0, \ldots, a_{n-1}] := f^k_{\mathcal{A}}(t_1[a_0, \ldots, a_{n-1}], \ldots, t_k[a_0, \ldots, a_{n-1}]).$$

Now we can define the relation $\mathcal{A} \models \varphi[a_0, \ldots, a_{n-1}]$:

a) If R^k is an k-ary relation symbol, and if $t_1, \ldots, t_k$ are terms, then let

$$\begin{aligned} &\mathcal{A} \models R^k t_1 \ldots t_k[a_0, \ldots, a_{n-1}] \\ &\qquad \iff (t_1[a_0, \ldots, a_{n-1}], \ldots, t_k[a_0, \ldots, a_{n-1}]) \in R^k_{\mathcal{A}}. \end{aligned}$$

If t_1 and t_2 are terms, let

$$\mathcal{A} \models t_1 = t_2[a_0, \ldots, a_{n-1}] \iff t_1[a_0, \ldots, a_{n-1}] = t_2[a_0, \ldots, a_{n-1}].$$

b) If φ and ψ are formulae of L, let

$$\mathcal{A} \models (\neg\varphi)[a_0, \ldots, a_{n-1}] \iff \neg(\mathcal{A} \models \varphi[a_0, \ldots, a_{n-1}]),$$
$$\mathcal{A} \models (\varphi \vee \psi)[a_0, \ldots, a_{n-1}] \iff \mathcal{A} \models \varphi[a_0, \ldots, a_{n-1}] \vee \mathcal{A} \models \psi[a_0, \ldots, a_{n-1}]$$

and

$$\mathcal{A} \models \exists v_n \varphi[a_0, \ldots, a_{n-1}] \iff \exists a \in A(\mathcal{A} \models \varphi[a_0, \ldots, a_{n-1}, a]).$$

---

[2] or $(A, R, \ldots)$

[3] See the footnote on page 178.

With the usual abbreviations we obtain the assertions

$$\mathcal{A} \models (\varphi \wedge \psi)[a_0, \ldots, a_{n-1}], \qquad \mathcal{A} \models (\varphi \implies \psi)[a_0, \ldots, a_{n-1}],$$
$$\mathcal{A} \models (\varphi \iff \psi)[a_0, \ldots, a_{n-1}] \text{ and } \mathcal{A} \models \forall v_n \varphi[a_0, \ldots, a_{n-1}],$$

where, for example, the last one is the assertion $\forall a \in A\ \mathcal{A} \models \varphi[a_0, \ldots, a_{n-1}, a]$.

If φ is a formula of the language L, whose free variables are occuring in $(v_0, \ldots, v_{n-1})$ then we say that **the formula φ holds in $\mathcal{A}$** iff $\mathcal{A} \models \forall v_0 \ldots \forall v_{n-1} \varphi$ is provable in ZFC.

If $\alpha, \beta \leq \omega$, and if $\mathcal{A} = (A, (g_k : k < \alpha), (Q_k : k < \beta))$ and $\mathcal{B} = (B, (f_k : k < \alpha), (R_k : k < \beta))$ are structures for L, then $\mathcal{A}$ is called a **substructure of** $\mathcal{B}$, written as $\mathcal{A} \subseteq \mathcal{B}$ iff A is a subset of B and for all k the following holds: If n_k is the arity of f_k, R_k, then $g_k = f_k \restriction A^{n_k}$, $Q_k = R_k \cap A^{n_k}$ respectively; in particular, for $n_k = 0$, this means $f_k = g_k$.

If $\mathcal{A}$ is a substructure of $\mathcal{B}$, then $\mathcal{A}$ is called an **elementary substructure of** $\mathcal{B}$, written as $\mathcal{A} \preceq \mathcal{B}$, iff, for any formula φ of the language L whose free variables occur in the sequence $(v_0, \ldots, v_{n-1})$, we have

$$\forall a_0 \in A \ldots \forall a_{n-1} \in A(\mathcal{A} \models \varphi[a_0, \ldots, a_{n-1}] \iff \mathcal{B} \models \varphi[a_0, \ldots, a_{n-1}]).$$

Lemma 4.2.2 and the **Löwenheim-Skolem theorem**, Theorem 4.2.3, still hold for languages with countably many function and relation symbols. The proofs run completely analogous to those of Section 4.2.

Definition *A structure $(A, (f_k : k < \omega))$ with countably many functions f_k is called a* **Jónsson algebra** *iff it has no proper substructure of the same cardinality.* J(κ) *is the assertion "There exists a Jónsson algebra of cardinality κ".*

In this section we will deal with the question for which infinite cardinals κ we can prove J(κ). *κ, λ and μ are syntactical variables for* **infinite** *cardinal numbers.*

Example Let $f_n = \{(m, n) : m \in \omega\}$ for each $n \in \omega$. Then $(\omega, (f_n : n \in \omega))$ is obviously a Jónsson algebra. This proves J$(\aleph_0)$.

We prove first a criterion which yields a necessary and sufficient condition for the existence of Jónsson algebras and will be used frequently.

Definition *If $\mathcal{A}$ is a structure for a language L and $\varphi(v_0, \ldots, v_n)$ is a formula of L, then we say that a function f from A^n into A is a* **Skolem function for φ in $\mathcal{A}$** *iff, for all $a_1, \ldots, a_n \in A$,*

$$\mathcal{A} \models \exists v_0 \varphi[a_1, \ldots, a_n] \implies \mathcal{A} \models \varphi[f(a_1, \ldots, a_n), a_1, \ldots, a_n].$$

If $\mathcal{B}$ is a substructure of $\mathcal{A}$ such that, for each L-formula φ, there is at least one Skolem function for φ such that B is closed under this function, it is easy to check that $\mathcal{B}$ is an elementary substructure of $\mathcal{A}$.

Lemma 8.2.1 *Let λ be an infinite cardinal. Then the following propositions are equivalent:*

(i) $\mathrm{J}(\lambda)$.

(ii) *For any regular cardinal $\kappa \geq \lambda^+$ and any elementary substructure $(M, \in)$ of $(\mathrm{H}(\kappa), \in)$ satisfying $\lambda \in M$ and $|M \cap \lambda| = \lambda$, we have $\lambda \subseteq M$.*

(iii) *There is a regular cardinal $\kappa \geq \lambda^+$ and a structure $\mathcal{H} = (\mathrm{H}(\kappa), \in, \ldots)$, such that, for all $\mathcal{M} \preceq \mathcal{H}$,*

$$\lambda \in M \wedge |M \cap \lambda| = \lambda \implies \lambda \subseteq M.$$

Proof First we show that (ii) follows from (i). Let $\kappa \geq \lambda^+$ be a regular cardinal and $(M, \in)$ be an elementary substructure of $(\mathrm{H}(\kappa), \in)$ such that $\lambda \in M$ and $|M \cap \lambda| = \lambda$. By Lemma 4.1.4, it is clear that any function from λ^n into λ and thus any Jónsson algebra on λ is a member of $\mathrm{H}(\kappa)$. Using the absoluteness for $\mathrm{H}(\kappa)$ of the formula $\varphi(x, y) \equiv$ "x is a Jónsson algebra on y", we get $\mathrm{H}(\kappa) \models$ "There is a Jónsson algebra on λ". Now we obtain, together with Lemma 4.2.13, a Jónsson algebra $\mathcal{A} = (\lambda, (f_k : k < \omega))$ on λ which is a member of M. Then $(f_k : k \in \omega) \in M$ and thus $f_k \in M$ for every $k \in \omega$ since $\omega \subseteq M$.

Fix $k \in \omega$. Since $\mathrm{H}(\kappa) \models f_k : \lambda^{n_k} \longrightarrow \lambda$, $M \preceq \mathrm{H}(\kappa)$ and $f_k, n_k, \lambda \in M$, we can conclude that $M \models f_k : \lambda^{n_k} \longrightarrow \lambda$. This implies $f_k(\vec{\alpha}) \in \lambda \cap M$ for all $\vec{\alpha} \in (\lambda \cap M)^{n_k}$, hence $(B, (f_k \restriction B^{n_k} : k < \omega))$ is a substructure of $\mathcal{A}$ where $B := \lambda \cap M$. Since $\mathcal{A}$ is a Jónsson algebra, we can infer from $|B| = |\lambda \cap M| = \lambda$ that $\lambda = B = \lambda \cap M \subseteq M$.

Clearly (ii) implies (iii). There remains to be shown that (iii) implies (i). So let us assume (iii) and let $\mathcal{H}$ be a structure with universe $\mathrm{H}(\kappa)$ for an appropriate language L, for which $\in$ is interpreted as usual. By the Löwenheim-Skolem theorem, there is an elementary substructure $\mathcal{M}$ of $\mathcal{H}$ such that $\lambda \in M$, $\lambda \subseteq M$ and $|M| = \lambda$. Choose a bijection g from λ onto M. Clearly g is a binary relation on M, and we identify it with the interpretation of a binary relation symbol of the language L_g which results from the language L by adjoining the symbol g to its alphabet.

Claim The structure $\mathcal{M}' := (M, \in, g, \ldots)$ has no proper elementary substructure of cardinality λ.

Proof Consider an elementary substructure $\mathcal{N}' := (N, \in, g, \ldots)$ of $\mathcal{M}'$ such that $|N| = \lambda$. Let $\varphi(x)$ be the L_g-formula "x is the least ordinal such that $\mathrm{dom}(g) \subseteq x$". An easy calculation shows that $\mathcal{M}' \models \varphi(x)[\lambda]$, hence $\mathcal{M}' \models \exists x\, \varphi(x)$, and from Lemma 4.2.2 and $\mathcal{N}' \preceq \mathcal{M}'$ we can infer that there is $a \in N$ such that $\mathcal{M}' \models \varphi(a)$. Now it is easy to see that $a = \lambda$, hence $\lambda \in N$. $\mathcal{M}' \models \forall x \exists y\, g(y) = x$ gives $\mathcal{N}' \models \forall x \exists y\, g(y) = x$, and thus $g[N \cap \lambda] = N$. Therefore we have $|N \cap \lambda| = |g[N \cap \lambda]| = |N| = \lambda$. If we put $\mathcal{N} := (N, \in, \ldots)$, then $\mathcal{N} \preceq \mathcal{M}$ and consequently $\mathcal{N} \preceq \mathcal{H}$. With assumption (iii) we can conclude that $\lambda \subseteq N$, and thus we have $M = \mathrm{ran}(g) = \mathrm{ran}(g \restriction N \cap \lambda) = N$ which proves the claim.

Now we choose, for each formula φ of L_g, a Skolem function f_φ for φ in $\mathcal{M}'$. Then $(M, (f_\varphi : \varphi \in L_g))$ is a Jónsson algebra. To see this, let $(N, (f_\varphi : \varphi \in L_g))$ be a substructure of $(M, (f_\varphi : \varphi \in L_g))$ such that $|M| = |N|$. Then the structure $(N, \in, g, \ldots)$ is closed under Skolem functions and is thus an elementary substructure of $(M, \in, g, \ldots)$. Now the claim yields $N = M$.

Theorem 8.2.2 $\mathrm{J}(\lambda) \Longrightarrow \mathrm{J}(\lambda^+)$

Proof We want to apply Lemma 8.2.1. So let us assume that $M \preceq \mathrm{H}(\lambda^{++})$, $\lambda^+ \in M$, and $|M \cap \lambda^+| = \lambda^+$. We will show that λ^+ is a subset of M and fix $\beta \in \lambda^+$.

Clearly, there is some set $C \in [M \cap \lambda^+]^\lambda$. Since λ^+ is regular, we get $\alpha' = \sup(C \cup \{\beta\}) \in \lambda^+$. From $|M \cap \lambda^+| = \lambda^+$ we conclude that there is an ordinal $\alpha \in M \cap \lambda^+$ with $\alpha > \alpha'$, and hence α satisfies $\alpha > \beta$ and $|M \cap \alpha| = \lambda$. From $\lambda^+ \in M$ and Example 4.2.15 we can infer that $\lambda \in M$. By Lemma 4.2.13, we obtain a bijection g from λ onto α which is a member of M. Example 4.2.15 implies that $g \restriction \lambda \cap M$ is a bijection from $\lambda \cap M$ onto $\alpha \cap M$. So we get $|\lambda \cap M| = |\alpha \cap M| = \lambda$.

Now $\mathrm{J}(\lambda)$ and Lemma 8.2.1 yield $\lambda \subseteq M$, hence $\alpha = g[\lambda] = g[\lambda \cap M] = \alpha \cap M$. So we have shown that $\beta \in \alpha \subseteq M$, which gives $\lambda^+ \subseteq M$, as desired.

Theorem 8.2.3 (Tryba, Woodin) *If λ is a regular uncountable cardinal number and if there exists a nonreflecting stationary subset of λ, then* $\mathrm{J}(\lambda)$.

Proof Again we want to apply Lemma 8.2.1. In assertion (iii) we choose $\kappa = \lambda^+$ and assume that $M \preceq \mathrm{H}(\lambda^+)$, $\lambda \in M$ and $|M \cap \lambda| = \lambda$. We must show that $\lambda \subseteq M$.

By Lemma 1.8.14, there exists a nonreflecting stationary subset S_1 of λ. Then $S_1 \in \mathrm{H}(\lambda^+)$, and the considered property of S_1 can be expressed by a formula which is absolute for $\mathrm{H}(\lambda^+)$. Therefore we can find, using Lemma 4.2.13, a nonreflecting stationary subset S of λ which is a member of M. Now Solovay's theorem, Theorem 1.8.13, says that there is a partition of S into λ stationary sets; it can be regarded as a function from λ into $\mathcal{P}(S)$. So let $\varphi(x, y, z)$ be the formula "x is an injection with domain y such that $\mathrm{ran}(x)$ is a partition of z and all elements of $\mathrm{ran}(x)$ are stationary in y". The Lemmas 4.1.3 and 4.1.4 say that $\mathrm{H}(\lambda^+) \models \exists x\ \varphi[\lambda, S]$. Consequently, we can find such a function $g = (S_\xi : \xi < \lambda)$ which is a member of M. Let

$$C := \{\alpha < \lambda : \sup(M \cap \alpha) = \alpha\}.$$

We will see that C is a club in λ. If we can prove in addition, for any $\xi < \lambda$,

Claim 1 $C \cap S_\xi \subseteq M$,

then the rest of the proof is easy: Since C is a club and S_ξ is stationary in λ, we can choose an ordinal $\alpha \in C \cap S_\xi \subseteq M \cap S_\xi$. Since $g \in M$ and $\alpha \in M$, there is $\sigma \in M$ such that $\alpha \in S_\sigma$. From $S = \bigcup\{S_\zeta : \zeta \in \lambda\}$ and the fact that the sets S_ζ are pairwise disjoint, we can infer that $\sigma = \xi \in M$. So we have shown that $\lambda \subseteq M$. Now Lemma 8.2.1 yields the desired assertion.

For the proof proof of Claim 1 let us assume that its assertion is false. Then there is $\alpha \in (C \cap S_\xi) \setminus M$. Since $|M \cap \lambda| = \lambda$, the ordinal $\beta := \min(M \setminus \alpha)$ exists, and we have $\alpha < \beta < \lambda$. Note that β is a limit ordinal, since otherwise the immediate predecessor β' of β would be a member of M and would so be greater than α, contradicting the choice of β. Since S is nonreflecting and $S \in M$, there is a club C_β in β such that C_β is a member of M and S and C_β are disjoint.

Now we show that $C_\beta \cap \alpha$ is unbounded in α. Let $\sigma < \alpha$. Since $\alpha \in C$ and thus $\sup(M \cap \alpha) = \alpha$, there is $\delta \in M \cap \alpha$ such that $\delta > \sigma$. Since C_β is a club in β and $C_\beta \in M$, we obtain an ordinal $\epsilon \in [\delta, \beta) \cap C_\beta \cap M$. The choice of β gives $\epsilon < \alpha$.

Now we can conclude that $\alpha \in C_\beta$ since C_β is a club in β, which contradicts $\alpha \in S_\xi \subseteq S$ and $C_\beta \cap S = \emptyset$. This completes the proof of Claim 1. There remains the proof of

Claim 2 C is a club in λ.

For the proof of Claim 2, consider a limit ordinal $\gamma < \lambda$ such that $\gamma = \sup(C \cap \gamma)$. By the definition of C, we have $\bigcup(M \cap \alpha) = \alpha$ for all $\alpha \in C$. So we can conclude that

$$\sup(M \cap \gamma) = \bigcup(M \cap \bigcup(C \cap \gamma)) = \bigcup_{\alpha \in C \cap \gamma} \bigcup(M \cap \alpha) = \bigcup(C \cap \gamma) = \gamma,$$

i.e., $\gamma \in C$. Therefore C is closed in λ. Furthermore, if $\sigma \in \lambda$, then we can find, using $|M \cap \lambda| = \lambda$, a strictly increasing sequence $(\alpha_i : i < \omega)$ of elements of $M \cap \lambda$ such that $\sigma < \alpha_0$. Then $\alpha := \sup\{\alpha_i : i < \omega\} \in \lambda$, since $\mathrm{cf}(\lambda) > \aleph_0$, and we can conclude that $\alpha = \sup(M \cap \{\alpha_i : i < \omega\}) \leq \sup(M \cap \alpha) \leq \alpha$. This gives $\alpha \in C$ and $\alpha > \sigma$, and thus C is unbounded in λ.

Corollary 8.2.4 *For any regular cardinal number κ, we have* $\mathrm{J}(\kappa^+)$.

Proof This is an immediate consequence of Theorem 8.2.3, since we obtain from Lemma 1.8.14 a nonreflecting stationary subset of κ^+.

In the previous corollary we can omit the assumption that κ is regular if we add the generalized continuum hypothesis to our axioms of ZFC. To prove this, we need a preliminary combinatorial result.

Theorem 8.2.5 (Erdös, Hajnal, Rado) *If $2^\kappa = \kappa^+$, then there is a function $f : [\kappa^+]^2 \longrightarrow \kappa^+$ such that $f[[A]^2] = \kappa^+$ for any $A \in [\kappa^+]^{\kappa^+}$.*

Proof By assumption we have $|[\kappa^+]^\kappa| = (\kappa^+)^\kappa = (2^\kappa)^\kappa = 2^\kappa = \kappa^+$, and so there exists a bijection $g : \kappa^+ \longrightarrow [\kappa^+]^\kappa$.

Consider a fixed ordinal $\alpha \in [\kappa, \kappa^+)$ such that $[\alpha]^\kappa \cap g[\alpha] \neq \emptyset$. There are bijections $(\alpha_\beta : \beta < \kappa)$ and $((A^\alpha_\beta, \eta^\alpha_\beta) : \beta < \kappa)$ from κ onto α and onto $([\alpha]^\kappa \cap g[\alpha]) \times \alpha$, respectively. We define by recursion for $\beta < \kappa$

$$\sigma^\alpha_\beta := \min\{\sigma \in \kappa : \alpha_\sigma \in A^\alpha_\beta \setminus \{\alpha_{\sigma^\alpha_\gamma} : \gamma < \beta\}\}.$$

The minimum exists since A^α_β is a subset of α of cardinality κ, and it is obvious that the sequence $(\sigma^\alpha_\beta : \beta < \kappa)$ is injective.

Now we put, for any $\{\alpha, \delta\} \in [\kappa^+]^2$ with $\delta < \alpha$,

$$f(\{\alpha,\delta\}) := \begin{cases} \eta^\alpha_\gamma & \text{if } \alpha \geq \kappa \wedge [\alpha]^\kappa \cap g[\alpha] \neq \emptyset \wedge \exists \gamma \in \kappa\ \delta = \alpha_{\sigma^\alpha_\gamma}, \\ 0 & \text{otherwise.} \end{cases}$$

If $A \in [\kappa^+]^{\kappa^+}$ and $\eta < \kappa^+$, we take a member A' of $[A]^\kappa$ and choose an ordinal $\alpha > \max\{\sup(A'), g^{-1}(A'), \eta\}$ such that $\alpha \in A$. Then we have $\alpha \geq \kappa$, $A' \in [\alpha]^\kappa \cap g[\alpha]$, and $\eta \in \alpha$. Therefore there is $\gamma \in \kappa$ such that $(A', \eta) = (A^\alpha_\gamma, \eta^\alpha_\gamma)$. Putting $\delta := \alpha_{\sigma^\alpha_\gamma}$, we can conclude that $\delta \in A^\alpha_\gamma = A' \subseteq A$ and $\eta = \eta^\alpha_\gamma = f(\{\alpha, \delta\}) \in f[[A]^2]$, as desired.

Corollary 8.2.6 (GCH) $\Longrightarrow \forall \kappa \in \mathrm{ICN}\ \mathrm{J}(\kappa^+)$.

Proof Fix $\kappa \in$ ICN. Since $2^\kappa = \kappa^+$, we can choose a function $f : [\kappa^+]^2 \longrightarrow \kappa^+$ with the properties in Theorem 8.2.5. Let the function $g : \kappa^+ \times \kappa^+ \longrightarrow \kappa^+$ be defined by

$$g(\alpha,\beta) = \begin{cases} f(\{\alpha,\beta\}) & \text{if } \alpha \neq \beta, \\ 0 & \text{if } \alpha = \beta. \end{cases}$$

If $(A, g \restriction A^2)$ is a substructure of (κ^+, g) and $|A| = \kappa^+$, then $A \supseteq g[A^2] \supseteq f[[A]^2] = \kappa^+$, hence $A = \kappa^+$. Therefore (κ^+, g) is a Jónsson algebra.

If one replaces GCH by the stronger assumption V = L, then one can even show that, for any infinite cardinal κ, there is a Jónsson algebra of cardinality κ. However, we will not present the proof of this theorem since we are mainly interested in results which are provable in ZFC.

The preceding theorems say nothing about the provability in ZFC of $\mathrm{J}(\mu^+)$ for singular cardinal numbers μ. For such cardinal numbers, an application of the main lemma of pcf-theory will yield a sufficient criterion for the existence of a Jónsson algebra. In particular we will prove that there exists a Jónsson algebra on $\aleph_{\omega+1}$. We need two preliminary technical lemmas.

Definition *For any set A let $\xi_A : \bigcup_{n\in\omega}({}^{A^n}A) \longrightarrow \omega$ be that function which assigns to each function $f : A^n \longrightarrow A$ its arity $\xi_A(f) := n$. Further let $f''B := f[B^n]$, if $\xi_A(f) = n$ and $B \subseteq A$.*

Lemma 8.2.7 *If A is a set and F is a countable subset of $\bigcup\{{}^{A^n}A : n \in \omega\}$, then there is a countable set F' such that $F \subseteq F'$, $F' \subseteq \bigcup\{{}^{A^n}A : n \in \omega\}$, and F' is closed under the composition of functions (the definition should be clear from the proof).*

Proof We define by recursion a sequence $(F_i : i < \omega)$ of sets of functions. Put $F_0 := F$. If F_n is already defined, then take F_{n+1} as the union of the set F_n and the set of all functions

$$\begin{cases} A^k & \longrightarrow \quad A \\ (x_1, \ldots, x_k) & \longmapsto \quad f(f_1(\vec{x}_1), \ldots, f_{\xi_A(f)}(\vec{x}_{\xi_A(f)})) \end{cases}$$

such that $f, f_1, \ldots, f_{\xi_A(f)} \in F_n$, $k = \sum(\xi_A(f_i) : 1 \le i \le \xi_A(f))$, and

$$\vec{x}_i = (x_{1+\sum(\xi_A(f_j):j<i)}, \ldots, x_{\sum(\xi_A(f_j):j\le i)})$$

for all $i \in \{1, \ldots, \xi_A(f)\}$. Now we put $F' := \bigcup\{F_n : n \in \omega\}$.

If g results from the composition of the members $f, f_1, \ldots, f_k$ of F', then, for each $j \in \{1, \ldots, k\}$, there is i_j such that $f_j \in F_{i_j}$. Let $f \in F_{i_0}$. Taking $i := \max\{i_0, \ldots, i_k\}$, we get $\{f, f_1, \ldots, f_k\} \subseteq F_i$ and thus $g \in F_{i+1} \subseteq F'$. Corollary 1.6.6 says that $|[\omega]^{<\omega}| = \aleph_0$, and this gives $|F_n| \le \aleph_0$ for all natural numbers n. Therefore the set F' is countable.

Lemma 8.2.8 *If $\mathcal{A}$ is a structure for the (countable) language L, then there is a countable set $F \subseteq \bigcup\{{}^{A^n}A : n \in \omega\}$ such that, for all nonempty subsets B of A, the following holds: If one restricts all relations and functions of $\mathcal{A}$ to the set $\bigcup\{f''B : f \in F\}$, one obtains an elementary substructure of $\mathcal{A}$ whose universe $\bigcup\{f''B : f \in F\}$ is a superset of B.*

Proof For each L-formula φ, we choose a Skolem function f_φ for $\mathcal{A}$. Since L is countable, so is the set $G := \{f_\varphi : \varphi \text{ is an } L\text{-formula}\}$. Therefore Lemma 8.2.7 says that there is a countable superset F of G which is closed under compositions.

Now assume that $B \subseteq A$ is nonempty, and let $B' := \bigcup\{f''B : f \in F\}$. If $g : A^n \longrightarrow A$ is one of the Skolem functions in G and $x_1, \ldots, x_n \in B'$, then there is, for each $i \in \{1, \ldots, n\}$, a function $f_i \in F$ and $\vec{x}_i \in B^{\xi_A(f_i)}$ such that $x_i = f_i(\vec{x}_i)$. Therefore we have

$$x := g(x_1, \ldots, x_n) = g(f_1(\vec{x}_1), \ldots, f_n(\vec{x}_n)) = f(\vec{x}_1, \ldots, \vec{x}_n) \in f''B,$$

where $f : A^{\sum(\xi_A(f_i):1\le i\le n)} \longrightarrow A$ is a composition of the functions $g, f_1, \ldots, f_n$. So we get $x \in B'$, hence B' is closed under at least one Skolem function for each L-formula. This gives $B' \preceq A$. The formula $v_0 = v_1$ ensures that id_A is a member of G and thus of F, which yields $B \subseteq B'$.

Theorem 8.2.9 *Assume that λ is a singular cardinal. If there exists a cardinal $\mu < \lambda$ such that $\mathrm{J}(\kappa)$ holds for all $\kappa \in (\mu, \lambda)_{\mathrm{reg}}$, then there is a Jónsson algebra on λ^+.*

Proof Put $\Theta := (2^{2^\lambda})^+$. For the structure $\mathcal{A} := (\mathrm{H}(\Theta), \in)$ we choose a set F of functions as in Lemma 8.2.8. We want to apply Lemma 8.2.1 and consider an elementary substructure $(M, \in, (f : f \in F))$ of $(\mathrm{H}(\Theta), \in, (f : f \in F))$ such that $\lambda^+ \in M$ and $|M \cap \lambda^+| = \lambda^+$. Example 4.2.15 tells us that $\lambda \in M$. If we can show that $\lambda^+ \subseteq M$, then Lemma 8.2.1 implies $\mathrm{J}(\lambda^+)$.

First let us note that $\lambda^+ \subseteq M$ if we know that $\lambda \subseteq M$: If $\alpha \in [\lambda, \lambda^+)$, then $|M \cap \lambda^+| = \lambda^+$ yields an ordinal $\beta \in M \cap \lambda^+$ such that $\alpha < \beta$. Since $|\beta| = \lambda$ and $\lambda, \beta \in M$, there is a bijection e from λ onto β which is a member of M, and from $\lambda \subseteq M$ we can conclude that $\beta = e[\lambda] = e[\lambda \cap M] \subseteq M$ and thus $\alpha \in M$, as desired. We will prove the assertion $\lambda \subseteq M$ with the help of pcf-theory.

By Lemma 3.3.5, there is a progressive set a of regular cardinals satisfying $|\mathsf{a}| = \mathrm{cf}(\lambda)$ and $\sup(\mathsf{a}) = \lambda$ and an ultrafilter D on a such that $\lim_D(\mathsf{a}) = \lambda$ and $\mathrm{cf}(\prod \mathsf{a}/D) = \lambda^+$. Let $\bar{h} = (h_\xi : \xi < \lambda^+)$ be a sequence of members of $\prod \mathsf{a}$ which is cofinal in $\prod \mathsf{a}$ modulo D. We choose a cardinal μ corresponding to the assumption of the theorem and assume without loss of generality, using Lemma 3.3.5, that $\mathsf{a} \subseteq [\mu, \lambda)_{\mathrm{reg}}$. By the choice of Θ, the sets a, D, $\bar{h}$ and μ are members of $\mathrm{H}(\Theta)$. Now let ψ be that formula with the parameter $\lambda \in M$ which formalizes the assertion that the objects a, D, $\bar{h}$ and μ with the above properties exist. As in Example 4.2.15 f) one can show, using $\Theta > 2^{\sup(\mathsf{a})}$, that ψ is absolute for $\mathrm{H}(\Theta)$. Consequently, since M is an elementary substructure of $\mathrm{H}(\Theta)$, we can assume without loss of generality that these sets are members of M. Let

$$A := \{\kappa \in M \cap \mathsf{a} : \sup(M \cap \kappa) = \kappa\}.$$

If $\kappa \in A$, then we can infer from Lemma 8.2.1 and $\mathrm{J}(\kappa)$ that $\kappa \subseteq M$. Now, for the proof of $\lambda \subseteq M$, it suffices to show the following

Claim A is unbounded in λ.

Proof We assume, contrary to our hopes, that the assertion of the claim is false, and let

$$\lambda' := \min\{\nu \in \lambda \cap \mathrm{CN} : A \subseteq \nu\}.$$

By our assumption, this minimum exists, and we have $\lambda' < \lambda$.

Subclaim 1 $\neg\ (M \cap \mathsf{a} \subseteq \lambda')$.

Assume that the assertion of the subclaim is false. First we note that $(M, \in) \models \exists x (x \in y)[\mathsf{a}]$, and thus we have $M \cap \mathsf{a} \neq \emptyset$, hence $\lambda' > 0$ by our assumption that $M \cap \mathsf{a} \subseteq \lambda'$. So there exists $\kappa \in A$. We can conclude that $\mathrm{cf}(\lambda) = |\mathsf{a}| < \min(\mathsf{a}) \leq \kappa \subseteq M$, hence $\mathrm{cf}(\lambda) \subseteq M$ and $\mathrm{cf}(\lambda) \in M$. If $\varphi(x, y, z)$ is the formula "$y, z \in \mathrm{ON}$ and $x : y \longrightarrow z$ is strictly increasing and $\bigcup \mathrm{ran}(x) = z$", then we obtain, using $\mathrm{H}(\Theta) \models \exists x \varphi[\mathrm{cf}(\lambda), \lambda]$, a strictly increasing function $e : \mathrm{cf}(\lambda) \longrightarrow \lambda$ which is a member of M and whose range is cofinal in λ. Since $\mathrm{cf}(\lambda) \subseteq M$, we get $\mathrm{ran}(e) \subseteq M$, and thus $M \cap \lambda$ is cofinal in λ. Furthermore, $(M, \in) \models$ "x is cofinal in y"$[\mathsf{a}, \lambda]$ since $\sup(\mathsf{a}) = \lambda$ and $\mathsf{a}, \lambda \in M$. Therefore $M \cap \mathsf{a}$ is cofinal in $M \cap \lambda$ and thus cofinal in λ. This contradicts $M \cap \mathsf{a} \subseteq \lambda' < \lambda$.

Now let

$$M' := \bigcup \{f''(M \cup \mathsf{a}) : f \in F\}.$$

The choice of F yields $M \cup \mathsf{a} \subseteq M'$ and $M' \preceq \mathrm{H}(\Theta)$. We will show that there is a final segment $\mathsf{a} \cap (\lambda'', \lambda)_{\mathrm{reg}}$ of a such that the restriction of $\chi_{M'}$ to $\mathsf{a} \cap (\lambda'', \lambda)_{\mathrm{reg}}$, extended to a function g on a by taking the value zero for cardinals not greater than λ'', has the property that $g \in \prod \mathsf{a}$. Together with the fact that $\bar{h}$ is cofinal in $\prod \mathsf{a}$, this will lead to a contradiction as follows. Let

$$g(\kappa) := \begin{cases} \sup(M' \cap \kappa) & \text{if } \kappa \in (\lambda'', \lambda)_{\mathrm{CN}}, \\ 0 & \text{otherwise,} \end{cases}$$

for every $\kappa \in \mathsf{a}$. Let us assume that $g \in \prod \mathsf{a}$. Then the choice of $\bar{h}$ yields an ordinal $\alpha < \lambda^+$ such that $g \leq_D h_\alpha$. Since $M \cap \lambda^+$ is unbounded in λ^+, we can find an ordinal $\beta \in M' \cap (\alpha, \lambda^+)$, and so we have $g <_D h_\beta$, hence $\mathrm{B}(g, h_\beta) \in D$. From $\lim_D(\mathsf{a}) = \lambda$ we get $\mathsf{a} \cap (\lambda'', \lambda)_{\mathrm{reg}} \in D$, and in particular we obtain a member κ of the set $\mathrm{B}(g, h_\beta) \cap [\lambda'', \lambda)_{\mathrm{CN}}$. $\kappa \in \mathsf{a}$ and $\mathsf{a} \subseteq M'$ yields $\kappa \in M'$, and $\beta, \bar{h} \in M'$ gives $h_\beta \in M'$, hence $h_\beta(\kappa) \in M' \cap \kappa$, from which we get the contradiction $g(\kappa) < h_\beta(\kappa) \leq \sup(M' \cap \kappa) = g(\kappa)$. Thus we have seen that it suffices to prove the following

Subclaim 2 $\exists \lambda'' \in [\lambda', \lambda)_{\mathrm{CN}}\ \forall \kappa \in \mathsf{a} \cap (\lambda'', \lambda)_{\mathrm{CN}}\ (\sup(M' \cap \kappa) < \kappa)$.

Proof Assume that this assertion is false. Then we note first that $\lambda \subseteq M'$. Namely, if $\alpha \in \lambda$, then we can choose $\lambda'' \in [\lambda', \lambda)_{\mathrm{CN}}$ with $\lambda'' > \alpha$. By our supposition, we obtain a cardinal $\kappa \in \mathsf{a} \cap (\lambda'', \lambda)_{\mathrm{CN}}$ such that $\sup(M' \cap \kappa) = \kappa$. Since $\kappa \in \mathsf{a} \subseteq M'$, we infer from $\mathrm{J}(\kappa)$ and Lemma 8.2.1 that $\kappa \subseteq M'$, hence $\alpha \in M'$. So we get $\lambda \subseteq M'$.

By Subclaim 1, there exists a cardinal $\kappa \in M \cap \mathsf{a} \cap [\lambda', \lambda)_{\mathrm{CN}}$. Then $\kappa \notin A$, hence $\sup(M \cap \kappa) < \kappa$. We will show that $\sup(M' \cap \kappa) \leq \sup(M \cap \kappa)$. On the other hand, we have $\kappa < \lambda \subseteq M'$, hence $\kappa \subseteq M'$. This will yield $\kappa = \sup(M' \cap \kappa) \leq \sup(M \cap \kappa)$, contradicting $\sup(M \cap \kappa) < \kappa$, and complete the proof of Subclaim 2 and thus of the theorem.

So fix $\alpha \in M' \cap \kappa$. By the definition of M', there is a function $f : \mathrm{H}(\Theta)^n \longrightarrow \mathrm{H}(\Theta)$ which is a member of F, and there are $s'_1, \ldots, s'_n \in M \cup \mathsf{a}$ such that $\alpha = f(s'_1, \ldots, s'_n)$. Put $S := \{s_1, \ldots, s_m\} := \{s'_1, \ldots, s'_n\} \cap M$. Then

$|f''(S \cup \mathsf{a}) \cap \kappa| \leq |\mathsf{a}| < \min(\mathsf{a}) \leq \kappa$, hence $f''(S \cup \mathsf{a}) \cap \kappa$ is bounded in κ. Let $\varphi \equiv \varphi(x, y, z, v_1, \ldots, v_m)$ be the formula

$$x \in y \wedge \forall w_1 \ldots \forall w_n (\bigwedge_{i=1}^{n} (w_i \in z \vee \bigvee_{j=1}^{m} w_i = v_j) \wedge f(w_1, \ldots, w_n) \in y \Rightarrow f(w_1, \ldots, w_n) \in x).$$

Using $(\mathrm{H}(\Theta), \in, F) \models \exists x \, \varphi[\kappa, \mathsf{a}, s_1, \ldots, s_m]$ and $(M, \in, F) \preceq (\mathrm{H}(\Theta), \in, F)$, we obtain an ordinal $\alpha_{S,f} \in M \cap \kappa$ which is a bound for $f''(S \cup \mathsf{a})$ in $M \cap \kappa$. Now $\alpha \in f''(S \cup \mathsf{a})$ gives $\alpha < \alpha_{S,f} \leq \sup(M \cap \kappa)$, and altogether we get $\sup(M' \cap \kappa) \leq \sup(M \cap \kappa)$, as desired.

Corollary 8.2.10 *There exists a Jónsson algebra on* $\aleph_{\omega+1}$.

Proof In the example preceding Lemma 8.2.1, we have proved $\mathrm{J}(\aleph_0)$. From Theorem 8.2.2 we get $\mathrm{J}(\aleph_n)$ for all $n \in \omega$. Now the previous theorem, 8.2.9, guarantees for $\lambda = \aleph_\omega$ and $\mu = \aleph_0$ the existence of a Jónsson algebra on $\lambda^+ = (\aleph_\omega)^+ = \aleph_{\omega+1}$.

8.3 A Partition Theorem of Todorčević

Definition *If* λ *is an infinite cardinal, then* $\mathrm{p}(\lambda)$, *also written as* $\lambda \not\rightarrow [\lambda]^2_\lambda$, *is the following assertion: There is a function* $c : [\lambda]^2 \longrightarrow \lambda$ *such that, for any subset* X *of* λ *of cardinality* λ, $\{c(x) : x \in [X]^2\} = \lambda$.

Remark Theorem 8.2.5 says that $\mathrm{p}(\lambda)$ holds if $\lambda = 2^\kappa = \kappa^+$ for some cardinal κ. Furthermore, we can conclude from $\mathrm{p}(\lambda)$ that there is a Jónsson algebra on λ, namely, the algebra (λ, g) where, choosing c as in the definition of $\mathrm{p}(\lambda)$, the function $g : \lambda \times \lambda \longrightarrow \lambda$ is defined by $g(\alpha, \beta) := c(\{\alpha, \beta\})$ for $\alpha \neq \beta$, and $g(\alpha, \alpha) := 0$. To see that (λ, g) is a Jónsson algebra, assume that $(X, g \restriction X \times X)$ is a substructure of (λ, g) such that $|X| = \lambda$. Then $X \supseteq \{c(x) : x \in [X]^2\} = \lambda$.

Lemma 8.3.1 *Let* λ *be an infinite cardinal and* $c : [\lambda^+]^2 \longrightarrow \lambda$ *be a function such that, for any subset* X *of* λ^+ *of cardinality* λ^+, *we have* $\{c(x) : x \in [X]^2\} = \lambda$. *Then* $\mathrm{p}(\lambda^+)$.

Proof To realize $\lambda^+ \not\rightarrow [\lambda^+]^2_{\lambda^+}$, we choose for each $\beta < \lambda^+$ an injection $i_\beta : \beta \longrightarrow \lambda$, and define the function $c^* : [\lambda^+]^2 \longrightarrow \lambda^+$ by $c^*(\alpha, \beta) = i_\beta^{-1}(c(\alpha, \beta))$, if $\alpha < \beta < \lambda^+$ and $c(\alpha, \beta) \in \mathrm{ran}(i_\beta)$, $c^*(\alpha, \beta) = 0$ otherwise.

Consider $X \subseteq \lambda^+$ with $|X| = \lambda^+$, and $\delta < \lambda^+$. We search for $\alpha, \beta \in X$ such that $\alpha < \beta$ and $c^*(\alpha, \beta) = \delta$. If $Y := \{\beta \in X : \delta < \beta\}$, then $|Y| = \lambda^+$ and $\{i_\beta(\delta) : \beta \in Y\} \subseteq \lambda$. Since λ^+ is regular, there is an ordinal $\eta < \lambda$ such that $|\{\beta \in Y : i_\beta(\delta) = \eta\}| = \lambda^+$. By assumption, we can find $\alpha, \beta \in Y \subseteq X$ such that $\alpha < \beta$ and $c(\alpha, \beta) = \eta = i_\beta(\delta)$. So we get $\delta = i_\beta^{-1}(c(\alpha, \beta)) = c^*(\alpha, \beta)$, as desired.

Theorem 8.3.2 *Assume that there is a strictly increasing sequence* $(\kappa_n : n \in \omega)$ *of regular cardinals such that* $\mathrm{p}(\kappa_n)$ *holds for all* $n \in \omega$, *and let* $\mathsf{a} := \{\kappa_n : n \in \omega\}$ *and* $\lambda := \sup(\mathsf{a})$. *If* $\lambda^+ \in \mathrm{pcf}(\mathsf{a})$, *then* $\mathrm{p}(\lambda^+)$.

Proof Let $h := (\kappa_n : n \in \omega)$ and, without loss of generality, $\kappa_0 > \omega$. Since $\lambda^+ \in \mathrm{pcf}(\mathsf{a})$, we know from Lemma 3.3.1 that there is an ultrafilter D on ω such that $\mathrm{cf}(\prod h/D) = \lambda^+$. Clearly D is nonprincipal. We choose a sequence $(f_\xi : \xi < \lambda^+)$ of members of $\prod h$ which is cofinal in $\prod h$ modulo D and assume further that, for each $n \in \omega$, $c_n : [\kappa_n]^2 \longrightarrow \kappa_n$ is a function satisfying the property from the definition of $\mathrm{p}(\kappa_n)$. To apply Lemma 8.3.1, we define a function $c : [\lambda^+]^2 \longrightarrow \lambda$ and show that, for every subset X of λ^+ of cardinality λ^+, c takes on $[X]^2$ each value $\delta < \lambda$:

> If $\alpha < \beta < \lambda^+$ and n is the least natural number k such that $f_\alpha(k) \neq f_\beta(k)$, then we put $c(\alpha, \beta) := c_n(f_\alpha(n), f_\beta(n))$.

Now consider a fixed subset X of λ^+ satisfying $|X| = \lambda^+$ and $\delta < \lambda$. Then $Z = \{f_\xi : \xi \in X\}$ is unbounded in $\prod h$ modulo D. Since each cardinal κ_m is regular, there must be $n \in \omega$ such that $\delta < \kappa_n$ and $|\{f_\xi(n) : \xi \in X\}| = \kappa_n$. Otherwise, we could easily get a bound for Z: Since D is a nonprincipal ultrafilter, we have $d := \{n \in \omega : \delta < \kappa_n\} \in D$. The supposition $\sup\{f_\xi(n) : \xi \in X\} < \kappa_n$ for all $n \in d$ leads to $g \in \prod h$, where g is defined by

$$g(n) = \begin{cases} \sup\{f_\xi(n) + 1 : \xi \in X\} & \text{if } n \in d, \\ 0 & \text{if } n \in \omega \setminus d. \end{cases}$$

g would be an upper bound of Z modulo D in $\prod h$ which does not exist.

Since $|\prod\{\kappa_i : i < n\}| = \kappa_{n-1} < \kappa_n$, we obtain, together with the regularity of κ_n, a function $s \in \prod\{\kappa_i : i < n\}$ such that

$$|\{f_\xi(n) : s \subseteq f_\xi \text{ and } \xi \in X\}| = \kappa_n\,.$$

Using $\mathrm{p}(\kappa_n)$, we obtain ordinals $\alpha, \beta \in X$ such that $\alpha < \beta$, $c_n(f_\alpha(n), f_\beta(n)) = \delta$, $s \subseteq f_\alpha$, and $s \subseteq f_\beta$. So we have $n = \min\{k \in \omega : f_\alpha(k) \neq f_\beta(k)\}$, hence $c(\alpha, \beta) = \delta$, which proves the theorem.

Theorem 8.3.3 (Todorčević) *Assume that* λ *is a singular cardinal. If there is a cardinal* μ *with* $\mu < \lambda$ *such that* $\mathrm{p}(\kappa)$ *holds for every regular cardinal* $\kappa \in [\mu, \lambda)$, *then* $\mathrm{p}(\lambda^+)$.

Proof[4] By Lemma 8.3.1 it suffices to find a function $c : [\lambda^+]^2 \longrightarrow \lambda$ which maps, for any subset X of λ^+ of cardinality λ^+, $[X]^2$ onto λ. Lemma 3.3.5

[4]We follow the proof in [BM].

says that there is a set $\mathsf{a} \subseteq [\mu, \lambda)_{\text{reg}}$ of regular cardinals with $|\mathsf{a}| = \text{cf}(\lambda)$ and $|\mathsf{a}|^+ < \min(\mathsf{a})$ and an ultrafilter D on a such that $\lambda^+ = \text{cf}(\prod \mathsf{a}/D)$ and $\lim_D(\mathsf{a}) = \lambda$. Let

$$I := \text{I}_{\mathsf{b}}(\mathsf{a}).$$

With Lemma 3.4.9 we get $\mathcal{J}_{<\lambda}(\mathsf{a}) \subseteq I$. Since λ is singular, we have $\mathcal{J}_{<\lambda^+}(\mathsf{a}) = \mathcal{J}_{<\lambda}(\mathsf{a})$, hence $\mathcal{J}_{<\lambda^+}(\mathsf{a}) \subseteq I$.

Claim 1 $\text{tcf}(\prod \mathsf{a}/I \restriction \mathsf{b}) = \lambda^+$ for some $\mathsf{b} \in D \setminus I$.

$\lim_D(\mathsf{a}) = \lambda$ gives $I \cap D = \emptyset$ and thus $\mathcal{J}_{<\lambda^+}(\mathsf{a}) \cap D = \emptyset$.

Since $(\prod \mathsf{a}, <_{\mathcal{J}_{<\lambda^+}(\mathsf{a})})$ is λ^+-directed, we obtain by Lemma 3.4.16 a set $\mathsf{b} \in D$ such that $\text{tcf}(\prod \mathsf{a}/\mathcal{J}_{<\lambda^+}(\mathsf{a}) \restriction \mathsf{b}) = \lambda^+$. Then $\mathsf{b} \notin I$, and together with $\mathcal{J}_{<\lambda^+}(\mathsf{a}) \subseteq I$ we conclude, using Lemma 3.1.7, that $\text{tcf}(\prod \mathsf{a}/I \restriction \mathsf{b}) = \lambda^+$. This proves Claim 1.

Now fix $\mathsf{b} \in D \setminus I$ such that $\text{tcf}(\prod \mathsf{b}/I) := \text{tcf}(\prod \mathsf{b}/I \cap \mathcal{P}(\mathsf{b})) = \lambda^+$. For technical reasons we need the condition

$$(*) \qquad \sum(\nu : \nu \in \mathsf{b} \cap \kappa) < \kappa \text{ for all } \kappa \in \mathsf{b}.$$

If $\kappa \in \mathsf{b}$, then $\kappa \in \mathsf{a}$, and κ is regular. Thus $|\mathsf{b}| \leq |\mathsf{a}| < |\mathsf{a}|^+ < \min(\mathsf{a}) \leq \kappa$ yields $\sum(\nu : \nu \in \mathsf{b} \cap \kappa) < \kappa$.

Choose, for each $\kappa \in \mathsf{b}$, a function $c_\kappa : [\kappa]^2 \longrightarrow \kappa$ which witnesses $\text{p}(\kappa)$. Further let $(f_\xi : \xi < \lambda^+)$ be a sequence which is cofinal in $\prod \mathsf{b}$ modulo I. We define the desired function c as follows:

> If $\alpha < \beta < \lambda^+$, and if there is a greatest cardinal $\kappa \in \mathsf{b}$ such that $f_\alpha(\kappa) > f_\beta(\kappa)$, then we put $c(\alpha, \beta) := c_\kappa(f_\alpha(\kappa), f_\beta(\kappa))$. Otherwise, we let $c(\alpha, \beta) := 0$.

Consider $X \subseteq \lambda^+$ with $|X| = \lambda^+$, and let $\delta < \lambda$. We search for $\alpha, \beta \in X$ such that $\alpha < \beta$ and $c(\alpha, \beta) = \delta$. Choose, if possible, for each $\nu \in \mathsf{b}$ and each $\gamma < \nu$ a least ordinal $\alpha = \alpha(\nu, \gamma) \in X$ such that $f_\alpha(\nu) = \gamma$, and let the function $g \in \prod \mathsf{b}$ be defined by

$$g(\kappa) := \sup\{f_\alpha(\kappa) : \exists\gamma\exists\nu \in \mathsf{b}\ (\gamma < \nu < \kappa \land \alpha = \alpha(\nu, \gamma))\}.$$

Fix $\nu \in \mathsf{b}$. Then $|\{\alpha(\nu, \gamma) : \gamma < \nu\}| \leq \nu$. Further we conclude from $(*)$ that $\sum(\nu : \nu \in \mathsf{b} \cap \kappa) < \kappa$. Thus in the definition of $g(\kappa)$ we take the supremum of a set of cardinality less than κ, and we get $g(\kappa) \in \kappa$ and $g \in \prod \mathsf{b}$.

Claim 2 There is a set $Y \subseteq X$ with $|Y| = \lambda^+$, and a cardinal $\kappa_0 \in \mathsf{b}$ with $\kappa_0 > \delta$, such that $g(\kappa) < f_\xi(\kappa)$ for all $\xi \in Y$ and all $\kappa \geq \kappa_0$.

g is majorized under $<_I$ by a function f_{ξ_0} and thus by all functions f_ξ with $\xi \geq \xi_0$. Let $Y_1 := \{\xi \in X : \xi \geq \xi_0\}$. For each $\xi \in Y_1$, the set $B_\xi := \{\kappa \in \mathsf{b} : g(\kappa) \geq f_\xi(\kappa)\}$ is a member of I and thus strictly bounded in b by a cardinal $\mu_\xi \in \mathsf{b}$ satisfying $\mu_\xi > \delta$. Since $|Y_1| = \lambda^+$, we obtain by the pigeonhole principle a subset Y of Y_1 with $|Y| = \lambda^+$, such that $\mu_\xi =: \kappa_0 \in \mathsf{b}$ for all $\xi \in Y$. Now, if $\xi \in Y$ and $\kappa \geq \kappa_0$, then $\kappa \notin B_\xi$. This gives $g(\kappa) < f_\xi(\kappa)$, and Claim 2 is proved.

For each $\kappa \in \mathsf{b}$ and each $\xi < \lambda^+$, we put

$$A_\xi(\kappa) := \{f_\zeta(\kappa) : \zeta \in Y \wedge \xi < \zeta\}.$$

Claim 3 There is a cardinal $\kappa_1 \in \mathsf{b}$ with $\kappa_1 \geq \kappa_0$ such that, for each $\xi < \lambda^+$, the set $A_\xi(\kappa_1)$ is cofinal in κ_1.

We give an indirect proof for the claim and assume that, for each $\kappa \geq \kappa_0$ with $\kappa \in \mathsf{b}$, there is an ordinal $\xi(\kappa) < \lambda^+$ such that $A_{\xi(\kappa)}(\kappa)$ is bounded in κ. Put $\xi^* = \sup\{\xi(\kappa) : \kappa \geq \kappa_0\}$. Then $\xi^* < \lambda^+$, since $|\mathsf{b}| = \mathrm{cf}(\lambda) < \lambda^+$. Therefore $A_\xi(\kappa)$ is bounded in κ for all $\xi \geq \xi^*$ and all $\kappa \geq \kappa_0$ with $\kappa \in \mathsf{b}$. Assume that $\gamma(\kappa)$ is a bound for $A_{\xi^*}(\kappa)$ for each $\kappa \in \mathsf{b} \setminus \kappa_0$, and let $h(\kappa) := \gamma(\kappa) + 1$ for $\kappa \in \mathsf{b} \setminus \kappa_0$ and $h(\kappa) := 0$ for $\kappa \in \mathsf{b} \cap \kappa_0$. If $\xi > \xi^*$ and $\xi \in Y$, then $\{\kappa \in \mathsf{b} : f_\xi(\kappa) \geq h(\kappa)\} \subseteq \kappa_0 \cap \mathsf{b} \in I$, hence $f_\xi <_I h$. On the other hand, we have $|\{\xi \in Y : \xi > \xi^*\}| = \lambda^+$, and consequently $(f_\xi : \xi^* < \xi \wedge \xi \in Y)$ is a cofinal sequence in $\prod \mathsf{b}$ modulo I. This contradiction proves Claim 3.

Now we choose κ_1 according to Claim 3. Then in particular we get $|A_\xi(\kappa_1)| = \kappa_1 < \lambda$ for all $\xi < \lambda^+$. Furthermore, the sequence $(A_\xi(\kappa_1) : \xi < \lambda^+)$ is decreasing under the relation $\subseteq$. Therefore there is $\xi_1 < \lambda^+$, such that $A_\xi(\kappa_1) =: A$ for all $\xi \geq \xi_1$. Since $\delta < \kappa_0 \leq \kappa_1$, the choice of c_{κ_1} yields γ_1, $\gamma_2 \in A$ with $\gamma_1 < \gamma_2 < \kappa_1$ and $c_{\kappa_1}(\gamma_2, \gamma_1) = \delta$. From $\gamma_2 \in A = A_{\xi_1}(\kappa_1)$ we conclude that there is $\xi_2 \in Y$ satisfying $\xi_1 < \xi_2$ and $\gamma_2 = f_{\xi_2}(\kappa_1)$. So there exists $\alpha = \alpha(\kappa_1, \gamma_2) \in X$, and we have $\gamma_2 = f_\alpha(\kappa_1)$. Since $\gamma_1 \in A = A_{\xi_2}(\kappa_1)$, there is $\beta > \xi_2 \geq \alpha$ with $\beta \in Y$ such that $f_\beta(\kappa_1) = \gamma_1$.

It remains to be shown that κ_1 is the greatest cardinal $\kappa \in \mathsf{b}$ with $f_\alpha(\kappa) > f_\beta(\kappa)$, since in this case we can conclude that $\delta = c_{\kappa_1}(f_\alpha(\kappa_1), f_\beta(\kappa_1)) = c(\alpha, \beta)$, as desired. So let $\kappa > \kappa_1$. Then $\kappa_0 \leq \kappa_1 < \kappa$. From the definition of g, $\beta \in Y$ and Claim 2 we infer that $f_\alpha(\kappa) \leq g(\kappa) < f_\beta(\kappa)$. This completes, together with Lemma 8.3.1, the proof of the theorem.

8.4 Cofinalities of Partial Orderings $([\lambda]^{\leq\kappa}, \subseteq)$

In this section we will investigate the cofinalities of partial orderings of the form $([\lambda]^{\leq\kappa}, \subseteq)$. We will prove the following theorem: If κ is an infinite cardinal number, α is an ordinal such that $1 \leq \alpha < \kappa$, and $\lambda := \kappa^{+\alpha}$, then

$$\mathrm{cf}([\lambda]^{\leq\kappa}, \subseteq) = \max \mathrm{pcf}(\{\kappa^{+(\beta+1)} : \beta < \alpha\}).$$

First we present some simple properties of these cofinalities.

Lemma 8.4.1 *If λ and κ are cardinals such that $\omega \le \kappa \le \lambda$, then*

$$\max\{\mathrm{cf}([\lambda]^{\le\kappa}, \subseteq), 2^\kappa\} = \lambda^\kappa.$$

Proof Obviously it suffices to prove the relation "$\ge$". Assume that $\mathcal{S}$ is a subset of $[\lambda]^{\le\kappa}$ which is cofinal in $[\lambda]^{\le\kappa}$ under $\subseteq$. Then $[\lambda]^{\le\kappa} = \bigcup\{[S]^{\le\kappa} : S \in \mathcal{S}\}$, and Corollary 1.6.6 gives

$$\lambda^\kappa = |[\lambda]^{\le\kappa}| \le \sum_{S\in\mathcal{S}} |[S]^{\le\kappa}| \le |\mathcal{S}| \cdot 2^\kappa.$$

Hereby we get immediately the desired assertion.

Lemma 8.4.2 *For any cardinal λ,* $\mathrm{cf}([\lambda]^{\le\lambda}, \subseteq) = 1$.

Lemma 8.4.3 *Assume that λ is an infinite cardinal and $\kappa \in [1, \lambda)_{\mathrm{CN}}$. Then*

$$\mathrm{cf}([\lambda]^{\le\kappa}, \subseteq) \ge \lambda.$$

Proof Let us assume, to get a contradiction, that there is a set $\mathcal{P}$ which satisfies $\mathcal{P} \subseteq [\lambda]^{\le\kappa}$ and $|\mathcal{P}| < \lambda$ and is cofinal in $[\lambda]^{\le\kappa}$ under $\subseteq$. Then $|\bigcup\mathcal{P}| < \lambda$, and so there exists $x \in \lambda\setminus\bigcup\mathcal{P}$. However, there is *no* $P \in \mathcal{P}$ such that $\{x\} \subseteq P$.

Lemma 8.4.4 *If κ and λ are cardinals such that $\omega \le \kappa \le \lambda$, then*

$$\mathrm{cf}([\lambda^+]^{\le\kappa}, \subseteq) \le \max\{\lambda^+, \mathrm{cf}([\lambda]^{\le\kappa}, \subseteq)\}.$$

Proof Choose, for each $\alpha \in [\lambda, \lambda^+)_{\mathrm{ON}}$, a set $\mathcal{P}_\alpha$ cofinal in $[\alpha]^{\le\kappa}$ under $\subseteq$ such that $|\mathcal{P}_\alpha| = \mathrm{cf}([\lambda]^{\le\kappa}, \subseteq)$, and put

$$\mathcal{P} := \bigcup\{\mathcal{P}_\alpha : \alpha \in [\lambda, \lambda^+)_{\mathrm{ON}}\}.$$

Obviously we have $|\mathcal{P}| = \max\{\lambda^+, \mathrm{cf}([\lambda]^{\le\kappa}, \subseteq)\}$. We show that $\mathcal{P}$ is cofinal in $[\lambda^+]^{\le\kappa}$.

If $X \subseteq \lambda^+$ and $|X| \le \kappa$, then $X \subseteq \alpha$ for some $\alpha \in [\lambda, \lambda^+)_{\mathrm{ON}}$, since λ^+ is regular. Now there exists a set $P \in \mathcal{P}_\alpha \subseteq \mathcal{P}$ such that $X \subseteq P$.

Corollary 8.4.5 *If λ is an infinite cardinal and $1 \le n < \omega$, then*

$$\mathrm{cf}([\lambda^{+n}]^{\le\lambda}, \subseteq) = \lambda^{+n}.$$

Lemma 8.4.6 *Assume that κ and λ are cardinals such that $\omega < \kappa^+ < \lambda$. Then*

$$\mathrm{cf}([\lambda]^{\le\kappa}, \subseteq) \le \mathrm{cf}([\lambda]^{\le\kappa^+}, \subseteq).$$

Proof Choose $\mathcal{P}$ cofinal in $[\lambda]^{\leq\kappa^+}$ under $\subseteq$, and let without loss of generality $|P| = \kappa^+$ for all $P \in \mathcal{P}$. Further assume that, for each $P \in \mathcal{P}$, $(x_\alpha^P : \alpha < \kappa^+)$ is a bijection from κ^+ onto P. We put $M_\alpha^P := \{x_\beta^P : \beta < \alpha\}$ for each $P \in \mathcal{P}$ and each $\alpha < \kappa^+$, and

$$\mathcal{M} := \{M_\alpha^P : P \in \mathcal{P} \wedge \alpha < \kappa^+\}.$$

Then $|\mathcal{M}| \leq \max\{|\mathcal{P}|, \kappa^+\}$, and from Lemma 8.4.3 and $\kappa^+ < \lambda$ we can conclude that $|\mathcal{M}| \leq |\mathcal{P}|$.

We show that $\mathcal{M}$ is cofinal in $[\lambda]^{\leq\kappa}$. Fix $X \subseteq \lambda$ with $|X| \leq \kappa$. Then $X \subseteq P$ for some $P \in \mathcal{P}$. Since κ^+ is regular, there exists $\alpha < \kappa^+$ such that $X \subseteq M_\alpha^P \in \mathcal{M}$.

Remark The previous lemma is false for $\kappa^+ = \lambda$, as can easily be inferred from Lemmas 8.4.2 and 8.4.3.

Theorem 8.4.7 *Assume that κ is an infinite cardinal, α is an ordinal such that $1 \leq \alpha < \kappa$, and $\lambda := \kappa^{+\alpha}$. Then*

$$\mathrm{cf}\,([\lambda]^{\leq\kappa}, \subseteq) = \max \mathrm{pcf}(\{\kappa^{+(\beta+1)} : \beta < \alpha\}).$$

Proof[5] Put $\mathsf{a} := \{\kappa^{+(\beta+1)} : \beta < \alpha\}$. Obviously, a is progressive and thus an interval of regular cardinals, since any limit cardinal between κ and λ has a cofinality less than α.

We prove first the relation "$\geq$". Let $\mathcal{P}$ be cofinal in $[\lambda]^{\leq\kappa}$. Let

$$f_P(\nu) := \sup(P \cap \nu)$$

for each $P \in \mathcal{P}$ and any $\nu \in \mathsf{a}$. $|P| \leq \kappa < \min(\mathsf{a})$ gives $f_P \in \prod \mathsf{a}$ for all $P \in \mathcal{P}$. By Theorem 3.4.21 it suffices to show that $\mathrm{cf}\,(\prod \mathsf{a}, \leq_{\mathsf{a}}) \leq |\mathcal{P}|$. So fix $f \in \prod \mathsf{a}$. Since $|\mathsf{a}| < \kappa$, we have $\mathrm{ran}(f) \in [\lambda]^{\leq\kappa}$, and thus $\mathrm{ran}(f) \subseteq P$ for some $P \in \mathcal{P}$. So we get

$$f(\nu) \leq \sup(P \cap \nu) = f_P(\nu)$$

for all $\nu \in \mathsf{a}$, hence $f \leq f_P$.

Now we turn to the proof of the reverse inequality. Without loss of generality we can assume that $\alpha \geq \omega$, since otherwise $\max \mathrm{pcf}(\mathsf{a}) = \max(\mathsf{a}) = \lambda$ and $\mathrm{cf}\,([\lambda]^{\leq\kappa}, \subseteq) = \lambda$ by Corollary 8.4.5. If $\mu := |\alpha|^+$, then $\mu \in (|\mathsf{a}|, \min(\mathsf{a}))_{\mathrm{reg}}$.

Choose a regular cardinal $\Theta > 2^{2^{\sup(\mathsf{a})}}$ and a control function $\mathcal{F} = (\mathcal{F}_{\mathsf{b}} : \mathsf{b} \subseteq \mathsf{a})$ for a generated by $\bar{h}$, where $\bar{h}$ is a special sequence for a and μ. We define

$$\mathcal{G} := \{f \in \mathcal{F}_{\mathsf{a}} : \forall \nu \in \mathsf{a}\ \mathrm{cf}\,(f(\nu)) = \mu\}.$$

[5]In this proof we use ideas of S. Shelah communicated in an e-mail to one of the authors.

For each $\xi < \lambda$ with $\mathrm{cf}(\xi) = \mu$, we choose a club C_ξ in ξ of order type μ. Furthermore, we choose for each $f \in \mathcal{G}$ an elementary substructure N_f of $\mathrm{H}(\Theta)$ such that $|N_f| = \kappa$ and

$$\kappa \cup \{\kappa\} \cup \mathsf{a} \cup \bigcup\{C_{f(\nu)} : \nu \in \mathsf{a}\} \subseteq N_f.$$

Lemma 6.1.4 says that $|\mathcal{G}| \leq \max \mathrm{pcf}(\mathsf{a})$. So it is enough to show that the set

$$\{N_f \cap \lambda : f \in \mathcal{G}\}$$

is cofinal in $[\lambda]^{\leq\kappa}$.

For this, fix $X \subseteq \lambda$ with $|X| \leq \kappa$. Further let $(M_i : i \leq \mu)$ be an approximation sequence in $\mathrm{H}(\Theta)$ such that $|M_i| = \kappa$ for all $i < \mu$ and

$$X \cup \kappa \cup \{\kappa\} \cup \mathsf{a} \cup \{\bar{h}, \mathsf{a}\} \subseteq M_0.$$

Theorem 6.1.5 tells us that $f := \chi_{M_\mu} \restriction \mathsf{a} \in \mathcal{F}_{\mathsf{a}}$. By Lemma 4.3.3, there is, for each $\nu \in \mathsf{a}$, a club $D_\nu \subseteq M_\mu \cap \nu$ of order type μ in $f(\nu)$ – in particular we get $f \in \mathcal{G}$. Since μ is uncountable, we can assume without loss of generality that $D_\nu \subseteq C_{f(\nu)}$; otherwise we can take the intersection of the clubs. Now we put $\mathsf{a}^* := \{\kappa\} \cup \mathsf{a}$ and $D_\kappa := \emptyset$. Then $\mathsf{a}^* \setminus \{\min(\mathsf{a}^*)\}$ is an interval of regular cardinals and $(D_\nu : \nu \in M_\mu \cap \mathsf{a}^*)$ is a sequence of subsets of N_f characterizing M_μ over a^*. From Lemma 6.2.2 we can conclude that $X \subseteq M_\mu \cap \lambda \subseteq N_f \cap \lambda$, as desired.

Remark Lemma 8.4.1 says that $\aleph_\omega^{\aleph_1} = 2^{\aleph_1} \cdot \mathrm{cf}([\aleph_\omega]^{\leq\aleph_1}, \subseteq)$. Therefore we could determine $\aleph_\omega^{\aleph_1}$ if we could "compute" $2^{\aleph_1}$ and $\mathrm{cf}([\aleph_\omega]^{\leq\aleph_1}, \subseteq)$. The first cardinal cannot be estimated by the results of P. J. Cohen, but Shelah gives an estimate for the second cardinal: Theorem 8.4.7 tells us that $\mathrm{cf}([\aleph_\omega]^{\leq\aleph_1}, \subseteq) = \max \mathrm{pcf}(\{\aleph_{n+2} : n < \omega\})$. By Theorem 3.3.6, $\mathrm{pcf}(\{\aleph_{n+2} : n < \omega\})$ is an interval of regular cardinals, and from Corollary 7.2.7 we get $|\mathrm{pcf}(\{\aleph_{n+2} : n < \omega\})| < \aleph_{\omega_4}$. So we can conclude that $\mathrm{cf}([\aleph_\omega]^{\leq\aleph_1}, \subseteq) < \aleph_{\omega_4}$. Altogether we can infer that

$$\aleph_\omega^{\aleph_1} < \max\{(2^{\aleph_1})^+, \aleph_{\omega_4}\}.$$

On the other hand, Lemma 8.4.6 says that $\mathrm{cf}([\aleph_\omega]^{\leq\aleph_0}, \subseteq) \leq \mathrm{cf}([\aleph_\omega]^{\leq\aleph_1}, \subseteq)$, hence

$$\mathrm{cf}([\aleph_\omega]^{\leq\aleph_0}, \subseteq) < \aleph_{\omega_4}.$$

Therefore, if we want to estimate $\aleph_\omega^{\aleph_0}$, we have to estimate $2^{\aleph_0}$ which is impossible, and the cardinal $\mathrm{cf}([\aleph_\omega]^{\leq\aleph_0}, \subseteq)$ which is possible by Shelah's results.

Chapter 9

The Cardinal Function $\mathrm{pp}(\lambda)$

In this chapter, we introduce Shelah's cardinal function $\mathrm{pp}_\kappa(\lambda)$ whose properties we now summarize. If λ is a singular cardinal, then the definition of $\mathrm{pp}_\kappa(\lambda)$ gives $\mathrm{pp}_\kappa(\lambda) \leq \lambda^\kappa$ for any cardinal κ with $\mathrm{cf}(\lambda) \leq \kappa < \lambda$. For singular cardinals λ with uncountable cofinality and $\kappa = \mathrm{cf}(\lambda)$ which are in addition κ-strong, we get $\mathrm{pp}_\kappa(\lambda) = \lambda^\kappa$. In particular $\mathrm{pp}_\kappa(\lambda) = 2^\lambda$ holds for every strong limit cardinal λ with uncountable cofinality κ.

When $\kappa = \mathrm{cf}(\lambda)$, we put $\mathrm{pp}(\lambda) := \mathrm{pp}_\kappa(\lambda)$. So the function $\mathrm{pp}(\lambda)$ takes as values, for certain cardinals, the values of the Gimel function and of the continuum function, respectively. In analogy to the Galvin-Hajnal theorem we will prove that

$$\mathrm{pp}(\aleph_\delta) < \aleph_{(|\delta|^{\mathrm{cf}(\delta)})^+}$$

for singular cardinals $\aleph_\delta$. Analogously to Silver's theorem we get: If λ is a singular cardinal with uncountable cofinality, and if the set $\{\mu < \lambda : \mathrm{pp}(\mu) = \mu^+\}$ is stationary in λ, then $\mathrm{pp}(\lambda) = \lambda^+$.

Assume that κ and λ are cardinals and $\mathrm{cf}(\lambda) \leq \kappa < \lambda$. Then λ will be called pp_κ-strong if $\mathrm{pp}_\kappa(\nu) < \lambda$ for all singular cardinals $\nu < \lambda$ with $\mathrm{cf}(\nu) \leq \kappa < \nu$. If we rewrite the lemma of Galvin and Hajnal for $\mathrm{pp}(\lambda)$, it reads as follows: If λ is a singular cardinal with uncountable cofinality κ which is pp_κ-strong, $(\lambda_\xi : \xi < \kappa)$ is a normal function cofinal in λ, and $\Phi \in {}^\kappa\mathrm{ON}$ is a function such that $\mathrm{pp}_\kappa(\lambda_\xi) = \lambda_\xi^{+\Phi(\xi)}$ for all $\xi < \kappa$, then $\mathrm{pp}(\lambda) \leq \lambda^{+\|\Phi\|_{I_d(\kappa)}}$.

9.1 pp(λ) and the Theorems of Galvin, Hajnal and Silver

As usual, we use κ and λ in this section as syntactical variables for infinite cardinals.

We remind the reader on the definition of the cardinals $\mathrm{pp}_\kappa(\lambda)$ and $\mathrm{pp}^+_\kappa(\lambda)$. Assume that λ is a singular cardinal such that $\mathrm{cf}(\lambda) \leq \kappa < \lambda$, and let A be the set

$$\{\mathrm{cf}(\prod \mathsf{a}/D) : \mathsf{a} \subseteq \mathrm{Reg} \ \wedge \ \sup(\mathsf{a}) = \lambda \ \wedge \ |\mathsf{a}| \leq \kappa \\ \wedge \ D \text{ ultrafilter on } \mathsf{a} \ \wedge \ D \cap \mathrm{I_b}(\mathsf{a}) = \emptyset\}.$$

Then $\mathrm{pp}_\kappa(\lambda) = \sup(A)$ and $\mathrm{pp}^+_\kappa(\lambda) = \sup^+(A)$, where $\sup^+(A)$ is the cardinal $\min\{\kappa \in \mathrm{CN} : \forall \nu \in A \ \nu < \kappa\}$. If $\kappa = \mathrm{cf}(\lambda)$, then $\mathrm{pp}(\lambda) = \mathrm{pp}_\kappa(\lambda)$ and $\mathrm{pp}^+(\lambda) = \mathrm{pp}^+_\kappa(\lambda)$.

Let κ and λ be given as above. Then we get $\mathrm{cf}(\prod \mathsf{a}/D) \leq |{}^\kappa\lambda| = \lambda^\kappa$ for all a and D in question. If μ is a cardinal with $\mu < \lambda$ and a satisfies the properties in the definition of A, then we know from Lemma 3.3.5 that we can assume without loss of generality that $\mu < \min(\mathsf{a})$; in particular we can always assume that a is progressive.

Now we choose a as a set of regular cardinals of cardinality $\mathrm{cf}(\lambda)$ and cofinal in λ and D as an ultrafilter of final segments on a. Then Lemma 3.3.4 says that $\mathrm{cf}(\prod \mathsf{a}/D) > \lambda$. So we can conclude that

$$\lambda < \mathrm{pp}_\kappa(\lambda) \leq \lambda^\kappa.$$

The following Lemma yields a characterization of the cardinal $\mathrm{pp}_\kappa(\lambda)$ with the help of ideals.

Lemma 9.1.1 *If* $\mathrm{cf}(\lambda) \leq \kappa < \lambda$, *then*

$$\mathrm{pp}_\kappa(\lambda) = \sup\{\mathrm{tcf}(\prod f/I) : f \in {}^\kappa\lambda \ \wedge \ \mathrm{ran}(f) \subseteq \mathrm{Reg} \ \wedge \ \lim_I(f) = \lambda\}.$$

Proof Put $\mathrm{pp}^*_\kappa(\lambda) := \sup\{\mathrm{tcf}(\prod f/I) : f \in {}^\kappa\lambda \wedge \mathrm{ran}(f) \subseteq \mathrm{Reg} \wedge \lim_I(f) = \lambda\}$. If, according to the definition of $\mathrm{pp}_\kappa(\lambda)$, $\mathsf{a} \subseteq \lambda \cap \mathrm{Reg}$, $\sup(\mathsf{a}) = \lambda$, $|\mathsf{a}| \leq \kappa$, D is an ultrafilter on a, and $D \cap \mathrm{I_b}(\mathsf{a}) = \emptyset$, then we can assume without loss of generality that $\kappa < \min(\mathsf{a})$. Let f be a surjection from κ onto a and J be the dual ideal of the ultrafilter D. Note that f is progressive and $I := \{f^{-1}[X] : X \in J\}$ is an ideal on κ such that $J = f[I]$. Then we have $\lim_I(f) = \lambda$ and, using Lemma 3.3.1, $\mathrm{cf}(\prod \mathsf{a}/D) = \mathrm{tcf}(\prod \mathsf{a}/J) = \mathrm{tcf}(\prod f/I)$. So we get $\mathrm{pp}_\kappa(\lambda) \leq \mathrm{pp}^*_\kappa(\lambda)$.

Conversely, if $f \in {}^{\kappa}\lambda$ is a function satisfying $\operatorname{ran}(f) \subseteq \mathrm{Reg}$ and $\lim_I(f) = \lambda$, then $c := \{\xi < \kappa : f(\xi) \leq \kappa\}$ is a member of I. Putting $g := f \restriction (\kappa \setminus c)$ and $I' := I \cap \mathcal{P}(\kappa \setminus c)$, we obtain $\operatorname{tcf}(\prod f/I) = \operatorname{tcf}(\prod g/I')$ from Lemma 3.1.8. For $\mathsf{a} := \operatorname{ran}(g)$, we can conclude that $\mathsf{a} \subseteq \mathrm{Reg}$, $\sup(\mathsf{a}) = \lambda$, and $|\mathsf{a}| \leq \kappa$; $\mathrm{I_b}(\mathsf{a}) \subseteq g[I']$ follows from $\lim_I(f) = \lambda$. By construction, g is progressive, and consequently we can infer from Lemma 3.3.1 that $\operatorname{tcf}(\prod g/I') = \operatorname{tcf}(\prod \operatorname{ran}(g)/g[I'])$. Now consider an arbitrary ultrafilter D on a such that $D \cap g[I'] = \emptyset$. Then we have $D \cap \mathrm{I_b}(\mathsf{a}) = \emptyset$ and $\operatorname{tcf}(\prod f/I) = \operatorname{tcf}(\prod g/I') = \operatorname{tcf}(\prod \operatorname{ran}(g)/g[I']) = \operatorname{cf}(\prod \mathsf{a}/D)$. This yields $\mathrm{pp}^*_\kappa(\lambda) \leq \mathrm{pp}_\kappa(\lambda)$, as desired.

Theorem 9.1.2 *Assume that λ is a cardinal,* $\operatorname{cf}(\lambda) \leq \kappa < \lambda$, *and λ is κ-strong and not a fixed point of the aleph function. Then*

$$\mathrm{pp}_\kappa(\lambda) = \lambda^\kappa.$$

Proof If $\lambda =: \aleph_\delta$, then the assumption gives $\delta < \aleph_\delta$. Furthermore, the cardinal $\mu := (|\delta|^\kappa)^+$ is less than λ and κ-strong.

Let $\mathsf{a} := [\mu, \lambda)_{\mathrm{reg}}$. Then $|\mathsf{a}| \leq |\delta| < \mu$, and thus a is a progressive interval of regular cardinals such that $\min(\mathsf{a})$ is κ-strong and $\sup(\mathsf{a}) = \lambda$. From Theorem 6.2.4 we infer that $\sup \mathrm{pcf}_\kappa(\mathsf{a}) = \lambda^\kappa$. Now we show that $\sup \mathrm{pcf}_\kappa(\mathsf{a}) = \mathrm{pp}_\kappa(\lambda)$, which will prove the theorem. Note that we can infer from the fact that λ is κ-strong that $\max \mathrm{pcf}(\mathsf{d}) < \lambda$ for every $\mathsf{d} \in \mathrm{I_b}(\mathsf{a})$ with $|\mathsf{d}| \leq \kappa$: If $\nu < \lambda$ satisfies $\mathsf{d} \subseteq \nu$, then $|\prod \mathsf{d}| \leq \nu^{|\mathsf{d}|} < \lambda$.

For the proof of the relation "$\leq$", let $\mathsf{c} \in [\mathsf{a}]^{\leq\kappa}$. If $\mathsf{c} \in \mathrm{I_b}(\mathsf{a})$, then $\max \mathrm{pcf}(\mathsf{c}) < \lambda < \mathrm{pp}_\kappa(\lambda)$. So we can assume that $\sup(\mathsf{c}) = \lambda$. Consider a fixed ultrafilter D on c. We distinguish two cases. If $D \cap \mathrm{I_b}(\mathsf{c}) \neq \emptyset$, choose a set $\mathsf{d} \in D \cap \mathrm{I_b}(\mathsf{c})$ and observe that, using Lemma 3.1.8, we have $\operatorname{cf}(\prod \mathsf{c}/D) = \operatorname{cf}(\prod \mathsf{d}/D \cap \mathcal{P}(\mathsf{d})) < \lambda$. If $D \cap \mathrm{I_b}(\mathsf{c}) = \emptyset$, then $\operatorname{cf}(\prod \mathsf{c}/D)$ is a member of that set in the definition of $\mathrm{pp}_\kappa(\lambda)$ whose supremum is $\mathrm{pp}_\kappa(\lambda)$. So we have shown that $\sup \mathrm{pcf}_\kappa(\mathsf{a}) \leq \mathrm{pp}_\kappa(\lambda)$.

For the proof of the relation "$\geq$" assume that $\mathsf{c} \subseteq \mathrm{Reg}$, $|\mathsf{c}| \leq \kappa$, $\sup(\mathsf{c}) = \lambda$, and D is an ultrafilter on c such that $D \cap \mathrm{I_b}(\mathsf{c}) = \emptyset$. Let $\mathsf{b} := \mathsf{c} \cap \mu$ and $\nu \in \mathrm{pcf}(\mathsf{b})$. Taking $\mathrm{pcf}(\mathsf{c}) = \mathrm{pcf}(\mathsf{c} \setminus \mathsf{b}) \cup \mathrm{pcf}(\mathsf{b})$ and $\max \mathrm{pcf}(\mathsf{b}) < \lambda$ into account, we can assume without loss of generality that $\mathsf{c} \subseteq [\mu, \lambda)_{\mathrm{reg}} = \mathsf{a}$, which gives $\mathrm{pp}_\kappa(\lambda) \leq \sup \mathrm{pcf}_\kappa(\mathsf{a})$.

This raises the question of whether Theorem 9.1.2 is true without the assumption that λ is not a fixed point of the aleph function. We can answer it positively if we accept the assumption that λ has uncountable cofinality.

Theorem 9.1.3 *Let λ be a singular cardinal. If λ has the uncountable cofinality κ and is κ-strong, then*

$$\mathrm{pp}_\kappa(\lambda) = \lambda^\kappa.$$

Proof At the beginning of this section we noted that $\mathrm{pp}_\kappa(\lambda) \leq \lambda^\kappa$. For the proof of the reverse inequality, consider a cardinal $\mu \in (\lambda, \lambda^\kappa]_{\mathrm{CN}}$. Without loss of generality we can assume that μ is regular, since any singular cardinal is the supremum of a set of regular cardinals. By Lemma 3.1.15, there is a function $f \in {}^\kappa\lambda$ and an ideal I on κ such that $\lim_I(f) = \kappa$ and $\mu = \mathrm{tcf}(\prod f/I)$. Theorem 3.1.13 says that $\mathrm{tcf}(\prod \mathrm{cf} \circ f/I) = \mu$. To show that $\lim_I(\mathrm{cf} \circ f) = \lambda$, let us suppose that there is a cardinal $\nu < \lambda$ such that $X := \{i < \kappa : \mathrm{cf}(f(i)) \leq \nu\} \in I^+$. The assumption that λ is κ-strong yields

$$\mu = \mathrm{tcf}(\prod \mathrm{cf} \circ f/I) = \mathrm{tcf}(\prod_{i \in X} \mathrm{cf}(f(i))/I \cap \mathcal{P}(X)) \leq |{}^X\nu| \leq \nu^\kappa < \lambda.$$

This contradiction shows that $\lim_I(\mathrm{cf} \circ f) = \lambda$. Now we can infer from Lemma 9.1.1 that $\mu \leq \mathrm{pp}_\kappa(\lambda)$, hence $\lambda^\kappa \leq \mathrm{pp}_\kappa(\lambda)$.

In the following we will prove generalizations of the Galvin-Hajnal theorem and of Silver's theorem.

Theorem 9.1.4 (Shelah) *If $\aleph_\delta$ is a singular cardinal, then*

$$\mathrm{pp}(\aleph_\delta) < \aleph_{(|\delta|^{\mathrm{cf}(\delta)})^+}.$$

Proof We use Shelah's interval technique and claim that

$$[\aleph_0, \mathrm{pp}^+(\aleph_\delta))_{\mathrm{reg}} \subseteq \bigcup \{\mathrm{pcf}(\mathsf{a}) : \mathsf{a} \subseteq \aleph_\delta \cap \mathrm{Reg} \wedge |\mathsf{a}| \leq \mathrm{cf}(\delta) \wedge \mathsf{a} \text{ is progressive}\}.$$

So let $\mu \in [\aleph_0, \mathrm{pp}^+(\aleph_\delta))_{\mathrm{reg}}$. If $\mu < \aleph_\delta$, then there is a finite progressive subset a of $\aleph_\delta \cap \mathrm{Reg}$, namely $\mathsf{a} := \{\mu\}$, such that $|\mathsf{a}| \leq \mathrm{cf}(\delta)$ and $\mu \in \mathrm{pcf}(\mathsf{a})$. So let $\mu \in (\aleph_\delta, \mathrm{pp}^+(\aleph_\delta))_{\mathrm{reg}}$. By Lemma 3.3.5, there is a progressive subset a of $\aleph_\delta \cap \mathrm{Reg}$ such that $|\mathsf{a}| \leq \mathrm{cf}(\aleph_\delta) = \mathrm{cf}(\delta)$ and $\mu \in \mathrm{pcf}(\mathsf{a})$, as desired.

Now $|\aleph_\delta \cap \mathrm{Reg}| \leq |\delta|$ gives $|[\aleph_\delta \cap \mathrm{Reg}]^{\leq \mathrm{cf}(\delta)}| \leq |\delta|^{\mathrm{cf}(\delta)}$. From Corollary 3.4.14 we conclude that $|\mathrm{pcf}(\mathsf{a})| \leq 2^{|\mathsf{a}|} \leq 2^{\mathrm{cf}(\delta)}$, and thus we get $|[\aleph_0, \mathrm{pp}^+(\aleph_\delta))_{\mathrm{reg}}| \leq |\delta|^{\mathrm{cf}(\delta)} \cdot 2^{\mathrm{cf}(\delta)} = |\delta|^{\mathrm{cf}(\delta)}$. This yields $\mathrm{pp}(\aleph_\delta) < \aleph_{(|\delta|^{\mathrm{cf}(\delta)})^+}$.

Remark If $\aleph_\delta$ is singular, $\mathrm{cf}(\delta)$-strong and not a fixed point of the aleph function, then we can infer from Theorem 9.1.2 and Theorem 9.1.4 that

$$\aleph_\delta^{\mathrm{cf}(\delta)} < \aleph_{(|\delta|^{\mathrm{cf}(\delta)})^+}.$$

If $\aleph_\delta = \delta$, this assertion holds trivially. Consequently, Theorem 9.1.4 is a generalization of the Galvin-Hajnal theorem.

In the rest of this section we prove Shelah's generalization of Silver's theorem.

Lemma 9.1.5 *Let* λ *be a singular cardinal number. If* $\mathrm{pp}(\lambda) < \lambda^{+\mathrm{cf}(\lambda)^+}$, *then* $\mathrm{pp}(\lambda) = \mathrm{pp}_\kappa(\lambda)$ *for all* κ *with* $\mathrm{cf}(\lambda) \leq \kappa < \lambda$.

Proof Consider κ and λ such that $\mathrm{cf}(\lambda) \leq \kappa < \lambda$ and $\mathrm{pp}(\lambda) < \lambda^{+\mathrm{cf}(\lambda)^+}$. By definition we have $\mathrm{pp}(\lambda) \leq \mathrm{pp}_\kappa(\lambda)$. To get a contradiction, let us assume that there is a cardinal $\kappa \in [\mathrm{cf}(\lambda), \lambda)_{\mathrm{CN}}$ for which equality does not hold. Then we can find a subset a of $\lambda \cap \mathrm{Reg}$ with $\sup(\mathsf{a}) = \lambda$ and $|\mathsf{a}| \leq \kappa$ and an ultrafilter D on a, such that $D \cap \mathrm{I_b}(\mathsf{a}) = \emptyset$ and $\theta := \mathrm{pp}(\lambda) < \mathrm{cf}(\prod \mathsf{a}/D)$. As usual we can assume without loss of generality that $\kappa < \min(\mathsf{a})$.

Corollary 3.4.11 says that $D \cap \mathcal{J}_{<\theta^+}(\mathsf{a}) = \emptyset$, since $\mathrm{cf}(\prod \mathsf{a}/D) \geq \theta^+$, hence $\mathcal{J}_{<\theta^+}(\mathsf{a}) \neq \mathcal{P}(\mathsf{a})$. If $\rho < \lambda$ is a cardinal, then $\mathsf{a} \setminus \rho \in D$ and thus $\theta < \mathrm{cf}(\prod(\mathsf{a} \setminus \rho)/D)$, and the same argument yields $\mathcal{J}_{<\theta^+}(\mathsf{a} \setminus \rho) \neq \mathcal{P}(\mathsf{a} \setminus \rho)$. Therefore, for any $\mathsf{b} \in \mathcal{J}_{<\theta^+}(\mathsf{a})$, the set $\mathsf{a} \setminus \mathsf{b}$ is unbounded in λ: Since $\mathsf{b} \setminus \rho$ is a member of $\mathcal{J}_{<\theta^+}(\mathsf{a} \setminus \rho)$, $\mathsf{b} \setminus \rho$ is a proper subset of $\mathsf{a} \setminus \rho$, and so there is $\nu \in \mathsf{a} \setminus \mathsf{b}$ such that $\rho \leq \nu$.

Choose a sequence $(\mathsf{b}_\mu : \mu \in \mathrm{pcf}(\mathsf{a}))$ of generators for a, let $\mathsf{b}_\mu := \emptyset$ if $\mu \notin \mathrm{pcf}(\mathsf{a})$, and define

$$B := [\,[\lambda^+, \mathrm{pp}^+(\lambda))_{\mathrm{reg}}\,]^{<\aleph_0} = \{X \subseteq [\lambda^+, \mathrm{pp}^+(\lambda))_{\mathrm{reg}} : X \text{ is finite }\}.$$

The assumption $\mathrm{pp}(\lambda) < \lambda^{+\mathrm{cf}(\lambda)^+}$ gives $|B| = |[\lambda^+, \mathrm{pp}^+(\lambda))_{\mathrm{reg}}| \leq \mathrm{cf}(\lambda)$.

Consider $X \in B$. Then we have $\bigcup_{\mu \in X} \mathsf{b}_\mu \in \mathcal{J}_{<\theta^+}(\mathsf{a})$, and we can choose, as shown above, a set c_X which is unbounded in λ and satisfies $\mathsf{c}_X \subseteq \mathsf{a} \setminus \bigcup_{\mu \in X} \mathsf{b}_\mu$ and $|\mathsf{c}_X| = \mathrm{cf}(\lambda)$. Then also the set $\mathsf{c} := \bigcup_{X \in B} \mathsf{c}_X$ satisfies $|\mathsf{c}| = \mathrm{cf}(\lambda)$. Clearly c is unbounded in λ. Furthermore the construction gives $\mathsf{c} \setminus \rho \notin \mathcal{J}_{<\theta^+}(\mathsf{a})$ for all cardinals $\rho < \lambda$, since otherwise, by Corollary 5.3.4, there would be $X_0 \in B$ such that $\mathsf{c} \setminus \rho \subseteq \bigcup_{\mu \in X_0} \mathsf{b}_\mu$, and thus c_{X_0} would be bounded in λ.

Now we can conclude from Exercise 2.1.10 that there exists an ultrafilter D' on a such that $\mathsf{c} \setminus \rho \in D'$, for all cardinals $\rho < \lambda$, and $D' \cap \mathcal{J}_{<\theta^+}(\mathsf{a}) = \emptyset$. But then we get $\mathrm{cf}(\prod \mathsf{c}/D' \cap \mathcal{P}(\mathsf{c})) \geq \theta^+ \geq \mathrm{pp}^+(\lambda)$ and $D' \cap \mathrm{I_b}(\mathsf{c}) = \emptyset$, contradicting the definition of $\mathrm{pp}(\lambda)$.

Theorem 9.1.6 (Shelah) *If* λ *is a singular cardinal with uncountable cofinality, and if the set* $\{\nu < \lambda : \mathrm{pp}(\nu) = \nu^+\}$ *is stationary in* λ, *then* $\mathrm{pp}(\lambda) = \lambda^+$.

Proof Let a be a progressive set of regular cardinals such that $\sup(\mathsf{a}) = \lambda$ and $|\mathsf{a}| = \mathrm{cf}(\lambda)$. To prove that $\mathrm{cf}(\prod \mathsf{a}/D) = \lambda^+$ for any ultrafilter D on a with $D \cap \mathrm{I_b}(\mathsf{a}) = \emptyset$, we will show that $\mathrm{tcf}(\prod \mathsf{a}/\mathrm{I_b}(\mathsf{a})) = \lambda^+$.

Since $\sup(\mathsf{a}) = \lambda$, it is easy to see that the set $\{\rho < \lambda : \sup(\mathsf{a} \cap \rho) = \rho\}$ is a club of cardinals in λ. By Theorem 7.2.4, there is a club C_0 of cardinals in λ such that $\mathrm{otp}(C_0) = \mathrm{cf}(\lambda)$ and

$$(1) \qquad \mathrm{tcf}(\prod C_0^+/\mathrm{I_b}(C_0^+)) = \lambda^+.$$

Then $C := \{\rho < \lambda : \sup(\mathsf{a} \cap \rho) = \rho \wedge \rho \neq 0\} \cap C_0$ is also a club in λ with $|C| = \mathrm{cf}(\lambda)$, and from $C^+ \notin \mathrm{I_b}(C_0^+)$ and (1), together with Lemma 3.1.8, we can conclude that

(2) $$\mathrm{tcf}(\prod(C^+/\mathrm{I_b}(C^+)) = \lambda^+.$$

By assumption, the set $S := \{\nu \in C : \mathrm{pp}(\nu) = \nu^+\}$ is stationary in λ. From $S^+ \notin \mathrm{I_b}(C^+)$ we infer with (2) that

(3) $$\mathrm{tcf}(\prod S^+/\mathrm{I_b}(S^+)) = \lambda^+.$$

Now, if $\nu \in S$, then $\nu = \sup(\mathsf{a} \cap \nu)$, and from $\mathrm{cf}(\nu) \leq |\mathsf{a} \cap \nu| \leq |\mathsf{a}| = \mathrm{cf}(\lambda) < \min(\mathsf{a}) < \nu$ and $\mathrm{pp}(\nu) = \nu^+ < \nu^{+\mathrm{cf}(\nu)^+}$ we conclude, using Lemma 9.1.5, that $\mathrm{pp}_{|\mathsf{a}\cap\nu|}(\nu) = \nu^+$.

For any ultrafilter D on $\mathsf{a} \cap \nu$ with $D \cap \mathrm{I_b}(\mathsf{a} \cap \nu) = \emptyset$, we have $\mathrm{cf}(\prod(\mathsf{a} \cap \nu)/D) = \nu^+$, and thus Theorem 3.4.18 tells us that $\mathrm{tcf}(\prod(\mathsf{a}\cap\nu)/\mathrm{I_b}(\mathsf{a}\cap\nu)) = \nu^+$.

We put $I_\nu := \{X \subseteq \mathsf{a} : X \cap \nu \in \mathrm{I_b}(\mathsf{a} \cap \nu)\}$ for each $\nu \in S$. Then Lemma 3.1.8 says that $\mathrm{tcf}(\prod \mathsf{a}/I_\nu) = \nu^+$, hence $S^+ \subseteq \mathrm{pcf}(\mathsf{a})$. $|S| = |C| = \mathrm{cf}(\lambda) < \min(\mathsf{a})$ gives $|S^+| < \min(\mathsf{a})$. Taking

$$I^* := \{X \subseteq \mathsf{a} : \{\nu^+ \in S^+ : X \notin I_\nu\} \in \mathrm{I_b}(S^+)\},$$

we can infer from (3) and Corollary 3.3.8 that

$$\mathrm{tcf}(\prod S^+/\mathrm{I_b}(S^+)) = \lambda^+ = \mathrm{tcf}(\prod \mathsf{a}/I^*).$$

It remains to show that $I^* \subseteq \mathrm{I_b}(\mathsf{a})$. If $X \subseteq \mathsf{a}$ and $X \notin \mathrm{I_b}(\mathsf{a})$, then $\{\rho < \lambda : \sup(X \cap \rho) = \rho\}$ is a club in λ. Therefore the set $\{\nu \in S : \sup(X \cap \nu) = \nu\} = \{\nu \in S : X \notin I_\nu\}$ is stationary in λ. Consequently, the set $\{\nu^+ \in S^+ : X \notin I_\nu\}$ is unbounded in S^+, and thus $X \notin I^*$.

Remark Let us demonstrate why the preceding theorem "generalizes" Silver's theorem. If λ is a singular cardinal with uncountable cofinality and is not a limit point of fixed points of the aleph function, then there is some final segment E of λ containing no fixed points of the aleph function. Assume that $S := \{\nu < \lambda : 2^\nu = \nu^+\}$ is stationary in λ. Then λ is a strong limit cardinal. Furthermore the set L of all singular strong limit cardinals in E is a club in λ, and thus $L \cap S$ is stationary in λ. For any cardinal $\nu \in L \cap S$ we get $\mathrm{pp}(\nu) = 2^\nu$, using Theorem 9.1.2 and Lemma 1.6.15 d). Now Theorem 9.1.6 says that $\mathrm{pp}(\lambda) = \lambda^+$. Since λ is a strong limit cardinal, we get $\mathrm{pp}(\lambda) = \lambda^{\mathrm{cf}(\lambda)}$ from Theorem 9.1.3, and $\lambda^{\mathrm{cf}(\lambda)} = 2^\lambda$ from Theorem 1.6.15 d). In particular we have $2^\lambda = \lambda^+$. In this sense, Theorem 9.1.6 is a generalization of Silver's theorem.

9.2 The Lemma of Galvin and Hajnal for pp(λ)

In this section we will prove a theorem that is analogous to Corollary 2.3.4: If λ is a singular cardinal with uncountable cofinality κ and pp_κ-strong, $(\lambda_\xi : \xi < \kappa)$ is a normal sequence cofinal in λ, and $\Phi \in {}^\kappa\mathrm{ON}$ is an ordinal function such that $\mathrm{pp}(\lambda_\xi) = \lambda_\xi^{+\Phi(\xi)}$ for all $\xi < \kappa$, then $\mathrm{pp}(\lambda) \le \lambda^{+\|\Phi\|_{\mathrm{I_d}(\kappa)}}$. The proof of this theorem of Shelah[1] needs several preparatory considerations.

Lemma 9.2.1 *Assume that μ and λ are singular cardinals of cofinality $\le \kappa$, where κ is a regular cardinal and $\kappa < \mu < \lambda$. Further let $\lambda \le \mathrm{pp}_\kappa(\mu)$. Then $\mathrm{pp}_\kappa(\lambda) \le \mathrm{pp}_\kappa(\mu)$.*

Proof Fix $\nu \in \mathrm{Reg}$ such that $\lambda < \nu < \mathrm{pp}_\kappa^+(\lambda)$. From Lemma 3.3.5 we obtain a progressive set $\mathsf{b} \subseteq \lambda$ of regular cardinals satisfying $|\mathsf{b}| \le \kappa$, $\sup(\mathsf{b}) = \lambda$, and $\mu < \min(\mathsf{b})$, and an ultrafilter D on b such that $\mathrm{cf}(\prod \mathsf{b}/D) = \nu$. In the same way we can infer from $\lambda \le \mathrm{pp}_\kappa(\mu)$ that, for each $\rho \in \mathsf{b}$, there exists a progressive set $\mathsf{a}_\rho \subseteq \mu \cap \mathrm{Reg}$ with $|\mathsf{a}_\rho| \le \kappa < \min(\mathsf{a}_\rho)$ and $\sup(\mathsf{a}_\rho) = \mu$, and an ultrafilter D'_ρ on a_ρ, such that $\mathrm{cf}(\prod \mathsf{a}_\rho/D'_\rho) = \rho$ and $D'_\rho \cap \mathrm{I_b}(\mathsf{a}_\rho) = \emptyset$. Since also $D'_\rho \cap \mathrm{I_b}(\mu) = \emptyset$, we can extend each of the ultrafilters D'_ρ to an ultrafilter D_ρ on $\mathsf{c} := \bigcup\{\mathsf{a}_\rho : \rho \in \mathsf{b}\}$, and we can easily verify that $D_\rho \cap \mathrm{I_b}(\mathsf{c}) = \emptyset$. Furthermore, we have $\mathsf{b} \subseteq \mathrm{pcf}(\mathsf{c})$, $|\mathsf{b}| \le \min(\mathsf{c})$, and c is a progressive set. Now Corollary 3.3.8 says that $D^* := \{X \subseteq \mathsf{c} : \{\rho \in \mathsf{b} : X \in D_\rho\} \in D\}$ is an ultrafilter on c such that $\mathrm{cf}(\prod \mathsf{c}/D^*) = \mathrm{cf}(\prod \mathsf{b}/D) = \nu$. No member of D^* is bounded in c, since this holds for all D_ρ and $\emptyset \notin D$. Furthermore we have $\mathsf{c} \subseteq \mu \cap \mathrm{Reg}$, $|\mathsf{c}| \le \kappa \cdot \kappa = \kappa$, and $\sup(\mathsf{c}) = \mu$. So we can conclude that $\nu \le \mathrm{pp}_\kappa(\mu)$.

Lemma 9.2.2 *Assume that λ is a singular cardinal with uncountable cofinality κ, $(\lambda_\xi : \xi < \kappa)$ is a normal sequence of cardinals which is cofinal in λ, and $I \supseteq \mathrm{I_b}(\kappa)$ is a normal ideal on κ. If $f \in {}^\kappa\mathrm{Reg}$ is a function with $\lim_I(f) = \lambda$, then the set $\{\xi < \kappa : f(\xi) \le \lambda_\xi\}$ is a member of I.*

Proof Lemma 1.8.6 says that we can infer from $I \supseteq \mathrm{I_b}(\kappa)$ that $I \supseteq \mathrm{I_d}(\kappa)$. Let us assume that the set $X := \{\xi < \kappa : \kappa < f(\xi) \le \lambda_\xi\}$ is not a member of I. If C is a club in κ, then $\kappa \setminus C$ is not stationary and is thus a member of I. This gives $X \cap C \notin I$ since $X \subseteq (X \cap C) \cup (\kappa \setminus C)$, hence in particular $S := X \cap \mathrm{Lim} \notin I$. Fix $\xi \in S$. With the continuity of the sequence $(\lambda_\xi : \xi < \kappa)$, we can conclude that $\mathrm{cf}(\lambda_\xi) \le \xi < \kappa < \lambda_\xi$, hence λ_ξ is singular. Since $f(\xi)$ is regular, we get $f(\xi) < \lambda_\xi$, and since ξ is a limit ordinal, there exists an ordinal $j(\xi) < \xi$ such that $f(\xi) < \lambda_{j(\xi)}$. Now the function $j : S \longrightarrow \kappa$ is regressive. Therefore we obtain, by Lemma 1.8.9, an ordinal $\alpha < \kappa$ and a set $S_0 \subseteq S$ such that $S_0 = \{\xi \in S : j(\xi) = \alpha\} \notin I$. Consequently, the set $\{\xi < \kappa : f(\xi) \le \lambda_\alpha\}$ is not a member of I. This contradicts the assumption that $\lim_I(f) = \lambda$.

[1] Our proof elaborates Shelah's representation in [Sh5], VIII, §1.

Remember that, for any ordinal function $g \in {}^A\mathrm{ON}$,

$$\mathrm{pcf}(g) = \{\lambda : \exists I (I \text{ is an ideal on } A \wedge \lambda = \mathrm{tcf}(\prod g/I))\}.$$

If in addition g is progressive, then Lemma 3.3.1 says that $\mathrm{pcf}(g) = \mathrm{pcf}(\mathrm{ran}(g))$.

Lemma 9.2.3 *Assume that $g \in {}^A\mathrm{ON}$ is a progressive function with $\mathrm{ran}(g) \subseteq \mathrm{Reg}$ and I is an ideal on A. Further assume that $\lambda = \max \mathrm{pcf}(g \restriction B)$ for all $B \in I^+$. Then $\lambda = \mathrm{tcf}(\prod g/I)$.*

Proof It suffices to show that $\lambda = \mathrm{tcf}(\prod \mathrm{ran}(g)/g[I])$. From $A \notin I$ and the assumption we can infer that $\lambda = \max \mathrm{pcf}(g)$, hence $\lambda = \max \mathrm{pcf}(\mathrm{ran}(g))$ since g is progressive. We put $\mathsf{a} := \mathrm{ran}(g)$ and want to apply Theorem 3.4.18. For this we suppose, contrary to our hopes, that there is an ultrafilter D on a such that $D \cap g[I] = \emptyset$ and $\mu := \mathrm{cf}(\prod \mathsf{a}/D) \neq \lambda$. Then we get $\mu < \lambda$ and $D \cap \mathcal{J}_{<\mu^+}(\mathsf{a}) \neq \emptyset$. So we choose a set $\mathsf{b} \in D \cap \mathcal{J}_{<\mu^+}(\mathsf{a})$ and can conclude that $\max \mathrm{pcf}(\mathsf{b}) < \mu^+ \leq \lambda$. The set $B := g^{-1}[\mathsf{b}]$ is not a member of I, since $\mathsf{b} \notin g[I]$. This yields $\mathrm{pcf}(g \restriction B) = \mathrm{pcf}(g[B]) = \mathrm{pcf}(\mathsf{b})$, and thus $\max \mathrm{pcf}(\mathsf{b}) = \max \mathrm{pcf}(g \restriction B) < \lambda$, contradicting the assumption.

Theorem 9.2.4 *Assume that a is a set of regular cardinals with $|\mathsf{a}| \leq \min(\mathsf{a})$, $\kappa > \aleph_0$ is regular, and $(\mathsf{b}_\xi : \xi < \kappa)$ is a sequence of subsets of a with $\mathsf{a} = \bigcup_{\xi<\kappa} \mathsf{b}_\xi$ which is increasing under inclusion. Further let $\lambda \in \mathrm{pcf}(\mathsf{a}) \setminus \bigcup_{\xi<\kappa} \mathrm{pcf}(\mathsf{b}_\xi)$. Then there is an ordinal $\xi_0 < \kappa$ and a sequence $(\lambda_\xi : \xi_0 \leq \xi < \kappa)$ satisfying $\lambda_\xi \in \mathrm{pcf}(\mathsf{b}_\xi)$ and $\lambda_\xi > \kappa^+$ for all ξ with $\xi_0 \leq \xi < \kappa$, and a subset A of κ which is unbounded in κ, such that $\lambda = \mathrm{tcf}(\prod_{\xi \in A} \lambda_\xi / \mathrm{I_b}(A))$.*

If $\kappa = \aleph_0$, then there is a countable set $\mathsf{b} \subseteq \bigcup\{\mathrm{pcf}(\mathsf{b}_\xi) : \xi < \kappa\}$ such that $\lambda \in \mathrm{pcf}(\mathsf{b})$.

Proof Let a, κ, $(\mathsf{b}_\xi : \xi < \kappa)$, and λ satisfy the assumptions of the theorem. Clearly a is infinite since $\lambda \in \mathrm{pcf}(\mathsf{a}) \setminus \mathsf{a}$.

Claim 1 We can assume without loss of generality that

a) $\max\{|\mathsf{a}|^+, \aleph_2\} < \min(\mathsf{a})$.
b) $\lambda = \max \mathrm{pcf}(\mathsf{a})$.
c) a has no maximum.
d) $\mathsf{b}_\xi \neq \emptyset$ for all $\xi < \kappa$.

Proof Put $\mathsf{a}' := \mathsf{a} \setminus \{|\mathsf{a}|, |\mathsf{a}|^+, \aleph_2\}$. Then a' is progressive, and it is obvious that $\lambda \in \mathrm{pcf}(\mathsf{a}')$ and thus $J_{<\lambda^+}(\mathsf{a}') \setminus J_{<\lambda}(\mathsf{a}') \neq \emptyset$. We choose a set $\mathsf{a}'' \subseteq \mathsf{a}'$ such that $\mathsf{a}'' \in J_{<\lambda^+}(\mathsf{a}') \setminus J_{<\lambda}(\mathsf{a}')$. From Lemma 3.4.19 and $\max \mathrm{pcf}(\mathsf{a}'') < \lambda^+$ we get $\lambda = \max \mathrm{pcf}(\mathsf{a}'')$. Let $\rho := \min\{\nu \in \mathrm{ICN} : |\mathsf{a}'' \setminus \nu| < \aleph_0\}$. Then we have $\sup(\mathsf{a}'' \cap \rho) = \rho$. Since $\mathrm{pcf}(\mathsf{a}'') = \mathrm{pcf}(\mathsf{a}'' \cap \rho) \cup \mathrm{pcf}(\mathsf{a}'' \setminus \rho) = \mathrm{pcf}(\mathsf{a}'' \cap \rho) \cup (\mathsf{a}'' \setminus \rho)$

and $\lambda \in \mathrm{pcf}(\mathsf{a}'') \setminus \mathsf{a}''$, we get $\lambda = \max \mathrm{pcf}(\mathsf{a}'' \cap \rho)$, and $\mathsf{a}'' \cap \rho$ has no maximum. It is easy to see that it suffices to prove the theorem for $\mathsf{a}'' \cap \rho$ and $(\mathsf{b}'_\xi : \xi < \kappa)$, where $\mathsf{b}'_\xi := \mathsf{b}_\xi \cap \mathsf{a}'' \cap \rho$. Since the sequence $(\mathsf{b}_\xi : \xi < \kappa)$ is increasing, this holds also for the sequence $(\mathsf{b}'_\xi : \xi < \kappa)$, and by assumption there is $\xi_0 < \kappa$ such that $\mathsf{b}'_\xi \neq \emptyset$ for all ξ with $\xi_0 \leq \xi < \kappa$. This proves Claim 1.

Next we put $\mu := \max\{|\mathsf{a}|^+, \aleph_2\}$. Then $\mu \in (|\mathsf{a}|, \min(\mathsf{a}))_{\mathrm{reg}}$. For $S := \sum_{\mu,\aleph_0}$ we further assume that $(E_\beta : \beta \in S)$ is a $\diamondsuit_{\mathrm{Club}}$-sequence for S according to Lemma 4.5.3, and, according to Lemma 6.1.1, $\bar{f} = (\bar{f}^{\mathsf{b}} : \mathsf{b} \subseteq \mathsf{a})$ is a special sequence for a and all cardinals $\rho \in (|\mathsf{a}|, \min(\mathsf{a}))_{\mathrm{reg}}$. We choose a regular cardinal $\Theta > 2^{2^{\sup(\mathsf{a})}}$ and an approximation sequence $(N_i : i \leq \mu)$ in $\mathrm{H}(\Theta)$ such that

$$\mathsf{a} \cup \{\mathsf{a}, \lambda, (\mathsf{b}_\xi : \xi < \kappa), (E_\beta : \beta \in S), \bar{f}\} \subseteq N_0$$

and $|N_i| \leq \mu$ for all $i < \mu$. Let $N := N_\mu$. From Lemma 4.3.3 we get $\mu \subseteq N$, and thus, using $|\mathsf{a}| < \mu$, we can assume without loss of generality that $|\mathsf{a}| \subseteq N_0$. Example 4.2.15 yields $|\mathsf{a}|^+ \in N$. Clearly $\aleph_2 \in \mathrm{H}(\Theta)$, hence $\aleph_2 \in N$ and thus $\mu \in N$. The supposition $|\mathsf{a}| < \kappa$ leads, together with the fact that κ is regular, to $\mathsf{a} \subseteq \bigcup\{\mathsf{b}_\xi : \xi < \sigma\}$ for some $\sigma < \kappa$. Then $\mathsf{a} \subseteq \mathsf{b}_\sigma$, which contradicts $\lambda \in \mathrm{pcf}(\mathsf{a}) \setminus \mathrm{pcf}(\mathsf{b}_\sigma)$. Therefore $\kappa \leq |\mathsf{a}|$, and we also get $\kappa + 1 \subseteq N_0$.

Claim 2 There is a sequence $h := (((\mathsf{c}_{\xi,k}, \lambda_{\xi,k}) : k \leq n(\xi)) : \xi < \kappa)$ such that

(i) For all $\xi < \kappa$

(a) $n(\xi) < \omega$ and $\mathsf{b}_\xi = \mathsf{c}_{\xi,0} \supseteq \ldots \supseteq \mathsf{c}_{\xi,n(\xi)} = \emptyset$.

(b) If $k < n(\xi)$, then $\mathsf{c}_{\xi,k+1} \in J_{<\lambda_{\xi,k}}(\mathsf{a})$, where $\lambda_{\xi,k} = \max \mathrm{pcf}(\mathsf{c}_{\xi,k})$.

(c) If $k < n(\xi)$ and $\theta \in \mathsf{c}_{\xi,k} \setminus \mathsf{c}_{\xi,k+1}$, then $\chi_N(\theta) = f^{\mathsf{c}_{\xi,k}}_{\chi_N(\lambda_{\xi,k})}(\theta)$.

(ii) $h \in N$.

Proof By an iterated application of Lemma 6.1.3 it is easy to find finite sequences with the properties in (i). The crucial point is the proof of (ii).

First we define by recursion for each $n < \omega$ sets $T_n \subseteq {}^{2n}\mu$ of strictly increasing sequences, and, for each $\eta \in T_n$ and $\xi < \kappa$, sets $\mathsf{c}^\eta_\xi \subseteq \mathsf{b}_\xi$ and clubs C^η_ξ and C^η in μ as follows.

Put $T_0 := \{\emptyset\}$ and $\mathsf{c}^\emptyset_\xi := \mathsf{b}_\xi$ for all $\xi < \kappa$. Note that $\mathsf{c}^\emptyset_\xi \in N$ for all $\xi < \kappa$. Fix $n < \omega$, and assume that T_n and $\mathsf{c}^\eta_\xi \in N$ are defined for all $\eta \in T_n$ and all $\xi < \kappa$. Now consider $\xi < \kappa$.

If $\eta \in T_n$, we can apply Lemma 6.1.3 to c^η_ξ and $(N_i : i \leq \mu)$ instead of b and $(N_i : i \leq \mu)$ and obtain clubs C^η_ξ in μ which have that properties with respect to $\mathfrak{c}_\xi^\eta$ which are satisfied by the club C in the lemma with respect to $\mathfrak{b}$. Then the set $C^\eta := \bigcap_{\xi<\kappa} C_\xi^\eta$ is also a club in μ, since $\kappa \le |\mathfrak{a}| < \mu$. Now we put

$$T_{n+1} := \{\eta^\frown(i,j) : \eta \in T_n \wedge i,j \in C^\eta \wedge \eta(2n-1) < i < j\},$$

and get from the lemma, for any $\eta^\frown(i,j) \in T_{n+1}$ and $\xi < \kappa$, that

$$\mathfrak{c}_\xi^{\eta^\frown(i,j)} := \mathrm{B}[f^{\mathfrak{c}_\xi^\eta}_{\chi_{N_j}(\lambda_{\xi,\eta})}, \chi_{N_i}] \cap \mathfrak{c}_\xi^\eta \;\in\; N \cap J_{<\lambda_{\xi,\eta}}(\mathfrak{c}_\xi^\eta),$$

where $\lambda_{\xi,\eta}$ is the maximum of $\mathrm{pcf}(\mathfrak{c}_\xi^\eta)$. Furthermore, for any $\theta \in \mathfrak{c}_\xi^\eta \setminus \mathfrak{c}_\xi^{\eta^\frown(i,j)}$, we have $\chi_N(\theta) = f^{\mathfrak{c}_\xi^\eta}_{\chi_N(\lambda_{\xi,\eta})}(\theta)$. In addition we obtain from Lemma 6.1.3 and Lemma 4.2.14 a formula φ such that $\mathfrak{c}_\xi^{\eta^\frown(i,j)}$ is the unique member u of $\mathrm{H}(\Theta)$ satisfying $\mathrm{H}(\Theta) \models \varphi(u, \mathfrak{c}_\xi^\eta, \bar{f}, N_i, N_j)$. This completes the definition of the desired sets.

Next we want to find a club F in μ such that $\eta \restriction 2n \in T_n$ for all strictly increasing functions $\eta : \omega \to F$ and all $n < \omega$.

Consider $n < \omega$ and $\alpha < \mu$. Then $|T_n \cap {}^{2n}\alpha| \le |\alpha|^{2n} < \mu$, hence $F_{n,\alpha} := \bigcap_{\eta \in T_n \cap {}^{2n}\alpha} C^\eta$ is a club in μ. For each $n \in \omega$ we put

$$F_n := \Delta_{\alpha<\mu}(F_{n,\alpha} \cap \{\delta < \mu : \delta \in \mathrm{Lim}\}).$$

Now, if $\eta \in T_n$ and $i,j \in F_n$ satisfy $\beta := \eta(2n-1) < i < j$, then $\eta^\frown(i,j) \in T_{n+1}$. To see this, we first observe that $i,j \in \mathrm{Lim}$, $\eta \in T_n \cap {}^{2n}(\beta+1)$, and $\beta+1 < i < j$. So the definition of F_n yields $i,j \in F_{n,\beta+1}$; the definition of $F_{n,\beta+1}$, together with $\eta \in T_n \cap {}^{2n}(\beta+1)$, gives $i,j \in C^\eta$. Thus, by definition, $\eta^\frown(i,j) \in T_{n+1}$. By induction it is easy to see that $F := \bigcap_{n<\omega} F_n$ satisfies our claims.

$(E_\beta : \beta \in S)$ is a $\diamondsuit_{\text{Club}}$-sequence for S, and thus the set $\{\beta \in S : E_\beta \subseteq F\}$ is stationary in μ. Consequently, there is $\alpha \in S$ such that $E_\alpha \subseteq F$. Since $\mu \subseteq N$, we have $\alpha \in N$ and thus $E_\alpha \in N$. Furthermore E_α has order type ω, and the unique order isomorphism q from ω onto E_α is a member of $\mathrm{H}(\Theta)$, hence of N.

From $\mathrm{ran}(q) = E_\alpha \subseteq F$ and the fact that q is strictly increasing we can now infer that $q \restriction 2n \in T_n$ for all $n \in \omega$. This enables us to define the desired set $\mathfrak{c}_{\xi,k}$. We put $\mathfrak{c}_{\xi,k} := \mathfrak{c}_\xi^{q\restriction 2k}$ and $\lambda_{\xi,k} := \max \mathrm{pcf}(\mathfrak{c}_{\xi,k})$. Then by construction, for ξ fixed, $\mathfrak{c}_{\xi,k+1}$ is the unique member u of $\mathrm{H}(\Theta)$ such that $\mathrm{H}(\Theta) \models \varphi(u, \mathfrak{c}_{\xi,k}, \bar{f}, N_{q(2k)}, N_{q(2k+1)})$. Furthermore, Lemma 3.4.20 together with $\mathfrak{c}_{\xi,k} \neq \emptyset$ gives $\lambda_{\xi,k+1} < \lambda_{\xi,k}$, using $\mathfrak{c}_\xi^{q\restriction 2(k+1)} \in J_{<\lambda_{\xi,q\restriction 2k}}(\mathfrak{c}_\xi^{q\restriction 2k})$. Consequently, there must be $k \in \omega$ such that $\mathfrak{c}_{\xi,k} = \emptyset$, and we can define the natural number $n(\xi) := \min\{k \in \omega : \mathfrak{c}_{\xi,k} = \emptyset\}$. The construction shows that, for each $\xi < \kappa$, the sequence $(\mathfrak{c}_{\xi,k} : k \le n(\xi))$ satisfies (i).

Since $(N_i : i \leq \alpha) \in N$ and $E_\alpha \subseteq \alpha$, we get $g := (N_{q(n)} : n < \omega) \in N$. With the above property of the formula φ we can conclude that the sequence $h_1 := ((\mathsf{c}_{\xi,n} : n < \omega) : \xi < \kappa)$ is a member of N. Namely, h_1 is the unique member of $\mathrm{H}(\Theta)$ such that

$$\mathrm{H}(\Theta) \models \mathrm{Func}(h_1) \wedge \mathrm{dom}(h_1) = \kappa \wedge \forall \xi < \kappa(\mathrm{Func}(h_1(\xi)) \wedge \mathrm{dom}(h_1(\xi)) = \omega$$

$$\wedge h_1(\xi)(0) = \mathsf{b}_\xi \wedge \forall n \in \omega(\varphi(h_1(\xi)(n+1), h_1(\xi)(n), \bar{f}, g(2n), g(2n+1)))),$$

and thus $h_1 \in N$ by Lemma 4.2.13 since the parameters of the defining formula are members of N. It is easy to see that the function $h_2 := ((\mathsf{c}_{\xi,k} : k \leq n_\xi) : \xi < \kappa)$ is definable in $\mathrm{H}(\Theta)$ with the parameter h_1, hence $h_2 \in N$.

From Example 4.2.15, together with the assumption on Θ, we know that

$$x \subseteq \mathsf{a} \Longrightarrow \mathrm{pcf}(x) \in \mathrm{H}(\Theta) \wedge \forall y \in \mathrm{H}(\Theta)(y = \mathrm{pcf}(x) \Longleftrightarrow \mathrm{H}(\Theta) \models y = \mathrm{pcf}(x)).$$

So we can conclude for $h_3 := ((\lambda_{\xi,k} : k \leq n_\xi) : \xi < \kappa)$, where $\max \mathrm{pcf}(\emptyset) := 0$, that

$$h_3 = \{(\xi, x) : \mathrm{H}(\Theta) \models \xi < \kappa \wedge \mathrm{Func}(x) \wedge \mathrm{dom}(x) = \mathrm{dom}(h_2(\xi)) \wedge$$

$$\forall k \in \mathrm{dom}(x)(x(k) = \max \mathrm{pcf}(h_2(\xi)(k)))\}.$$

This gives $h_3 \in N$. Now it is easy to define the function h in $\mathrm{H}(\Theta)$ with the parameters $h_2, h_3 \in N$, which proves (ii) and completes the proof of Claim 2.

$h_3 \in N$ yields immediately that $(\mathsf{d}_\xi : \xi < \kappa) \in N$, where $\mathsf{d}_\xi := \{\lambda_{\xi,k} : k < n(\xi)\}$ for all $\xi < \kappa$.

Claim 3 If $A \subseteq \kappa$ is unbounded in κ, then $\lambda = \max \mathrm{pcf}(\bigcup_{\xi \in A} \mathsf{d}_\xi)$.

Proof To get a contradiction, we assume that there is an unbounded subset A of κ such that $\max \mathrm{pcf}(\bigcup_{\xi \in A} \mathsf{d}_\xi) < \lambda$. Let $\mathsf{d}' := \bigcup_{\xi \in A} \mathsf{d}_\xi$. We have $\lambda_{\xi,k} \in \mathrm{pcf}(\mathsf{b}_\xi) \subseteq \mathrm{pcf}(\mathsf{a})$, hence $\kappa^+ < \lambda_{\xi,k}$ for all ξ and k. So we get $\mathsf{d}' \subseteq \mathrm{pcf}(\mathsf{a})$ and $|\mathsf{d}'|^+ < \min(\mathsf{d}')$. Now Corollary 7.1.19 says that $\mathrm{pcf}(\mathsf{d}') \subseteq \mathrm{pcf}(\mathsf{a})$.

Since $(\mathsf{d}_\xi : \xi < \kappa) \in N$, we can assume in addition that A is a member of N, hence $\mathsf{d}' \in N$. We want to choose a special sequence $\bar{g} = (\bar{g}^{\mathsf{e}} : \mathsf{e} \subseteq \mathsf{d})$ for appropriate d and μ. Of course we want to ensure the existence of such a sequence which in addition is a member of N. A natural candidate for d is d', but unfortunately we do not know that d' is infinite. So let $\mathsf{d} := \mathsf{d}'$ if d' is infinite. If $|\mathsf{d}'| < \aleph_0$ and $\kappa > \aleph_0$, then there exists $\xi \in A$ such that $\aleph_0 \leq |\mathsf{b}_\xi|$. Let $\mathsf{d} := \mathsf{d}' \cup \mathsf{b}_\xi$. Then we have $|\mathsf{d}| = |\mathsf{b}_\xi| < |a|^+ < \min(\mathsf{a}) \leq \min(\mathsf{d})$, hence d is an infinite, progressive set satisfying $\max \mathrm{pcf}(\mathsf{d}) < \lambda$. Furthermore, $\mathsf{b}_\xi \in N$ yields $\mathsf{d} \in N$. If $|\mathsf{d}'| < \aleph_0$, $\kappa = \aleph_0$, and at least one b_ζ is infinite for some $\zeta \in A$, then the same procedure works. Finally we consider the case that $|\mathsf{d}'| < \aleph_0$, $\kappa = \aleph_0$, and b_ξ is finite for all $\xi \in A$. Fix $\zeta \in A$. Then $\mathrm{pcf}(\mathsf{b}_\zeta) = \mathsf{b}_\zeta$ and thus

$\lambda_{\zeta,0} = \max(\mathsf{b}_\zeta)$. Since $\mathsf{a} = \bigcup_{\xi \in A} \mathsf{b}_\xi$ and a has no maximum, there exists $\xi \in A$ such that $\zeta < \xi$ and $\max(\mathsf{b}_\zeta) < \max(\mathsf{b}_\xi)$. From this we get $|\mathsf{d}'| \geq \aleph_0$, and we can can infer that this case cannot occur.

Next we choose a special sequence $\bar{g} = (\bar{g}^{\mathsf{e}} : \mathsf{e} \subseteq \mathsf{d})$ for d and μ. Lemma 6.1.2 says that all such sequences are members of $\mathrm{H}(\Theta)$ if only $\Theta > 2^{2^{\mathrm{sup}(\mathsf{a})}}$. So we know why we have chosen Θ so large. The previous considerations show, together with $\mathsf{d} \in N$ and $\mu \in N$, that we can assume without loss of generality that $\bar{g} \in N$. Again we obtain with Lemma 6.1.3 a sequence $((\mathsf{e}_l, \rho_l) : l \leq m) \in N$, such that $\mathsf{d} = \mathsf{e}_0 \supseteq \ldots \supseteq \mathsf{e}_m = \emptyset$, $\rho_l = \max \mathrm{pcf}(\mathsf{e}_l)$, and $\chi_N(\rho) = g^{\mathsf{e}_l}_{\chi_N(\rho_l)}(\rho)$ for any $\rho \in \mathsf{d}$ and $l < m$ with $\rho \in \mathsf{e}_l \setminus \mathsf{e}_{l+1}$.

Consider $\theta \in \mathsf{a} \subseteq N$. Then $\xi := \min\{\zeta \in A : \theta \in \mathsf{b}_\zeta\} \in N$. Furthermore we have $k := \min\{k < n(\xi) : \theta \notin \mathsf{c}_{\xi,k+1}\} \in N$, and we can conclude from Claim 2 that

$$\chi_N(\theta) = f^{\mathsf{c}_{\xi,k}}_{\chi_N(\lambda_{\xi,k})}(\theta).$$

Since $\lambda_{\xi,k} \in \mathsf{d} \cap N$ and thus $l := \min\{j < m : \lambda_{\xi,k} \notin \mathsf{e}_{j+1}\} \in N$, we infer that

$$\chi_N(\lambda_{\xi,k}) = g^{\mathsf{e}_l}_{\chi_N(\rho_l)}(\lambda_{\xi,k}).$$

These considerations show that the function $\chi_N \restriction \mathsf{a}$ is definable in $\mathrm{H}(\Theta)$ with the parameters a, $\bar{f}$, A, $(((\mathsf{c}_{\xi,k}, \lambda_{\xi,k}) : k \leq n(\xi)) : \xi < \kappa)$, $\bar{g}$, $(\mathsf{e}_l : l \leq m)$, and $(\chi_N(\rho_l) : l \leq m)$. Therefore, putting

$$N^+ := \mathrm{Skol}_\Theta(N \cup \{\chi_N(\rho_l) : l \leq m\}),$$

we can conclude from Lemma 4.2.14 and Lemma 4.4.2 that $\chi_N \restriction \mathsf{a} \in N^+$. Now we have $\rho_l \in N$, and our supposition gives

$$\rho_l = \max \mathrm{pcf}(\mathsf{e}_l) \leq \max \mathrm{pcf}(\mathsf{d}) < \lambda$$

for all $l \leq m$. So we get $\chi_N(\rho_l) < \chi_N(\lambda)$, and Corollary 4.4.4 tells us that $\chi_N(\lambda) = \chi_{N^+}(\lambda)$. On the other hand, we have $\chi_N \restriction \mathsf{a} \in N^+$ and, using $|N| < \min(\mathsf{a})$, $\chi_N \restriction \mathsf{a} \in \prod \mathsf{a}$. Therefore there is $\alpha < \lambda$ such that $\alpha \in N^+$ and $\chi_N \restriction \mathsf{a} <_{\mathsf{a}} f^{\mathsf{a}}_\alpha$. Then $\alpha \in N^+ \cap \lambda$ gives $\alpha < \chi_{N^+}(\lambda) = \chi_N(\lambda)$, and from Lemma 6.1.3 we can infer that

$$\chi_N \restriction \mathsf{a} <_{\mathsf{a}} f^{\mathsf{a}}_\alpha <_{\mathcal{J}_{<\lambda}(\mathsf{a})} f^{\mathsf{a}}_{\chi_N(\lambda)} \leq_{\mathsf{a}} \chi_N \restriction \mathsf{a}.$$

This contradiction completes the proof of Claim 3.

We can already complete the proof of the assertion of the theorem for countable κ: If we choose $\mathsf{b} := \bigcup\{\mathsf{d}_\xi : \xi < \kappa\}$, then $\mathsf{b} \subseteq \bigcup\{\mathrm{pcf}(\mathsf{b}_\xi) : \xi < \kappa\}$ and, by Claim 3, $\lambda = \max \mathrm{pcf}(\mathsf{b})$.

For $\xi < \kappa$ and $n \geq n_\xi$, let $\lambda_{\xi,n} := \lambda_{\xi,n(\xi)-1}$. Obviously the proof of the theorem will be finished if we can show the following claim.

Claim 4 There is $n^* < \omega$ and an unbounded subset A of κ such that $\lambda = \operatorname{tcf}(\prod_{\xi \in A} \lambda_{\xi,n^*}/\mathrm{I_b}(A))$.

Proof Since $\kappa^+ < \lambda_{\xi,n}$ for all $\xi < \kappa$ and $n \in \omega$, the function $(\lambda_{\xi,n} : \xi < \kappa)$ is progressive. Therefore it suffices to prove, by Claim 3 and Lemma 9.2.3, that there is $n^* \in \omega$ and an unbounded subset A of κ such that $\lambda \in \operatorname{pcf}\{\lambda_{\xi,n^*} : \xi \in B\}$ for all $B \in \mathrm{I_b}(A)^+$. Let us assume, contrary to our hopes, that this assertion is false.

First we obtain with $\kappa > \aleph_0$ an $n' \in \omega \setminus \{0\}$ such that $A_0 := \{\xi < \kappa : n(\xi) = n'\}$ is unbounded in κ. If the set A_k is already defined and unbounded in κ, we can choose, by the assumption of the indirect proof, a set $A_{k+1} \subseteq A_k$ which is unbounded in κ and satisfies $\lambda \notin \operatorname{pcf}\{\lambda_{\xi,k} : \xi \in A_{k+1}\}$. With $A_{n'} \subseteq A_0$ and by the choice of n' we can conclude that

$$\bigcup_{\xi \in A_{n'}} \mathsf{d}_\xi = \bigcup_{k<n'} \{\lambda_{\xi,k} : \xi \in A_{n'}\} \subseteq \bigcup_{k<n'} \{\lambda_{\xi,k} : \xi \in A_{k+1}\}.$$

This inclusion also holds for the sets of possible cofinalities of these sets. But Claim 3 says that $\lambda \in \operatorname{pcf}(\bigcup_{\xi \in A_{n'}} \mathsf{d}_\xi)$, and so we can infer from $\operatorname{pcf}(a_1 \cup a_2) = \operatorname{pcf}(a_1) \cup \operatorname{pcf}(a_2)$ that there is $k < n'$ such that $\lambda \in \operatorname{pcf}(\{\lambda_{\xi,k} : \xi \in A_{k+1}\})$. This contradicts the choice of A_{k+1}.

We can slightly improve the previous result.

Lemma 9.2.5 *Assume that* a *is a progressive set of regular cardinals,* $\kappa > \aleph_0$ *is a regular cardinal, and* $(\mathsf{b}_\xi : \xi < \kappa)$ *is a sequence of subsets of* a *which is increasing under inclusion, such that* $\mathsf{a} = \bigcup\{\mathsf{b}_\xi : \xi < \kappa\}$ *and* $\lambda \in \operatorname{pcf}(\mathsf{a}) \setminus \bigcup_{\xi<\kappa} \operatorname{pcf}(\mathsf{b}_\xi)$. *Then there is a club* C^* *in* κ *and a sequence* $(\mu_\xi : \xi \in C^*)$ *of regular cardinals such that* $\mu_\xi \in \operatorname{pcf}(\mathsf{b}_\xi) \setminus \bigcup_{\zeta<\xi} \operatorname{pcf}(\mathsf{b}_\zeta)$ *and* $\lambda = \operatorname{tcf}(\prod_{\xi \in C^*} \mu_\xi/\mathrm{I_b}(C^*))$.

Proof With the previous theorem we obtain an unbounded subset A of κ and, without loss of generality, a progressive sequence $(\lambda_\xi : \xi < \kappa)$ of regular cardinals such that $\lambda_\xi \in \operatorname{pcf}(\mathsf{b}_\xi)$ and $\lambda_\xi > \kappa^+$ for all $\xi < \kappa$ and $\lambda = \operatorname{tcf}(\prod_{\xi \in A} \lambda_\xi/\mathrm{I_b}(A))$. We will use Lemma 3.3.1 without mentioning. Let

$$C := \{\xi < \kappa : \sup(A \cap \xi) = \xi\}$$

be the union of $\{0\}$ and the set of limit points of A. We know that C is a club in κ.

Fix $\xi \in C$. Extending the ultrafilter on $A \cap \xi$ which is generated by $\{\xi \setminus \alpha : \alpha < \xi\}$, we obtain an ultrafilter D_ξ on A such that $A \cap \xi \in D_\xi$ and $\sup(X \cap \xi) = \xi$ for all $X \in D_\xi$. In particular we have $D_\xi \cap \mathrm{I_b}(A \cap \xi) = \emptyset$. The cardinal

$$\mu_\xi := \operatorname{cf}(\prod_{\zeta \in A} \lambda_\zeta/D_\xi) = \operatorname{cf}(\prod_{\zeta \in A \cap \xi} \lambda_\zeta/D_\xi \cap \mathcal{P}(A \cap \xi))$$

is a member of $\operatorname{pcf}\{\lambda_\zeta : \zeta \in A \cap \xi\}$. Since $\{\lambda_\zeta : \zeta \in A \cap \xi\}$ is progressive, we can infer from Corollary 7.1.21 that $\operatorname{pcf}\{\lambda_\zeta : \zeta \in A \cap \xi\} \subseteq \operatorname{pcf}(\mathsf{b}_\xi)$. So we get $\mu_\xi \in \operatorname{pcf}(\mathsf{b}_\xi)$.

Now we show that $\operatorname{tcf}(\prod_{\xi\in C} \mu_\xi / \mathrm{I}_\mathrm{b}(C)) = \lambda$. It is natural to apply Lemma 3.3.7. For this, put $f := (\lambda_\xi : \xi \in A)$, $g := (\mu_\xi : \xi \in C)$, and let, for $\xi \in C$, I_ξ be the dual ideal of D_ξ and

$$I^* := \{X \subseteq A : \{\xi \in C : X \notin I_\xi\} \in \mathrm{I}_\mathrm{b}(C)\}.$$

Then $|C| < \min(f[A])$, and we know that $\operatorname{tcf}(\prod f/I_\xi) = \operatorname{cf}(\prod f/D_\xi) = \mu_\xi = g(\xi)$ for every $\xi \in C$. From Lemma 3.3.7 we can conclude that $\operatorname{tcf}(\prod g/\mathrm{I}_\mathrm{b}(C)) = \operatorname{tcf}(\prod f/I^*)$. If we can prove that $\mathrm{I}_\mathrm{b}(A) \subseteq I^*$, then we get

$$\lambda = \operatorname{tcf}(\prod f/\mathrm{I}_\mathrm{b}(A)) = \operatorname{tcf}(\prod f/I^*) = \operatorname{tcf}(\prod g/\mathrm{I}_\mathrm{b}(C)) = \operatorname{tcf}(\prod_{\xi\in C} \mu_\xi/\mathrm{I}_\mathrm{b}(C)).$$

To show that $\mathrm{I}_\mathrm{b}(A) \subseteq I^*$, fix $X \subseteq A$ bounded in κ. Then $X \subseteq A \cap \zeta$ for some $\zeta \in C$. By the choice of the sequence $(D_\xi : \xi \in C)$, we have $X \notin D_\xi$ for all $\xi \in C$ with $\xi > \zeta$. Consequently, the set $\{\xi \in C : X \in D_\xi\}$ is a member of $\mathrm{I}_\mathrm{b}(C)$, hence $X \in I^*$. This completes the proof of $\operatorname{tcf}(\prod_{\xi\in C} \mu_\xi/\mathrm{I}_\mathrm{b}(C)) = \lambda$.

For all $\xi < \kappa$, the set $\{\zeta \in C : \mu_\zeta \in \operatorname{pcf}(\mathsf{b}_\xi)\}$ is bounded in C, since otherwise $\lambda \in \operatorname{pcf}\{\mu_\zeta : \zeta \in C \wedge \mu_\zeta \in \operatorname{pcf}(\mathsf{b}_\xi)\}$ and thus $\lambda \in \operatorname{pcf}(\mathsf{b}_\xi)$.

So we can choose, for each $\zeta < \kappa$, a club $C_\zeta \subseteq \{\xi \in C : \mu_\xi \notin \operatorname{pcf}(\mathsf{b}_\zeta)\}$, for example a final segment of C, and put $C^* := \triangle_{\zeta<\kappa} C_\zeta$. Now C^* has the desired properties: If $\xi \in C^*$, then $\mu_\xi \notin \operatorname{pcf}(\mathsf{b}_\zeta)$ for all $\zeta < \xi$ and $\mu_\xi \in \operatorname{pcf}(\mathsf{b}_\xi)$, as demonstrated above. Furthermore we have $C^* \subseteq C$, $\mathrm{I}_\mathrm{b}(C) \cap \mathcal{P}(C^*) = \mathrm{I}_\mathrm{b}(C^*)$, and $C^* \notin \mathrm{I}_\mathrm{b}(C)$, hence $\mathrm{I}_\mathrm{b}(C) \restriction C^*$ is an ideal on C. Now we can infer from Lemma 3.1.8 that

$$\operatorname{tcf}(\prod_{\xi\in C^*} \mu_\xi/\mathrm{I}_\mathrm{b}(C^*)) = \operatorname{tcf}(\prod_{\xi\in C} \mu_\xi/(\mathrm{I}_\mathrm{b}(C) \restriction C^*)) = \operatorname{tcf}(\prod_{\xi\in C} \mu_\xi/\mathrm{I}_\mathrm{b}(C)) = \lambda,$$

which proves the lemma.

Lemma 9.2.6 (Jech) *Assume that λ is a singular cardinal of uncountable cofinality κ and $(\lambda_\xi : \xi < \kappa)$ is a normal sequence of infinite cardinals which is cofinal in λ. Further let $I \supseteq \mathrm{I}_\mathrm{b}(\kappa)$ be a normal ideal on κ. If $f \in {}^\kappa\mathrm{Reg}$ satisfies $\lim_I(f) = \lambda$, $\Phi \in {}^\kappa\mathrm{ON}$ is a function such that $f(\xi) \leq \lambda_\xi^{+\Phi(\xi)}$ for all $\xi < \kappa$, and $\operatorname{tcf}(\prod f/I)$ exists, then $\operatorname{tcf}(\prod f/I) \leq \lambda^{+\|\Phi\|_I}$.*

Proof Let us assume that the assertion is false. The Lemmas 1.8.7 and 2.2.2 show that $<_I$ is a well-founded relation. Therefore we can choose a function $\Phi \in {}^\kappa\mathrm{ON}$ which is minimal under $<_I$ with respect to the property that there is a function $f \in {}^{\kappa}\mathrm{Reg}$ satisfying $f(\xi) \leq \lambda_{\xi}^{+\Phi(\xi)}$ for all $\xi < \kappa$, $\lim_I(f) = \lambda$, and $\mathrm{tcf}(\prod f/I) > \lambda^{+\|\Phi\|_I}$. Furthermore we can assume without loss of generality that $\lambda_\xi < f(\xi) = \lambda_{\xi}^{+\Phi(\xi)}$ and thus in particular $\Phi(\xi) \neq 0$ for all $\xi < \kappa$. Namely, Lemma 9.2.2 says that the set $B := \{\xi < \kappa : f(\xi) \leq \lambda_\xi\}$ is a member of I. Putting $f'(\xi) := \lambda_\xi^+$ for $\xi \in B$ and $f'(\xi) := f(\xi)$ for $\xi \in \kappa \setminus B$, and defining the function $\Phi' \in {}^{\kappa}\mathrm{ON}$ by $f'(\xi) =: \lambda_{\xi}^{+\Phi'(\xi)}$, we get $\Phi' \leq_I \Phi$. Consequently, Φ' is also minimal under $<_I$, and f' and Φ' have the desired properties.

Next we prove that $\|\Phi\|_I$ is a successor ordinal; in particular, $\lambda^{+\|\Phi\|_I}$ will be regular. Assume that this is not the case. Then the set $X := \{\xi < \kappa : \Phi(\xi)$ is a limit ordinal$\}$ is not a member of I, since otherwise, together with Lemma 2.2.3, the function $\Psi \in {}^{\kappa}\mathrm{ON}$, given by

$$\Phi(\xi) = \begin{cases} \Psi(\xi) & \text{if } \xi \in X, \\ \Psi(\xi) + 1 & \text{if } \xi \in \kappa \setminus X, \end{cases}$$

would satisfy $\Psi <_I \Phi$ and $\|\Psi\|_I + 1 = \|\Phi\|_I$, contradicting our assumption that $\|\Phi\|_I$ is not a successor ordinal. So we get $X \in I^+$.

Since, for any $\xi \in X$, the cardinal $f(\xi) = \lambda_{\xi}^{+\Phi(\xi)}$ is regular, we have $f(\xi) = \mathrm{cf}(\lambda_{\xi}^{+\Phi(\xi)}) \leq \mathrm{cf}(\Phi(\xi)) \leq \Phi(\xi)$. Any modulo $I \restriction X$ cofinal sequence of members of $\prod f$ is strictly increasing under $<_{I \restriction X}$, and thus Lemma 2.2.3 says that $\|\Phi\|_{I \restriction X} \geq \mathrm{tcf}(\prod f/I \restriction X)$. Together with the assumption that $\mathrm{tcf}(\prod f/I) > \lambda^{+\|\Phi\|_I}$ we can conclude that $\|\Phi\|_{I \restriction X} \geq \mathrm{tcf}(\prod f/I \restriction X) = \mathrm{tcf}(\prod f/I) > \lambda^{+\|\Phi\|_I} \geq \|\Phi\|_I$. Lemma 2.2.9 b) tells us that $\kappa \setminus X \in I^+$, and so we can infer from Lemma 2.2.9 c) that $\|\Phi\|_{I \restriction (\kappa \setminus X)} = \|\Phi\|_I$, since $\|\Phi\|_I = \|\Phi\|_{I \restriction (X \cup \kappa \setminus X)} = \min\{\|\Phi\|_{I \restriction X}, \|\Phi\|_{I \restriction (\kappa \setminus X)}\}$. However, the function $\Psi \in {}^{\kappa}\mathrm{ON}$ defined above satisfies $\|\Phi\|_{I \restriction (\kappa \setminus X)} = \|\Psi\|_{I \restriction (\kappa \setminus X)} + 1$, hence $\|\Phi\|_I = \|\Psi\|_{I \restriction (\kappa \setminus X)} + 1$. This contradicts our assumption, and we have shown that $\|\Phi\|_I$ is a successor ordinal.

By the main lemma of pcf-theory, Theorem 3.2.6, and by Theorem 3.1.13 we obtain a function $g \in {}^{\kappa}\mathrm{Reg}$ such that $g <_I f$, $\lim_I(g) = \lambda$, and $\mathrm{tcf}(\prod g/I) = \lambda^{+\|\Phi\|_I}$. Lemma 9.2.2 tells us that $\{\xi < \kappa : g(\xi) \leq \lambda_\xi\} \in I$. Putting

$$h(\xi) := \begin{cases} g(\xi) & \text{if } g(\xi) > \lambda_\xi, \\ \lambda_\xi^+ & \text{if } g(\xi) \leq \lambda_\xi, \end{cases}$$

we get $h =_I g$ and thus $h <_I f$. If the function $\Psi \in {}^{\kappa}\mathrm{ON}$ is given by $h(\xi) = \lambda_{\xi}^{+\Psi(\xi)}$, then clearly $\Psi <_I \Phi$. But this contradicts the choice of Φ, since also $\lim_I(h) = \lambda$ and $\mathrm{tcf}(\prod h/I) = \mathrm{tcf}(\prod g/I) = \lambda^{+\|\Phi\|_I} > \lambda^{+\|\Psi\|_I}$.

Definition *Assume that λ is a singular cardinal and κ is a cardinal with* $\mathrm{cf}(\lambda) \leq \kappa < \lambda$. *We say that λ is* **$\mathbf{pp}_\kappa$-strong** *iff* $\mathrm{pp}_\kappa(\nu) < \lambda$ *for all* $\nu < \lambda$ *with* $\mathrm{cf}(\nu) \leq \kappa < \nu$.

Lemma 9.2.7 *Assume that λ is a singular cardinal and κ is a cardinal such that* $\mathrm{cf}(\lambda) \leq \kappa < \lambda$, *and let λ be* pp_κ*-strong. If* $\mathsf{a} \subseteq \mathrm{Reg}$ *satisfies* $\sup(\mathsf{a}) = \lambda$ *and* $|\mathsf{a}| \leq \kappa < \min(\mathsf{a})$ *and D is an ultrafilter on* a *with* $\mathrm{cf}(\prod \mathsf{a}/D) > \lambda$, *then* $D \cap \mathrm{I_b}(\mathsf{a}) = \emptyset$ *and, in particular,* $\mathrm{cf}(\prod \mathsf{a}/D) \leq \mathrm{pp}_\kappa(\lambda)$.

Therefore we have

$$\mathrm{pp}_\kappa(\lambda) = \sup\{\max \mathrm{pcf}(\mathsf{a}) : \mathsf{a} \subseteq \mathrm{Reg} \wedge |\mathsf{a}| \leq \kappa < \min(\mathsf{a}) \wedge \sup(\mathsf{a}) = \lambda\}.$$

Proof Let λ, κ, a and D satisfy the assumptions of the lemma. Assume, to get a contradiction, that there is $\nu < \lambda$ such that $\mathsf{a} \cap \nu^+ \in D$, and put $\mu := \min\{\nu < \lambda : \mathsf{a} \cap \nu^+ \in D\}$. Then we have $\mu = \lim_D(\mathsf{a})$ and $\kappa < \mu$, and from Lemma 3.3.4 we can infer that μ is a limit cardinal and $\mathsf{a} \cap \mu \in D$. The choice of μ gives $\mu = \sup(\mathsf{a} \cap \mu)$ and thus $\mathrm{cf}(\mu) \leq |\mathsf{a} \cap \mu| \leq |\mathsf{a}| \leq \kappa < \mu$. Furthermore, $D' := \{X \subseteq \mathsf{a} \cap \mu : X \in D\}$ is an ultrafilter on $\mathsf{a} \cap \mu$ which satisfies $D' \cap \mathrm{I_b}(\mathsf{a} \cap \mu) = \emptyset$ by the choice of μ. So we conclude from our supposition that $\mathrm{cf}(\prod \mathsf{a}/D) = \mathrm{cf}(\prod(\mathsf{a} \cap \mu)/D') \leq \mathrm{pp}_\kappa(\mu) < \lambda$, and this contradicts the assumption. This yields $D \cap \mathrm{I_b}(\mathsf{a}) = \emptyset$.

Lemma 9.2.8 *Assume that λ is a* $\mathrm{pp}_{\mathrm{cf}(\lambda)}$*-strong singular cardinal and* $\kappa := \mathrm{cf}(\lambda) > \aleph_0$. *Then there is a normal sequence* $(\lambda_\xi : \xi < \kappa)$ *of singular cardinals which is cofinal in λ such that* $\lambda_0 > \kappa$ *and* $\mathrm{pp}_\kappa(\nu) < \lambda_\xi$ *for all ξ with* $0 < \xi < \kappa$ *and all cardinals* $\nu < \lambda_\xi$ *satisfying* $\mathrm{cf}(\nu) \leq \kappa < \nu$.

Proof First we show that, for any cardinal $\lambda' < \lambda$,

$$p(\lambda') := \sup\{\mathrm{pp}_\kappa(\nu) : \nu \in \mathrm{ICN} \wedge \mathrm{cf}(\nu) \leq \kappa < \nu < \lambda'\} < \lambda.$$

Without loss of generality we can assume that $p(\lambda') > \lambda'$. If the above assertion is false, then there is a cardinal $\nu < \lambda'$ such that $\mathrm{cf}(\nu) \leq \kappa < \nu$ and $\lambda' < \mathrm{pp}_\kappa(\nu)$. Choose ν minimal with respect to this property. Then we can conclude that $\mathrm{pp}_\kappa(\nu') \leq \mathrm{pp}_\kappa(\nu)$ for all $\nu' < \lambda'$ with $\mathrm{cf}(\nu') \leq \kappa < \nu'$: If $\nu' < \nu$, then $\mathrm{pp}_\kappa(\nu') \leq \lambda'$ by the choice of ν. If $\nu < \nu'$, then $\nu < \nu' < \lambda' < \mathrm{pp}_\kappa(\nu)$, and Lemma 9.2.1 gives $\mathrm{pp}_\kappa(\nu') \leq \mathrm{pp}_\kappa(\nu)$. So we get $p(\lambda') = \mathrm{pp}_\kappa(\nu)$, and since λ is pp_κ-strong, we conclude that $p(\lambda') = \mathrm{pp}_\kappa(\nu) < \lambda$, contradicting our supposition.

Now we can define by recursion the desired normal sequence $(\lambda_\xi : \xi < \kappa)$. Let $h : \mathrm{cf}(\lambda) \longrightarrow \lambda$ be a normal function such that $\sup(\mathrm{ran}(h)) = \lambda$. Choose λ_0 singular such that $\kappa < \lambda_0 < \lambda$ and put, inevitably, $\lambda_\xi := \sup\{\lambda_\zeta : \zeta < \xi\}$ for limit ordinals ξ. For the successor step, define $\lambda_{\xi,0} := \max\{\lambda_\xi, h(\xi)\}$, $\lambda_{\xi,n+1} := \max\{\lambda_{\xi,n}^+, p(\lambda_{\xi,n})\}$ for each $n \in \omega$, and $\lambda_{\xi+1} := \sup_{n<\omega} \lambda_{\xi,n}$. Then $\lambda_{\xi+1} < \lambda$, since $\mathrm{cf}(\lambda) > \aleph_0$. If $\mathrm{cf}(\nu) \leq \kappa < \nu$ and $\nu < \lambda_{\xi+1}$, that means $\nu < \lambda_{\xi,n}$ for some $n \in \omega$, then, by definition, $\mathrm{pp}_\kappa(\nu) \leq p(\lambda_{\xi,n}) \leq \lambda_{\xi,n+1} < \lambda_{\xi+1}$.

Lemma 9.2.9 *Assume that λ is a singular cardinal of uncountable cofinality κ which is pp_κ-strong. Further let $(\lambda_\xi : \xi < \kappa)$ be a normal sequence of singular cardinals which is cofinal in λ. If $\mu \in (\lambda, \mathrm{pp}^+(\lambda))_{\mathrm{CN}}$ is regular, then there is a club C in κ and a function $f \in {}^C(\mathrm{Reg} \cap \lambda)$, such that $\mathrm{tcf}\,(\prod f/\mathrm{I_b}(C)) = \mu$ and $\lambda_\xi < f(\xi) < \mathrm{pp}^+_\kappa(\lambda_\xi)$ for all $\xi \in C$.*

Proof Without loss of generality let $\lambda_\xi > \kappa$ for all $\xi < \kappa$. By Lemma 3.3.5, there is a set a of regular cardinals satisfying $\sup(\mathsf{a}) = \lambda$, $|\mathsf{a}| \leq \kappa < \min(\mathsf{a})$, and $\mu \in \mathrm{pcf}(\mathsf{a})$. The assumption on λ and Lemma 9.2.7 imply that we can assume without loss of generality that $\mu = \max \mathrm{pcf}(\mathsf{a})$. To see this, let $\mu = \mathrm{cf}\,(\prod \mathsf{a}/D)$ for some ultrafilter D on a. We can replace, if necessary, a by a generator b_μ of $\mathcal{J}_{<\mu^+}(\mathsf{a})$ over $\mathcal{J}_{<\mu}(\mathsf{a})$. Corollary 5.3.5 says that $\mathsf{b}_\mu \in D$, hence $\sup(\mathsf{b}_\mu) = \lambda$ since, by Lemma 9.2.7, $D \cap \mathrm{I_b}(\mathsf{a}) = \emptyset$, and from Lemma 5.3.1 we infer that $\max \mathrm{pcf}(\mathsf{b}_\mu) = \mu$. Furthermore we can assume without loss of generality that $\sup(\mathsf{a} \cap \lambda_\xi) = \lambda_\xi$, hence $\mathrm{cf}\,(\lambda_\xi) \leq \kappa < \lambda_\xi$, and $\mathrm{pp}_\kappa(\nu) < \lambda_\xi$ for all $\xi < \kappa$ and all $\nu < \lambda_\xi$ such that $\mathrm{cf}\,(\nu) \leq \kappa < \nu$, that means that λ_ξ is pp_κ-strong. Otherwise we could thin out the given sequence $h := (\lambda_\xi : \xi < \kappa)$ with the help of Lemma 9.2.8. If $(\nu_\xi : \xi < \kappa)$ is the sequence of singular cardinals yielded by this lemma, the set $E := \{\xi < \kappa : \lambda_\xi = \nu_\xi \wedge \sup(\mathsf{a} \cap \lambda_\xi) = \lambda_\xi\}$ is a club in κ, and we can replace h by the sequence $(\lambda_\xi : \xi \in E)$.

Now fix $\xi < \kappa$ and put $\mathsf{a}_\xi := \mathsf{a} \cap \lambda_\xi$. Note that $\lambda_\xi < \max \mathrm{pcf}(\mathsf{a}_\xi)$. From Lemma 9.2.7 we infer that $\max \mathrm{pcf}(\mathsf{a}_\xi) \leq \mathrm{pp}_\kappa(\lambda_\xi) < \lambda$ for all $\xi < \kappa$. Consequently we can apply Lemma 9.2.5 to obtain a club C_1 in κ and a function $f_1 := (\mu_\xi : \xi \in C_1)$ satisfying $\mu_\xi \in \mathrm{pcf}(\mathsf{a}_\xi) \setminus \bigcup\{\mathrm{pcf}(\mathsf{a}_\zeta) : \zeta < \xi\}$ for all $\xi \in C_1$ and $\mathrm{tcf}\,(\prod f_1/\mathrm{I_b}(C_1)) = \mu$.

Let $C := C_1 \cap \mathrm{Lim}$, $f := f_1 \restriction C$ and fix $\xi \in C$. First we note that $\mu_\xi \notin \bigcup\{\mathsf{a}_\zeta : \zeta < \xi\} = \mathsf{a}_\xi$. We claim that $\mu_\xi \geq \lambda_\xi$. Otherwise we could infer from $\sup(\mathsf{a}_\xi) = \lambda_\xi$ that $\mu_\xi < \sup(\mathsf{a}_\zeta) = \lambda_\zeta$ for some $\zeta < \xi$. From $\mu_\xi \notin \mathrm{pcf}(\mathsf{a}_\zeta)$ we can conclude that $\mu_\xi \in \mathrm{pcf}(\mathsf{a}_\xi \setminus \mathsf{a}_\zeta)$, hence $\lambda_\zeta \leq \min(\mathsf{a}_\xi \setminus \mathsf{a}_\zeta) \leq \mu_\xi$, contradicting $\mu_\xi < \lambda_\zeta$. Furthermore we have $\mu_\xi < \lambda$, since $\mu_\xi \in \mathrm{pcf}(\mathsf{a}_\xi)$ and $\max \mathrm{pcf}(\mathsf{a}_\xi) < \lambda$.

To apply Lemma 9.2.2, we extend the function f to κ, putting $f(\xi) := \lambda_\xi^+$ for all $\xi \in \kappa \setminus C$. From $\lambda_\xi \leq f(\xi)$ for all $\xi < \kappa$ and $\mathrm{I_b}(\kappa) \subseteq \mathrm{I_d}(\kappa)$ we get $\lim_{\mathrm{I_d}(\kappa)}(f) = \lambda$. Now Lemma 9.2.2 tells us that $\{\xi < \kappa : f(\xi) \leq \lambda_\xi\} \in \mathrm{I_d}(\kappa)$. So we can assume without loss of generality that $\lambda_\xi < f(\xi) = \mu_\xi$ for all $\xi \in C$. Next we claim that $\mu_\xi < \mathrm{pp}^+_\kappa(\lambda_\xi)$. Namely, there is an ultrafilter D on a_ξ such that $\mu_\xi = \mathrm{cf}\,(\prod \mathsf{a}_\xi/D) > \lambda_\xi$. Now Lemma 9.2.7, together with the fact that λ_ξ is pp_κ-strong, implies that $\mu_\xi = f(\xi) < \mathrm{pp}^+_\kappa(\lambda_\xi)$. Finally we can conclude from $C \notin \mathrm{I_b}(C_1)$ and $\mathrm{tcf}\,(\prod f_1/\mathrm{I_b}(C_1)) = \mu$ that $\mathrm{tcf}\,(\prod f/\mathrm{I_b}(C)) = \mu$. This completes the proof of the lemma.

Theorem 9.2.10 (Shelah) *Assume that $\aleph_0 < \kappa = \mathrm{cf}\,(\lambda) < \lambda$, λ is a pp_κ-strong cardinal, $(\lambda_\xi : \xi < \kappa)$ is a normal sequence of singular cardinals which is* *cofinal in* λ, *and* $\Phi \in {}^{\kappa}\mathrm{ON}$ *is a function such that* $\mathrm{pp}_{\kappa}(\lambda_\xi) = \lambda_\xi^{+\Phi(\xi)}$ *for all* $\xi < \kappa$. *Further let the norm of* Φ *be determined by the ideal* $\mathrm{I_d}(\kappa)$, *i.e., let* $\|\Phi\| := \|\Phi\|_{\mathrm{I_d}(\kappa)}$. *Then*

$$\mathrm{pp}(\lambda) \leq \lambda^{+\|\Phi\|}.$$

Proof Consider $\mu \in (\lambda, \mathrm{pp}^{+}(\lambda))_{\mathrm{reg}}$. It suffices to show that $\mu \leq \lambda^{+\|\Phi\|}$. By Lemma 9.2.9, we obtain a club C in κ and a function $f \in {}^{C}\mathrm{Reg}$ such that $\mathrm{tcf}(\prod f/\mathrm{I_b}(C)) = \mu$ and $\lambda_\xi < f(\xi) < \mathrm{pp}^{+}_{\kappa}(\lambda_\xi)$ for all $\xi \in C$. This gives $f(\xi) \leq \lambda_\xi^{+\Phi(\xi)}$ for all $\xi \in C$. To apply Lemma 9.2.6, we extend the function f to κ, putting $f(\xi) := \lambda_\xi^{+}$ for every $\xi \in \kappa \setminus C$. From Corollary 1.8.10 we can conclude that $I := \mathrm{I_d}(\kappa) \restriction C$ is a normal ideal on κ satisfying $\mathrm{I_b}(\kappa) \subseteq \mathrm{I_d}(\kappa) \subseteq I$. Obviously we have $\lim_I(f) = \lambda$, and thus Lemma 9.2.6 yields $\mathrm{tcf}(\prod f/I) \leq \lambda^{+\|\Phi\|_I}$ if $\mathrm{tcf}(\prod f/I)$ exists. But

$$\mathrm{tcf}\left(\prod f/I\right) = \mathrm{tcf}\left(\prod f \restriction C/\mathrm{I_d}(C)\right) = \mathrm{tcf}\left(\prod f \restriction C/\mathrm{I_b}(C)\right) = \mu$$

shows that $\mathrm{tcf}(\prod f/I)$ exists, and we can infer that $\mu \leq \lambda^{+\|\Phi\|_I}$.

Now the supposition $\|\Phi\| < \|\Phi\|_I$ gives $\mathrm{I_d}(\kappa) \neq \mathrm{I_d}(\kappa) \restriction C$, hence $\kappa \setminus C \notin \mathrm{I_d}(\kappa)$, contradicting the fact that C is a club in κ. Therefore we get $\|\Phi\| = \|\Phi\|_I$ and thus $\mu \leq \lambda^{+\|\Phi\|}$, as desired.

Bibliography

[Ba] H. Bachmann, *Transfinite Zahlen*, Springer-Verlag, Berlin, Heidelberg, New York 1967.

[BM] M. R. Burke, M. Magidor, *Shelah's pcf theory and its applications*, Ann. Pure Appl. Logic 50 (1990), 207–254.

[Dr] F. R. Drake, *Set Theory. An Introduction to Large Cardinals*, Studies in Logic and the Foundations of Mathematics 76, North-Holland, Amsterdam 1974.

[Ea] W. B. Easton, *Powers of regular cardinals*, Ann. Math. Logic 1 (1970), 139–178.

[En] H. B. Enderton, *Elements of Set Theory*, Academic Press, New York 1977.

[EHM] P. Erdős, A. Hajnal, A. Máté, R. Rado, *Combinatorial Set Theory: Partition Relations for Cardinals*, Studies in Logic and the Foundations of Mathematics 76, North-Holland, Amsterdam 1984.

[GH] F. Galvin, A. Hajnal, *Inequalities for cardinal powers*, Ann. of Math. 101 (1975), 491–498.

[Je1] T. Jech, *The Axiom of Choice*, Studies in Logic and the Foundations of Mathematics 75, North-Holland, Amsterdam 1973.

[Je2] T. Jech, *Set Theory*, 2nd edition, Springer-Verlag, Berlin 1997.

[Je3] T. Jech, *Shelah's theorem on* $2^{\aleph_\omega}$, Bull. London Math. Soc. 24 (1992), 127–139.

[Je4] T. Jech, *A variation on a theorem of Galvin and Hajnal*, Bull. London Math. Soc. 25 (1993), 97–103.

[Je5] T. Jech, *Singular cardinals and the pcf theory*, Bull. Symbolic Logic 1 (1995), 408–424.

[JS] T. Jech, S. Shelah, *On a conjecture of Tarski on products of cardinals*, Proc. Amer. Math. Soc. 112 (1991), 1117–1124.

[JW1] W. Just, M. Weese, *Discovering Modern Set Theory. I*, The Basics, Graduate Studies in Mathematics, vol. 8, American Mathematical Society, 1996.

[JW2] W. Just, M. Weese, *Discovering Modern Set Theory. II*, Set-Theoretic Tools for Every Mathematician, Graduate Studies in Mathematics, vol. 18, American Mathematical Society, 1997.

[Ku] K. Kunen, *Set Theory. An Introduction to Independence Proofs*, Studies in Logic and the Foundations of Mathematics 102, North-Holland, Amsterdam 1980.

[Le] A. Levy, *Basic Set Theory*, Springer-Verlag, Berlin, Heidelberg, New York 1979.

[Ma] M. Magidor, *On the singular cardinals problem I*, Israel J. Math. 28 (1977), 1–31.

[RR] H. Rubin, J. E. Rubin, *Equivalents of the Axiom of Choice, II*, Studies in Logic and the Foundations of Mathematics 102, North-Holland, Amsterdam 1985.

[Sh1] S. Shelah, *Proper Forcing*, Lecture Notes in Mathematics 940, Springer-Verlag, Berlin 1982.

[Sh2] S. Shelah, *Successors of singulars, cofinalities of reduced products of cardinals and productivity of chain conditions*, Israel J. Math. 62 (1988), 213–256.

[Sh3] S. Shelah, *Products of regular cardinals and cardinal invariants of products of Boolean algebras*, Israel J. Math. 70 (1990), 129–187.

[Sh4] S. Shelah, *Cardinal arithmetic for sceptics*, Bull. Am. Math. Soc. 26 (1992), 197–210.

[Sh5] S. Shelah, *Cardinal Arithmetic*, Oxford Logic Guides 29, Clarendon Press, Oxford 1994.

[Sh6] S. Shelah, *The pcf theorem revisited*, The mathematics of Paul Erdös, II, 420–459, Algorithms Combin. 14, Springer-Verlag, Berlin 1997.

[Si] J. Silver, *On the singular cardinals problem*, in Proc. Internat. Congress of Math., Vancouver 1974, vol. I, 265–268.

[We] E. Weitz, *Untersuchungen über die Grundlagen der pcf-Theorie von Saharon Shelah*, Dissertation, Hannover 1996.

List of Symbols

Index

Printed in the United States
By Bookmasters